# MICROTUBULES AND MICROTUBULE INHIBITORS

# MICROTUBULES AND MICROTUBULE INHIBITORS

Proceedings of the International Symposium on Microtubules and Microtubule Inhibitors.
Beerse, Belgium, 2-5 September, 1975.
Organized by the Janssen Research Foundation under the auspices of the Belgian Society for Cell Biology.

*Editors:*
M. Borgers
M. De Brabander

1975

NORTH-HOLLAND PUBLISHING COMPANY - AMSTERDAM · OXFORD
AMERICAN ELSEVIER PUBLISHING COMPANY, INC. - NEW YORK

North-Holland ISBN: 0 7204 0395 2
American Elsevier ISBN: 0 444 11041 0

Published by:

North-Holland Publishing Company - Amsterdam
North-Holland Publishing Company, Ltd. - Oxford

Sole distributors for the U.S.A. and Canada:

American Elsevier Publishing Company, Inc.
52 Vanderbilt Avenue
New York, N.Y. 10017

PRINTED IN THE NETHERLANDS

# PREFACE

This symposium which gathers microtubule specialists from many laboratories, who we are happy to welcome here in Belgium, has been organized by the Janssen Research Foundation under the auspices of the Belgian Society for Cell Biology. I wish to thank warmly Dr. Paul A.J. Janssen for his support and for the enthusiasm with which he has accepted the organization of this international meeting.

In 1961 Slautterback coined the word "microtubule". Since then new papers on these organelles have been accumulating at a nearly exponential rate. However, microtubules were known before they got their name : the neurofibrils, the mitotic spindle, the marginal bundle of erythrocytes had been described long before as fibrillar protoplasmic structures. The first suggestion that they might be tubular was made by K. Porter at the 8th Congress of the Society for Cell Biology, in 1954. W. Bernhard and E. de Harven in 1956, and several others brought confirmation. However, I do not think that microtubules became a fashionable topic only because they had received a name. The observation that they played a part in so many different cell functions, and that mitosis was only one of these ; the discovery that they were present in all eukaryotic cells, and that they may have originated at the same time in evolution as the cell nucleus ; the demonstration of their chemical composition and the possibility of assembling at will microtubules in vitro, all these facts have led to an explosion of discoveries. This explains why, one year after the excellent meeting of the New York Academy of Sciences, it was thought useful to bring together again a group of workers in this field. The important problems considered in this book range from the structure of the colchicine molecule to the mechanisms of mitosis, from the relation between tubulin and actin to the control of cell secretion, to mention only a few topics. They extend also into the field of therapeutics, and this explains the support the Janssen Research Foundation has given to the organization of this meeting.

Microtubules illustrate perfectly well the close relations between fundamental and applied biology. Should I recall that colchicine, already at the end of the eighteenth century, was considered a miracle drug ? It is through the study of its toxicological side-effects that the Italian author, Pernice, in 1889, published the first description of mitosis in the intestine of the dog, in a paper which would be forgotten for sixty years. It is a consequence of the observations of Dixon and

Malden, in 1906, on colchicine leukocytosis in mammals that in 1933, F. Lits, at that moment a young student in medicine, studied the effects of colchicine in the mouse, and found that it modified in a spectacular way the mitotic activity. It was in the same laboratory in Brussels in 1937, that my father, with Havas and Lits demonstrated the action of colchicine on plant meristems. The application of colchicine in the production of polyploid and amphidiploid plants is still playing a considerable role in the fight of humanity for providing cereals with greater nutritive yield. Now, through the work on microtubules, colchicine has again become a tool for the analysis of cell functions - movement, intra-cytoplasmic transport, secretion, phagocytosis, etc. And new therapeutic applications of this drug are reported - in psoriasis, periodic fever, amyloidosis, histiocytosis - while vincristine has, apart from cancer chemotherapy, interesting effects in the treatment of thrombocytopenia.

Microtubule research has still much to teach us : several fundamental problems remain clouded in mystery. The most extraordinary aspect of these organelles is their nearly continuous assembly and disassembly : what may be the methods used by the cell to grow microtubules at the proper moment and the proper place. This appears so closely related to the determinism of cell shape, that a better understanding of microtubule regulation - and this involves of course also centrioles and cilia - should lead to a better understanding of the origin of the highly assymetrical cell forms which lead the eukaryotes to the incredibly varied shapes of life on this earth.

P. DUSTIN

Beerse (Belgium), September 1975.

# CONTENTS

# INTRODUCTORY LECTURE

*Microtubules and Microtubule Inhibitors, eds. M. Borgers and M. de Brabander*



# MICROTUBULES AND EVOLUTION

Lynn Margulis
Department of Biology
Boston University

Summary

The enigma of the origin of microtubules both phylogenetically (in the origins of mitotic cells) and ontogenetically (in the development of individuals) will be placed in the context of the adaptive radiation of eukaryotic cells. This event, which probably occurred in the late Precambrian is thought to have resulted in the polyphyletic origin of mitosis and meiosis. The exogenous serial endosymbiotic theory and the endogenous theory of the origin of microtubule organizing centers (MTOC's, Pickett-Heaps[1], 1974) are compared.

The multiple roles of microtubules in the evolutionary diversification of protists (e.g., organisms formerly classified as protozoa, nucleate algae, flagellate fungi, slime molds, slime net amoebae and their relatives) are truly amazing. For example, microtubules as cell structural elements in cilia and flagella, tracks along which pigment granules, organelles and endosymbionts are translocated and structures involved in mitotic motility are well known. But microtubule-based structures may also be used as cell anchoring devices, "teeth", "legs", "fishing rods", "stirrers", cell elongation machines, as well as instruments in the more serious business of mitotic and meiotic sexual reproduction in protists. It is likely that the evolution of microtubule diversity in protists preadapted these proteinaceous structures for their physiological roles in animals and plants.

## I. Introduction

Because of their ubiquity in eukaryotes and their presumed role in many morphogenetic and physiological processes, the literature on microtubules has been extensively reviewed (Olmsted and Borisy[2], 1973; McIntosh[3], 1975; Stephens[4], 1971; Margulis[5], 1973; Soifer[6], 1975, and so forth). Rather than perpetuate the tradition of reviewing microtubule literature as fast as it appears, it is my purpose here merely to raise some questions concerning the origin and evolution of the microtubular systems of eukaryotes. It will probably be another decade until meaningful answers will be available.

The major issues raised in this paper concern the mysterious origin of the microtubule system both phylogenetically (in the evolutionary history of the populations of organisms in which it arose) and ontogenetically (in the development of microtubule-based structures in the life history of individuals). Some quick comments on the subsequent diversification of the microtubule system in protists are also included.

## II. The Mysterious Origin of the Microtubule System

There are two articulated and mutually exclusive views concerning the origins of the ubiquitous microtubule systems of eukaryotes: the endogenous theory (Pickett-Heaps[1], 1974) and the exogenous (Sagan[7], 1967; Margulis[8], 1970). The former considers microtubules to have arisen during the origin of the mitotic process in those prokaryotes that eventually became eukaryotes. In this view the origin of microtubules, organized from microtubule-organizing centres (MTOC's) preceded the origin of (9+2) flagella. Subsequent to the evolution of the motile (9+2) flagellum came the movement of flagellar basal bodies (centrioles) as passive organizing centres which were distributed to daughter cells in mitosis. The origin of flagella, is thought to have occurred after evolution of red algae (which lack centrioles and flagella at all stages of their life cycles) and before the origin of the flagellated algae (e.g., the chlorophytes, chrysophytes and so forth). The endogenous view, often with variations, is held by most biologists (e.g., Taylor[9], 1974).

Although I am very indebted to the original and important work of Pickett-Heaps (1971[10], 1974[1], 1975[11]) especially with respect to the clarification of the relationship between microtubular structures and their organizing centres, I disagree with his endogenous view of the origin of the microtubule system. I believe that all eukaryotes that contain bona fide microtubules originated from flagellated ancestors. The (9+2) flagella, which I take to be homologous in all organisms, preceded the origin of mitosis. The microtubules that originally were exclusively associated with organelles of motility (flagella, and their shorter relatives cilia) became reutilized in many other cellular processes including mitosis. I believe the protists (e.g., the nucleate algae, protozoans, flagellated fungi, slime net amoeba and so forth) represent organisms that are legacies from an event that classical evolutionists would call an "adaptive radiation". This event led to the origin and stabilization of many taxa starting with organisms that lack orthodox mitosis and ending with many that are clearly eumitotic. It most likely occurred in the late Precambrian prior to 750 million years ago (Knoll and Barghoorn[12], 1975; Margulis[13], 1974). Flagella, in this view, preadapted early eukaryotes for the polyphyletic origin of mitosis. Under continuous selection pressures to both increase and properly distribute DNA, mitosis arose independently in several lines of protists and was succeeded by the origin of meiosis in some of these. Luckily, the present biota contains enough relicts amongst living "lower eukaryotes" such that reconstruction of these events can begin to be approached (Kubai[14], 1975; Fuller[15], 1975; also see symposium on the origin of mitosis in eukaryotic microorganisms, BioSystems, December, 1975).

Every scholar who has in any way dealt with the issue since Henneguy and Lenhossék (1897, 1898; in Wilson[16], 1925) has become acutely aware that basal bodies and the centrioles of mitosis are homologous structures. An intimate relation exists in protists between the organelles of motility and feeding (flagella, axopods and so forth) and the mitotic spindle. Microtubules and the tubulin protein of which they are composed both in the mitotic spindle and in cilia and flagella are clearly homologous as judged even by analyses of amino acid sequences of tubulins from diverse sources (Luduena and Woodward[17], 1975). These new observations have simply strengthened the Henneguy and Lenhossék contention that centrioles from previous mitotic divisions often become basal bodies to cilia. The recognition of this classical homology between centrioles and basal bodies has been greatly enhanced by ultrastructural work which has revealed their 9-fold symmetry. A critical issue however, at this moment entirely unsolved, is from where (in the phylogenic sense) did the microtubule system itself originate? The endogenous view sees it as coming ultimately from the prokaryotes that became the eukaryote by differentiation of elements required for mitosis; I am unable to design experiments to test this idea. The exogenous view (one that we are involved in testing) is that microtubules and eventually the (9+2) substructure arose in a class of motile bacteria, probably spirochaete-like and that these bacteria at first had little or nothing to do with the prokaryote that eventually acquired mitochondria and became the nucleated eukaryote. That is, I believe that a class of mitochondria-containing nucleated organisms, non-phagocytotic but amoeba-like, acquired on their surfaces ectosymbiotic spirochaetes that happened to have very characteristic cross sections. What were originally scavenging symbionts, casually attached to the surface of their hosts eventually became microtubule-based organelles of motility. Analogies of such sequences of events of course are known; many protists regularly have attached spirochaetes (e.g., Pyrosonympha, Smith et al.[18], 1975; Kentrophoros, Raikov[19], 1974; see Margulis[20], 1975a for review) and in some cases the surface spirochaetes are actually used as organelles of motility (e.g., Myxotricha, Cleveland and Grimstone[21], 1966). The ectosymbiotic association between spirochaete microbes and their nucleated hosts in the early eukaryotes, according to this view, went much further. The motile microbes that had once been spirochaetes were incorporated into the life of the host cell, the protein synthesizing systems of host and spirochaetes were merged, at least some of the genes that had once coded for spirochaete activities became incorporated into the nucleus, and MTOC's of many kinds evolved in many different cells. In this view the MTOC's represent nucleoprotein, parts of the replication and readout machinery of the original spirochaetes. When a new basal body arises from a preexisting one (as Allen[22], 1969 has shown in the development of the ciliate cortex) RNA at basal body sites is replicated and protein assembles accordingly so that a faithful replica of the parent basal body is developed. When several hundred basal bodies are produced from one (e.g., in the development of multiflagellated sperm of cycads or ginkgos) or when such 9-fold organized structures appear "de novo" as they do in amoeboflagellate transformation (Fulton[23], 1971; Fulton and Kowit[24], 1975) I think it is because RNA replicates relatively more times before the developmental stages set in. Thus, according to the exogenous theory MTOC's are fundamentally genetic entities that require both nucleic acid and protein assembly for their replication and phenotypic expression. Furthermore they are remnants of once free-living ectosymbiotic microorganisms. I say "RNA" explicitly here for three reasons (1) It has been shown rather conclusively that DNA does not accompany basal body "reproduction" in any of well-known systems (Fulton[23], 1971; Younger et al.[25], 1972) in spite of earlier reports to the contrary (Smith-Sonneborn and Plaut, 1967[26], 1969[27]) (2) RNA synthesis does seem to accompany centriole and/or basal body "reproduction" in several systems (Hartman, et al.[28], 1974; Laane and Haugli[29], 1974; Younger et al.[25], 1972) and at least in two cases this RNA may be localized with the lumen of the basal body itself (Hartman, et al.[28], 1974; Dippell, 1975, quoted in Sonneborn[30], 1974) (3) The assertion that an MTOC is ribonucleoprotein ultimately homologous with that from free-living spirochaetes may be outrageous but it has the distinct advantage of being explicit enough to be testable. Differences between the exogenous and endogenous views are presented in Table 1.

I leave it to those who espouse the endogenous view to assert predictions that follow therefrom. In the remainder of this paper I wish only to review briefly some of our new knowledge of microtubule-based processes in the context of the exogenous view.

III. Did microtubule-based processes precede the origin of mitosis?

It is my opinion that (9+2) based motility preceded the origin of mitosis. There are two cases, for example, of dinoflagellates that have no microtubules at all associated with nuclear division (they seem to segregate chromatin on membrane, Soyer[31], 1975) and certainly they are flagellated. It seems to me far more likely

Table 1
Comparison of Two Views of the Origin of Microtubule Systems

| Exogenous View (Margulis[8], 1970) | Endogenous View (Pickett-Heaps, 1974[1], 1975[11]) |
|---|---|
| 1. Complex (9+2) flagella preceded the evolution of the microtubule system in eukaryotes and were ultimately derived from eclosymbiotic surface microbes. | 1. Microtubules and their organizing centers preceded the origin of flagella and were ultimately derived endogenously perhaps from nuclear membrane cites. |
| 2. Microtubules will be found in spirochaetes, eventually a spirochaete with a (9+2) array of intracellular tubules may be found. | 2. Microtubules will be found in the cytoplasm of the prokaryotes that were ancestral to the eukaryotes. There is no reason to expect them in spirochaetes. |
| 3. MTOC's are replicative, RNA replication will accompany "reproduction" of centrioles, basal bodies, kinetochores and so forth. | 3. MTOC's are protein self assembly systems. |
| 4. All eumitotic organisms have flagellated ancestors; (9+2) organelles may be inducible in them. | 4. Some eumitotic organisms are primitively nonflagellated (red algae, higher fungi), thus (9+2) organelles could never be induced in them. |
| 5. The genes coding for the tubulin and microtubule associated proteins will be homologous to the genes coding for such protein in selected spirochaetes. | 5. The genes coding for tubulin and associated proteins are in the nucleus where they originated, such genes would only be expected among ancestral prokaryotes on the direct line to eukaryotes. |
| 6. The flagellum is monophyletic mitosis and meiosis are polyphyletic in several lines of protists. | 6. Mitosis in monophyletic; the flagellum subsequently evolved in a mitotic organism and was retained for its selective advantage. |
| 7. Once evolved mitosis was not lost; flagellates and amoeboids lacking mitosis are primitive. | 7. Mitosis was lost in several lines of flagellated protists. |

that mitosis sensu stricto never evolved in these flagellate forms rather than it evolved, led to flagella and was lost. There is good evidence that mitosis and nuclear development has taken its own idiosyncratic course in this group of dinoflagellate protists (Ris[32], 1975; Kubai[14], 1975). The exogenous view implies that eumitotic nonflagellated organisms such as the red algae and the ascomycetous fungi arose from flagellated ancestors that permanently withdrew their flagella. Such a proposition would be greatly strengthened if centriolar or related structures with 9-fold symmetry were inducible in these organisms. This result could not be predicted on the Pickett-Heaps[1] (1974) version of the endogenous theory. That mitosis evolved in the protists, or at least prior to its stabilization in animals and embryophyte green plants, is deducible from the variations upon it listed in Table 2. It is likely that it evolved polyphyletically in the fungi as well. The reader is encouraged to go to the original literature for details (see Fuller[15], 1975; Ris[32], 1975; Kubai[14], 1975, for such an éntre). Furthermore the variations on the mitotic theme listed here can be best understood by the concept that mitosis evolved polyphyletically, as suggested also by Goode[33] (1975).

Microtubules are involved in a rather large number of processes that seem to have nothing to do directly with mitosis, as can be seen in Fig. 1. Selection pressures similar to those which acted later on much larger and more complex diploid multicellular organisms resulted in the convergent formation of microtubule-based stalks or holdfasts, "teeth" (e.g., Nassula), "feet" (e.g., hypotrich ciliates), and conduits for transport of various particles and substances (e.g., food, pigment granules, mitochondria and the like) (e.g., suctorian ciliates). Microtubules form the highly motile axostyle structure in flagellates such as Pyrsonympha; the role of these complex organelles in the vegetative cell is not known but the possibility that it helps to stir the cellulose digestion factory in the termite hindgut has been suggested (Langford[39], 1975)! Microtubule-based structures clearly function as the bone and muscle equivalents of animal tissue; perusal of recent books on protozoology will certainly confirm this notion (e.g., Sleigh[40], 1973; Grell[37], 1973).

Table 2
Variations of Standard Animal (metazoan) and Plant (metaphytan) Mitosis

| Variations | Organisms |
|---|---|
| 1. Nuclear division with no evidence of microtubules | Pelomyxa palustris[a]<br>Prorocentrum[d] |
| 2. Chromatin not condensing into chromosomes | Amoeba proteus[a]<br>Collozoum[r]<br>Saccharomyces[as] |
| 3. Kinetochores on nuclear membrane (or nuclear membrane pores) | Syndinium[d]<br>Trichonympha[z] |
| 4. No (or very undeveloped) kinetochores | Euglena[e]<br>Saccharomyces[as] |
| 5. Mitotic microtubules only inside nucleus | Chaos carolenensis[a]<br>Euglena[e]<br>Thalanophysidae[r]<br>Oxymonas[z]<br>Saccharomyces[as]<br>all micronuclei studied[ci] |
| 6. Microtubules in extranuclear channels only, chromatin apparently segregated on membrane | Cryptothecodinium[d]<br>Stentor coeruleus[ci] macronuclei |
| 7. Spindle tubules appear from granular masses which appear to originate from rhizoplast dissolution | Platymonas[pr] |

Table 2 cont.

| Variations | Organisms |
|---|---|
| 8. Nucleolar persistance | *Tetramitus*[z] |
| | *Naeglaria*[z] |
| | opalinids[z] |
| | *Polymyxa*[o] |
| | *Saprolegnia*[o] |
| | *Patellina*[f] |
| 9. No metaphase plate forms | *Euglena*[e] |
| | *Mucor*[zy] |
| | *Saccharomyces*[as] |
| | oomycetes |

Key to major groups ("phyla", Margulis[35], 1975b)

| Protists | Fungi |
|---|---|
| c - Caryoblastea | as - ascomycota |
| d - Dinoflagellata | zy - zygomycota |
| r - Radiolariata | |
| z - Zoomastigina | |
| pr - Prasinophyta | |
| a - asexual amoebae | |
| ci - Ciliophora | |
| f - Foraminiferan | |
| o - Oomycota | |
| e - Euglenophyta | |

See Heath[36], 1975; Grell[37-38], 1973, 1975; Ris[32], 1975; Fuller[15], 1975; and Symposium on the origin of mitosis in eukaryotic microorganisms, *Biosystems*, December, 1975, for detail.

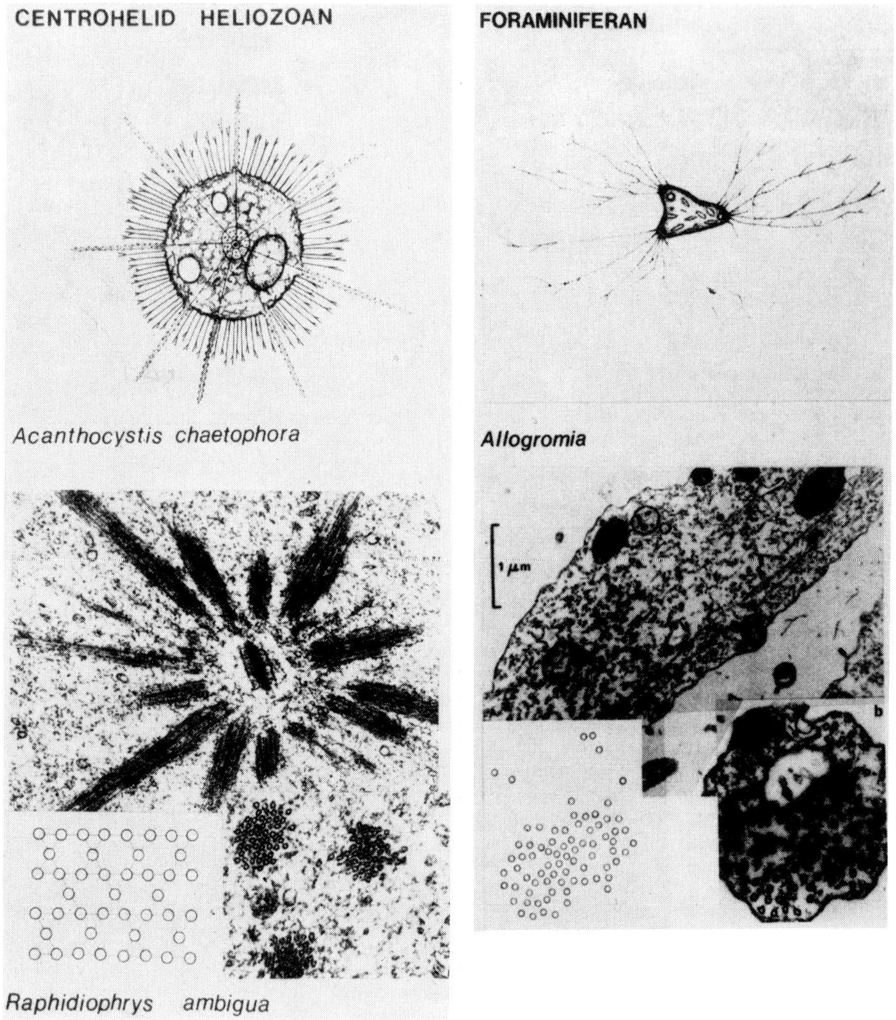


Fig. 1. Microtubular arrays in protists: heliozoan axopods (*Acanthocystis*, *Raphidiophrys*, *Echinosphaerium*); these are feeding and prey capture organelles; foraminiferan rhizopods (these are active in food capture, locomotion, as intracellular conduits for mitochondria and other particles and in calcium carbonate shell deposition); in polymastigotes microtubules comprise the axostyles such as this one in *Pyrsonympha* (they undergo vigorous nontranslational movements and may function as "stirrers"). In close relatives (*Saccinobaculus*, *Oxymonas*, *Notila*) the axostyles fuse during the sexual cycle and serve to unite the male and female gamete pronuclei (Cleveland[34], 1956). (Assembled from various sources)

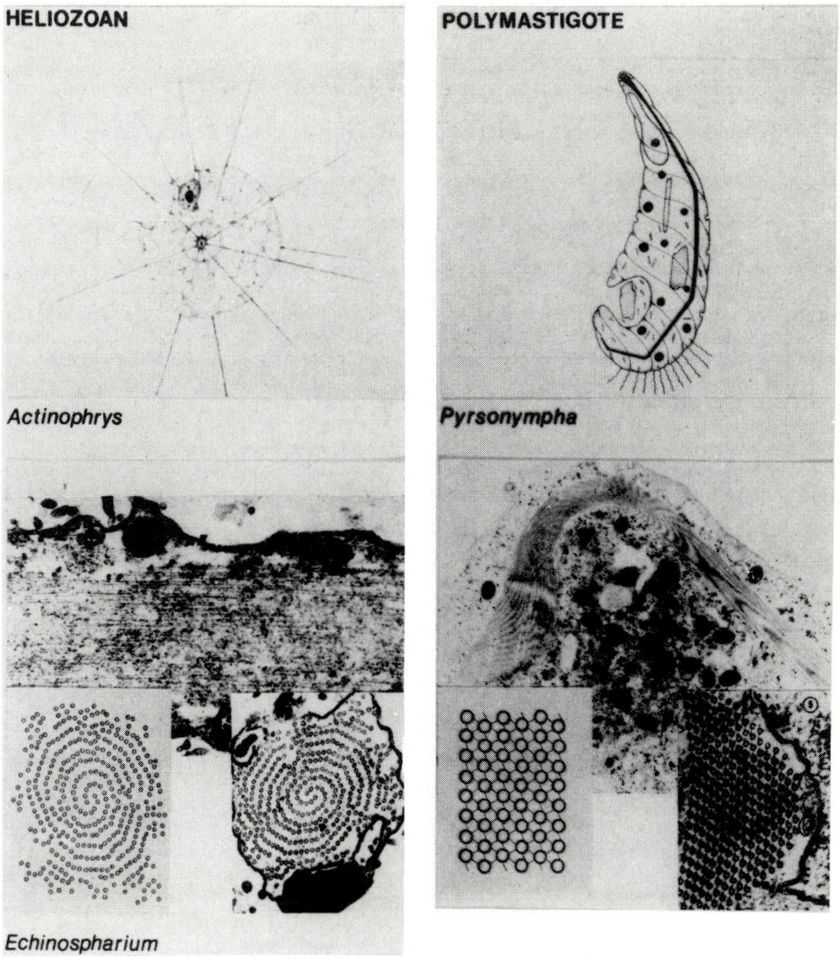
HELIOZOAN
Actinophrys
Echinospharium
POLYMASTIGOTE
Pyrsonympha

Fig. 1 cont.

In many organisms, however, exactly those microtubule-based organelles that, during active growing life, function as conduits for feeding and movement and so forth are withdrawn and apparently reutilized in mitotic-like cell division processes. Some amoeboflagellates and dinoflagellates can only divide after withdrawal of their flagella. Some meliozoans withdraw their axopods during cell division. In hypermastigotes the division figure superficially resembles that of a classical animal cell mitosis (with very large centrioles). However, ultrastructural studies indicate that the spindle (with its "long centriole") is a modification of the entire anterior end (the rostrum) of the flagellate. Elaborate microtubular and periodically striated bands appear (Fig. 2).

The evolution of structural microtubule systems may have preadapted certain flagellates for meiosis and fertilization as well. For example, Notila is a flagellate of the hindgut of wood roaches with a conspicuous axostyle, much like that of Pyrsonympha. During fusion of the gametes in fertilization the male pronucleus is brought to the female pronucleus by apposition of the axostyles to which these nuclei are attached (Cleveland[34], 1956). The relationship between mitotic and other microtubules is clearly intimate in many organisms. Probably the development of independence between the mitotic and the general somatic microtubules evolved several times. This certainly seems to be the case in the ciliate micronuclei. These undergo rather idiosyncratic mitotic divisions within closed nuclei independently of the somatic infraciliature (Davidson and LaFountain[41], 1975; Lewis[43], 1975).

This is of course not the place in which to summarize many years experience extending to the early days of cytology into the 19th century (Wilson[16], 1925) of tens of scientists concerning the fascinating origins of the mitotic process (could I if I would!). This brief overview may only serve to remind us of several concepts: (1) "Microtubule research began long before microtubules were discovered ..." (Mazia[44], 1975), (2) microtubule-based structures and processes are extremely

Fig. 2. Resolution of the hypermastigote "large mitotic apparatus" shows it to be a modification of the entire rostral region of the organisms; spindle microtubules form between the rostral rods. The light micrograph (left) is taken at a somewhat later stage than the electron micrograph (right). From L. R. Cleveland, A. V. Grimstone and I. R. Gibbons (see Grimstone and Gibbons[42], 1966).

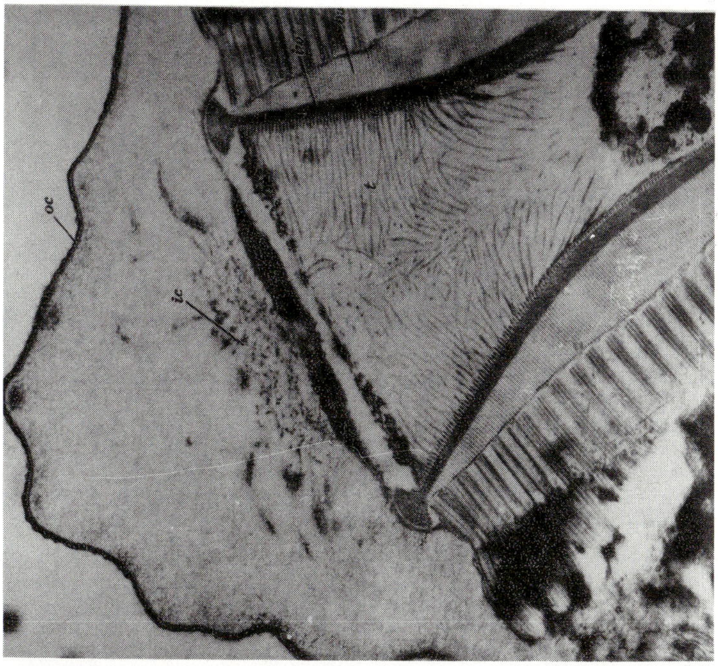
oc
ic
t

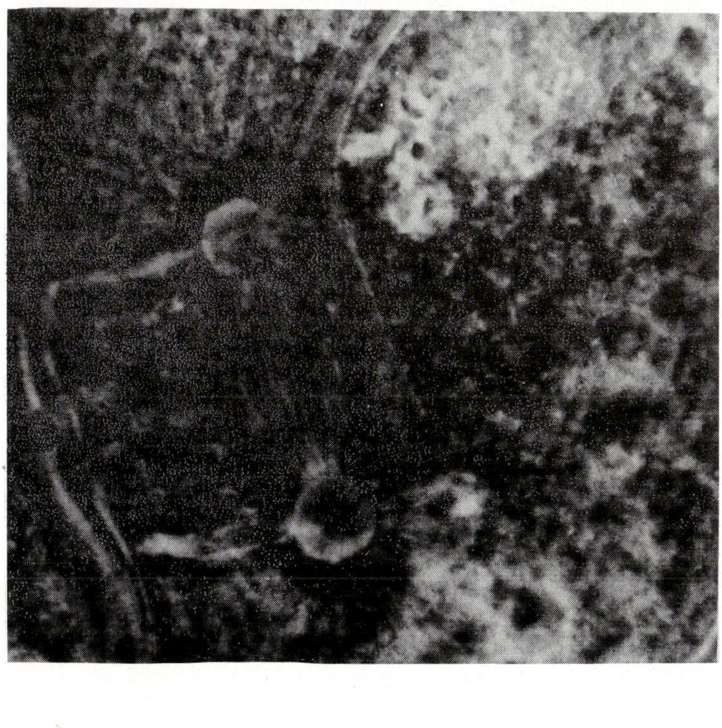

diverse in the "lower eukaryotes", (3) this diversification is intimately associated with the origin and evolution of mitosis and meiosis,* and (4) the origin of the predecessors of the microtubule system in prokaryotes stands as an exciting unsolved problem.

Acknowledgements

Research support from NASA NGR-004-025 (Microbial contributions to the Precambrian Earth) and from the Boston University Graduate School is gratefully acknowledged. I am indebted to N. Lowe for aid with the figures, to B. Miranda for editorial assistance, and to L. To and C. Bardele for useful discussions.

References

1. Pickett-Heaps, J. D., 1974. The evolution of mitosis and the eukaryotic condition. BioSystems 6: 37-48.

2. Olmsted, J. and Borisy, G. G. 1973. Microtubules. Ann. Rev. Biochem. 42: 507-540.

3. McIntosh, J. R. 1975. Introduction to microtubules. J. Supramol. Struct. 2: 385-392.

4. Stephens, R. E. 1971. Microtubules. Biological Macromolecules. 5: 355-391. Timasheff, S. N. and Fasman, G. D., eds. Marcel Dekker, Inc., New York, New York.

5. Margulis, L. 1973. Colchicine sensitive microtubules. Int. Rev. Cytol. 34: 333-361. G. Bourne and J. F. Danielli, eds. Academic Press, New York.

6. Soifer, D., ed. 1975. The Biology of Cytoplasmic Microtubules. Vol. 253. New York Academy of Sciences.

7. Sagan, L. 1967. Origin of mitosing cells. J. Theor. Biol. 14: 225-275. (=Margulis)

8. Margulis, L. 1970. Origin of Eukaryotic Cells. Yale University Press, New Haven, Conn.

9. Taylor, F. J. R. 1974. Implications and extensions of the serial endosymbiosis theory of the origin of eukaryotes. Taxon 23: 229-258.

10. Pickett-Heaps, J. D. 1971. The autonomy of the centriole: fact or fallacy. Cytobios 3: 205-214.

11. Pickett-Heaps, J. D. 1975. Aspects of spindle evolution. New York Acad. Sci. 253: 352-361.

12. Knoll, A., and Barghoorn, E. B. 1975. Precambrian eukaryotic organisms: A reassessment of the evidence. Science (in press)

* I believe this should be the basis of recognition of this group as the kingdom Protista (see Margulis[35], 1975b).

13. Margulis, L. 1974. On the evolutionary origin and possible mechanism of colchicine sensitive mitotic movements. BioSystems 6: 16-36.

14. Kubai, D. F. 1975. Evolution of the mitotic spindle. Int. Rev. Cytol. (in press)

15. Fuller, M. 1975. Mitosis in fungi. Int. Rev. Cytol. (in press)

16. Wilson, E. B. 1925. The Cell in Development and Heredity. p. 698-699. Macmillan, New York.

17. Luduena, R., and Woodward, D. 1975. α and β-tubulin: separation and partial sequence analysis. New York Acad. Sci. 253: 272-283.

18. Smith, H. E., Buhse, H. E., and Stanier, S. J. 1975. Possible formation and development of spirochaete attachment sites found on the surface of symbiotic polymastigote flagellates of the termite Reticultermes flavipes. BioSystems (in press)

19. Raikov, I. B. 1974. Étude ultrastructural des bactéries épizoïques et endozoïques de Kentrophoros latum Raikov, Cilié Holotriche mésopsammique. Cahiers de Biologie Marine 15: 379-393.

20. Margulis, L. 1975a. The genetic and evolutionary consequences of symbiosis. Parasitological Reviews: Experimental Parasitology. (in press)

21. Cleveland, L. R., and Grimstone, A. V. 1964. The fine structure of the flagellate Mixotricha paradoxa and its associated micro-organisms. Proc. Royal Soc. B 159: 668-686.

22. Allen, R. D. 1969. The morphogenesis of basal bodies and accessory structures of the cortex of the ciliated protozoan Tetrahymena pyriformis. J. Cell Biol. 40: 716-733.

23. Fulton, C. 1971. Centrioles: Results and Problems in Cell Differentiation. Vol. 2. Origin and Continuity of Cell Organelles. pp. 170-221. J. Reinert and H. Ursprung, eds., Springer-Verlag, New York.

24. Fulton, C. and Kowit, J. D. 1975. Programmed synthesis of flagellar tubulin during cell differentiation in Naeglaria. The Biology of Cytoplasmic Microtubules. New York Acad. Sci. 253: 318-332.

25. Younger, K. B., Banerjee, S., Kelleher, J. K., Winston, M. and Margulis, L. 1972. Evidence that synchronized production of new basal bodies is not associated with DNA synthesis in Stentor coeruleus. J. Cell Sci. 11: 621-637.

26. Smith-Sonneborn, J. and Plaut, W. 1967. Evidence for the presence of DNA in the pellicle of Paramecium. J. Cell Sci. 2: 225-234.

27. Smith-Sonneborn, J. and Plaut, W. 1969. Studies on the autonomy of pellicular DNA in Paramecium. J. Cell Sci. 5: 365-372.

28. Hartman, S. , Gurney, T. and Puma, T. 1974. Evidence for the association of RNA with the ciliary basal bodies of Tetrahymena. J. Cell Sci. 16: 241-259.

29. Laane, M. M. and Haugli, F. B. 1974. Division centres in mitotic nuclei of Physarum polycephalum plasmodia. Norweg. J. Botany 21: 309-318.

30. Sonneborn, T. M. 1974. Ciliate morphogenesis and its bearing on general cellular morphogenesis. Actualités Protozoologiques 1: 337-355.

31. Soyer, M. O. 1975. Chromosome division and organization in some dinoflagellates. BioSystems (in press)

32. Ris, H. 1975. Primitive mitotic systems. BioSystems (in press)

33. Goode, D. 1975. Evolution of mitosis in the Protozoa: The association of chromosomes, nuclear envelope, kinetochores, and microtubules. BioSystems (in press)

34. Cleveland, L. R. 1956. Brief accounts of the sexual cycles of the flagellates of Cryptocercus. J. Protozool. 3: 161-180.

35. Margulis, L. 1975b. The classification and evolution of prokaryotes and eukaryotes. Handbook of Genetics I: 1-41. R. King, ed. Plenum Press, N.Y.

36. Heath, I. B. 1975. The possible significance of variations in the mitotic systems of the aquatic fungi ("Phycomycetes"). BioSystems: (in press)

37. Grell, K. G. 1973. Protozoology. Springer-Verlag. Berlin, Heidelberg, N. Y.

38. Grell, K. G. 1975. Aulacantha scolymantha Haeckel: An enigmatic case of nuclear division. BioSystems (in press)

39. Langford, G. M. 1975. The of bending in the axostyle of Pyrsonympha vertens. BioSystems (in press)

40. Sleigh, M. 1973. The Biology of Protozoa. Edward Arnold, London.

41. Davidson, L. and LaFountain, J. 1975. Mitosis and early meiosis in Tetrahymena pyriformis and the evolution of mitosis in the phylum ciliophora. BioSystems (in press)

42. Grimstone, A. V. and Gibbons, I. R. 1966. The fine structure of the centriolar apparatus and associated structures in the complex flagellates Trichonympha and Pseudotrichonympha. Phil. Trans. Royal Soc. London, Series B 250: 215-242.

43. Lewis, L. 1975. The evolutionary significance of ultrastructural variations in the micronuclear spindle apparatus in the genus Paramecium. BioSystems (in press)

44. Mazia, D. 1975. Microtubule research in perspective. New York Acad. Sci. 253: 7-13.

# FUNDAMENTAL ASPECTS OF STRUCTURE AND BIOCHEMISTRY OF MICROTUBULES

*Microtubules and Microtubule Inhibitors, eds. M. Borgers and M. de Brabander*
© 1975, North-Holland Publishing Company - Amsterdam


# SUBSTRUCTURE AND SYMMETRY OF MICROTUBULES

Linda A. Amos

Medical Research Council Laboratory of Molecular Biology,
Cambridge, England.

## Summary

The substructure of microtubules has been investigated by applying both optical and digital techniques of image analysis to electron micrographs of negatively stained specimens. The surface lattice of flagellar tubules has been determined from the diffraction patterns of the electron microscope images, and a low resolution three-dimensional image of a singlet microtubule has been reconstructed, which provides an indication of the overall shape of the tubulin dimers.

The role of the microtubule as a support for other protein components is considered, in relation to its structure and symmetry. A model based on the close similarity between the symmetries of microtubules and F-actin is described, showing how actin filaments might, under some circumstances, form a paracrystalline coating around cytoplasmic tubules.

## Introduction

Microtubules observed in the electron microscope, either in section, or embedded intact in negative stain, are generally seen to be hollow tubular structures, about 24nm in diameter. Their walls appear to be made up of discrete globular subunits, arranged in longitudinal rows or protofilaments. Observations of cross-sectional views by a number of workers have shown that the number of protofilaments is thirteen in a wide variety of microtubules, including the A-tubules of flagellar outer doublets, as well as normal singlet tubules from different sources [1-6]. The B-tubule of flagellar doublets appears to possess only ten protofilaments of its own [3-7] and shares part of the wall of the A-tubule. The detailed packing of the subunits, which are arranged along the protofilaments with a longitudinal periodicity of 4nm, has been investigated by diffraction analysis of electron micrographs of negatively stained flagellar tubules [8-11] and of reconstituted brain tubules [7], and has been found to be identical in these cases, although the pairing of the subunits into 8nm dimers in the B-tubules of flagellar doublets appears to be different from the A-tubule arrangement [10].

All microtubules so far investigated are composed principally of equal proportions of two closely related 54 000 dalton chemical subunits, α- and β-tubulin, presumably associated in the form of a heterodimer[12-17]. It is tempting to identify the tubulin monomers with the 4nm globular subunits observed in the electron microscope, since they would be about the right size, but this identification has not been proved. Apart from tubulin, there are many other minor protein components associated with microtubules. Linck[18] has observed at least ten non-tubulin bands in SDS-acrylamide gels of purified flagellar outer doublet tubules, even after the removal of all the obvious projections and cross-bridges, and there is growing evidence for a number of proteins closely associated with neurotubules[19-21]. Microtubules are likely, therefore, to be structurally fairly complex.

Diffraction patterns of microtubules

The optical diffraction patterns from the electron microscope images of a wide variety of preparations show very similar major features, namely a pair of equatorial peaks at a reciprocal spacing of 1/5 $nm^{-1}$, which arise by diffraction from the longitudinal protofilaments, and a strong layer line at a height of about 1/4 $nm^{-1}$ above the equator. The 4nm layer line usually consists of two pairs of peaks, as seen in Figs. 1(b&d) and 3a. In addition, there are often signs of a layer line at a height of 1/8 $nm^{-1}$ above the equator whose intensity distribution is rather variable. It is usually completely absent from the diffraction patterns of highly purified brain tubules[7], although we have occasionally observed diffraction peaks in this region, as in the example shown in Fig. 1d. Other layer lines corresponding to longitudinal periodicities of up to 96nm are seen in the diffraction patterns of flagellar tubules and are believed to be due to the attachment of minor protein components.

The arrangement of subunits in a helical structure can be determined from its diffraction pattern as follows. Each $n$-fold family of helices in a helical surface lattice gives rise to two main diffraction peaks equidistant from the meridian of the pattern, one aris-

Fig. 1. Electron microscope images of singlet microtubules and their diffraction patterns. a) shows flagellar A-tubules near the tip of a flagellum. The B-tubules of the doublets have ended some distance below. Lengths of A-tubule included in the 3-D reconstruction are marked A and H. b) shows the optical diffraction pattern of tubule H. The diffraction pattern of A is shown in Fig. 3a. c) is a typical brain microtubule polymerised *in vitro* (preparation kindly provided by B.M.F. Pearse). d) is the diffraction pattern of c).

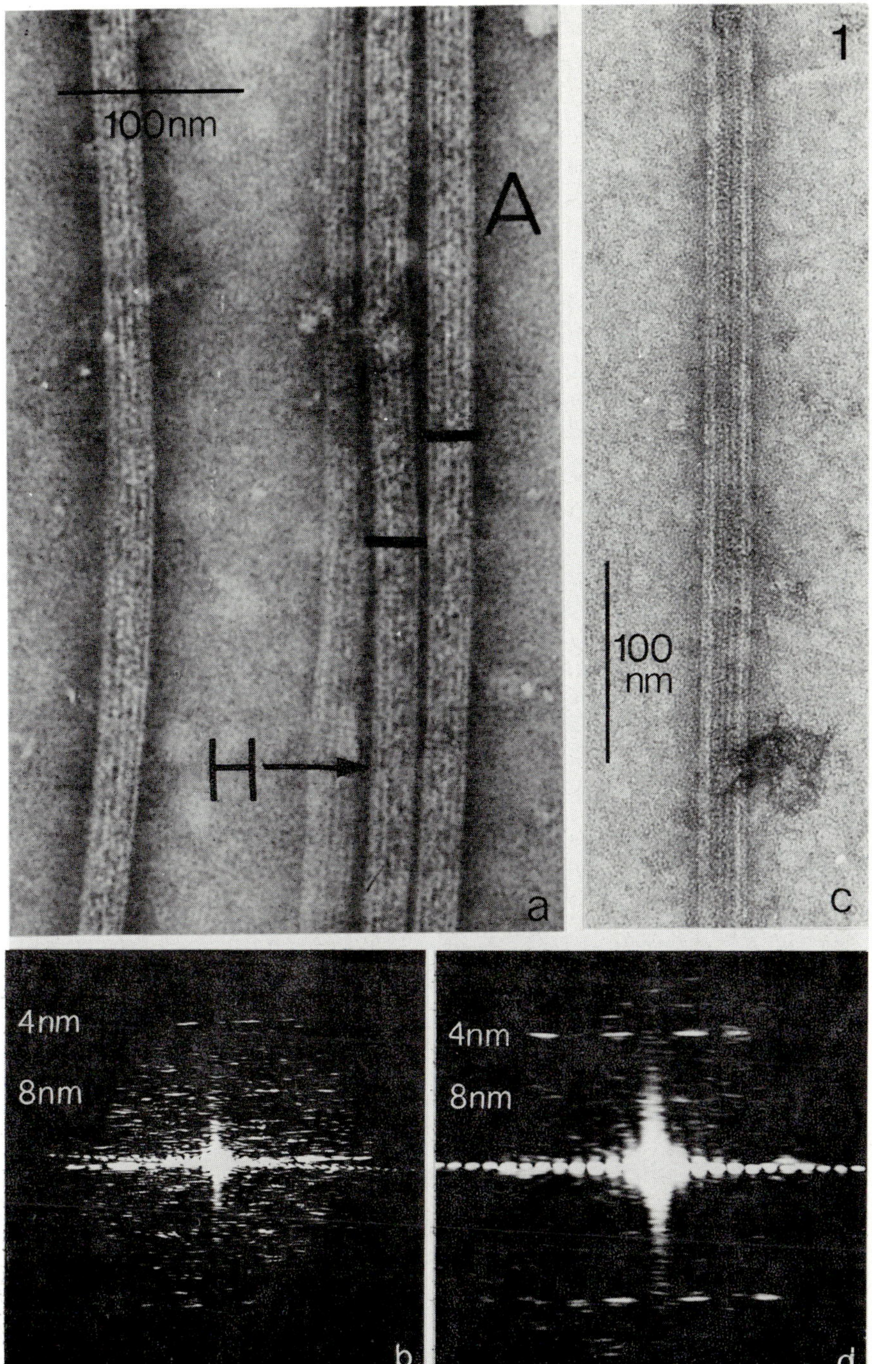
1
100nm
A
H
a
100
nm
c
4nm
8nm
b
4nm
8nm
d

ing from the "near" side of the helix, the other from the "far" side. The distance of each peak from the meridian depends upon the radius at which the corresponding helical family shows up most strongly and also upon the value of *n*. Approximate values for *n* on each layer line can thus be determined directly from a diffraction pattern of an electron microscope image[22]. The mathematical equivalent of an optical diffraction pattern, that is, a computed Fourier transform of the digitized image, provides even more information, since one obtains not only an array of amplitudes resembling the optical diffraction pattern, but also the relative phase of the diffracted wave at each point. The relative phases of pairs of peaks on either side of the meridian show whether *n* is odd or even. If the phase origin of the Fourier transformation is on the projected axis of the helical particle, the two peaks should be exactly in phase for an even value of *n*, or exactly out of phase in the case of an odd value of *n*. In practice, of course, the presence of noise in the electron microscope images makes this relationship somewhat imperfect.

The positions of the peaks on the 4nm layer line of microtubule diffraction patterns (see Fig. 1) limit the values of *n* to the ranges 2-4 and 6-12 respectively. The computed phases[10] of the peaks indicate an odd order (therefore 3) for the peaks nearer to the meridian and an even order for the outer pair of peaks. Since the number of longitudinal protofilaments is now definitely known to be thirteen,[5,6] the value of *n* for the outer peaks must be ten, in order to be consistent with a periodic helical surface lattice (see Fig. 2a). Subdivision of the reciprocal lattice to correspond to the introduction of an 8nm periodicity in the surface lattice leads to values of 5 and 8 for *n* on the 8nm layer line (c.f. Fig.2a). The diffraction peaks for such relatively close values of *n* tend to overlap on both sides of the layer line, giving rise to single long continuous peaks along the layer line. However, using information from both sides of the meridian, it is possible to separate the contributions from the two functions[10].

The hand of a symmetrical structure observed in the electron microscope, and hence the signs associated with each value of *n*, can be determined by tilting the particle in a known way and observing the relative displacements of the lattices on the near and far sides. We have determined the hands of the 3-start helices of both the A- and B-tubules of flagellar doublets to be left-handed, by tilting the tubules about an axis normal to the helix axis[11]. The hand of the lattice in reconstituted neurotubules also appears to be the same, according to the results of Erickson[7].

The final indexing of the diffraction pattern of a singlet microtubule is shown in Fig. 2a, and the surface lattice which would give rise to such a pattern is drawn in Fig. 2b.

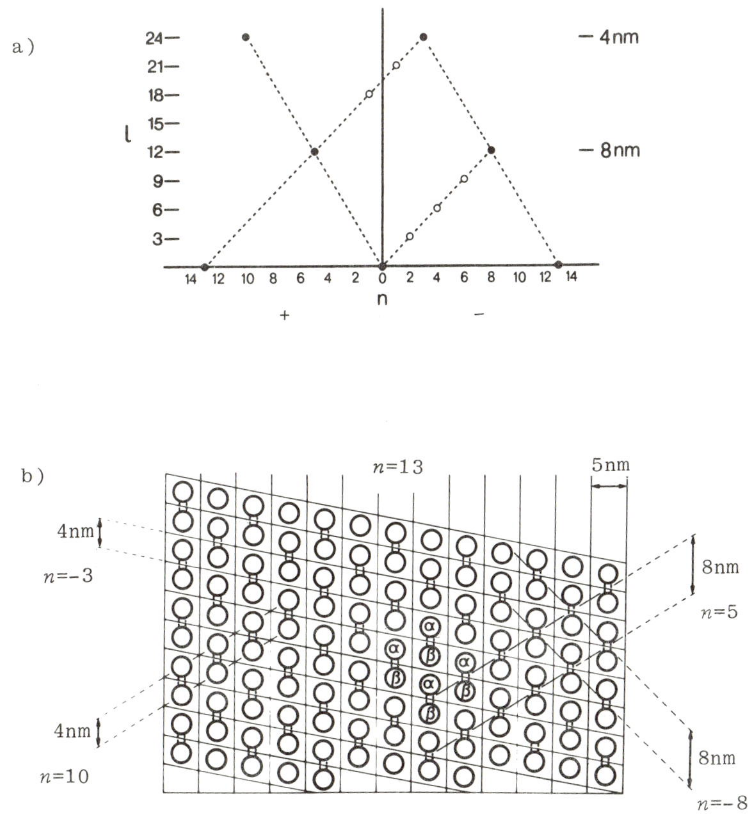


Fig. 2. a) the reciprocal lattice of a microtubule, represented by an $(n,l)$ plot[23], $n$ being the number of helices in a particular helical family and $l$ the layer line number in the diffraction pattern, based in this case on a possible 96nm longitudinal repeat. The reciprocal unit cell corresponding to the basic lattice of tubulin dimers is indicated by dotted lines. This unit cell would be further subdivided if other components were attached to the surface of the tubule in a helical arrangement with a longer longitudinal periodicity.

The $(n,l)$ plot for F-actin is extremely similar, except that it is of opposite hand.[32] If each of the labelled periodicities were increased by 10% and the signs reversed along the $n$-axis, the main points of the $(n,l)$ plot of F-actin would be represented by the open circles. b) the unrolled helical surface lattice of a microtubule, as seen from the outside, showing the main helical families, including the 8nm periodicities.

Just as one can introduce an 8nm periodicity into the basic 4nm surface lattice by proposing a difference between alternate subunits[8], even longer periodicities can be introduced, while still preserving overall helical symmetry. The way in which the $(n, l)$ plot can be subdivided to give a smaller reciprocal unit cell (and hence a larger unit cell in the surface lattice) is indicated in Fig. 2a. A possible way in which a microtubule might be decorated to give a helical structure with a longitudinal periodicity of about 32nm is discussed in more detail below.

Three-dimensional reconstruction of a microtubule image

Fig. 3c shows a model computed from the averaged data from three selected images of singlet flagellar A-tubules, following the method described in detail by DeRosier and Klug[25] and DeRosier and Moore[26]. The overlapping contributions on the equatorial and 8nm layer lines were separated as described by Amos and Klug[10]. Figs. 4 a-c show the reconstructed density in more detail, in the form of cylindrical and radial sections. At such low resolution, the boundaries between the protein and the contrasting stain around it are not very sharply defined, so the contour level chosen for construction of the model is one which leads to a reasonable degree of continuity between adjacent subunits. The volume enclosed is consistent with a molecular weight per 4nm unit of about 50 000 daltons and an assumed density of 800 dalton/$nm^3$. Further support for the assumption that each 4nm subunit corresponds to a similar molecular weight to that of the chemical monomer comes from the work of Gibbons[27]. He measured the amount of protein in flagellar axonemes as about $8 \times 10^{-15}$ gm/μm along the axoneme. Assuming that at least 50% of the protein is tubulin, the (9x23x240) 4nm units in a micron along the nine outer doublets and the (2x13x240) central pair 4nm units per micron would each

Fig. 3. Reconstructed images of flagellar A-tubules.
a) the optical diffraction pattern of tubule A in Fig.1a. The phases in the computed Fourier transform of the four peaks on the 4nm layer line, from left to right, were $235^\circ$, $210^\circ$, $35^\circ$, $240^\circ$.[10] On the 8nm layer line the phase of the left-hand peak ranged from $275^\circ$ to $335^\circ$, that of the right-hand peak from $70^\circ$ to $135^\circ$.
b) left: an enlarged image of tubule A, and right: an optically filtered image of this tubule. Because there is an overlap of two functions with similar values of $n$ on the 8nm layer line (see Fig. 2), optical filtering is successful only in a few favourable cases.
c) a model of the three-dimensional reconstructed image of a tubule. The two kinds of morphological unit are labelled α and β, although they may not represent α- and β-tubulin.

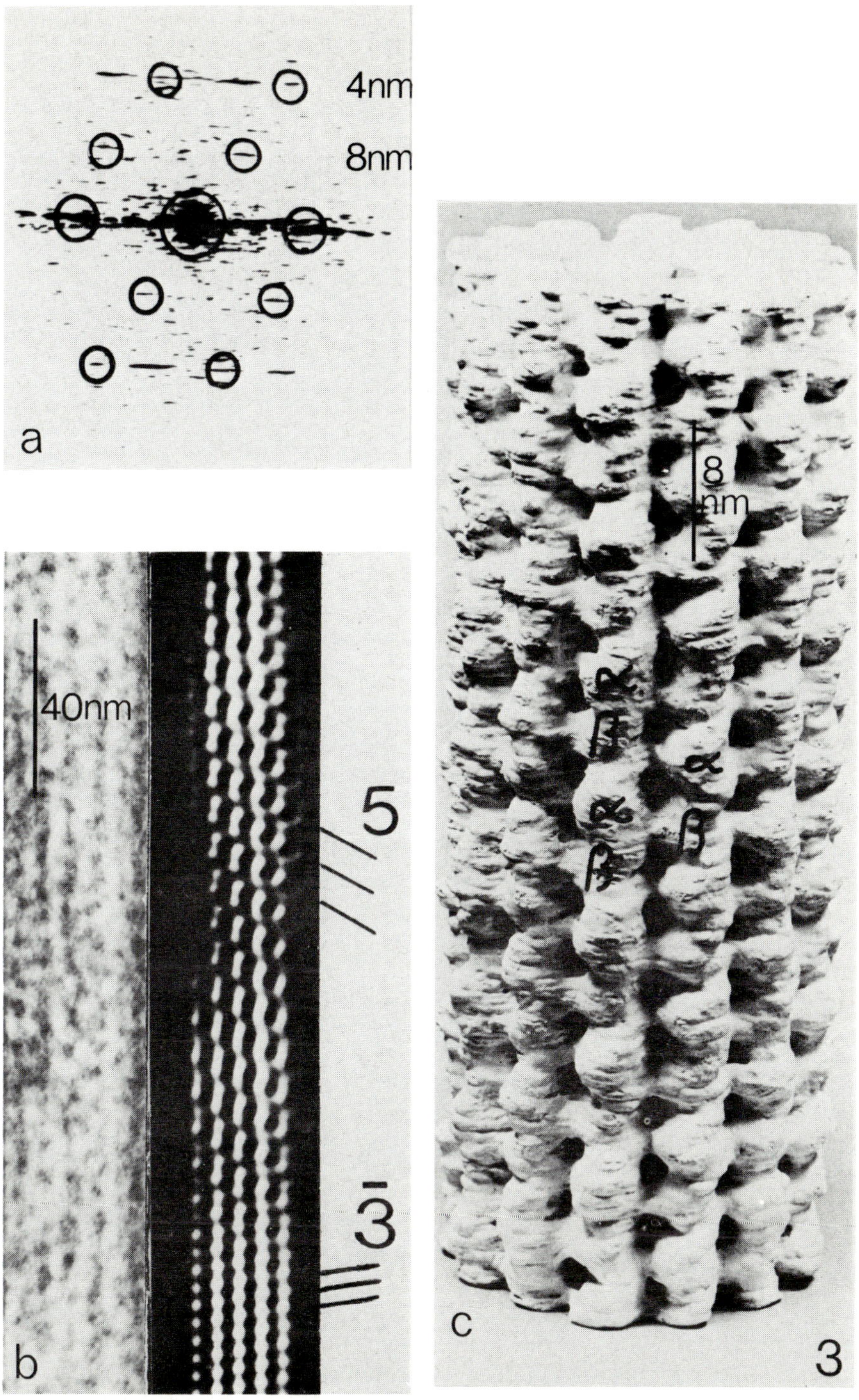
4nm
8nm
a
40nm
5
3̄
b
8
nm
α
β
α
β
α
β
c
3

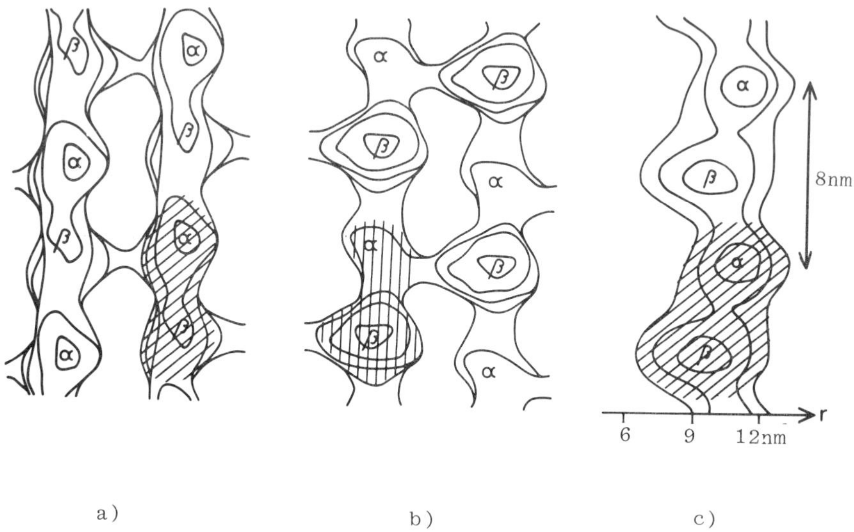


Fig. 4. Sections through the three dimensional reconstructed density map of a flagellar A-tubule. a) and b) represent superimposed cylindrical sections through two protofilaments of the reconstructed image. A single contour has been drawn for each radius. a) shows the surface as viewed from the outside, with sections between radii of 9.6nm and 12nm. b) shows the view from inside, with sections between radii of 7.2nm and 9.6nm. What is thought to be an individual heterodimer is shown shaded. Alternate 4nm morphological units are labelled α and β.

c) is a single radial section through the centre of a protofilament which shows the radial tilting of the tubulin dimers, the main feature of the reconstruction which distinguishes one type of 4nm subunit from the other.

have a molecular weight of about 45 000 daltons; this value is sufficiently close to the molecular weight of the tubulin monomer to suggest a one-to-one correspondence.

However, although the two halves of the dumb-bells in Fig. 3c are tentatively labelled α and β, they may not actually represent the two species of chemical monomer. The most likely alternative is that the pairs of 4nm units correspond to the tubulin dimers, but that they are divided down their long axes in some way into monomers. This would provide a possible explanation for the observations of Warner and Meza[28] and Erickson[7] of an apparent longitudinal splitting of the protofilaments, especially in disintegrating tubules. Unfortunately, no specimen has yet been found which shows the longitudinal splitting while still preserving an 8nm periodicity.

Thus, at the resolution limit of the investigation described here, the A-tubule of flagellar doublets appears to be made up of slightly tilted dumb-bell-shaped dimers, joined end-to-end to form longitudinal protofilaments (Figs.3c & 4). Adjacent protofilaments are staggered in such a way as to produce 5-start and 8-start helical families of dimers, of opposite hand; whilst a 3-start helical family of 4nm minor pitch can be drawn through the individual subunits. Purified brain microtubules and other cytoplasmic microtubules are probably almost identical in their basic structure[7], even though they often do not show much evidence of an 8nm periodicity. Since α and β-tubulin subunits are equal in molecular weight and apparently similar in sequence[16,17], one would not expect that the difference between alternate subunits would necessarily be apparent at the available resolution. The manifestation of the 8nm periodicity may reflect a perturbation in the positions of alternate subunits due to the attachment of some other component of the tubule.

## Microtubules as skeletal structures

An obvious role for flagellar microtubules is to provide a support for other components which project from them, although of course they may also play a more active role. However, since it seems to be fairly common for non-flagellar microtubules also to have other material associated with their walls *in vivo* (see Roberts[30] for a review), it is possible that a major role of microtubules in general is to act as a cellular skeleton. The question arises as to what sort of structure they might be supporting in an intra-cellular environment. One protein which it has been suggested is associated with

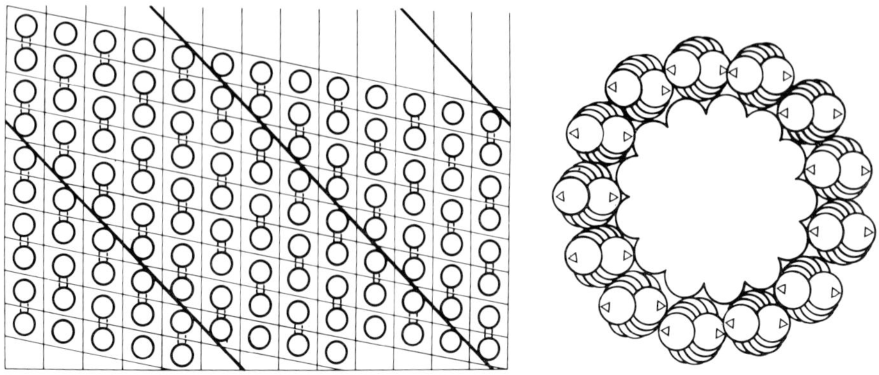

a) b)

Fig. 5. a) The helical surface lattice of a microtubule, as in Fig. 2b. The strong black lines indicate the direction of a 2-start helical family with a longitudinal repeat of 32nm, obtained by selecting every fourth helix of the 8-start helical family of dimers in Fig. 2b.

b) A diagram showing the end-on view of thirteen double-stranded actin filaments, packed to form a cylinder. The small triangles show the positions of equivalent points on the actin subunits. Adjacent filaments are exactly in register, but rotated relative to the cylinder surface by $2\pi/13$. Seen from the side, this rotation has the same effect as an axial shift of 2/13 of the actin longitudinal repeat (see Fig. 6).

---

Fig. 6. An idealised model of a microtubule coated with actin filaments packed side-by-side, so that where they come in close contact at every half turn, adjacent strands in different helices have a half staggering similar to that between the two strands of the same filament. As a result, groups of subunits along the filaments form approximately hexagonal close-packed arrays. A full complement of 13 actin filaments around the tubule model would produce two helices of close-packed actin subunits along the directions of the two-start family of the microtubule superlattice (indicated by the two black wires). This type of half-staggered packing has been observed in some actin paracrystals, as shown for example in the micrographs of Spudich *et al*.33. The resulting proportions of tubulin and actin in such a complex would be 4 tubulin dimers to 13 actin monomers, that is, approximately equal parts by weight.

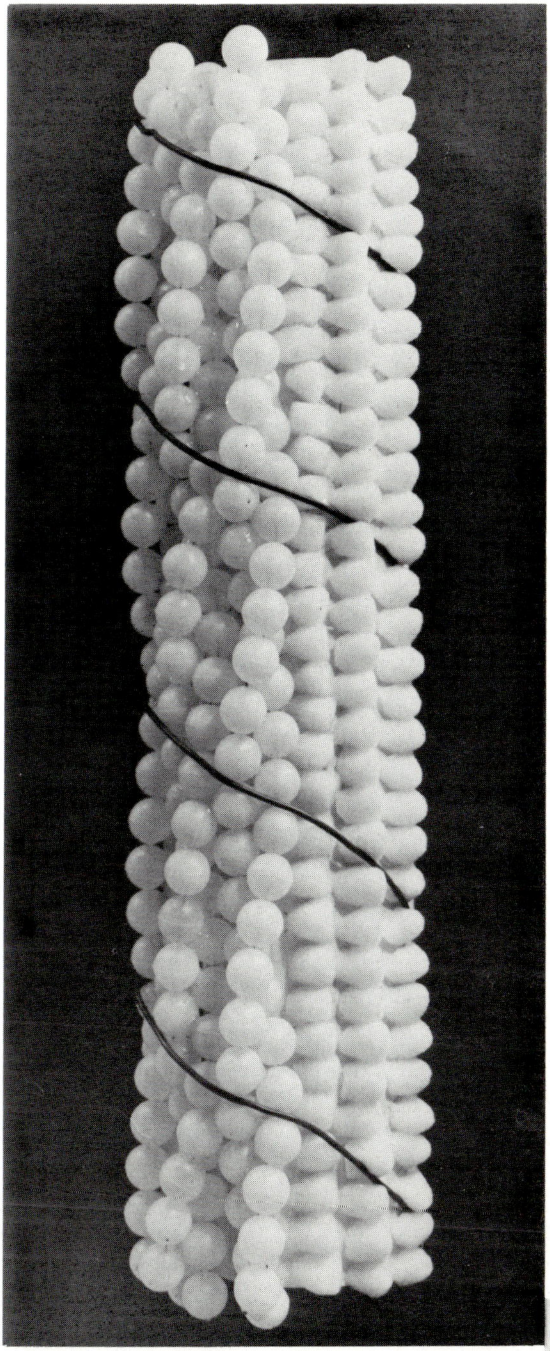

6

microtubules is actin, which would presumably be accompanied by myosin. Indeed, Forer and Behnke[31] have presented some evidence for an association *in vivo* between actin filaments and cytoplasmic microtubules. It is interesting therefore that the symmetries of these two structures seem to be fairly closely matched. F-actin is a right handed[32], double-stranded helix with thirteen subunits per turn and a longitudinal repeat distance of about 35nm[24]. Experiments with models have shown that it is possible to take thirteen actin filaments and to pack them neatly in parallel around a cylinder, with their axes parallel to the cylinder axis. The relative diameters of F-actin (about 8nm) and microtubules (about 24nm) would allow thirteen filaments to pack snugly around the circumference of a microtubule, as shown in Fig. 5b. The subunits in the two strands of an actin filament are half-staggered, and when a number of filaments are lined up alongside one another, it is possible for subunits in adjacent strands of different filaments to be half-staggered also. When actin filaments are packed in this manner around a cylinder, as in Fig. 6, the actin subunits in closest contact with the cylinder lie on a left-handed two-start helical family with a minor pitch equal to the 35nm F-actin repeat. This helical family fairly closely matches the two left-handed helices of the microtubule superlattice drawn in Fig. 5a.

It is not clear, however, that the axial spacings of the two lattices would match perfectly: the nominal 4nm periodicity of the basic microtubule lattice would give two-start helices with an axial periodicity of 32nm, as opposed to an F-actin repeat which is closer to 35nm. This 10% difference might mean that the complex could form with stretches of actin filament bulging out between regular points (every 32nm) of close contact with the tubule wall. Alternatively, the subunit spacings in the two polymers might well be somewhat flexible *in vivo* and might adjust to match one another perfectly during association.

At present, this type of interaction between microtubules and actin is simply speculation. However, such a close similarity in helical symmetry seems unlikely to have arisen by sheer coincidence: there certainly seems to be no reason why the actin filament should have evolved this particular symmetry within the myofilament lattice of a striated muscle. It is possible, therefore, that under some circumstances microtubules, by virtue of their intrinsic polarity, may provide a directional support system for transient actin-myosin interactions. This possible role for cytoplasmic microtubules as "polarised skeletal" structures is discussed in more detail by Roberts[30].

Acknowledgement

I am very grateful to Dr A. Klug and other colleagues for advice and stimulating discussions.

References

1. Ledbetter, M.C. and Porter, K.R., 1964, Science 144, 872.
2. Phillips, D.M., 1966, J. Cell Biol. 31, 635.
3. Ringo, D.L., 1967, J. Ultrastruct. Res. 17, 266.
4. Futaesaku, Y., Mizuhira, V. and Nakamura, H., 1972, J. Histochem. Cytochem. 20, 155.
5. Warner, F.D. and Satir, P., 1973, J. Cell Biol. 63, 35.
6. Tilney, L.G., Bryan, J., Bush, D.J., Fujiwara, K., Mooseker, M.S., Murphy, D.B. and Snyder, D.H., 1973, J. Cell Biol. 59, 267.
7. Erickson, H.P., 1974, J. Cell Biol. 60, 153.
8. Grimstone, A.V. and Klug, A., 1966, J. Cell Sci. 1, 351.
9. Chasey, D., 1972, Exptl. Cell Research 74, 140.
10. Amos, L.A. and Klug, A., 1974, J. Cell Sci. 14, 523.
11. Linck, R.W. and Amos, L.A., 1974, J. Cell Sci. 14, 551.
12. Witman, G.B., 1970, J. Cell Biol. 47, 229a.
13. Bibring, T. and Baxandall, J., 1971, J. Cell Biol. 48, 324.
14. Bryan, J. and Wilson, L., 1971, Proc. Nat. Acad. Sci. 68, 1762.
15. Fine, R.E., 1971, Nature New Biol. 233, 283.
16. Bryan, J., 1972, J. Mol. Biol. 66, 157.
17. Luduena, R.F. and Woodward, D.O., 1973, Proc. Nat. Acad. Sci., 70, 4594.
18. Linck, R.W., 1975, submitted to J. Cell Sci.
19. Borisy, G.G., Olmsted, J.B., Marcum, J.H. and Allen, C., 1974, Fed. Proc. 33, 167.
20. Jacobs, M., Smith, H. and Taylor, E.W., 1974, J. Mol. Biol. 89, 455.
21. Weingarten, M.D., Lockwood, A.H., Hwo,S-Y. and Kirschner, M.W., 1975, Proc. Nat. Acad. Sci. 72, 1858.
22. Klug, A. and Berger, J.E., 1964, J. Mol. Biol. 47, 384.
23. Klug, A., Crick, F.H.C. and Wyckoff, H.W., 1958, Acta Cryst. 11, 199.
24. Hanson, J. and Lowy, J., 1963, J. Mol. Biol. 6, 46.
25. DeRosier, D.J. and Klug, A., 1968, Nature 217, 130.
26. DeRosier, D.J. and Moore, P.B., 1970, J. Mol. Biol. 52, 355.
27. Gibbons, I.R., 1965, Archs. Biol., Liège 76, 317.

28. Warner, F.D., 1972, in *Advances in Cell and Molecular Biology* 2, 192. (ed. E.J. DuPraw) New York Acad. Press.

29. Warner, F.D. and Meza, I., 1974, J. Cell Sci. 15, 495.

30. Roberts, K., 1974, Prog. in Biophys. & Mol. Biol. 28, 371.

31. Forer, A. and Behnke, O., 1972, Chromosoma 39, 145.

32. Depue, R.H. and Rice, R.V., 1965, J. Mol. Biol. 12, 302.

33. Spudich, J.A., Huxley, H.E. and Finch, J.T., 1972, J. Mol. Biol. 72, 619.

*Microtubules and Microtubule Inhibitors, eds. M. Borgers and M. de Brabander*



# THE SLEEVE ELEMENT OF MICROTUBULES

H. Stebbings and C.E. Bennett

Department of Biological Sciences,

University of Exeter

## Summary

This study of the microtubules in nutritive tubes of the telotrophic ovary of Notonecta glauca glauca employed freeze-etching and electron microscope histochemical techniques. The normal appearance of these microtubules in sections stained with uranyl acetate and lead citrate is of a cylinder with outer and inner diameters of 22nm and 12nm respectively, and this is surrounded by an electron lucent zone of 66nm diameter.

Dialised iron and ruthenium red, which are both cationic stains with an affinity for mucopolysaccharides and other anionic surfaces, increased the apparent diameter of the microtubules to 32nm. The fact that freeze-etched microtubules also had an outer diameter of 32nm, suggested that the dialised iron and the ruthenium red were not simply adding on to the negatively charged surface of the microtubules at 22nm, but that they were staining some previously invisible component immediately surrounding the microtubules. This is most likely to be a mucopolysaccharide. The zone between 32nm and 66nm does not stain, and may be the result of electrostatic repulsion by the negatively charged microtubules.

## Introduction

Microtubules, although often labile structures, have been shown consistently to be preserved after fixation with glutaraldehyde. Indeed, using the standard techniques of electron microscopy, microtubules within a wide variety of plant and animal cells have been shown to have the same morphology, appearing as circular profiles of between 18-34nm diameter in transverse section, and as two parallel electron dense lines in longitudinal section[1,2].

A conspicuous feature of certain microtubules is an electron clear zone around their periphery into which cellular inclusions such as ribosomes do not encroach. Initially this region was thought to be a shrinkage artifact of fixation[3], but it has subsequently been shown to be a real component which is not normally stained. A part of the clear zone immediately surrounding neurotubules has been shown to stain with lanthanum hydroxide[4] and ruthenium red[5], and filamentous elements associated with certain neurotubules are especially pronounced after staining with lanthanum[6] and Alcian blue[7].

A massive system of microtubules has been discovered within the ovaries of

This work was supported by S.R.C. Research Grant, Nos. B/RG/5743.6

hemipteran insects[8]. Such ovaries are known as 'telotrophic' since they possess nutritive cells which comprise a trophic region at their anterior ends. The trophic region is in cytoplasmic continuity with the oocytes via nutritive tubes, and ribosomes pass back along microtubular tracts within the nutritive tubes to each of the developing oocytes in the chain.

In this study, we have employed histochemistry and freeze-etching to re-investigate the complete structure and arrangement of the cytoplasmic microtubules within the telotrophic ovary of Notonecta glauca glauca and have paid particular attention to the extent and nature of the zone immediately surrounding each microtubule which may be an integral 'sleeve element'. By examining the site of probable interaction between the microtubules and the ribosomes, we also hoped to cast light on the role of the microtubules in the transport of the ribosomes.

## Materials and Methods

The animals used in this study were of the species Notonecta glauca glauca (Linn.) and were collected from ponds on the University campus at Exeter.

### Controls:

Single ovarioles were freshly dissected into a 3.0% solution of glutaraldehyde in 0.06M phosphate buffer at pH 7.3 and 18°C. After 15m in glutaraldehyde, the ovarioles were washed briefly in phosphate buffer and transferred to 1.0% osmium tetroxide buffered to pH 7.3 with veronal acetate[9]. Ovarioles were then dehydrated through an ethanol series.

### Freeze-etching:

For freeze-etching, ovaries were removed and small pieces of ovarioles placed in 20% glycerol in Ringer's solution for 10m. The 'neck' regions of ovarioles were then excised and mounted on gold collared specimen holders before freezing in Freon 22. The freeze-etching method was according to the now standard technique of Moor and Mühlethaler[10]. The etched specimens were shadowed with platinum/carbon in a "Balzers" 360M freeze-etcher. The replicas of the frozen fractured surfaces were cleaned using commercial bleach (Chloros, ICI) for 3h followed by 70% sulphuric acid for 6h, and then mounted on Formvar coated grids.

Prior to staining with Alcian blue, ruthenium red, iron and thorium, ovaries were removed, pinned out onto dental wax under insect Ringer, and perfused and fixed for 1h with 3.0% glutaraldehyde in the buffer to be used in the particular staining procedure.

To allow the penetration of the stains into the nutritive tubes, and thus to the vicinity of the microtubules, individual ovarioles were cut transversely just posterior to the trophic region, across the 'neck' of the ovarioles through which many nutritive tubes pass. The cuts were made with a steel blade using the minimum of pressure and a slight rolling action to reduce the possibility of sealing the nutritive tubes.

Alcian blue (after Hayat[11]):

After glutaraldehyde fixation in 0.06M Sorenson's buffer at pH 6.5, the tissue was transferred to a mixture of 3.0% glutaraldehyde in 0.06M Sorenson's buffer at pH 6.5 and 1.0% Alcian blue for a further 1h. The tissue was then washed for 1h in several changes of distilled water and postfixed for 1h in 1.0% osmium tetroxide buffered to pH 6.5 with veronal acetate. The tissue was then dehydrated through an ethanol series.

Ruthenium red (after Tani and Ametani[12]):

After fixation in glutaraldehyde buffered to pH 7.3 with 0.2M cacodylate buffer, the tissue was transferred to a solution of 3.0% glutaraldehyde buffered to pH 7.3 with cacodylate buffer containing 0.1% ruthenium red for 1h. The tissue was then washed in several changes of cacodylate buffer over a period of 1h prior to postfixation for 3h with 1.0% osmium tetroxide buffered to pH 7.3 with cacodylate buffer and containing ruthenium red. The tissue was then dehydrated through an ethanol series.

Dialised iron (after Rinehart and Abul-Haj[13]):

After glutaraldehyde fixation in 0.06M Sorenson's buffer at pH 7.3, the tissue was washed several times over a period of 1h in Sorenson's buffer alone. Dialised iron stain was prepared by the method of Rinehart and Abul-Haj (1951). One part of glacial acetic acid was added to three parts of the stock iron solution to achieve a pH below 2.0 and this was allowed to act on the tissue for 12h at room temperature. The tissue was then washed in distilled water for 1h prior to postfixation with 1.0% osmium tetroxide buffered to pH 7.3 with veronal acetate, at $4^{o}C$ for 1h. This was followed by dehydration in an ethanol series.

Thorium (after Rambourg and Leblond[14]):

After glutaraldehyde fixation in 0.06M Sorenson's buffer at pH 7.3, the tissue was washed in several changes of Sorenson's buffer alone over a period of 1h. It was then postfixed for 1h in 1.0% osmium tetroxide buffered to pH 7.3 with veronal acetate and washed for a further 1h in distilled water. 1.0% colloidal thorium, "Thorotrast" was prepared in 3.0% acetic acid (pH 2.5) and the tissue stained for 24h. The tissue was then rinsed in 3.0% acetic acid for 1h followed by dehydration in an ethanol series.

All ovarioles were passed through propylene oxide, and embedded in Araldite. Silver to grey sections were cut with glass knives on a Sorvall MT2-B Ultramicrotome, mounted on grids without supporting films and in some cases their contrast was enhanced by counter-staining with a saturated solution of uranyl acetate in 50% ethanol for 20m and lead citrate for 2m.

Sections and freeze-etched replicas were examined with an AEI EM 801 electron microscope.

## Results

Nutritive tubes extend backwards from the trophic region to each of the developing oocytes within an ovariole (Fig. 1). Transverse sections of the 'neck' regions of ovarioles will therefore reveal up to twenty nutritive tubes - each cut transversely. The tubes are roughly circular in cross section with diameters of between 10-20µm.

After fixation with glutaraldehyde and osmium, and staining with uranyl acetate and lead citrate, the nutritive tubes showed the same morphology as described by Macgregor and Stebbings[8]. Each tube was packed with longitudinally arranged microtubules. The microtubules appeared as electron dense circular profiles in transverse section with an outer diameter of approximately 22nm and an inner electron clear core of approximately 12nm diameter (Fig. 2). The microtubules rarely touched each other and there was a distinct electron clear zone of approximately 66nm diameter around each of the microtubules, into which the ribosomes which also pack the nutritive tubes, did not encroach (Diag. 1, Fig. 2). The spatial relationships between the microtubules and the ribosomes within nutritive tubes was not easily discernible in longitudinal sections since the dimensions of the microtubules and the ribosomes are such that they may be superimposed even in the thinnest sections.

In transverse fractures of nutritive tubes the microtubules also showed as circular profiles, but with a diameter of approximately 32nm and an inner core of 14nm (Diag. 1, Fig. 3). Each microtubule was surrounded by a clear zone of approximately 66nm diameter and there were projections from the microtubules into this

---

Fig. 1. Longitudinal section of an ovariole showing parts of the nutritive tubes (arrowed) which connect the trophic region (tr) to each of the string of oocytes.

Fig. 2. Electron micrograph of transverse section of part of a nutritive tube which had been fixed in glutaraldehyde and osmium, and stained with uranyl acetate and lead citrate. The transversely sectioned microtubules are surrounded by distinct electron clear zones (arrows), and are rarely seen to touch each other.

Fig. 3. Electron micrograph of part of a transversely fractured nutritive tube, showing microtubules which are also fractured transversely.

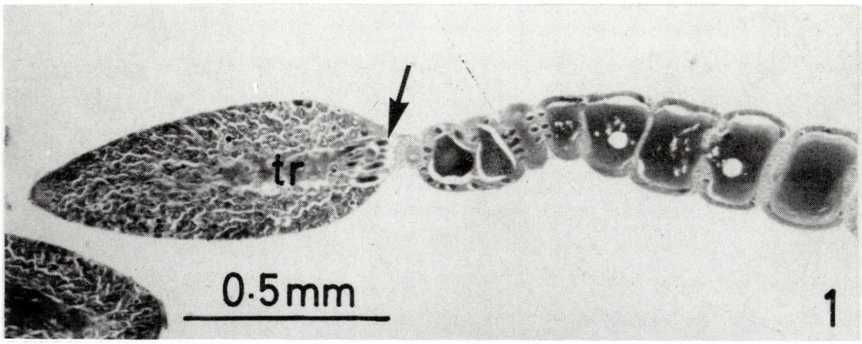
tr
0.5mm
1

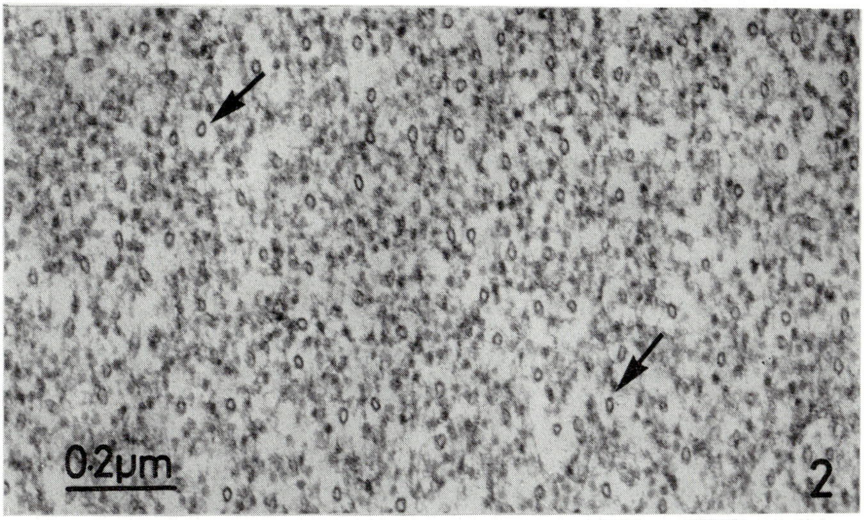
0.2μm
2

0.2μm
3

region.

Alcian blue penetrated into the nutritive tubes approximately 15μm from their cut ends during the one hour staining period. Thus, observation of longitudinal sections allowed a direct comparison between the Alcian blue - stained, and unstained regions of the nutritive tubes. In transverse section the microtubules appeared as relatively electron lucent circles 25nm in diameter and occasionally

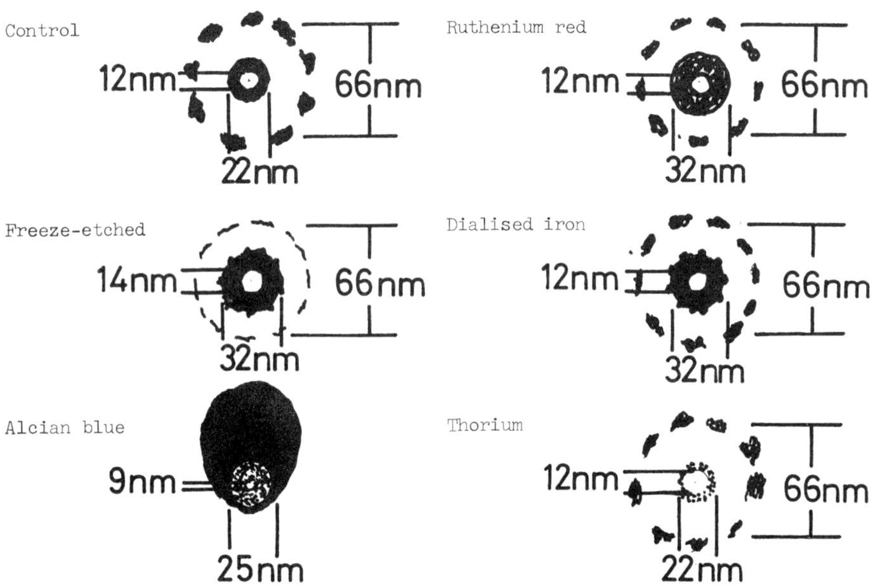


Diag. 1. Summary of microtubule images and their dimensions after the different preparative techniques.

---

Figs. 4-8. Electron micrographs of transverse and longitudinal sections of microtubules within nutritive tubes.

Fig. 4. Transverse sections of microtubules stained with Alcian blue.

Fig. 6. Longitudinal sections of microtubules stained with Alcian blue. cf. Fig. 5 of unstained microtubules further up the same nutritive tube where the stain had not penetrated.

Fig. 7. Transverse sections of microtubules stained with ruthenium red. The outer and inner circumferences of the microtubules are heavily stained (arrows).

Fig. 8. Longitudinal sections of microtubules stained with ruthenium red.

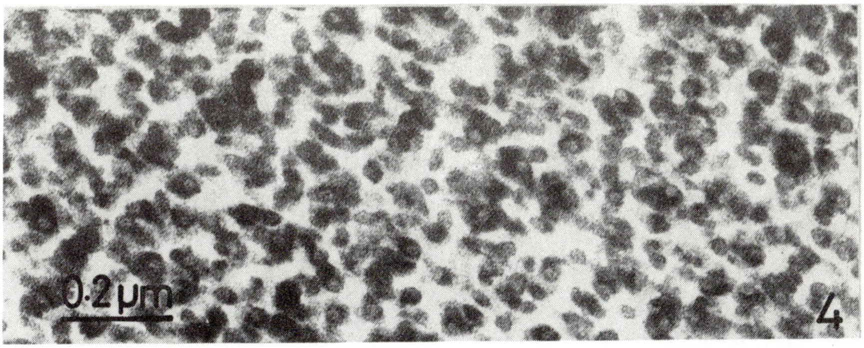
0·2 µm
4

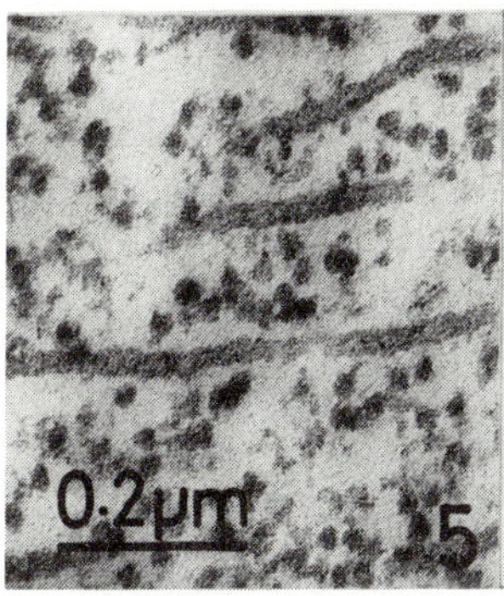
0·2 µm
5

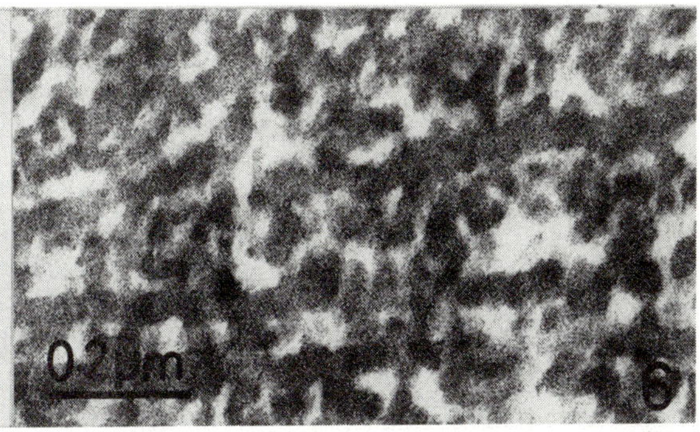
0·2 µm
6

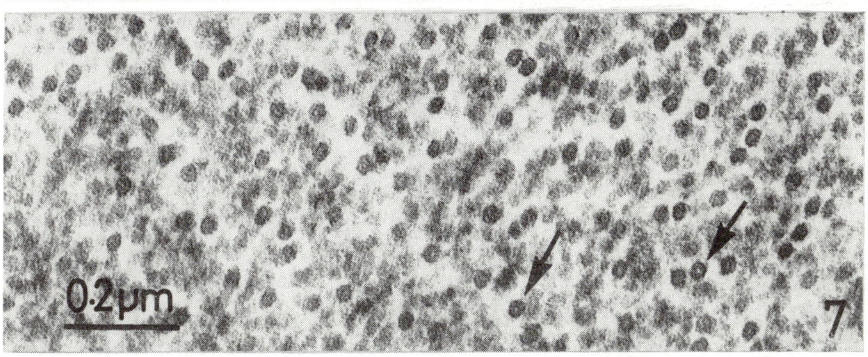
0·2 µm
7

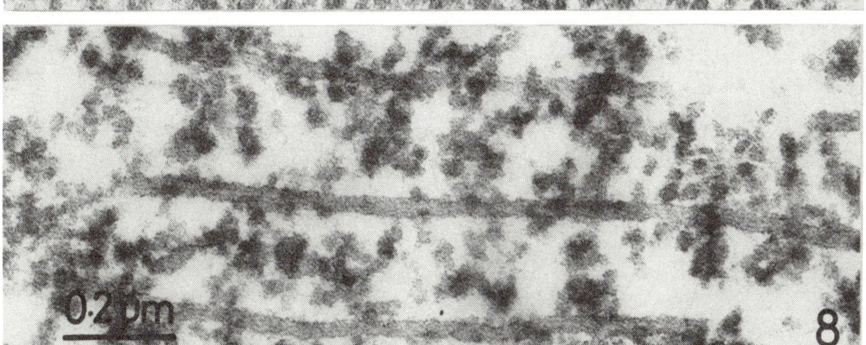
0·2 µm
8

an inner core of approximately 9nm in diameter which showed less electron density could be discerned (Diag. 1, Fig. 4). Asymmetrically positioned electron dense aggregates could be seen extending up to 41nm around and ensheathing many of the microtubules. These stained aggregates, which coat the microtubules, can be seen in longitudinal section in Fig. 6, and compared with the unstained image further up the nutritive tube in Fig. 5.

Fixation of ovarioles in the presence of ruthenium red increased the electron contrast of the microtubules which appeared as two concentric electron dense circles in transverse section. The outer circle had a diameter of approximately 32nm and the inner core a diam. of approximately 12nm (Diag. 1, Fig. 7). Each tubule was surrounded by a distinct clear zone 66nm in diameter and wispy material associated with the surfaces of the microtubules ran through this zone. In longitudinal sectionsstained with ruthenium red, the surface structure of sub-units can be observed.

There was a positive reaction between the dialised iron stain and the outer circumference of each of the microtubules (Figs. 9 and 10). This increased their outer diameters to approximately 32nm (Diag. 1, Fig. 9). The electron clear core of the microtubules was approximately 12nm in diameter. Each tubule was surrounded by an electron clear zone of 66nm in diameter.

Thorium did not stain the microtubules (Figs. 11 and 12), but was deposited in the form of fine particles within the matrices of the nutritive tubes.

## Discussion

The passage of ribosomes within insect telotrophic ovarioles, from the trophic regions to the developing oocytes has been well documented (Macgregor and Stebbings[8]). The nutritive tubes which connect the trophic regions to the oocytes are strongly birefringent and these have been shown to be packed with approximately 30,000 microtubules.

Distinctly oriented microtubules have been regularly described within tracts of cytoplasmic movement, but although their presence has been shown to be essential for movement to take place, it is not clear whether the microtubules cause or

---

Figs. 9-12. Electron micrographs of transverse and longitudinal sections of microtubules within nutritive tubes.

Figs. 9 and 10. Transverse and longitudinal sections of microtubules stained with dialised iron. Dense particles of iron can be seen, bound to the microtubular surfaces.

Figs. 11 and 12. Transverse and longitudinal sections of microtubules after staining with thorium. The microtubules themselves (arrowed) did not stain with thorium which aggregated in clumps within the matrices of the nutritive tubes.

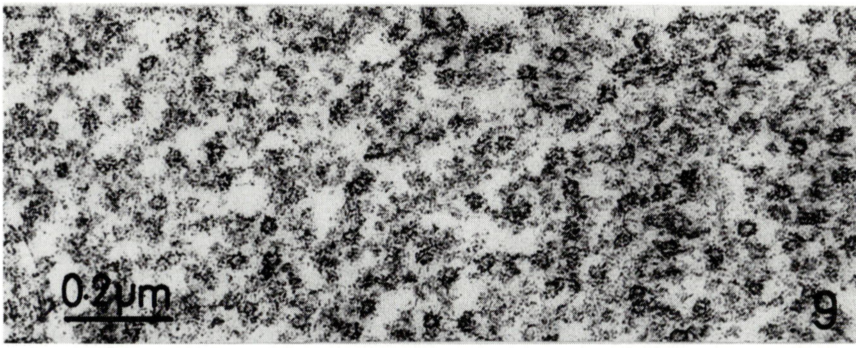
0.2μm
9

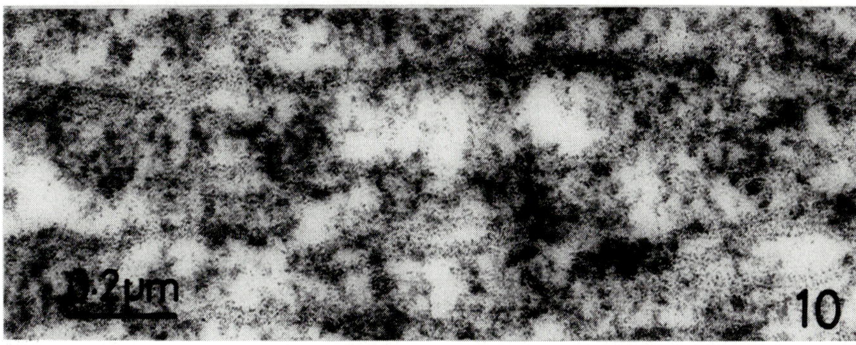
0.2μm
10

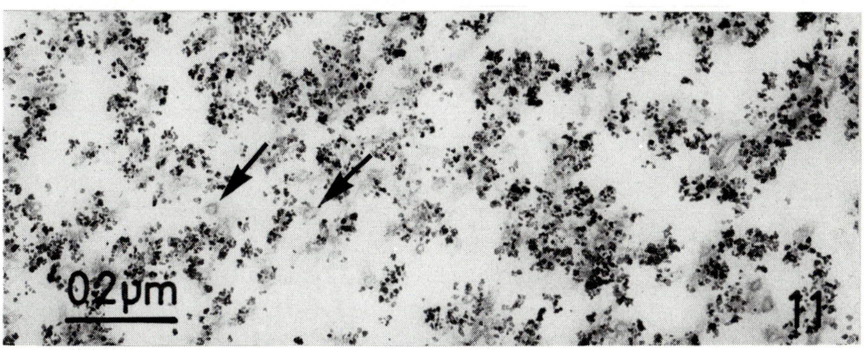
0.2μm
11

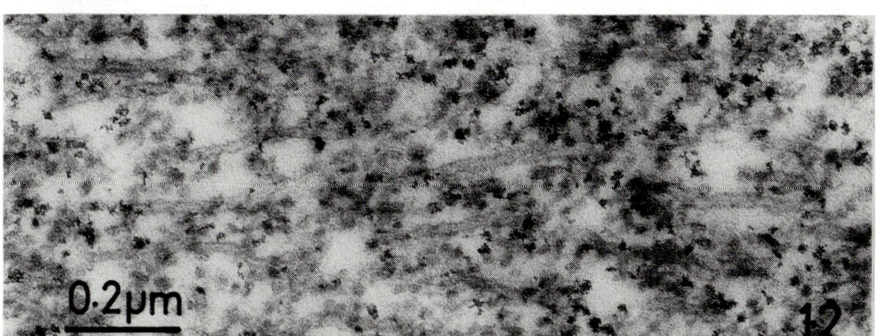
0.2μm
12

merely facilitate the motility.

We considered it significant that within systems, such as the nutritive tubes, where microtubules are arranged parallel to the direction of cytoplasmic movement, the tubules are invariably surrounded by a distinct zone which does not stain with conventional E.M. preparative procedures. Since this region is likely to be the site of any interaction between the microtubular complex and the moving ribosomes, we decided to examine the structure of the system after freeze-etching and different E.M. histochemical techniques.

The gross morphology of the nutritive tubes was the same after fixation and staining with Alcian blue, ruthenium red, dialised iron and thorium, as with conventional preparation. After freeze-etching, on the other hand, rows of microtubules, which were not a feature of fixed and stained preparations, were observed. The reason for this is unclear. However, the appearance and in particular the dimensions of the microtubules, as visualized by the different preparative procedures, showed significant similarities and differences.

The most consistent feature was a distinct zone of approximately 66nm diameter around individual microtubules from which the ribosomes were excluded. This was clearly apparent after freeze-etching and after each of the different staining procedures, except Alcian blue. None of the techniques stained and thus revealed the nature of the entire 66nm zone.

In transverse sections of conventionally prepared ovarioles, the microtubules appeared as electron dense circles, 22nm in diameter, with electron clear cores, 12nm in diameter. The microtubules in transverse fractures also appeared as circles, but with a larger diameter of 32nm and an inner core of 14nm.

As with certain neurotubules, regions associated with microtubules in the nutritive tubes stained with cationic stains. These stains are thought to possess some specificity for mucopolysaccharide, or at least for polyanionic molecules having a high charge density. Care must be taken in interpreting the nature of the stained regions, since as Burton and Hinkley[15] have pointed out, the binding of cationic stains by microtubules could be accounted for by the fact that some tubulins are themselves strongly anionic.

Our results clearly showed the both ruthenium red and dialised iron stained the microtubules in such a way that their outer diameters of 32nm corresponded to those of freeze-fractured microtubules. This suggests that a previously invisible component (between 22nm and 32nm), probably composed of mucopolysaccharide, is visualized by these completely different techniques.

Alcian blue also bound as electron opaque aggregates to the outer surfaces of the microtubules. Such aggregates are generally 30-50nm in diameter and are thought to be formed by an interaction between Alcian blue cations and an electronegative acid mucopolysaccharide-protein complex.

In any event, the periphery of the microtubules is certainly strongly anionic in character, and the fact that the microtubules only rarely come into contact with each other, and that ribosomes position themselves in 'halos' around microtubules, could possibly be explained in part by an electrostatic repulsion between fixed negative charges on the surfaces of these cytoplasmic components.

We are in the process of investigating this by introducing 'foreign' particles of designated charge and size into microtubular motile systems and observing the behaviour of these and particularly their positioning in relation to the microtubules. The insect ovarian system which we have chosen is ideally suitable for such a study because of the enormous numbers of microtubules arranged within distinct tracts, because its size lends itself to manipulative experiments such as those involving microinjection, and because the ovaries can be cultured in vitro[16].

References

1. Olmsted, J.B. and Borisy, G.G., 1973, A. Rev. Biochem. 42, 507.
2. Hepler, P.K. and Palevitz, B.A., 1974, A. Rev. Pl. Physiol. 25, 309.
3. Maser, M.D. and Philpott, C.W., 1964, Anat. Rec. 150, 365.
4. Lane, N.J. and Treherne, J.E., 1970, J. Cell Sci. 7, 217.
5. Tani, E. and Ametani, T., 1970, J. Cell Biol. 46, 159.
6. Burton, P.R. and Fernandez, H.L., 1973, J. Cell Sci. 12, 567.
7. Hinkley, R.E., 1973, J. Cell Sci. 13, 753.
8. Macgregor, H.C. and Stebbings, H., 1970, J. Cell Sci. 6, 431.
9. Palade, G.E., 1952, J. exp. Med. 95, 285.
10. Moor, H. and Mühlethaler, K., 1963, J. Cell Biol. 17, 609.
11. Hayat, M.A., 1972, in: Basic electron microscopy techniques (Van Nostrand Reinhold Co.) p. 75.
12. Tani, R. and Ametani, T., 1970, J. Cell Biol. 46, 159.
13. Rinehart, J.F. and Abul-Haj, S.K., 1951, A.M.A. Arch. Pathol. 52, 189.
14. Rambourg, A. and Leblond, C.P., 1967, J. Cell Biol. 32, 27.
15. Burton, P.R. and Hinkley, R.E., 1974, J. Submicr. Cytol. 6, 311.
16. Stebbings, H. and Ratcliffe, N.A., 1973, Nature, New Biol. 242, 253.

*Microtubules and Microtubule Inhibitors, eds. M. Borgers and M. de Brabander*



# CROSS-LINKED TUBULIN

R. Luduena, L. Wilson and E. M. Shooter

Departments of Pharmacology, Genetics, and Biochemistry,

Stanford University School of Medicine, Stanford, California

Summary

The structure and stability of the tubulin dimer have been studied by chemical cross-linking in vitro. Samples of embryonic chick brain tubulin were cross-linked with a variety of bifunctional reagents and the results were analyzed by acrylamide gel electrophoresis in a system capable of resolving the αα, αβ and ββ dimers. Optimal results were obtained with dimethyl-3,3'-(tetramethylene-dioxy)dipropionimidate (DTDI). Tubulin cross-linked with DTDI migrated largely as the αβ dimer, indicating that tubulin consists mainly of heterodimers. The αβ dimer undergoes spontaneous conformational changes and aggregation, resulting in the appearance of cross-linked αα and ββ dimers. These processes are inhibited by colchicine or vinblastine. Tubulin from sea urchin outer doublets and from vinblastine crystals also consists mainly of αβ dimers. DTDI can thus be used as a probe of the structure and stability of tubulin.

## 1. Introduction

Microtubules from a number of sources contain two kinds of 55,000 molecular weight monomeric proteins: α- and β-tubulin, differing from each other in their electrophoretic mobilities, amino acid sequences, and states of phosphorylation[1-6]. The functional subunit of microtubules is a 110,000 molecular weight dimer which, unlike either of the monomers, has the ability to bind to colchicine, vinblastine and GTP at distinct sites, to polymerize onto pre-formed microtubules, and to assemble into 36 S "rings"[7-10]. These properties are each subject to decay, at different rates, once microtubules are dissociated into their subunits[9,11].

There has been some controversy about the composition of the tubulin dimer, which could theoretically exist as either heterodimers (αβ), homodimers (αα and ββ) or both heterodimers and homodimers (αα, αβ, and ββ). Bryan and Wilson[3], observing that α- and β-tubulin occur in almost equal quantities in every microtubule system, proposed that tubulin is made of heterodimers. Witman et al[12], however, showed that in the partition protofilaments of Chlamydomonas outer doublets, one form of tubulin monomer was present in great excess over the other, suggesting that the tubulin in this area consisted of homodimers. Resolution of the question is complicated by the fact that the tubulin dimer, held together by non-covalent bonds, dissociated into monomers under the conditions present in any analytical system which could presumably distinguish between the possible dimers[13]. Furthermore, α- and β-tubulin are very close in molecular weight[13,14] and systems such as gel filtration and sedimentation equilibrium, whose conditions are gentle enough to keep the dimer intact, are unable to identify the type of dimer present.

Our strategy, therefore, has been to treat solubilized tubulin dimers with cross-linking reagents to form covalent bonds between the two monomers within the dimer and then to identify the cross-linked dimer on acrylamide gels. We have found that the tubulin dimer exists largely as the heterodimer αβ and is stabilized by colchicine and vinblastine, and that the decay of the αβ dimer is accompanied by the randomization of the cross-linking.

## 2. Materials and Methods

Dimethylsuberimidate dihydrochloride (DMSI) was purchased from the Eastman Chemical Co. Dimethyl-3,3'-(tetramethylenedioxy)dipropionimidate dihydrochloride (DTDI) was the kind gift of Dr. George Stark. DTDI has the formula $[ClH_2N = C(OCH_3)-(CH_2)_2O(CH_2)_2]_2$. Methyl-4-mercaptobutyrimidate (MMBI) was from the Pierce Chemical Co.; phenylenedimaleimide (PDM) was from the Aldrich Chemical Co. N,N'-hexamethylene(bis)iodoacetamide ($IC_6I$) was synthesized by the method of Ozawa[15]. Vinblastine was purchased from the Eli Lilly Co. and colchicine from Sigma.

Solubilized tubulin from 13-16-day-old chick embryo brains was purified by chromatography on DEAE-cellulose according to the method of Eipper[5]. Outer doublet microtubules from the sperm of the sea urchin *Strongylocentrotus purpuratus* were purified[16] and partially solubilized by incubation with 0.05 M sodium pyrophosphate-2.5 mM $MgCl_2$ (pH 7.0) (PPMg) for 1 hour at 1°C, followed by centrifugation at 27,000 g. The supernatant was used for cross-linking. Vinblastine crystals from sea urchin embryos were prepared[17], centrifuged and resuspended in PPMg for cross-linking. α- and β-tubulin were separated from each other by the procedure of Luduena and Woodward[6].

MMBI, DTDI, and DMSI were dissolved in PPMg and the pH readjusted to 7.0. $IC_6I$ was dissolved in methanol and PDM in dimethylformamide. Freshly purified tubulin (final concentration 0.2-0.3 mg/ml[18]) was mixed with the cross-linker solution and reacted for 3 hours at 24°C. The tubulin was then dialyzed against PPMg and reduced and carboxymethylated[19]. MMBI-cross-linked dialyzed tubulin was treated with $H_2O_2$ according to Traut et al[20]. αα, αβ, and ββ tubulin dimers used as standards for the mobility of cross-linked tubulin were prepared by reacting α- and β-tubulin with DMSI by the procedure of Carpenter and Harrington[21]. All cross-linked tubulin samples were analyzed by electrophoresis on acrylamide gels in the system of Yang and Criddle[22].

---

Fig.1. Separation of αα, αβ, and ββ dimers on Yang-Criddle gels. Samples of cross-linked tubulin were run on 4% gels, which were then stained with Coomassie blue. Samples were as follows: 1. α- and β-tubulin mixed before cross-linking; 2. α- and β-tubulin mixed after cross-linking; 3. Cross-linked β-tubulin; 4. Cross-linked α-tubulin. Origins are at the top

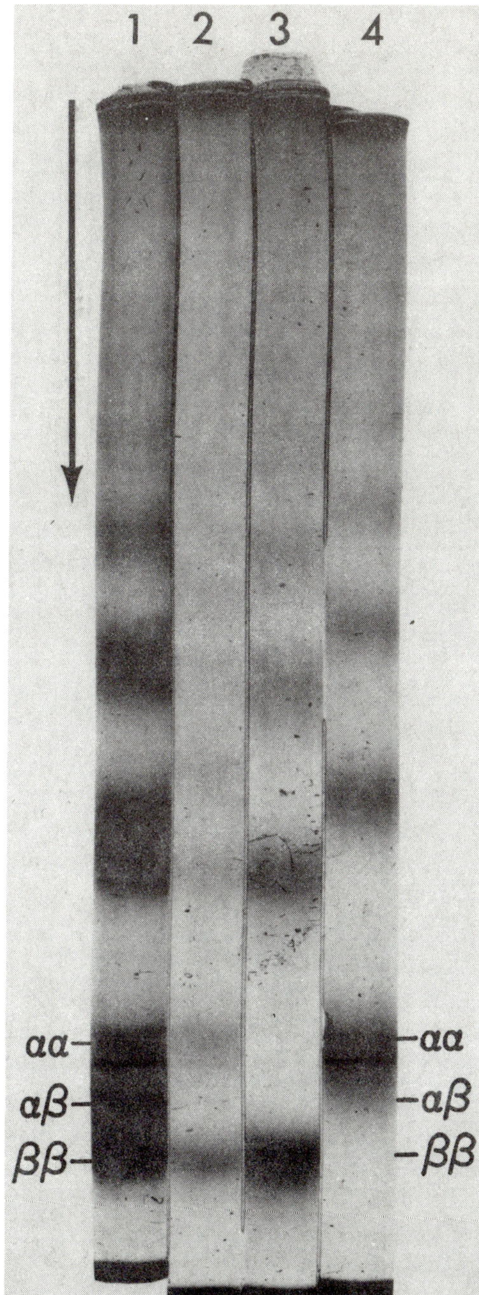
1
2
3
4
αα
αβ
ββ
αα
αβ
ββ

3. Separation of αα, αβ, and ββ dimers

To correctly identify the cross-linked dimer, it is necessary to use an acrylamide gel system on which the αα, αβ, and ββ dimers migrate at different rates. The gel system in which these three species are best separated is that of Yang and Criddle[22], in which the α- and β-tubulin monomers are also well-resolved[6] (Fig. 1). The tubulin samples used for these gels were cross-linked by a procedure[21] which generates a series of oligomers including the dimers. On 4% gels tubulin monomers migrate at the dye front. The αα and ββ dimers migrate at distinctly different rates (gels 2-4); mixing α- and β-tubulin prior to cross-linking results in the appearance of an αβ dimer band between αα and ββ (gel 1).

In most tubulin preparations which have not been treated with cross-linker a fraction (1-2% in chick brain tubulin, 10% in sea urchin sperm outer doublets) runs as a dimer. Whether this indicates the presence of a cross-linking enzyme *in vivo* or whether it is an artifact due to irreversible aggregation is not clear. This "background" dimer is about 40% αα, 40% αβ, and 20% ββ. Unless otherwise indicated, the results of the cross-linking given in this paper have not been corrected for "background" dimer.

4. Cross-linking embryonic chick brain tubulin *in vitro*

When embryonic chick brain tubulin is cross-linked with $IC_6I$, MMBI, or PDM, the results are the following. Reaction with 10 mM MMBI produces a low yield (6.0%) of a dimer which runs on 4% gels as a single broad band with little or no evidence of more than one component. Concentrations of $IC_6I$ above 0.171 mM generate higher yields (10-15%) of a dimer which also can not be resolved into components. PDM at 25 μM or higher also cross-links an unresolvable dimer at high yield. These results can be explained by the fact that tubulin monomers are poorly resolved on gels unless they have been reduced and carboxymethylated[14]; this must be true of the dimers also. Use of $IC_6I$, PDM, or MMBI, which bind to sulfhydryl groups, precludes subsequent complete carboxymethylation and may thus prevent resolution of a dimer cross-linked with these reagents. At concentrations of 0.171 mM or lower, $IC_6I$ generates a dimer consisting of three components with mobilities corresponding to those of αα, αβ, and ββ. The dimers are very poorly resolved, however. 12.6 μM PDM forms a cross-linked dimer containing 4 poorly resolved bands; under these same conditions PDM also generates a third monomer band migrating ahead of β-tubulin, suggesting that the fourth dimer band is an artifact and making the cross-linking results difficult to interpret.

The di-imidate reagents are much more promising. DMSI and DTDI form cross-links 11.4 A and 16.5 A long, respectively, between the ε-amino groups of lysine residues[23,24]. When freshly purified tubulin is cross-linked with DTDI or DMSI a dimer is formed in high yield and is resolvable into three components, co-migrating exactly with the $\alpha\alpha$, $\alpha\beta$, and $\beta\beta$ dimers. The cross-linking reaction gives optimal results in PPMg, pH 7.0-7.5, at cross-linker concentrations of 30 mM or lower. Raising the pH or the cross-linker concentration increases the yield of dimer but causes extensive blurring of its components. Under such conditions a majority of the di-imidate molecules bind to the protein with one imidate group only, leaving the other group free to hydrolyze[24]. This leads to extensive charge heterogeneity causing blurring of the dimer bands in this gel system which is sensitive to charge[14]. As expected the $\alpha$ and $\beta$ monomer bands are also blurred when tubulin is cross-linked at high pH or high cross-linker concentration. Paradoxically, therefore, the cross-linking reaction gives the clearest results under conditions where the reaction goes poorly.

The yield of cross-linked dimer at pH 7.0 is 10% or less. Although this represents only a small fraction of the total tubulin present, it probably provides a valid basis from which to draw conclusions about the rest of the tubulin population, because under the same set of conditions, both DTDI and DMSI cross-link relatively small amounts (12-28%) of other multimeric proteins (results not shown).

Analysis on 4% gels shows that DMSI cross-links $\alpha\alpha$, $\alpha\beta$, and $\beta\beta$ with approximate relative yields of 25%, 50%, and 25%, respectively (Fig. 2, gel 3). Since $\alpha$- and $\beta$-tubulin contain almost equal numbers of lysine residues[3], these results do not necessarily reflect the actual composition of the population of tubulin dimers, but may arise from non-specific cross-linking between dimers colliding randomly or forming aggregates. DTDI consistently cross-links a dimer which is about 60% $\alpha\beta$. If DTDI cross-linking is intra-dimer, then this result suggests that the tubulin dimer is largely $\alpha\beta$; however, if the tubulin is aggregating, then DTDI may be cross-linking non-specifically, but in a different manner than DMSI.

5. Evidence for the heterodimer model

If tubulin is largely a heterodimer, then the $\alpha\alpha$ and $\beta\beta$ dimers that are cross-linked may arise from aggregation. Similarly, if tubulin is mostly homodimers, then aggregation would lead to the appearance of more $\alpha\beta$. There are two ways of distinguishing these possibilities.

a. Effect of tubulin-specific drugs. If tubulin is cross-linked in the presence of drugs which stabilize it, then the cross-linking results should show a lower relative yield of those species of dimer which are formed during decay. Colchicine decreases tubulin aggregation and vinblastine stabilizes colchicine binding to tubulin[8,11,25]. When tubulin is cross-linked with DTDI in the presence of either of these drugs, there is a significant increase in the yield of $\alpha\beta$

relative to αα and ββ (see Table 1). In the presence of both colchicine and vinblastine 81% of the DTDI-cross-linked tubulin is αβ. (Fig. 2, gel 1). When this figure is corrected for the "background" dimer, about 88% of the cross-linked tubulin is αβ. Analogues of vinblastine have effects on the yield of cross-linked αβ that correlate with their known abilities to bind to tubulin (Luduena et al, manuscript in preparation). When DMSI is used as the cross-linker, vinblastine

Table 1

Effect of drugs on the cross-linking of tubulin by DTDI

| Drug added | Relative yield of αβ dimer (%) |
|---|---|
| None | 58.8 ± 0.9 |
| 0.1 mM colchicine | 64.7 ± 0.2 |
| 0.025 mM vinblastine | 65.4 ± 0.5 |
| 0.1 mM colchicine + 0.025 mM vinblastine | 81 |

reduces the total yield of dimer by over 50%, but has very little effect on the relative yield of αβ. This is consistent with a model whereby DMSI cross-links mainly within an aggregate. In the presence of vinblastine less of this aggregate would form and less would be cross-linked by DMSI, but the pattern of cross-linking would not change. DTDI, on the other hand, is probably the correct length to form an intra-dimer cross-link, but could also cross-link within an aggregate. Therefore, any agent which diminishes aggregation should make DTDI cross-linking more specific. Concentrations of vinblastine slightly higher than those used here are known to form an 8-9 S tubulin aggregate, perhaps a tetramer[26]. If vinblastine is forming such an aggregate in this system, the cross-linking may not be affected by it, since the structure of the 8-9 S tubulin may not allow inter-dimer cross-linking. On the other hand, the aggregate formed by tubulin decay would probably be a large non-specific one, containing numerous possibilities for inter-dimer cross-linking.

b. Effect of tubulin decay. If tubulin decay is accompanied by the formation of an aggregate, and if aliquots are cross-linked at a sequence of time points, then there should be an increasing yield of cross-linked dimer with time. This is observed with both DMSI and DTDI (Fig. 3). Vinblastine strongly stabilizes against this decay. In addition, the amounts of cross-linked oligomers heavier than dimers increase with time. All this is evidence that aggregation is occurring.

Fig. 2. Analysis of dimers cross-linked *in vitro* on 4% Yang-Criddle gels. Embryonic chick brain tubulin was incubated with 0.1 mM colchicine for 1 hour at 37°C, then cross-linked, reduced and carboxymethylated, and run on gels as in Fig. 1. Samples were cross-linked as follows: 1. with DTDI in the presence of 0.025 mM vinblastine; 2. with DTDI (no vinblastine); 3. with DMSI (no vinblastine).

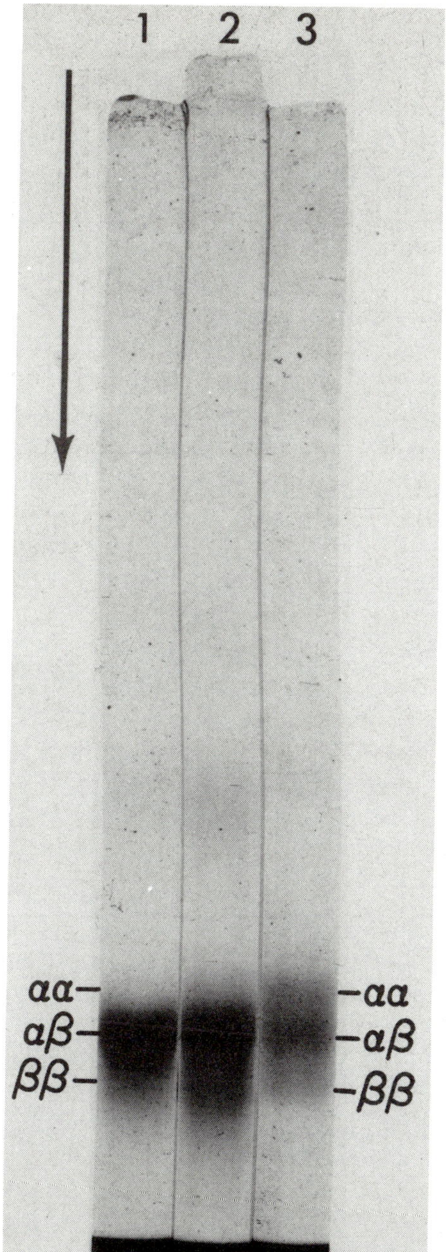

1
2
3
αα
αβ
ββ
αα
αβ
ββ

Aggregation should also lead to the formation of non-specific inter-dimer bonds. That is, if tubulin is largely αβ, then with time the yield of cross-linked αβ relative to αα and ββ should decrease. This is exactly what happens (Fig. 4). As aggregation proceeds, DTDI starts cross-linking more αα and ββ, causing a lower relative yield of αβ. This strongly suggests that the tubulin dimer originally exists mostly as αβ and that the cross-linking of αα and ββ is largely caused by aggregation.

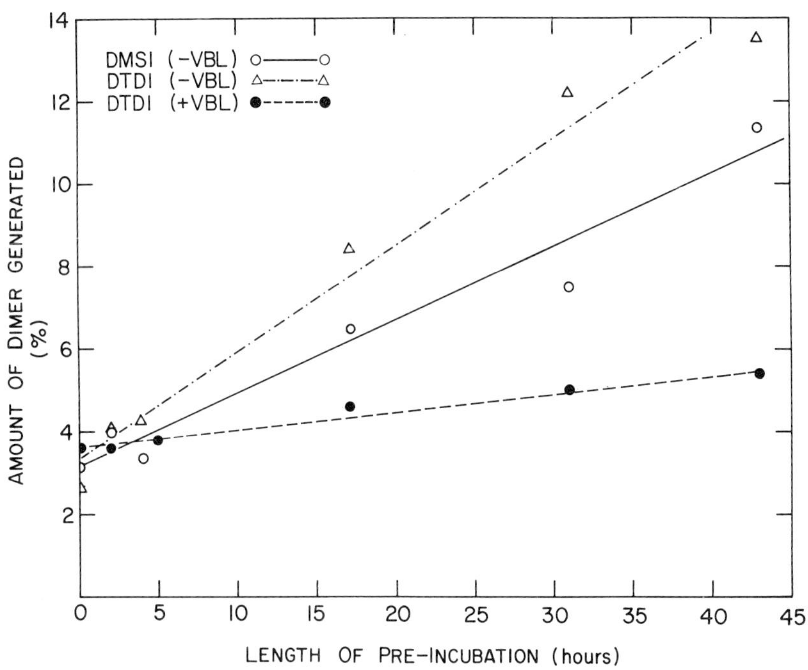


Fig. 3. Effect of tubulin decay on the amount of dimer generated by cross-linking. Samples of embryonic chick brain tubulin were incubated with 0.1 mM colchicine for 43 hours at 37°C. At the indicated times aliquots were removed and cross-linked with 15 mM DMSI or DTDI. One sample of tubulin was incubated with 0.025 mM vinblastine. Cross-linked samples were reduced and carboxymethylated and run on 7% Yang-Criddle gels. Gels were stained with fast green and scanned.

When DMSI is used to cross-link tubulin in this decay experiment, no change in the relative yield of αβ is observed with time, although the absolute yield of dimer increases (Figs. 3 and 4). This is explained if DMSI mainly forms inter-dimer cross-links, presumably within an aggregate. As aggregation proceeds, there should be more opportunities for inter-dimer cross-linking, but the relative yields of the three cross-linked dimers should not change. Recent studies using fluorescence-labeling and circular dichroism have indicated that the tubulin dimers undergo spontaneous conformational changes and also form aggregates. These processes are slowed up in the presence of colchicine or vinblastine[27,28].

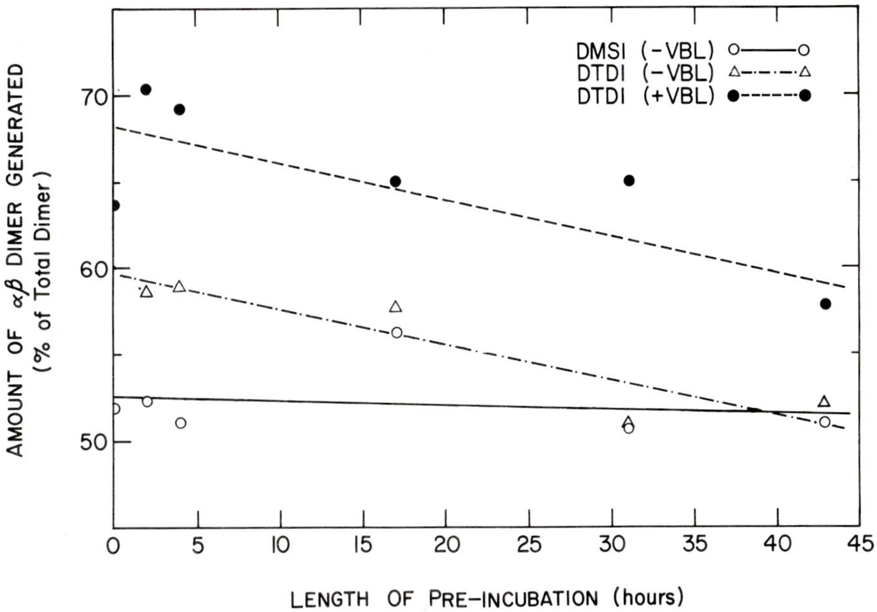


Fig. 4. Effect of tubulin decay on the relative yield of the αβ dimer. Samples of tubulin were incubated, cross-linked, and carboxymethylated as in Fig. 3. Samples were run on 4% Yang-Criddle gels, stained with fast green and scanned.

6. Cross-linking tubulin from other sources

DTDI cross-links intact vinblastine-tubulin crystals and solubilized tubulin from sea urchin outer doublets in a manner similar to the way it cross-links chick brain tubulin (see Table 2). DMSI, however, cross-links differently in that vinblastine increases the yield of αβ. This may reflect a slight difference in conformation of sea urchin tubulin resulting in more intra-dimer cross-linking with DMSI.

The vinblastine-tubulin crystals are a potential system for X-ray crystallographic analysis. Such studies would be immensely useful for extending our knowledge of microtubule structure and function, provided that the tubulin in these crystals is in the same conformation as in the soluble dimer and in the intact microtubule. Drug-binding and cross-linking may be used as probes of the

Table 2

Cross-linking of sea urchin tubulin with DTDI and DMSI

| Source of tubulin | Cross-linker | Vinblastine | Relative yield of αβ(%) |
|---|---|---|---|
| Outer doublet | DTDI | No | 61 |
| Outer doublet | DTDI | Yes | 73 |
| Vinblastine crystal | DTDI | Yes | 65 |
| Outer doublet | DMSI | No | 48 |
| Outer doublet | DMSI | Yes | 61 |
| Vinblastine crystal | DMSI | Yes | 62 |

conformational states of these different forms of tubulin. Both the crystals and the solubilized outer doublet tubulin bind to vinblastine, colchicine, and GTP, suggesting that the conformation around these binding sites is similar in the two systems[29,30]. Intact microtubules do not bind either colchicine or vinblastine[30,31], but this does not necessarily mean that the binding sites for these drugs are not there; they may simply be masked. Since the pattern of cross-linking in the crystals is similar to that of outer doublet tubulin, this may indicate that the conformations of the crystalline and soluble tubulin are alike.

## 7. Conclusion

The data presented here suggest that the population of soluble tubulin dimers consists mostly of heterodimers, which are likely to form large aggregates. Both colchicine and vinblastine decrease aggregation and stabilize the αβ dimer. Certain lysine residues in the intact αβ dimer are apparently positioned so that DTDI can form an intra-dimer bond linking the two monomers together. The possibility cannot be eliminated that some of the soluble tubulin is in the form of homodimers although they are probably in the minority. Similarly, until more extensive cross-linking studies are done on intact microtubules, it is still possible that microtubule polymerization is accompanied by extensive rearrangement of the tubulin, and conceivably portions of the microtubule could be composed of homodimers as Witman et al suggest[12]. Nevertheless, cross-linking with DTDI shows that the soluble colchicine or vinblastine-binding tubulin dimers are almost entirely αβ. Cross-linking with DTDI may also be a useful tool for testing the stabilization, decay, and conformational states of various forms of tubulin.

We gratefully acknowledge the advice and assistance of Linda Grisham, Daniel Chin, Michael Marcum, George Stark, and Jeremy Thorner. This work was supported by a grant from the Jane Coffin Childs Memorial Fund for Medical Research, by USPHS grants NS09335 and NS04270, and by American Cancer Society Grant CI-95.

References

1. Feit, H., Slusarek, L. and Shelanski, M.L., 1971, Proc. Nat. Acad. Sci. USA 68, 2028.

2. Everhart, L.P., 1971, J. Mol. Biol. 61, 745.

3. Bryan, J. and Wilson, L., 1971, Proc. Nat. Acad. Sci. USA 68, 1762.

4. Fine, R.E., 1971, Nature New Biol. 233, 283.

5. Eipper, B.A., 1972, Proc. Nat. Acad. Sci. USA 69, 2283.

6. Luduena, R.F. and Woodward, D.O., 1973, Proc. Nat. Acad. Sci. USA 70, 3594.

7. Weisenberg, R.C., Borisy, G.G. and Taylor, E.W., 1968, Biochemistry 7, 4466.

8. Wilson, L., Bamburg, J.R., Mizel, S.B., Grisham, L.M. and Creswell, K.M., 1974, Fed. Proc. 33 (2), 158.

9. Erickson, H.P., 1974, J. Supramolec. Struct. 2, 393.

10. Weingarten, M.D., Lockwood, A.H., Hwo, S.-Y. and Kirschner, M.W., 1975, Proc. Nat. Acad. Sci. USA 72, 1858.

11. Wilson, L. and Bryan, J., 1974, Adv. in Cell and Molec. Biol. 3, 21.

12. Witman, G.B., Carlson, K. and Rosenbaum, J.L., 1972, J. Cell Biol. 54, 540.

13. Lee, J.C., Frigon, R.P. and Timasheff, S.N., 1973, J. Biol. Chem. 248, 7253.

14. Luduena, R.F. and Woodward, D.O., 1975, Ann. N. Y. Acad. Sci. 253, 272.

15. Ozawa, H., 1967, J. Biochem. 62, 531.

16. Stephens, R.E., 1970, J. Mol. Biol. 47, 353.

17. Bryan, J., 1971, Exptl. Cell Res. 66, 129.

18. Lowry, D.H., Rosebrough, N.J., Farr, A.L. and Randall, R.J., 1951, J. Biol. Chem. 193, 265.

19. Crestfield, A.A., Moore, S. and Stein, W.H., 1963, J. Biol. Chem. 238, 622.

20. Traut, R.R., Bollen, A., Sun, T.-T., Hershey, J.W.B., Sundberg, J. and Pierce, L.R., 1973, Biochemistry 12, 3266.

21. Carpenter, F.H. and Harrington, K.T., 1972, J. Biol. Chem. 247, 5580.

22. Yang, S. and Criddle, R.S., 1970, Biochemistry 9, 3063.

23. Weast, R.C. (ed.), 1974, in Handbook of Chemistry and Physics, 55th ed., Cleveland, Ohio, CRC Press.

24. Davies, G.E. and Stark, G.R., 1970, Proc. Nat. Acad. Sci. USA 66, 651.

25. Wilson, L., 1970, Biochemistry 9, 4999.

26. Weisenberg, R.C. and Timasheff, S.N., 1970, Biochemistry 9, 4110.

27. Becker, J.S., Oliver, J.M. and Berlin, R.D., 1975, Nature 254, 152.

28. Ventilla, M. Cantor, C.R., and Shelanski, M., 1972, Biochemistry 11, 1554.

29. Bryan, J., 1972, Biochemistry 11, 2611.

30. Wilson, L., 1975, Ann. N. Y. Acad. Sci. 253, 213.

31. Wilson, L. and Meza, I., 1973, J. Cell Biol. 58, 709.

*Microtubules and Microtubule Inhibitors, eds. M. Borgers and M. de Brabander*



# THE EFFECT OF PRESSURE ON NEURONAL MICROTUBULES

Yves Engelborghs, Karel A.H. Heremans and
Johan Hoebeke*

Department of Chemistry, Katholieke
Universiteit te Leuven, B-3030 Heverlee
*Janssen Pharmaceutica, B-2340 Beerse,
Belgium.

The influence of pressure on the polymerization of microtubules was studied. It is shown that the pressure effect is linked to the temperature. At temperatures higher than 25°C only a small pressure effect is observed, while at lower temperatures a complete dissociation can be obtained with moderate pressures.

The equilibrium data can be analysed in terms of a nucleation and propagation equilibrium constant. It is shown that temperature and pressure mainly influence the nucleation equilibrium.

## 1. Introduction

Reversible pressure effects on different important cell functions have been observed long before the existence of microtubules was known and their involvement in cellular processes was recognized. In vitro experiments on pure microtubules showed them to be pressure sensitive, and the common view is that their disintegration is responsible for the effect of pressure on mitotic spindle, cell rounding and motility, integrity of the heliozoan axopodia etc.[1-5]

O'Connor et al.[6] however found no effect of pressure on the stability of microtubules of neuronal origin, in vivo as well as in vitro. These authors therefore state that the sensitivity towards pressure of microtubules from non-neuronal origin may have been overgeneralized.

In this paper we report on the influence of pressure on the polymerization of rat-brain microtubules. We show that the pressure effect is linked to the temperature. These results explain the apparent discrepancy mentioned above.

We analysed the equilibrium data not only in terms of a propagation equilibrium constant, as other authors did,[3,7] but also in terms of a nucleation equilibrium constant. We show that the pressure effect is essentially linked to the nucleation equilibrium.

## 2. Materials and methods

Rat-brain tubulin was prepared according to Shelanski et al.[8] After the third polymerization cycle the protein was dialyzed against a hundredfold volume excess of MES buffer at pH 6.4 with the following composition : 50 mM-MES, 70 mM-KCl, 1 mM-EGTA and 0.5 mM-$MgCl_2$. After dialysis GTP was added up to 1 mM. Protein concentration was determined with the method of Lowry[9]. The purity of our preparation was checked by polyacrylamide gel electrophoresis and was found to be over 90 %, and the integrity of the microtubules was checked by electron microscopy.

Light scattering was measured at 90° (wavelength 540 nm) with the equipment developed by Putzeys and Vaneghem. An essential feature of the instrument is that the cell containing the protein solution is fitted in a cell holder which has a screwed-on cover providing support for a pressure transducing membrane. More details and theoretical background of the method are dealt with elsewhere.[10,11]

In the course of our pressure studies we found that complete reversibility of the polymerization was maintained up to 1000 atm. The increase of the pH of the MES buffer with increase in pressure is very small and was neglected.

## 3. Results and Discussion

Fig. 1 shows the light scattering as a function of total tubulin concentration. The total light scattering shows a biphasic increase with protein concentration. The first part is the extrapolation of the data obtained at 15°C, 500 atm and higher, and therefore represents the light scattering of the completely dissociated state. The steeper part is due to the contribution of the polymers which are formed above a certain critical concentration. The data of Fig. 1 clearly show the linkage between the pressure effect and temperature. A pressure of 500 atm is sufficient to obtain a complete dissociation at 15°C, even at the highest concentrations used, while at 35°C only a very small effect is noticed.

It is thus clear that at 15°C and lower the neuronal microtubules have the same pressure sensitivity as these of non-neuronal origin. The conclusion of O'Connor et al.[6] about the difference of the two types of tubules is thus invalid.

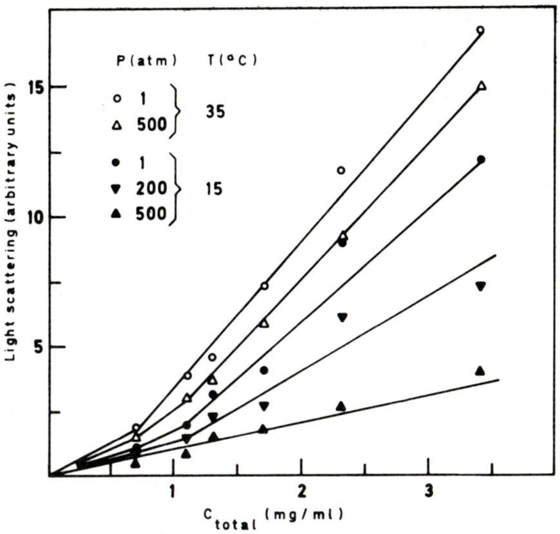


Fig. 1. The effect of pressure on the light scattering of microtubules at increasing total concentrations. Pressure was applied to solutions equilibrated at 35°C and at 15°C. For the sake of clarity, only a limited number of data are given. The measurements are corrected for window effects and turbidity.

Due to the presence of a critical concentration, the polymerization can be compared with a condensation or crystallization phenomenon. As shown by Gaskin et al.[7], the equilibrium data can then be analysed in terms of this critical concentration which is the reciprocal of the propagation equilibrium constant. The pressure-induced depolymerization of spindle microtubules was studied in an analogous way.[3]

With the model of Oosawa[12] it is however possible to analyse the data in more detail i.e. in terms of the equilibrium constants for propagation and nucleation.

The overall nucleation equilibrium is defined as follows :

$$nT \xrightarrow{AK^{n-1}} T_n \qquad (1)$$

T are the protomers, regardless whether they are present as 6S protomers or associated into 36S rings, n is the number of protomers involved in the nucleus $T_n$, $AK^{n-1}$ is the overall nucleation equilibrium constant, K is the propagation equilibrium constant, which is assumed to be independent on the degree of polymerization from n on, A is a dimensionless equilibrium constant which corresponds to the destabilisation of the nucleus relative to a n-mer in the polymer.

The propagation equilibrium is defined as the addition of a protomer to an existing polymer.

Gaskin et al.[7] have shown that the light scattering of microtubules solutions is nearly independent on the length of the tubules and is inversely proportional to the third power of the wavelength. Therefore the light scattering is a good approximation of the total number concentration of protomers present in the microtubules : $c_p = \sum_{i=2}^{\infty} ic_i$, where $c_i$ is the concentration of the i-mer. According to Oosawa[12] the relation between $c_i$ and $c_1$, the protomer concentration is :

$$c_i = AK^{-1}(Kc_1)^i \tag{2}$$

As this mathematical representation is only valid from n on,

$$c_p = \sum_{i=2}^{n-1} ic_i + AK^{-1} \sum_{i=n}^{\infty} i(Kc_1)^i \tag{3}$$

The second term can be simplified by series expansion[12] if the summation is started at i=2 therefore :

$$c_p = Ac_1/(1 - Kc_1)^2 - AK^{-1} \sum_{i=2}^{n-1} i(Kc_1)^i + \sum_{i=2}^{n-1} ic_i \tag{4}$$

We found that the inverse dependence of the scattering on the third power of the wavelength is valid down to 8°C, while an inverse fourth power dependence was only found as low as 4°C. This proves that the concentration of the low molecular weight intermediates is very small and therefore the second and third term of equation (4) can be neglected. The first part can then be rearranged to

$$(c_1/c_p)^{1/2} \simeq A^{-1/2}.(1 - Kc_1) \tag{5}$$

In a previous report[13] we showed that the influence of temperature can be analysed in this way and the enthalpy and entropy changes can be determined for the propagation equilibrium and for the destabilization of the nucleus.

If we assume that the steepest slope in Fig. 1 corresponds to the specific light scattering of the polymers, $c_p$ and $c_1$ can easily be determined and the data plotted according to equation 5, as shown in

Fig. 2.

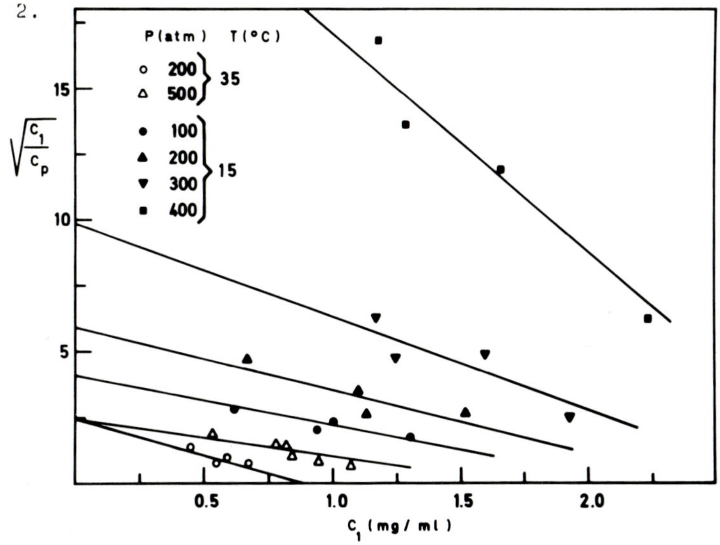


Fig. 2. Linearization of the data of Fig. 1 according to equation (5).

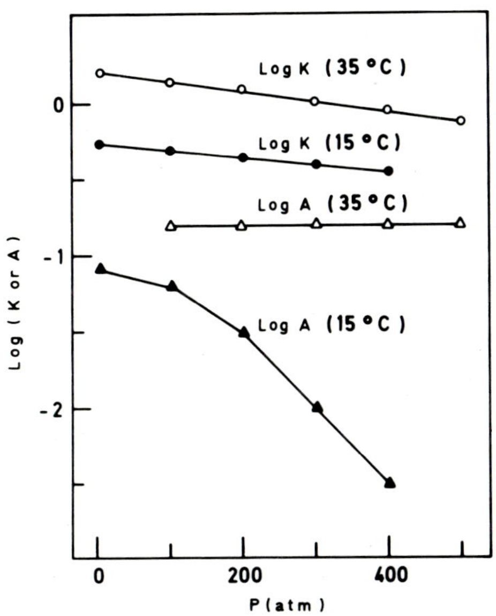


Fig. 3. The logarithm of the equilibrium constants determined from the plot of Fig. 2, as a function of pressure. The dimensions of K are $(mg/ml)^{-1}$ and A is dimensionless.

Fig. 3 shows the logarithm of the equilibrium constants as determined by equation 5, plotted against pressure. The associated volume changes can be calculated by the relation :

$$d \ln K / dP = - \Delta V / RT \quad (6)$$

and are shown in the table.

| Temperature | 35°C | 15°C |
|---|---|---|
| K | 26 | 21 |
| A | 0 | 49 ( 1 - 100 atm)<br>135 (200 - 400 atm) |

Volume changes associated with the equilibrium constants for propagation and for the destabilization of the nucleus. The volume changes ($cm^3$/mole) are determined with equation (6) from the data of figure 3.

These data, and Fig. 3 clearly show that the pressure dependence of the propagation equilibrium is small at both temperatures. The pressure dependence for the destabilization of the nucleus is negligible at 35°C but very large at 15°C. It is interesting to note that temperature [13] as well as pressure mainly influence the nucleation equilibrium.

A link between the effect of pressure and temperature has been observed for other protein-protein interacting systems such as flagellin[14], glutamate dehydrogenase[15] and casein[16]. It was also shown for the pressure induced disintegration of the heliozoan axopodia in the limited temperature range of 20 to 25°C, by Marsland et al.[5]

Small volume changes are usually associated with hydrophobic interactions, while ionic interactions are coupled with large volume changes due to electrostriction of the solvent molecules.[17-19] The fact that the volume changes associated with nucleation are so much larger at low than at high temperature, can be explained in different ways : either the number of protomers involved is larger at low temperatures or an additional conformational change is involved. The latter explanation seems more plausible as it is expected that n is only determined by the geometry of the nucleus. We assume that lowering the temperature decreases the stability of hydrophobic interactions which leads to a change in conformation and an exposure of charged groups.

Acknowledgement

The authors thank Greta Van Nijen for technical assistance with the preparation, Dr. N. Overbergh (Laboratorium voor macromoleculaire en organische scheikunde, K.U. Leuven) for electronmicroscopy and Prof. Leo C.M. De Maeyer for stimulating discussions. This work was supported by the "Fonds voor Kollectief Fundamenteel Onderzoek". Y.E. is "Aangesteld Navorser" of the "Nationaal Fonds voor Wetenschappelijk Onderzoek".

References

1. Tilney, L.G., Hiramoto, Y. and Marsland, D., 1966, J. Cell Biol. 29, 77 - 95.
2. Kitching, J.A., 1970, in High Pressure Effects on Cellular Process, 155 - 177, Ed. Zimmerman, A.M., Academic Press, New York.
3. Salmon, E.D., 1975, J. Cell. Biol. 66, 114 - 127.
4. Kennedy, J.R. and Zimmerman, A.M., 1970, J. Cell. Biol. 47, 568 - 576.
5. Marsland, D., Tilney, L.G. and Hirsfield, M., 1971, J. Cell. Physiol. 77, 187 - 194.
6. O'Connor, T.M., Houston, L.L. and Samson, F., 1974, Proc. Nat. Acad. Sci. USA 71, 4198 - 4202.
7. Gaskin, F., Cantor, C.R. and Shelanski, M.L., 1974, J. Mol. Biol. 89, 737 - 758.
8. Shelanski, M.L., Gaskin, F. and Cantor, C.R., 1973, Proc. Nat. Acad. Sci. USA 70, 765 - 768.
9. Lowry, O.H., Rosebrough, N.J., Farr, A.L. and Randall, R.J., 1951, J. Biol. Chem. 193, 265 - 275.
10. Vaneghem, J., 1966, Thesis, University of Leuven.
11. Heremans, K.A.H., 1975, Verh. Kon. Vlaamse Acad. Wet. België, in the press.
12. Oosawa, F., 1970, J. Theor. Biol. 27, 69 - 86.
13. Engelborghs, Y., Heremans, K.A.H., De Maeyer, L.C.M. and Hoebeke J., 1975, submitted for publication.
14. Gerber, B.R. and Noguchi, H., 1976, J. Mol. Biol. 26, 197 - 210.
15. Heremans, K.A.H., 1974, Proc. Int. Conf. High Pressure, Kyoto (The Physico-Chemical Society of Japan), 627 - 630.
16. Payens, T.A.J. and Heremans, K., 1969, Biopolymers 8, 335 - 345.
17. Heremans, K.A.H., Snauwaert, J., Vandersypen, H.A. and Van Nuland, Y., 1974, Proc. Int. Conf. High Pressure, Kyoto (The Physico-Chemical Society of Japan), 623 - 626.

18. Zipp, A. and Kauzmann, W., 1973, Biochemistry, 21, 4217 - 4228.

19. Suzuki, K. and Taniguchi, Y., 1972, in The Effects of Pressure on Organisms, 103 - 124, Cambridge, University Press.

*Microtubules and Microtubule Inhibitors, eds. M. Borgers and M. de Brabander*



# X-RAY ANALYSIS OF MICROTUBULE INHIBITORS*

T. N. Margulis
Department of Chemistry
University of Massachusetts
Boston, Massachusetts, U.S.A.

## Summary

The crystal and molecular structures of N-deacetylthiocolchicine and demethyl-iso-thiocolchicine have been determined. Substitution of $-SCH_3$ for $-OCH_3$ on the troponoid ring causes that ring to pucker, indicating less electron delocalization. Interchanging the positions of -C=O and $-OCH_3$ on the troponoid ring causes only small changes in the conformation of the other 7-membered ring. The conformations of the $-OCH_3$ groups on the benzenoid ring are different in these compounds and are each different from that of Colcemid. Both crystal structures are held together with hydrogen-bonds involving water of crystallization.

## 1. Introduction

There can be little doubt that a complete understanding of the nature of the interaction of microtubular inhibitors and tubulin will require a detailed knowledge of the structure of the inhibitors. X-ray diffraction is the only practical method that can provide the necessary parameters to precisely describe the geometry of a molecule the size of a typical inhibitor such as colchicine or podophyllotoxin. While it is true that these parameters (bond lengths, bond angles, torsion angles) refer to the extremely pure state of the single crystal there is reason to believe that many will remain essentially the same in solution. Certainly the structure found in the solid state is one that must be considered as a possible structure in solution. X-ray diffraction also provides information about how one molecule interacts with others of the same kind, and with solvent. For example, in both of the new structures described here water is found hydrogen-bonded to the colchicine derivatives.

It may be useful for workers who are preparing and testing new drugs to know what the crystallographer requires in order to carry

*This investigation was supported by Public Health Service Research Grant No. CA-17436-01 from the National Cancer Institute.

out a successful determination of the structure. Until about 15 years ago it would have been almost out of the question to determine the structure of a molecule the size of colchicine without the use of a heavy atom derivative. The heavy atom, often Br or I, provided a place in the molecule of high scattering power for X-rays which could be used as a handle to begin the resolution of the lighter atoms. Modern theory known as "direct methods" combined with larger and faster digital computers have made the requirement of a heavy atom less important. While it is probably still more certain that a solution of the crystal structure will be achieved if a heavy atom is present, direct methods are often successful and usually provide better molecular parameters than the heavy atom method since the atoms of interest (carbon, oxygen, nitrogen and hydrogen) are a larger fraction of the total scattering power of the molecule. The only critical requirement then is a good single crystal. This need not be large; a slowly grown crystal with linear dimensions of about 0.3-0.4 mm is about right.

The results of a diffraction study provide a list of coordinates of each atom, often including hydrogen, in the basic building block of the crystal--the unit cell. From these coordinates one can calculate bond lengths, bond angles, non-bonded distances (both inter- and intra-molecular), and torsion angles--the angle of twist about a bond. Information about the thermal motion of the atoms also comes out of an X-ray study but interpretation is not straightforward. The kind of precision obtained depends upon a number of things including crystal quality, amount of X-ray data obtained, presence or absence of a heavy atom, number of atoms in the molecule, and computational techniques employed. The agreement between the observed X-ray data and that calculated from the crystal structure is called the R-factor. While not on firm theoretical grounds, experience shows that in general low R's mean good structures, the best work giving values as low as 0.02 (meaning that the average difference between observed and calculated data is 2%). A rough measure of the accuracy of the crystal structure of an organic molecule is $\sigma$(C-C) the estimated standard deviation of a carbon-carbon single bond. In Table 1 are listed the mitotic spindle inhibitors for which crystal structures have been done, along with $\sigma$(C-C). In this paper I shall discuss only those which are derivatives of colchicine(I) both active and inactive.

Table 1

X-ray Studies of Mitotic Spindle Inhibitors

| Compound | σ(C-C),Å | Reference |
|---|---|---|
| colchicine--methylene bromide complex | -- | 1 |
| Colcemid (N-deacetyl-N-methylcolchicine) | 0.01 | 2 |
| N-deacetylthiocolchicine hydrochloride | 0.005 | This paper |
| demethyl-iso-thiocolchicine | 0.005 | This paper |
| 2-bromopodophyllotoxin | 0.04 | 3 |
| 5'-demethoxy-β-peltatin A methyl ether | 0.005 | 4 |
| melatonin | 0.004 | 5 |
| griseofulvin | 0.05 | 6 |
| leurocristine methiodide (related to vincristine and vinblastine) | 0.1 | 7 |

2. Colchicine(I) and Colcemid(II)

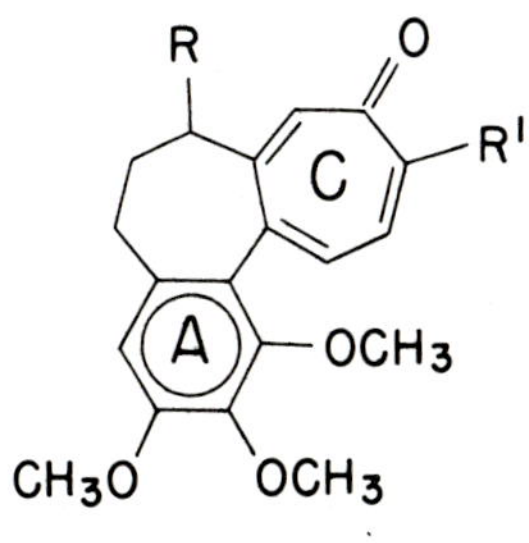


I R = $NHCOCH_3$, R' = $OCH_3$
II R = $NHCH_3$, R' = $OCH_3$
III R = $NH_3^+$, R' = $SCH_3$

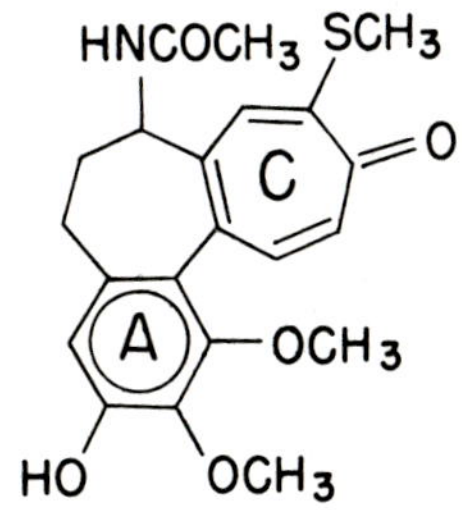


IV

In 1947 King et al.[1] produced the first X-ray diffraction study of a colchicine containing crystal. Using a complex of methylene bromide this work established the primary structure but the limited amount of X-ray data did not allow a complete 3-dimensional structure or provide details of molecular geometry.

Not until 1972 was a 3-dimensional study of a colchicine derivative carried out. The structure of Colcemid(II), N-deacetyl-N-methylcolchicine, also known as demecolcine, was solved by direct

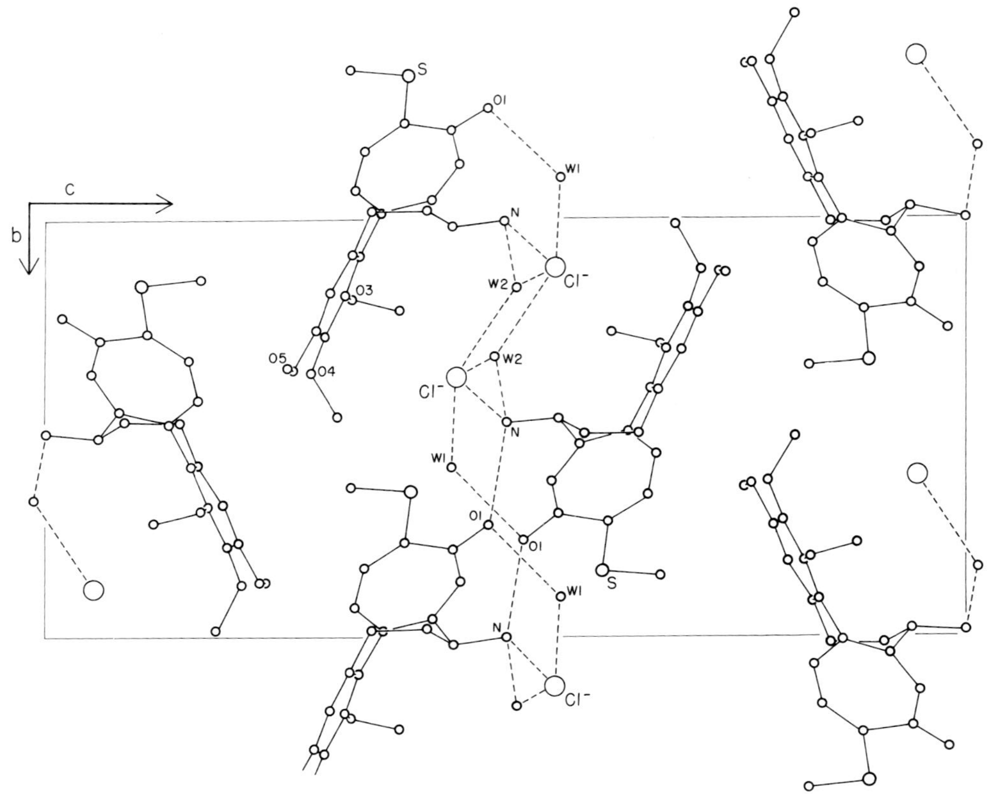


Fig. 1. Crystal structure of N-deacetylthiocolchicine(III). Dashed lines represent hydrogen-bonds. W1 and W2 are water molecules. Hydrogen atoms are not shown.

methods in my laboratory.[2] This study revealed several structural features since confirmed in other colchicine derivatives. The molecule is far from being planar. One way to describe it is in terms of two nearly planar fragments, the 7-membered troponoid ring (Ring C) and the 6-membered benzenoid ring (Ring A). The dihedral angle between these two fragments is 54°, whereas a planar molecule would have an angle of 0°. Inspection of the bond lengths in Ring C reveals an alternation consistent with a model in which the π electrons are only partially delocalized, i.e. the system is not really aromatic. Molecules of Colcemid are held together with hydrogen bonds from nitrogen of one molecule to carbonyl of another molecule, N-H----O=C. There is no intramolecular hydrogen bonding.

3. Thiocolchicines

When $-OCH_3$ on Ring C is replaced by $-SCH_3$ activity remains high and in some cases may even increase.[8,9] These compounds known as thiocolchicines, were prepared at Roussel Chemical by Velluz and Muller.[10-13] To examine what kinds of structural differences are caused by this substitution I have determined the crystal and molecular structure of N-deacetylthiocolchicine hydrochloride(III), a compound which has been mentioned as an anti-tumor agent.[14] Crystals of III, $C_{20}H_{24}NO_4S^+\ Cl^-\cdot 2H_2O$, are orthorhombic, space group $P2_12_12_1$, with a = 7.538, b = 11.663, c = 25.375 Å; 4 molecules per unit cell. The structure was determined by analysis of 1911 X-ray reflections and refined by the method of least squares, giving σ(C-C) = 0.005 Å and a conventional R of 0.035. Hydrogen atoms were found but are not included in the figures. The molecules are held together with hydrogen bonds to water and $Cl^-$ as shown in figure 1. There is also a direct link between molecules with an N-H----O=C hydrogen bond. The hydrogen bonding holds the molecules in sheets perpendicular to the c axis. There is no hydrogen bonding between sheets or between atoms in the same molecule. One molecule is illustrated in figure 2 along with the more important interatomic distances. The precision of this study is somewhat better than that of the Colcemid structure, and the bond lengths in the troponoid ring are in excellent agreement with those listed by Hamor and Derry[15] in their study of tropolone derivatives.

The troponoid ring is significantly changed by replacing $-OCH_3$ with $-SCH_3$. While the ring is essentially planar in II it adopts

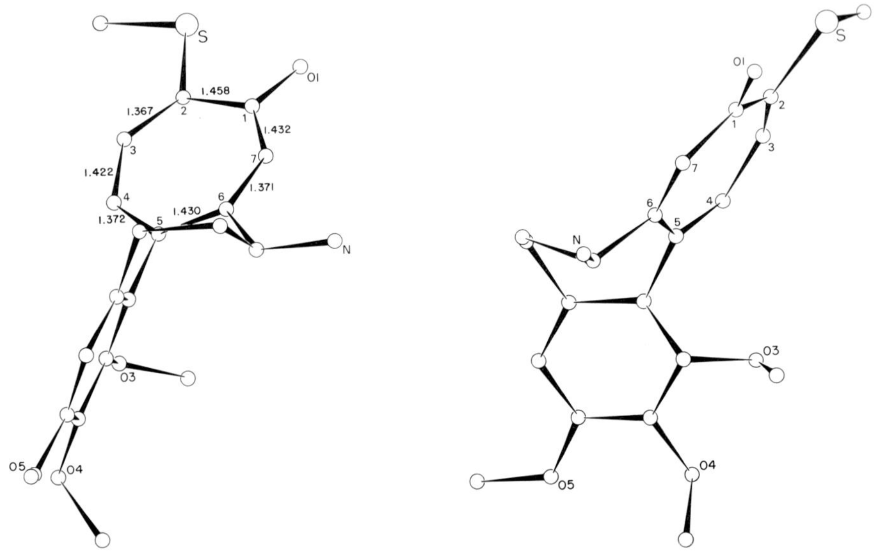


Fig. 2. Two views of a single molecule of N-deacetylthiocolchicine (III). Bond lengths are given in Å. Hydrogen atoms not shown. Atoms numbered 1-7 are carbon as are the unlabeled atoms.

the conformation of a shallow boat in III. The boat may be described by 3 planes:

| | |
|---|---|
| Plane 1 | C3, C4, C5, C6 |
| Plane 2 | C2, C3, C6, C7 |
| Plane 3 | C1, C2, C7, O1 |

The root-mean-square deviations of the atoms from their assigned planes are 0.0015, 0.0045, and 0.0035 Å for the three planes respectively. The dihedral angle between Plane 1 and Plane 2 is 6.7°, and between Plane 2 and Plane 3 is 5.1°. These may be compared with similar angles of 2 to 3° listed by Derry and Hamor for tropolones. An alternative way of describing this lack of planarity is in terms of the torsion angles in the ring. These will be 0° for a planar ring but vary from 0.5° to 8.5° in this ring. The puckering of the ring in this way is an indication of a less delocalized π electron system. It leads to an O1-O3 distance of 6.53 Å while this distance is 6.78 in Colcemid.

4. Isocolchicine

When the positions of the methoxy and carbonyl groups of Ring C are reversed, biological activity of all sorts is greatly diminished. The structure of demethyl-iso-thiocolchicine(IV) was determined in order to see what structural differences there are between the iso and ordinary forms of colchicine derivatives.

Crystals of IV, $C_{21}H_{23}NO_5S \cdot H_2O$, are monoclinic with a = 11.906, b = 8.713, c = 10.004, Å, β = 89.44°, and 2 molecules/cell. The structure was refined to an R of 0.04 with 2227 X-ray reflections yielding σ(C-C) = 0.005 Å. Hydrogen atoms were located but are not included in the figures.

A 3-dimensional network of hydrogen-bonds holds the structure together (figure 3). Water is hydrogen-bonded to both carbonyl groups and to nitrogen while the -OH group of Ring A is hydrogen-bonded to the oxygen of the acetyl group. The hydrogen-bonds are all between different molecules--there is no intra-molecular hydrogen-bonding. Figure 4 gives views of the individual molecules as well as some bond lengths.

The troponoid Ring C shows bond length alternation as found in II and III, but now long and short bonds are interchanged. For this reason the bond common to Ring C and the other 7-membered ring, Ring B is shorter than in II and III. This leads to small changes in the conformation of Ring B, with some torsion angles changing by as much as 10° in going from II to IV. However, these changes are no greater than changes found in going from II to III, both in the ordinary form.

As in III the troponoid ring in IV is markedly non-planar. Torsion angles in the ring are as large as 10.8°. Because the carbonyl oxygen atom, O1, is in a different position and because of the way Ring A is puckered, the O1-O3 distance is significantly shorter in IV (6.36 Å) than in II (6.78 Å) or III (6.53 Å). O1 is of course farther away from nitrogen, the distances being 4.83, 4.72, and 6.42 Å for II, III, and IV respectively.

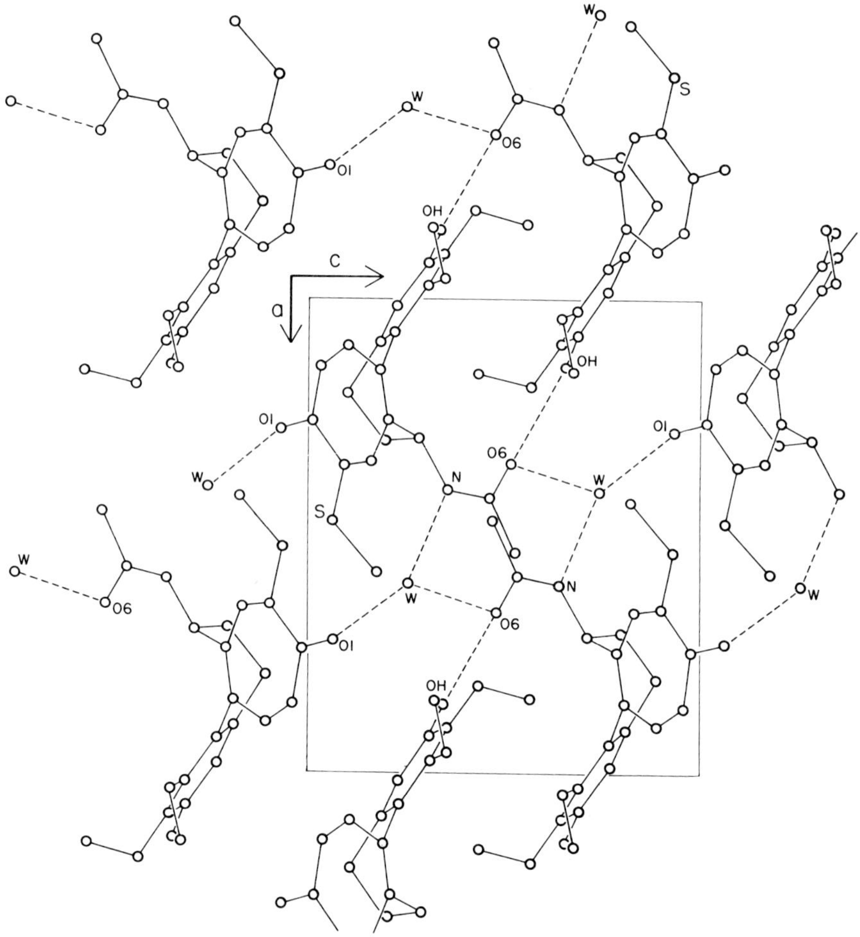


Fig. 3. The crystal structure of demethyl-iso-thiocolchicine(IV).

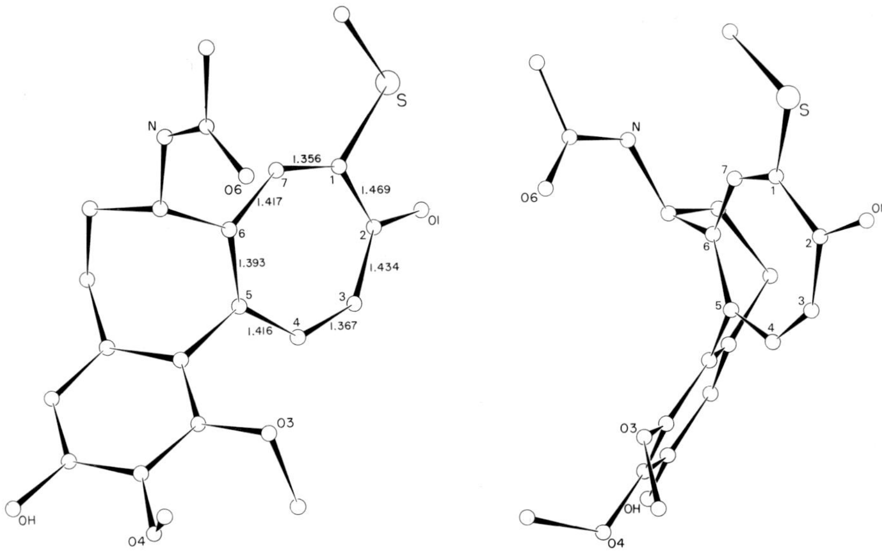


Fig. 4. Two views of a single molecule of demethyl-iso-thiocolchicine(IV). Bond lengths in Å. Hydrogen atoms not shown. Numbered atoms and unlabeled atoms are carbon.

5. Methoxy Groups

The largest difference, other than the change in double and single bonds, among these three colchicine derivatives is the orientation of the methoxy groups, $-OCH_3$, around Ring A. In general methoxy groups on aromatic rings tend to show two preferred conformations. Either the methyl carbon is in the plane of the aromatic ring or it is as far from that plane as possible. These two conformations correspond to torsion angles, τ, of 0° and 90°.

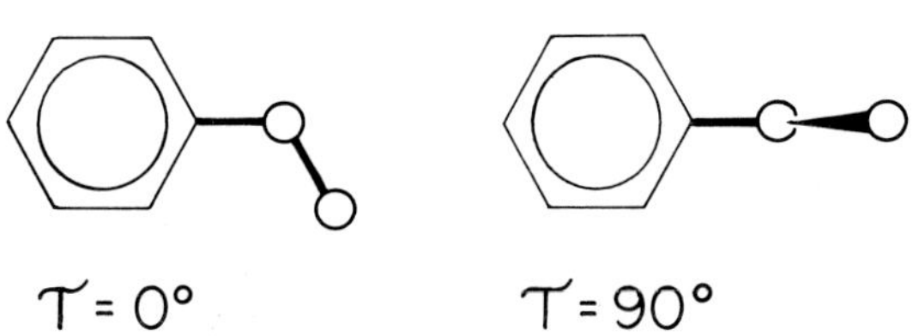

A torsion angle of -90° would correspond to a methyl group away from the plane but on the opposite side to that of $\tau = +90°$. In Table 2 are listed the torsion angles for compounds II, III, and IV, including those for $-SCH_3$ and $-OCH_3$ on Ring C. These latter are all close to 0° but the torsion angles on Ring A show great variation. In particular note that $\tau$ for the O3 methoxy is about +80° for 2 active compounds and -54° for the inactive one! The orientation of these groups is illustrated in Figure 5.

Table 2
Torsion angles (in degrees) of $-OCH_3$ and $-SCH_3$ about Ring A or C

| Compound | O3 | O4 | O5 | S or O1 |
|---|---|---|---|---|
| II | 81 | -79 | 8 | 1 |
| III | 79 | 71 | 47 | 1 |
| IV | -54 | 78 | -- | 2 |

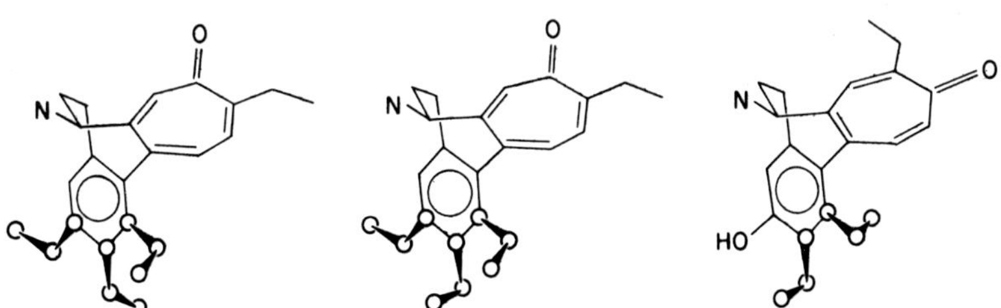


II III IV

Fig. 5. Conformation of $-OCH_3$ about Ring A. In each schematic drawing O3 is on the right, O4 in the center and O5 on the left.

It may be that this difference is just a result of crystal packing forces and has nothing to do with the orientation in solution. It is also true that differences in $-OCH_3$ orientation exist where biological activity is essentially the same, for example, $\tau$ at O4 is opposite in II and III. It could be further argued that a change on Ring C cannot cause a change on Ring A and the -OH group on Ring A is more likely to be the reason for any differences. Of course the Ring C change does cause a switching of double bonds which changes distances between O3 and carbon atoms of Ring C by .04 Å in going from III to IV.

I propose that the orientation of methoxy groups on Ring A, and in particular the orientation about O3 plays an important role in determining activity--that is whether or not a molecule can bind to tubulin or other receptors. Methoxy groups are prevalent in an enormous number of biologically active molecules, including not only all those listed in Table 1, but such diverse compounds as reserpine, mescaline, brucine, codeine, and quinine. What is the role of these groups? They are not chemically very active although their electron donating properties may activate the benzenoid ring. It seems unlikely that they form covalent bonds. I have suggested[2] that they may function as hydrogen-bond acceptors but so far this has not been the case in the crystal structures of any of the colchicine derivatives. I suggest that $-OCH_3$ groups provide specificity in colchicine derivatives by regulating the accessibility of a portion of the benzenoid ring. This could happen by a group blocking part of the ring until it was at the binding site and then turning to allow binding. Or the group stays rigid and simply allows access to the ring in some molecules, e.g. I, II, and III and not in others, e.g. IV. Estimates of the barrier to rotation of the $-OCH_3$ group give a value of about 5 kcal./mole.[16] Not very large but then neither is the energy involved in hydrogen-bonding which plays such a crucial role in biological structure and specificity.

References

1. King, M. F., de Vries, J. L. and Pepinsky, R., 1972, Acta Crystallogr. 5, 437.

2. Margulis, T. N., 1974, J. Amer. Chem. Soc. 96, 899.

3. Petcher, T. J., Weber, H. P., Kuhn, M. and von Wartburg, A., 1973, J. Chem. Soc. Perkin II, 288.

4. Bates, R. B. and Wood, J. B., 1972, J. Org. Chem. 37, 562.

5. Quarles, W. G., Templeton, D. H. and Zalkin, A., 1974, Acta Crystallogr. B30, 99.

6. Brown, W. A. C. and Sim, G. A., 1963, J. Chem. Soc., 1050.

7. Moncrief, J. W. and Lipscomb, W. N., 1966, Acta Crystallogr. 21, 322.

8. Deysson, G., 1968, Int. Rev. Cytol., vol. 24, eds. G. H. Bourne and J. F. Danielli (Academic Press, New York) p.99.

9. Fitzgerald, T. J., 1974, in: Antiinflammatory Agents, Chemistry and Pharmacology, vol. 1, eds. R. A. Scherrer and M. W. Whitehouse. (Academic Press, New York) p.295.

10. Velluz, L. and Muller, G., 1954, Bull. Soc. Chim., 755.

11. Velluz, L. and Muller, G., 1954, Bull. Soc. Chim., 1072.

12. Velluz, L. and Muller, G., 1955, Bull. Soc. Chim., 194.

13. Velluz, L. and Muller, G., 1955, Bull. Soc. Chim., 198.

14. Bernard, J., 1967, Lloydia, 30, 291.

15. Hamor, T. A. and Derry, J. E., 1973, Acta Crystallogr. B29, 2649.

16. Hehre, W. J., Radom, L. and Pople, J. A., 1972, J. Amer. Chem. Soc. 94, 1496.

Acknowledgement

I am indebted to Dr. André Courtin and Dr. G. Muller for their generous gifts of chemicals and to Dr. John Clark for growing the crystals.

*Microtubules and Microtubule Inhibitors, eds. M. Borgers and M. de Brabander*



# THE EFFECT OF MICROTUBULAR-ACTIVE AGENTS ON PHOSPHATIDYLINOSITOL PHOSPHODIESTERASE ACTIVITY

P.J. Quinn

Department of Biochemistry

Chelsea College, University of London

Summary

A phospholipase C-type enzyme specific for phosphatidylinositol is present in the soluble fraction of brain and thyroid. The enzyme is not associated with cytoplasmic microtubules in thyroid but phospholipase activity is complexed with different combinations of tubulin. A high molecular weight form consists of a colchicine-labile interaction between two tubulin dimers and one enzyme molecule. A lighter fraction consists of a combination of phospholipase with the β-subunit of tubulin. The products of phospholipase activity were found to be characteristic of each fraction; the high molecular weight form produces approximately equal amounts of inositol 1 phosphate and cyclic inositol 1:2 phosphate whereas the lighter fraction gives predominantly cyclic inositol 1:2 phosphate. Colchicine does not affect phospholipase activity but vinblastine and chlorpromazine both inhibit the enzyme and cause a corresponding reduction in the proportion of cyclic inositol 1:2 phosphate liberated. This suggests that the mode of inhibition of both drugs is different for high and low molecular weight enzyme components respectively. Chlorpromazine (2mM) selectively precipitates the high molecular weight form of the enzyme from brain post-microsomal supernatant suggesting that chlorpromazine possesses microtubular-active properties. This was checked by comparing the ability of chlorpromazine to selectively precipitate colchicine-binding microtubular protein from rat brain supernatant as well as from purified preparations of tubulin with that of vinblastine. Both drugs proved to be identical in these respects. The antipsychotic action of chlorpromazine may be due to an interaction of the drug with neurotubules.

## 1. Introduction

The structures formed when microtubular protein undergoes polymerisation fall broadly into two categories depending on their stability in the presence of low concentrations of colchicine. The tubules comprising the cytoplasmic microtubules of plant and animal cells, the mitotic spindle fibres which assemble during cell division, and the neurotubules located in axons and dendrites of the central nervous system are all referred to as labile structures and are depolymerised when treated with 1 to 5 μM colchicine. By contrast, polymerised microtubulin isolated from cilia and flagellae as well as a 36S component derived from neurotubulin are not disrupted by colchicine nor do they appear to bind the drug. Nevertheless, when such structures are depolymerised by other methods, colchicine binding activity, similar to that of the labile variety of tubulin, can be demonstrated. No evidence is yet available to suggest that the α and β subunits of the tubulin dimer interact differently but at least two types of interaction between dimers can be inferred depending upon the stability in the presence of colchicine.

Apart from polymerisation, microtubular protein is believed to bind to certain membrane fractions of brain, particularly the synaptosomal membrane[2], but the specificity of interactions with membranes is open to conjecture. Another type of interaction has been reported when microtubular protein of brain is phosphorylated via an intrinsic phosphokinase reaction[3]. Tubulin components in this case bind to other soluble proteins one of which has been identified as a phospholipase C-type enzyme specific for phosphatidylinositol[4]. The phospholipase enzyme appears to bind specifically to the phosphorylated β-subunit of tubulin. However, other combinations of the enzyme and microtubular protein seem likely. It may not be fortuitous therefore, that a variety of physiological processes, including secretion, that depend on an intact microtubular system[5] also involve a specific stimulation of phosphatidylinositol turnover[6]. The experiments described below were designed to establish the nature of the physical interactions between microtubular protein and phosphatidylinositol phosphodiesterase in the hope that the precise functional relationship between the two components would become clearer.

## 2. Enzyme binding to stable and labile tubulin

Phosphatidylinositol phosphodiesterase, like cytoplasmic microtubular protein, is found almost exclusively in the post-microsomal fraction of tissue homogenates. Brain appears to be exceptional in this regard because about half of the total enzyme activity is associated with the particulate fraction and can only be partially extracted with prolonged salt treatment[7]. The association of microtubular protein with the particulate fraction of brain has already been noted[2].

The thyroid gland exhibits a number of colchicine-sensitive functions and this tissue was selected for a preliminary study to establish whether or not cytoplasmic microtubules were associated with phospholipase activity. Washed pig thyroid slices were homogenised in buffers prepared in $H_2O$ or $D_2O$ at 23°C and centrifuged at 300 g. $D_2O$ is known to stabilize cytoplasmic tubules[8] and low centrifugal forces were used to prepare the extract to avoid sedimenting these structures. The extract prepared in $H_2O$ was incubated for 90 min at 37°C with 2.5 μM [$^3$H]-colchicine (1mCi/mmole) and the extract prepared in $D_2O$ was incubated without colchicine. Samples of each extract were fractionated by gel filtration on columns (2 x 20 cm) of Bio-Gel A-15m equilibrated with buffer prepared in $H_2O$ and calibrated with combinations of prokaryotic ribosomes and human γ-globulin. The elution profile of phospholipase activity, assayed at pH 5.4 against a substrate of [$^{32}$P]-phosphatidylinositol,and radioactivity is presented in figure 1. Colchicine binding to fractions of the extract prepared in $D_2O$ (figure 1a) was assayed as before except that bound radioactivity was precipitated by adding 2mM chlorpromazine and centrifuging at 34,000 g. Microtubular protein appears to be polymerised in the buffered $D_2O$ extract because only fractions eluting in the void volume of the column bind appreciable amounts of colchicine. Prior treatment of the $H_2O$ extract with colchicine (figure 1b) shifts the colchicine binding peak to fractions corresponding to the dimeric form of microtubular protein suggesting that the protein is polymerised in a labile manner. The distribution of phospholipase activity, on the other hand, does not change markedly, suggesting that the enzyme is not a constituent of cytoplasmic microtubules. This tends to be confirmed by the fact that fractions of the $D_2O$ extract containing enzyme activity do not bind significant amounts of colchicine although small changes in distribution of the enzyme may be due to colchicine-labile interactions.

The effect of colchicine on the distribution of enzyme activity was examined by density gradient fractionation. Washed thyroid slices were homogenised and a post-microsomal supernatant prepared at 4°C. Aliquots of this fraction were incubated for 90 min at 37°C with or without [$^3$H]-colchicine (5 μM, 0.5 mCi/μmole) and subjected to gel filtration through a small column of Sephadex G 150 to remove unbound colchicine. Phospholipase activities of both preparations and bound radioactivity of the labelled fraction eluted in the void volume of the column. Samples of these fractions were each layered over linear (5-22.5%,w/v) sucrose density gradients

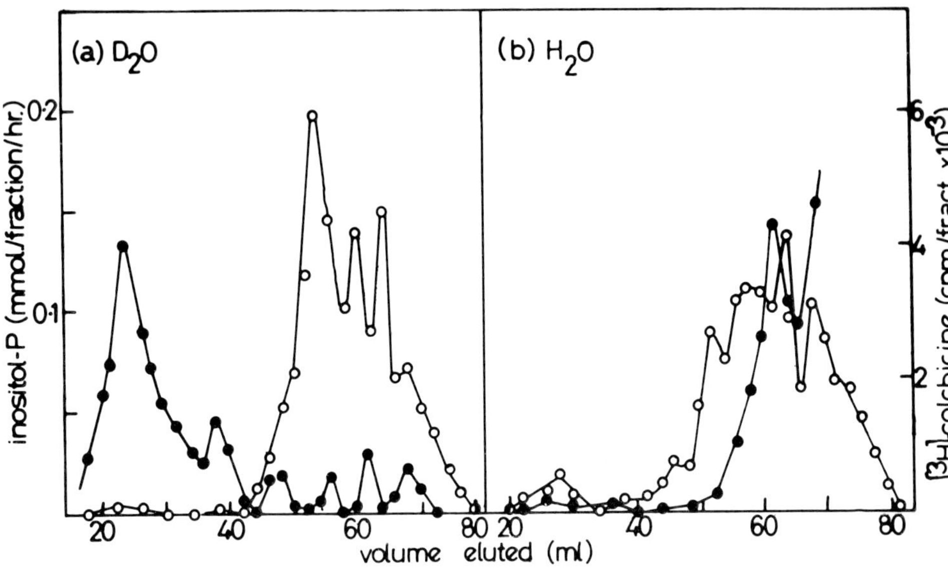


Fig. 1. Gel filtration of thyroid extracts prepared in (a) $D_2O$ and (b) $H_2O$. Phospholipase activity, O—O, and bound (a) or total (b) radioactivity, ●—●, of each fraction are shown.

and centrifuged for 17 hr at 47,000 r.p.m. The gradients were calibrated concurrently with γ-globulin and α-chymotrypsinogen. The gradients were fractionated commencing at the bottom of the tube and aliquots removed for phospholipase assay and radioactivity determination (figure 2a). In the control sample incubated without colchicine two peaks of phospholipase activity can be observed, one of high activity sedimenting at a rate corresponding to a molecular weight of about 220,000 and a lighter fraction of low activity at about 90,000 mol.wt. Treatment with colchicine decreases the apparent molecular weight of the heavy fraction yielding a new peak of about 140,000 mol.wt. but has no effect on the light fraction. Neither peak of enzyme activity is associated with the microtubular dimer of about 110,000 mol.wt. labelled with $[^3H]$-colchicine. The above experiment was repeated with rat brain post-microsomal supernatant except that unlabelled colchicine was used (figure 2b). Phospholipase activity in the control gradient was recovered in the same molecular weight fractions as in thyroid but the relative activity of the two peaks was found to be reversed. Three minor peaks of activity were also present, the lightest of which is believed to represent unbound enzyme (approx.mol.wt. 34,000).

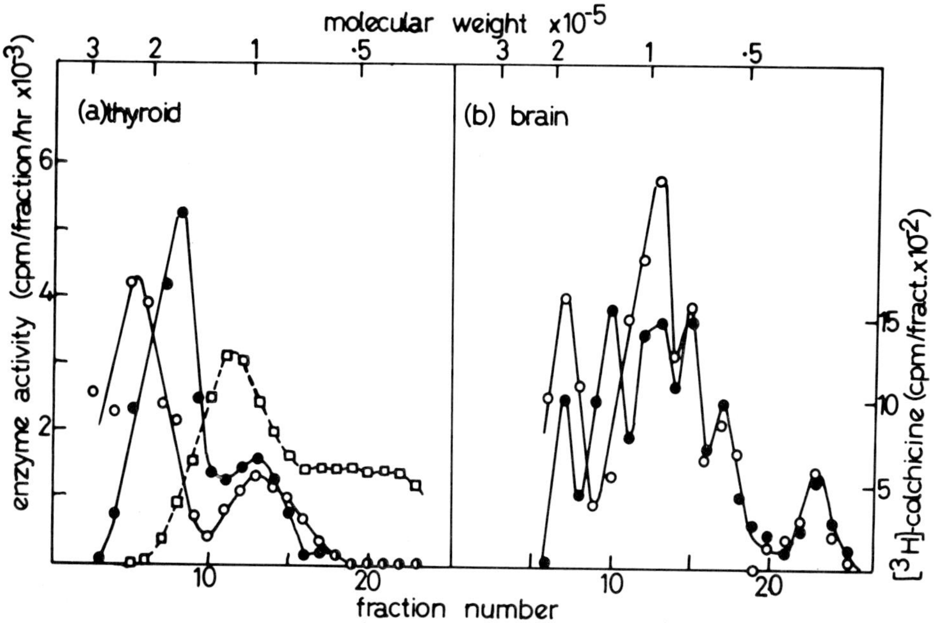


Fig. 2. Density gradient fractionation of phospholipase activity of thyroid (a) and brain (b) post-microsomal supernatants. Samples were incubated with, ●–●, or without, O–O, colchicine. The distribution of radioactivity, □--□, in gradient (a) is shown.

Colchicine treatment reduced phospholipase activity of both the 220,000 and 90,000 mol.wt. fractions and a new peak of activity of an apparent molecular weight 140,000 appeared. The phospholipase activity of the lighter fractions was not altered by colchicine treatment.

It may be concluded from these results that phosphatidylinositol phosphodiesterase is not a constituent of labile cytoplasmic microtubular structures, but that some enzyme activity is associated with a colchicine-labile component of tubulin. The molecular weight of this complex (220,000) suggests that it consists of a colchicine-sensitive interaction between two dimers of microtubular protein one of which interacts with the enzyme such that the colchicine-binding site is blocked or perturbed. Colchicine dissociates the complex into free dimer and enzyme-bound dimer; the latter, which does not bind colchicine, has an approximate molecular weight of 140,000. A complex of this type has already been reported in a purified preparation of microtubular protein from brain treated with colchicine[9]. It was found to consist of a dimer of microtubular protein combined with another protein of 33,000 mol.wt. The 90,000 mol.wt. fraction observed in both thyroid and brain is presumably identical with the complex reported to consist of an association between the β-subunit of tubulin and the phospholipase enzyme[4].

3. Analysis of phospholipase reaction products

Two water-soluble reaction products of phosphatidylinositol phosphodiesterase activity have been identified by high-voltage ionophoresis. They consist of D-myoinositol-1-phosphate and cyclic inositol 1:2 phosphate[10]. Furthermore, the proportion of each product released depends upon the pH of the reaction medium, the cyclic product predominating under acidic conditions. The gradient fractions, shown in figure 2, were assayed at pH 5.4 and the reaction products analysed to determine whether interaction of the enzyme with different combinations of microtubular protein affected the manner of substrate degradation. The results are presented in Table 1. Comparing firstly the two tissues, it can be seen that the proportion of cyclic inositol 1:2 phosphate released from assay of the unfractionated homogenate or soluble fractions is significantly greater in brain than thyroid. This can be explained by an examination of the products released by different complexes of the enzyme and relating this to the total activity of each recovered from the gradients (figure 2). Thus the heavier fractions (220,000 and 140,000 mol.wt.), which predominate in thyroid, yield less of the cyclic derivative than isositol-1-phosphate, whereas in brain the highest activity is observed in the lighter fraction (90,000 mol.wt.) which produces relatively more cyclic inositol 1:2 phosphate.

Table 1

Analysis of phospholipase reaction products from sucrose density gradient fractions

| Fraction | Colchicine | % cyclic 1:2 inositol phosphate | |
|---|---|---|---|
| | | Thyroid | Brain |
| Homogenate | - | 51 | 75 |
| Soluble | - | 45 | 69 |
| | + | 49 | 66 |
| 220,000 mol.wt. | - | 44 | 45 |
| 140,000 mol.wt. | + | 46 | 44 |
| 90,000 mol.wt. | - | 60 | 74 |
| | + | 54 | 72 |

It should be emphasized that colchicine does not alter the proportion of reaction products (table 1) nor the recovery of phospholipase activity from density gradients (figure 2). Furthermore, colchicine does not bind to the enzyme complex, and positive identification of microtubular protein cannot therefore be achieved by this method. For this reason, the effect of another microtubular-active drug, vinblastine, was investigated. The effect on phospholipase activity of incubating a rat brain soluble fraction for 30 min in the presence of varying concentrations of vinblastine is shown in figure 3 which also illustrates the concommittant changes in the proportion of cyclic inositol 1:2 phosphate released. Parallel experiments were performed with supernatant treated with colchicine or chlorpromazine. The latter was chosen because previous reports suggested that chlorpromazine was a potent inhibitor of the enzyme[11]. Colchicine, as expected, did not affect phospholipase activity nor the ratio of reaction products released. Vinblastine and chlorpromazine both inhibited enzyme activity and produced a proportional decrease in the per cent of cyclic inositol 1:2 phosphate recovered from the assay products. Because the reaction is almost completely inhibited by 1 mM of either drug it may be concluded that both the high and low molecular weight forms of the enzyme are inhibited. Nevertheless, if both complexes of the enzyme are inhibited equally one would expect the ratio of the reaction products to remain constant. Clearly this is not the case.

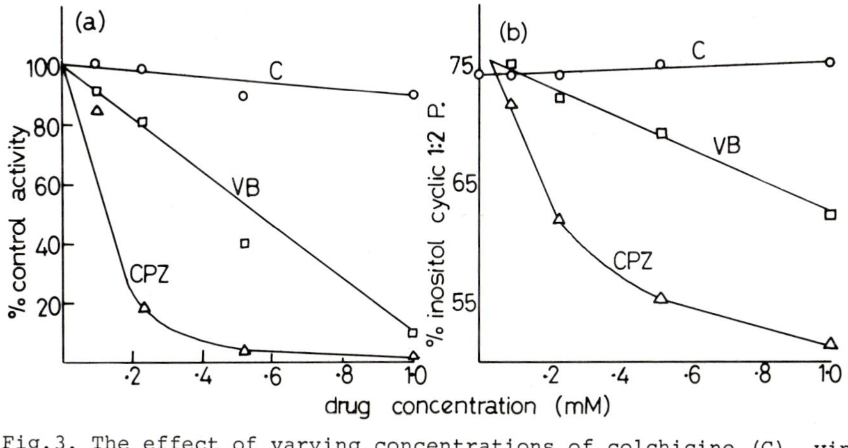


Fig.3. The effect of varying concentrations of colchicine (C), vinblastine (VB) and chlorpromazine (CPZ) on the activity of phosphatidylinositol phosphodiesterase activity of rat brain post-microsomal supernatant (a) and the proportion of cyclic inositol 1:2 phosphate recovered from the reaction products (b).

It is apparent that the 90,000 mol.wt. complex, which produces a higher proportion of the cyclic product, is inhibited to a greater extent than the heavier species of the enzyme and the effect of chlorpromazine in this respect is greater than that of vinblastine. This is most evident at drug concentrations of 1 mM where relatively more cyclic inositol 1:2 phosphate is recovered from the products of the sample treated with vinblastine compared to chlorpromazine.

How, then, do these drugs inhibit enzyme activity? Vinblastine is believed to link dimers of microtubular protein together in a manner not labile to colchicine[12]. The results shown in figure 3 are consistent with the notion that high molecular weight forms of the enzyme, consisting in part of microtubular dimers, are polymerised whereas the lighter fraction (90,000 mol.wt.) containing only β-subunit of tubulin is less affected. It is noteworthy that SDS-gel electrophoretograms of microtubular protein prepared by vinblastine precipitation almost invariably exhibit a prominent band corresponding to the molecular weight of the free enzyme (see, for example, figure 5). It has been suggested that chlorpromazine inhibits phospholipase activity by displacing calcium ions from the substrate surface[11]. Since vinblastine is also an amino compound it may be expected that it too will displace some calcium from the substrate. Calcium is required to assist the binding and penetration of the enzyme between the substrate molecules and, in addition, appears to participate more directly in the reaction[13]. If calcium displacement from the substrate surface was the only factor concerned with enzyme inhibition it would be necessary to postulate that the electrostatic requirements for enzyme-substrate interaction are different for each enzyme complex, otherwise the ratio of reaction products would remain unchanged at all levels of inhibition. An alternative explanation is that chlorpromazine also acts as a microtubular-active agent.

4. Further studies of the phospholipase complex

When chlorpromazine is added to the soluble fraction of tissue extracts a fine white precipitate forms which can be sedimented at 34,000g. To check whether specific complexes of the phospholipase enzyme were aggregated by chlorpromazine a density gradient analysis was performed as described in figure 2. A soluble fraction from rat brain, labelled with [$^3$H]-colchicine and filtered through Sephadex, was incubated for 30 min. with 2mM chlorpromazine and centrifuged for 30 min. at 34,000g. The supernatant from this step was fractionated by density gradient centrifugation and the resulting distribution of enzyme activity is presented in figure 4a.

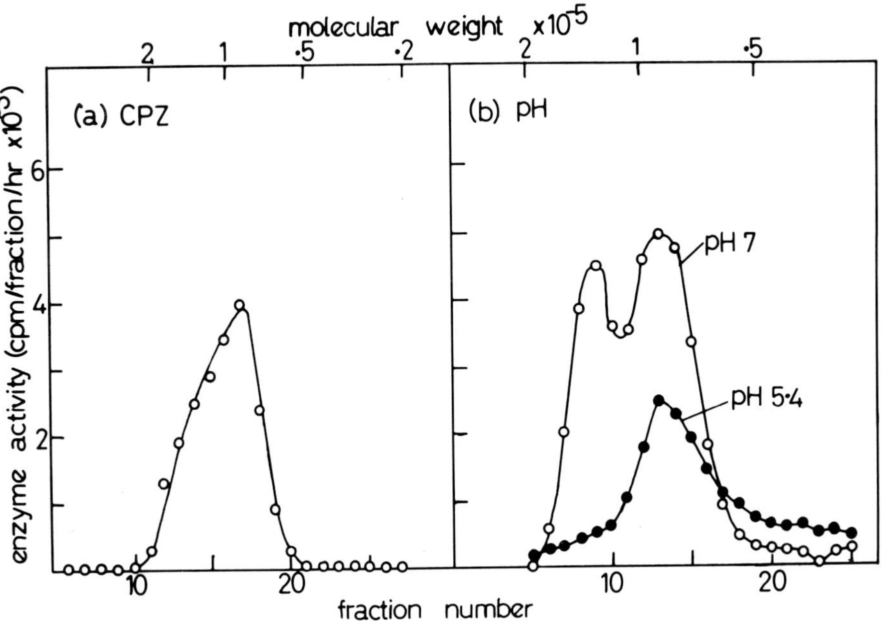


Fig. 4. Density gradient fractionation of phospholipase activity of rat brain post-microsomal supernatants. (a) after removal of a precipitate formed by treatment with 2mM chlorpromazine (CPZ) and (b) through gradients of pH 7 and pH 5.4.

Only a single peak of phospholipase activity is present (90,000 mol. wt.) suggesting that heavier species of enzyme are selectively removed by chlorpromazine. Thus chlorpromazine may inhibit phospholipase activity in two ways, one affecting both forms of the enzyme, for example by removing calcium ions bound to the substrate, and the other by complexing high molecular weight forms of the enzyme. The finding that all bound colchicine was precipitated by chlorpromazine (no radioactivity was recovered from the gradient), suggests that the drug may induce an interaction of the enzyme complex with colchicine-binding microtubular protein.

It was pointed out above that the ratio of reaction products depends on the assay pH. To examine the possibility that low pH causes a change in the molecular weight distribution of the enzyme, an additional experiment was done with $[^3H]$ -colchicine-labelled brain supernatant as described in figure 2 except that one sample was fractionated in a gradient prepared in buffer of pH 5.4. The phospholipase activity (assayed at pH 5.4) of the control (pH7), and low pH gradients are illustrated in figure 4b. Only about half of the phospholipase activity applied on the pH 5.4 gradient was recovered from individual fractions suggesting that the enzyme is unstable on prolonged exposure to acid conditions. Nevertheless, it is apparent from the distribution of the residual activity that phospholipase activity associated with the high molecular weight fraction is preferentially removed at pH 5.4. It is not possible to say whether the heavier fraction is completely inactivated or is converted to the lighter fraction but whichever is the case it is clear that cyclic inositol 1:2 phosphate will be the predominant product under acid conditions. The radioactivity profiles from these gradients (not shown) confirm earlier observations[14] that colchicine-binding activity is rapidly abolished at low pH.

5. The microtubular-active properties of chlorpromazine

One might conclude, especially from the results obtained in figure 4a, that chlorpromazine is a microtubular-active agent. Some preliminary experiments were performed to compare the action of chlorpromazine with that of the well-documented effect of vinblastine on precipitating microtubular protein[15]. A rat brain supernatant fraction was prepared as described by Shelanski et al[16], and tubulin was isolated from this by two cycles of polymerisation and depolymerisation. Samples of supernatant and tubulin were labelled with $[^3H]$-colchicine (5μM, 1.6 mCi/μmole) and aliquots incubated with 2mM chlorpromazine or vinblastine for 30 min. before centrifuging at 34,000g. The amount of colchicine precipitated in each fraction was the same for both drugs (table 2). SDS-disc gel electrophoresis of the untreated fractions and precipitates

Table 2

Comparison of the microtubular-active properties of vinblastine and chlorpromazine

| Fraction | Protein (mg) | Drug | Colchicine precipitated (pmoles) |
|---|---|---|---|
| Post-microsomal supernatant | 3.40 | vinblastine | 420 |
| | | chlorpromazine | 371 |
| Purified tubulin | 0.34 | vinblastine | 750 |
| | | chlorpromazine | 785 |

(figure 5) show that microtubular protein is selectively removed from the supernatant. Moreover, examination of gels of supernatants after incubation of purified tubulin with chlorpromazine or vinblastine (not shown) indicated that virtually all the protein had precipitated.

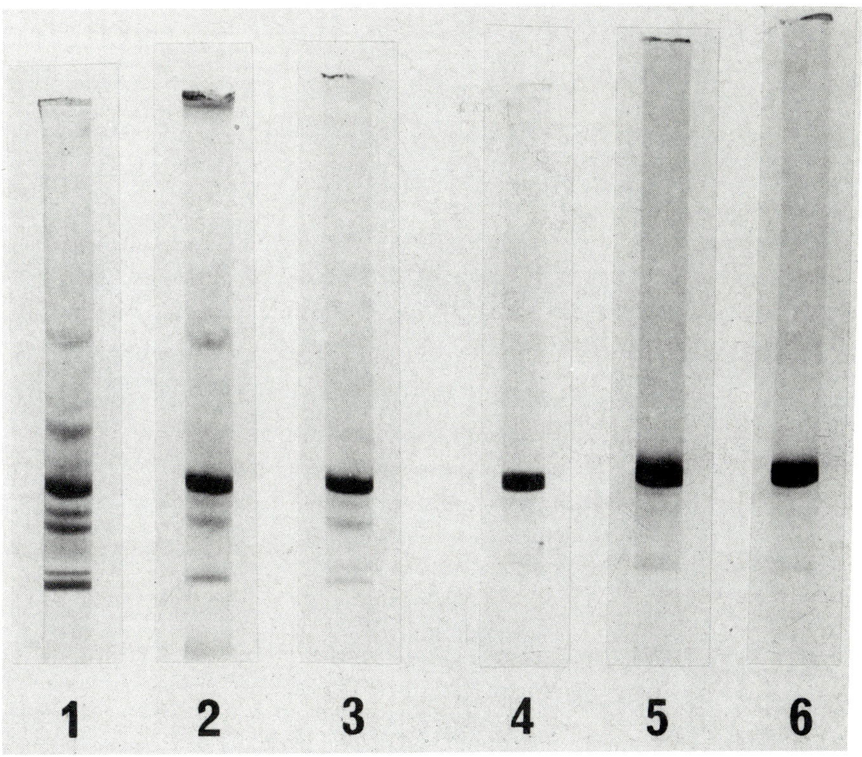


Fig.5. SDS-polyacrylamide gels of rat brain post-microsomal supernatant (1) and proteins precipitated from this extract by chlorpromazine (2) or vinblastine (3). Gel electrophoresis of purified tubulin[15] (4) and protein precipitated by chlorpromazine (5) and vinblastine (6) are also illustrated.

## 6. Conclusion

Phosphatidylinositol phosphodiesterase is not a component of cytoplasmic microtubules but is present in soluble fractions of brain and thyroid associated with microtubular protein. One form appears to be a colchicine-labile complex between the enzyme and two dimers of microtubular protein. Inhibition of phospholipase activity by chlorpromazine seems to be due in part to the formation of large inactive aggregate of this enzyme complex with microtubular protein. A possible effect of chlorpromazine on neurotubular function such as inhibition of axoplasmic transport[17] may have implications concerning the antipsychotic action of this drug.

## Acknowledgements

This investigation was aided and encouraged by J. McGuire, N. Freinkel and E.M. Naples.

## References

1. Wilson, L. and Meza, I., 1973, J.Cell Biol. 58, 709-19.
2. Stadler, J. and Franke, W.W., 1974, J.Cell Biol. 60, 297-303.
3. Murray, A.W. and Froscio, M., 1971, Biochem.Biophys.Res.Commun. 44, 1089-95.
4. Quinn, P.J., 1973, Biochem.J. 133, 273-81.
5. Roberts, K., 1974, Progr.Biophys.Mol.Biol. 28, 371-420.
6. Michell, R.H., 1975, Biochim.Biophys.Acta. 415, 81-147.
7. Keough, K.M.W. and Thompson, W., 1972, Biochim.Biophys.Acta. 270, 324-36.
8. Burgess, J. and Northcote, D.H., 1969, J.Cell Sci. 5, 433-51.
9. Weisenberg, R.C., Borisy, G.G. and Taylor, E.W., 1968, Biochemistry 7, 4466-79.
10. Dawson, R.M.C., Freinkel, N., Jungalwala, F.B. and Clarke, N., 1971, Biochem.J. 122, 605-7.
11. Jungalwala, F.B., Feinkel, N. and Dawson, R.M.C., 1971, Biochem. J. 123, 19-33.
12. Berry, R.W. and Shelanski, M.L., 1972, J.Mol.Biol. 71, 71-80.
13. Quinn, P.J. and Barenholz, Y., 1975, Biochem.J. 149, 199-208.
14. Wilson, L., 1970, Biochemistry 9, 4999-5007.
15. Wilson, L., Bryan, J., Ruby, A. and Mazia, D., 1970, Proc. Nat.Acad.Sci. USA. 66, 807-14.
16. Shelanski, M.L., Gaskin, F. and Cantor, C.R., 1973, Proc.Nat. Acad.Sci. USA. 70, 765-8.
17. Edström, A., Hansson, H.-A. and Norström, A., 1973, Z. Zellforsch. Mikro. Anat. 143, 71-91.

*Microtubules and Microtubule Inhibitors, eds. M. Borgers and M. de Brabander*



# INHIBITION OF TUBULIN ASSEMBLY BY RNA AND OTHER POLYANIONS. A MECHANISM.

J. Bryan, B.W. Nagle

Department of Biology

University of Pennsylvania, Philadelphia

I. Meza

Centro De Investigacion Del IPN

Mexico City, Mexico

Summary

Several types of non-neural cell extracts contain a heat stable, non-dialysable activity which will inhibit the spontaneous in vitro assembly of partially purified brain tubulin. This inhibitory activity in sea urchin extracts is sensitive to RNAses, but not other hydrolytic enzymes. Synthetic polynucleotides, purified RNA's from CHO cells, brain and sea urchin eggs all inhibit assembly. Half-maximal inhibition of tubulin assembly (at 1 mg/ml) for all the RNA's tested occurred between 0.24 and 0.33 mM added nucleic acid phosphate. Addition of exogenous RNA (or other polyanions) increased the apparent critical concentration of protein required for assembly. The addition of sonicated microtubule fragments to such mixtures of tubulin and a polyanion, which would not assemble spontaneously, resulted in tubule growth. The turbidity increase was a simple exponential function of time and the equilibrium values were independent of the numbers of fragments added. The initial rates of assembly were a linear function of the number of added fragments. Experiments with poly A-agarose and phosphocellulose indicate that added polyanions function by sequestering a heat stable, basic protein which is essential for neurotubule assembly. Tubulin treated with either of these insoluble polyanions would not assemble unless a protein fraction, eluted at higher salt concentration, was added. Reconstitution experiments indicate that there is some type of stoichiometric requirement for this factor and suggest that it partially determines the critical concentration and is essential for both nucleation and elongation of microtubules. The addition of this factor(s) to RNA inhibited solutions of tubulin or to DEAE purified tubulin produced assembly. These results further suggest that spontaneous assembly of microtubules in non-neural cell extracts may be blocked because the endogenous factors are complexed with RNA and other polyanions. This idea is supported by the observation that the ratio of tubulin to RNA is low in cultured cell extracts but very high in neural tissue extracts.

## 1. Introduction

Our current understanding of the mechanisms and control of microtubule assembly

(Abbreviations used: MES; morpholinoethane sulfonic acid, EGTA; ethylenebis (oxyethylene) tetraacetic acid, $C_c$; critical concentration, GTP; guanosine-t'-triphosphate, $C_1S$; brain tubulin (beef) purified by one cycle of assembly, $C_2S$; brain tubulin purified by two assembly cycles, PC-tubulin; phosphocellulose-stripped tubulin).

Supported by NSF grants GB 32287X and BMS 74-20302.

rests principally on information obtained by in vitro polymerization of neurotubulin. A direct extension of the conditions developed to obtain spontaneous neurotubule assembly to other systems has met with difficulty[1,2,3,4,5]. At least part of the difficulty appears to be the presence in non-neural cells of inhibitors of spontaneous tubulin assembly, since the addition of non-neural cell extracts to assembly competent neurotubulin will block spontaneous polymerization. In sea urchin eggs (Strongylocentrotus purpuratus), we have identified the principal inhibitor as RNA[6] and demonstrated that the mechanism of this inhibition is a general effect due to the sequestering, by added polyanions, of a heat stable essential protein factor[5,7]. In this paper we would like to review some of this evidence, argue that the essential factor is required for both nucleation and growth of microtubules and suggest that the partitioning of the factor between tubulin (or tubules) and RNA or other polyanions may serve as an efficient inhibitor of spontaneous tubulin assembly in vivo.

2. Evidence that RNA will inhibit spontaneous neurotubulin assembly.

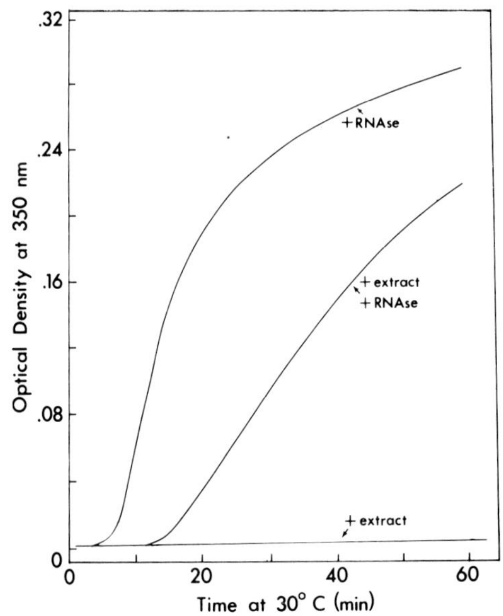


Fig. 1. Inhibition of microtubule assembly by a boiled, dialyzed extract of sea urchin eggs and relief of this inhibition by ribonuclease A. Microtubule assembly was monitored by development of turbidity at 350 nm. Each cuvette contained 1.0 ml of tubulin($C_1S$) at 2.0 mg/ml in 0.1 M MES, 0.5 mM Mg, 1 mM EGTA, 1 mM GTP and 2 M glycerol at pH 6.5. The samples labeled + extract contained 0.25 ml of a 30% extract of sea urchin eggs; the samples labeled + RNAse contained ribonuclease A at 10 μg/ml. Incubation was at 30°C. Sea urchin extracts were prepared by homogenizing 3 ml of S. purpuratus eggs in 7 ml cold MES-Mg, centrifuging at 100,000 g for 60 minutes, heating the supernatant at 100°C for 10 minutes and repeating the centrifugation. The final supernatant was dialyzed against MES-Mg with EGTA(1 mM) and used without further purification. (See references 4 and 5).

Several lines of evidence have been developed that indicate that RNA will inhibit tubulin assembly and that RNA is the heat stable inhibitor present in sea urchin egg extracts[4,5,6]. Figure 1 illustrates the inhibition of the spontaneous assembly of partially purified brain tubulin by a boiled sea urchin egg extract and the sensitivity of this inhibition to added RNAse A. Similar results can be obtained using RNAse N, but the inhibitory activity is resistant to DNAse, trypsin and several phospholipases. The addition of purified RNA's from sea urchin eggs, CHO cells or chick embryo brain all produce an inhibition of tubulin assembly (figure 2). Similarly, the addition of synthetic polynucleotides will inhibit the spontaneous formation of neurotubules. The data, presented in figure 2, indicate that there is little specificity in terms of base composition. Using the initial

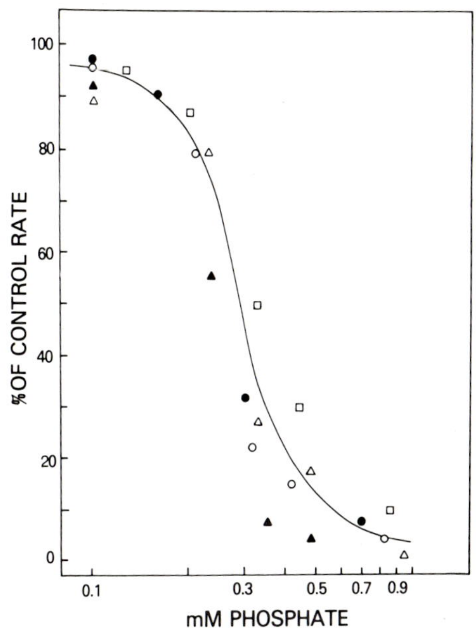


Fig. 2. Inhibition of microtubule assembly by purified RNA (sea urchin) and synthetic polynucleotides. The data are plotted as % of control rates. The measured rates were the maximum rates of OD (350 nm) increase obtained for each sample. Reaction mixtures contained tubulin ($C_1S$) at 1.5 mg/ml in MES-Mg with 1mM EGTA, 2 M glycerol and varying amounts of added RNA's. Incubation was at 30°C. The concentrations of RNA are given in mM phosphate. The concentrations were determined using the following millimolar extinction coefficients: (open squares) sea urchin, 8.5 (at 260 nm); (open circles) poly A, 10.4 (at 257 nm); (open triangles) poly U, 9.9 (at 260 nm); (closed circles) poly C, 10.1 (at 268 nm) and (closed triangles) poly G, 9.8 (at 254 nm). For clarity only one curve has been drawn through the points. (See reference 5 for further details).

rates of assembly as an indicator, half maximal inhibition for the RNA preparations tested occurs between 0.24 and 0.33 mM added nucleic acid phosphate. The observed inhibition is sharp, suggesting a markedly cooperative effect. Assuming a nominal polymer molecular weight of 100,000 (data from Sigma Chemical Co.) the inhibiting polymer concentrations are in the micromolar range. These numbers were obtained using 1 mg/ml of tubulin partially purified by one assembly cycle and measuring the effect on the second assembly cycle. In our general experience, however, the system becomes increasingly sensitive to polyanions with further purification. We can, for instance, routinely obtain half-maximal inhibition using second cycle tubulin at 50 μM added nucleic acid phosphate.

The inhibitory effect is not restricted to RNA; other polyanions including poly-L-glutamic acid, phosphocellulose, sulfo-propyl sephadex, carboxymethyl cellulose and an "insoluble RNA", polyadenylic acid agarose (poly A-agarose) all produce inhibition of neurotubulin assembly. These results argue that the inhibition is a general polyanion effect.

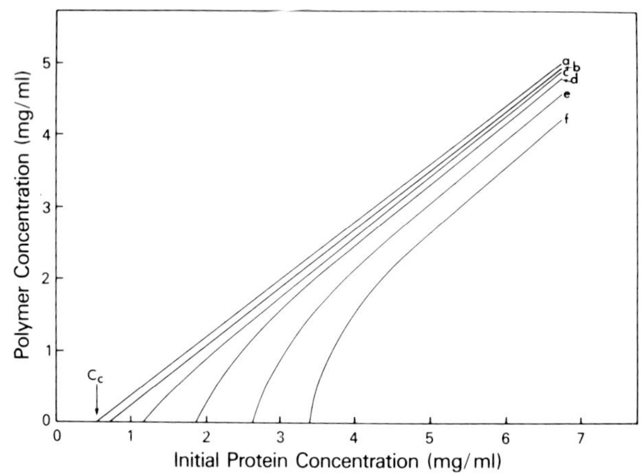


Fig. 3. Dependence of polymer formation on the initial protein concentration and on the concentration of boiled extract containing RNA. The polymer concentration was obtained by measuring the optical density at 350 nm and converting to mg/ml of polymer using the calibration procedures discussed in Bryan and Nagle.[4] The tubulin used was a $C_2S$ preparation in MES-Mg buffer with 1 mM GTP and 1 mM EGTA. The curves shown are: a. control; b. 10 μl; c. 20 μl; d. 50 μl; e. 100 μl and f. 200 μl of a boiled sea urchin extract containing 5.4 mg/ml of protein and 6 mM RNA as phosphate. The total reaction mixture was 1 ml. The extract was prepared as described in figure 1. The actual points have been left out for the sake of clarity, but each line represents at least six measurements. Temperature = 30°C. (See reference 4 for further details).

The results of a number of experiments on the effect of systematically changing the concentration of the sea urchin inhibitor and the concentration of tubulin are shown in figure 3. The results clearly demonstrate that increased concentrations of RNA affect the critical concentration ($C_c$) and also the extent of polymer formation. The data show that it is possible to have relatively concentrated tubulin solutions (2-4 mg/ml) which are below their $C_c$ and will not assemble spontaneously. These results were consistent with the idea that RNA sequesters either a molecule essential for assembly or a specific intermediate in the assembly pathway.

3. Separation of an essential heat stable protein factor(s) from tubulin.

The initial identification of the sea urchin inhibitor as an RNA and later as any polyanion suggested that treatment of assembly competent tubulin preparations with an insolubilized RNA or other cation exchange materials should block assembly. This was in fact the case[5,7] and in turn allowed a discrimination between the removal of a specific required aggregate of tubulin, such as a nucleating site, or the removal of an essential non-tubulin factor. The available evidence[5,7] indicates that a heat stable, trypsin sensitive non-tubulin factor is removed from neurotubulin preparations by cation exchangers which can be subsequently recovered by washing the exchanger at increased ionic strength. Weingarten _et al_[7] initially reported these observations and have termed the recovered factor(s) _tau_. The removal of this factor from tubulin using poly A-agarose seems to be markedly temperature sensitive as demonstrated in figure 4. This temperature sensitivity does not seem to be a function of the exchanger used, since we obtain qualitatively similar results using phosphocellulose and carboxymethyl cellulose. It is not yet clear if this temperature effect is related to the general temperature sensitivity of tubulin assembly or to some more complex process.

4. Reconstitution experiments

The results of an initial set of reconstitution experiments are shown in figure 5. Tubulin was stripped of the essential factor using phosphocellulose. For consistency, we will adhere to the terminology of Weingarten _et al_[7] and call this "PC-tubulin." Subsequent assembly of the PC-tubulin with and without added factor was monitored using turbidity and electron microscopy. The stripped samples alone produce no microtubules. The addition of factor(s) results in tubule assembly as judged by a turbidity increase and by electron microscopy. The fact that increasing concentrations of the factor(s) produce increased assembly is compatible with some type of stoichiometric requirement and argues

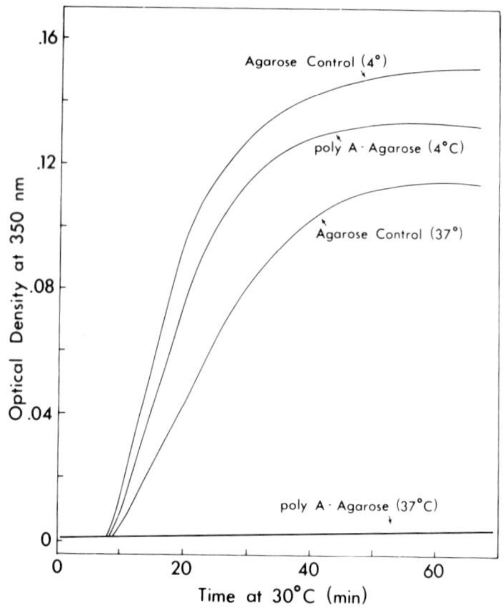


Fig. 4. Effects of treatment with agarose or poly A-agarose on the assembly reaction. Tubulin ($C_2S$) at 2 mg/ml in MES-Mg was mixed with an equal volume of packed agarose (Bio-Rad A-15 m) or poly A-agarose for 15 minutes at the temperature indicated. The mixture was then cooled on ice, centrifuged at 500 g for 10 minutes to remove the bulk of the agarose, the supernatants were collected and recentrifuged at 20,000 g for 15 minutes to remove fine particulates. These final supernatants were used to test for assembly after adding GTP to 1 mM. The control samples all had microtubules when checked in the EM using negatively stained preparations. There were no detectable microtubules present in the poly A-agarose (37°C) sample. Similar results were obtained with phosphocellulose and carboxymethylcellulose stripped tubulins. (See reference 5 for further details).

against a simple enzymatic or nucleation function for the factor. Our preliminary results using DEAE-purified tubulin dimer and a crude preparation of the stimulatory factor(s) confirm these results and argue that two measures of tubulin assembly, the $C_c$ and the extent of polymer formation are dependent upon both the tubulin dimer concentration and the concentration of factor(s). These results are also consistent with the data shown in figure 3. Here, an increased "inhibitor" concentration should produce a decrease in the activity of the factor with a resultant increase in the $C_c$ and this is, in fact, what we observe.

Only a preliminary characterization of the factor(s) has been done. The stimulatory activity is exceptionally heat stable and will withstand boiling for 5-15 minutes without loss of activity[5,7]. The activity is apparently a protein since it can be destroyed by the addition of trypsin but is insensitive to a series of other hydrolytic enzymes[5,7]. Slab gel electrophoresis indicates the presence of a number of bands in the factor containing fraction. The essential component(s) has not yet been purified.

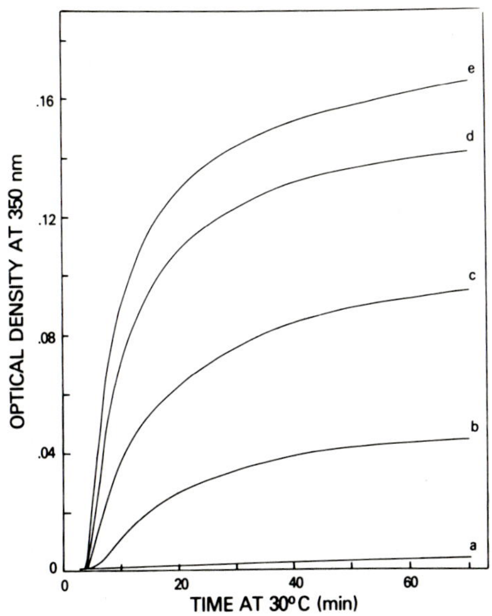


Fig. 5. Reconstitution experiments using PC-tubulin and factor(s). A $C_1S$ tubulin preparation was stripped using the protocol outlined in figure 4. The unbound tubulin peak was dialyzed against MES-Mg buffer containing 4 M glycerol for 6 hours at 4°C. The salt wash containing the factor(s) was dialyzed versus MES-Mg overnight at 4°C. PC-tubulin (400 µl at 2.8 mg/ml) was mixed with varying volumes of salt wash (2.0 mg/ml), GTP was added to 1 mM and the final volume adjusted to 800 µl with MES-Mg buffer. Samples were incubated at 30°C and the OD at 350 nm was recorded. Similar experiments using the parent unstripped $C_1S$ with increasing amounts of salt wash indicate no significant stimulatory effects on either the rates or extents of assembly. Curve a = either PC tubulin or factor(s) alone, curve b = 50 µl addition, curve c = 100 µl addition, curve d = 250 µl addition and curve e = 400 µl addition.

5. Some properties of polyanion inhibited neurotubulin solutions.

Since we think that <u>in situ</u> the microtubule assembling system must be exposed to various polyanions including RNA's, it seems reasonable that polyanion inhibited neurotubulin solutions can serve as useful models of tubule assembly. We find, in general, that such solutions will not assemble spontaneously if the concentrations of tubulin and polyanion are appropriately adjusted (figures 2 and 3). We have looked in these solutions for the "rings" found in control tubulin preparations. Their presence is determined by the pretreatment of the solutions. If poly A is added to a cold (0°) solution of tubulin containing rings, the rings remain intact. However, if this solution is warmed to 37°C for 2-5 minutes, the rings disappear and do not reappear upon cooling to 0°C. This result is consistent with the temperature sensitivity of inhibition shown in figure 4 and suggests that the factor is tightly complexed with tubulin in the rings at low temperature but more loosely associated at physiological temperatures. The data are in agreement with the results of Weingarten et al[7] which indicate the absence of rings in the stripped PC-tubulin solutions and their reformation upon addition of tau. These results are compatible with the notion that the rings are a low temperature aggregate of tubulin and the factor(s).

Although they will not assemble spontaneously, polyanion inhibited tubulin solutions will support microtubule growth. This is demonstrated directly in figure 6 by the addition of sonicated microtubule fragments to a tubulin solution inhibited by sea urchin RNA (see also reference 4). The initial rates of assembly are essentially a linear function of the numbers of fragments added and the equilibrium values are essentially independent of the numbers of fragments added[4]. The observed kinetics are consistent with the idea that spontaneous formation of nucleation centers is suppressed, that the number of growth points is constant and that elongation occurs by end-on addition of subunits[8]. We have done a more extensive investigation of the elongation reaction in tubulin solutions suppressed by addition of poly A, confirmed the sea urchin inhibitor results and made a preliminary search for "intermediates" in this nucleated assembly process using negative staining. In the electron microscope, there are no rings, sheets, or protofilament structures observed either before or immediately

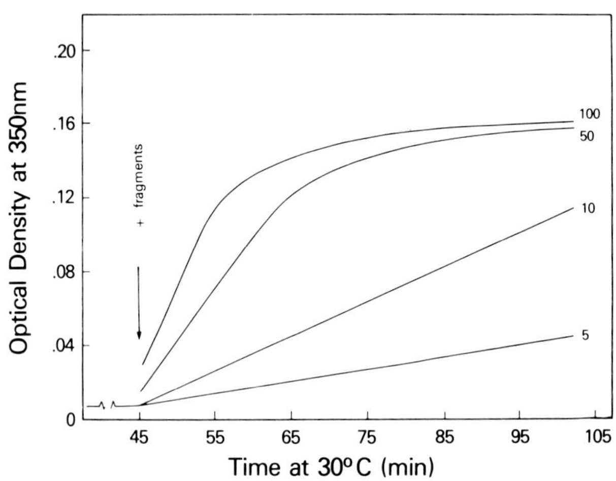


Fig. 6. The effect of adding sonicated microtubule fragments to suppressed brain tubulin solutions. The reaction mixture contained tubulin ($C_2S$) at 2.7 mg/ml, GTP at 1 mM, 0.2 ml of boiled dialyzed sea urchin extract (3 mg/ml of protein and 3.5 mM RNA as phosphate) in a total volume of 1.0 ml of MES-Mg buffer. The reaction mixtures were incubated at 30°C for 45 minutes before addition of the indicated volumes (in microliters) of sonicated fragments. The fragments were obtained by sonication of a control fully polymerized solution (2.7 mg/ml) for 15 seconds (Branson Model W, at lowest setting). The average fragment length was 0.45 μ.

after fragment addition. The only regular structures after fragment addition are microtubules. Within 1-2 minutes there are no longer any detectable short fragments (0.3-0.6 microns), only longer tubules. From these results and a more exhaustive analysis of the time course of this nucleated assembly, we conclude that the observed turbidity increase is due only to elongation or growth of the fragments. The failure to observe any "intermediates" is consistent with the idea that this elongation proceeds by end-on addition of individual subunits, presumably dimers or dimers and factor(s). While we observe no free "intermediates", our preliminary examination of the tubule ends does suggest that a significant number of the growing tubules (but not freshly sonicated fragments) have an end with a small length of unclosed tubule wall or sheet.

Direct evidence that the factor(s) is limiting in polyanion inhibited tubulin solutions comes from experiments involving the addition of the factor(s). One set of results is given in figure 7. Addition of exogenous factor produces assembly of tubules. The kinetics of assembly are markedly different from those shown in figure 6 for fragment addition and indicate that nucleation sites must be forming in the early stages of the reaction. Our preliminary morphological analyses of the early time periods in this assembly indicate the initial

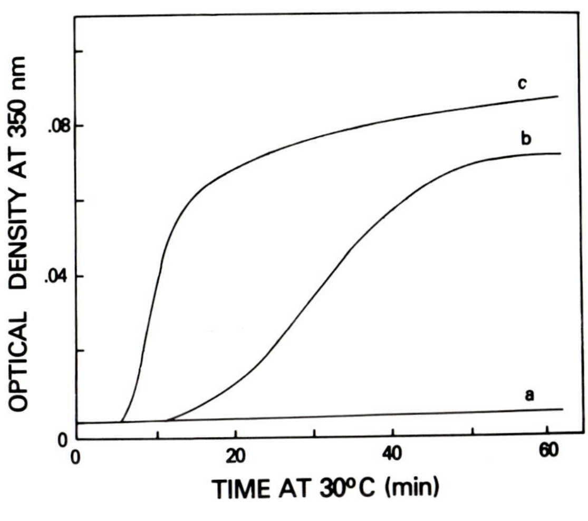


Fig. 7. Relief of poly A inhibition of tubulin assembly by addition of factor(s). Tubulin ($C_1S$) at a final concentration of 1.2 mg/ml was inhibited by addition of poly A (final concentration 0.28 mM as phosphate). Curve a = tubulin plus poly A, curve b = tubulin plus poly A plus phosphocellulose salt wash (0.4 mg/ml final concentration), curve c = control tubulin without additions. Each reaction mixture was buffered with MES-Mg plus 1 mM GTP and 2 M glycerol. Incubation was at 30°C.

formation of protofilaments and sheets of protofilaments followed by "wrapping" into closed tubes[9,10].

## 6. Discussion

The data presented are compatible with the idea that the formation of neurotubules requires the tubulin dimer and at least one other heat stable protein. The interaction between tubulin and this protein factor is temperature sensitive. At low temperatures tubulin and the factor aggregate to produce rings and spirals, at higher temperatures they form microtubules[5,7,9,10]. Although the relationship is not precisely defined, the reconstitution experiments indicate some type of a stoichiometric requirement for the factor and rule out, reasonably well, a merely enzymatic or nucleation function. These data argue that the factor is required during the growth process. This conclusion is supported by preliminary experiments which indicate that seeding of stripped tubulin with microtubule fragments will not produce elongation. The factor is also clearly required for nucleation since the reduction of the activity of the factor by polyanion addition blocks the nucleation pathway. This is demonstrated by the failure of polyanion inhibited tubulin preparations to spontaneously form microtubules and their ability to support tubule elongation if fragments of tubules are added as "nucleation" sites. Our preliminary search for "intermediates" in the elongation reaction suggests that growth occurs by addition of individual subunits and argues strongly that the protofilaments and sheets of protofilaments seen during the early assembly reactions are part of the nucleation pathway and that these structures are not essential intermediates in the elongation of tubules.

An extension of these in vitro results into an in vivo situation is made difficult by our lack of information on relative cytoplasmic concentrations of tubulin, factor(s) and polyanions. A conservative hypothesis would argue that cells normally have sufficiently high concentrations of both tubulin and tau (or other stimulatory proteins) that spontaneous assembly would be expected but that undirected assembly is blocked because cellular polyanions reduce or regulate the activity of the essential factors. The generation in situ of a nucleating site, for example a kinetochore, would then produce a 'directed' tubule growth analogous to what we observe upon addition of microtubule fragments. In this view, an interaction between polyanions and the tubulin - 'tau' system serves to lower the tau activity and provide a certain 'buffering' capacity for these factors. We are currently attempting to differentiate between this hypothesis and those involving actual fluctuations in the concentration of tau-like factors during the cell cycle.

References


1. Burns, R.G. and Starling, D., 1974, J. Cell Sci. 14, 411.
2. Farrell, K.W. and Burns, R.G., 1975, J. Cell Sci. 17, 669.
3. Bryan, J., 1975, Am. Zool. (in press).
4. Bryan, J. and Nagle, B.W., 1975, in: 26th Mosbach Colloquium, The Molecular Basis of Motility (in press).
5. Bryan, J., Nagle, B.W. and Doenges, K.H., 1975, Proc. Natl. Acad. Sci. (in press).
6. Bloom, G., Nagle, B.W. and Bryan, J., 1975, FEBS Abstracts.
7. Weingarten, M.D., Lockwood, A.H., Hwo, S. and Kirschner, M.W., 1975, Proc. Natl. Acad. Sci. 72, 1858.
8. Oosawa, F. and Higashi, S., 1967, in: Progress in Theoretical Biology, vol. 1, p. 79.
9. Kirschner, M.W., Williams, R.C., Weingarten, M.D. and Gerhart, J.C., 1974, Proc. Natl. Acad. Sci. 71, 1159.
10. Erickson, H.P., 1974, J. Supramol. Structure 2, 393.

*Microtubules and Microtubule Inhibitors, eds. M. Borgers and M. de Brabander*



# BIOCHEMICAL MECHANISMS OF ACTION OF MICROTUBULE INHIBITORS

L. Wilson, K. Anderson, L. Grisham, and D. Chin

Department of Pharmacology

Stanford University School of Medicine

Stanford, California 94305 USA

Summary

Several of the well known inhibitors of microtubule function exert their effects on microtubules through a nonstoichiometric "poisoning" of microtubule assembly. The vinca alkaloids possess the most striking poisoning effect of any inhibitors thus far investigated. There are two biologically specific vinblastine binding sites per mole of embryonic chick brain tubulin, with dissociation constants of approximately 3-6 x $10^{-6}$ M. The concentration of vinblastine required to inhibit cell growth in cultured EHB cells by 50% is 7.5 x $10^{-8}$ M; an almost 100-fold lower concentration than would be expected if vinblastine were working by means of a simple stoichiometric binding to tubulin. Vinblastine inhibits the initial rate of *in vitro* polymerization of embryonic chick brain tubulin by 50% at a concentration of 2.3 x $10^{-8}$ M, which is very similar to the concentration required to inhibit mitosis, but is such that only one tubulin-vinblastine complex could be present per 1400 free tubulin molecules in solution under the conditions employed. Podophyllotoxin exerts a similar, though slightly less potent poisoning effect. Griseofulvin, which does not bind to DEAE-cellulose-purified chick embryo brain tubulin, can prevent the *in vitro* polymerization of some microtubules in embryonic chick and bovine brain extracts, but many microtubules still polymerize. The biochemical explanation for this observation is not known, but this and other data suggest that several populations of microtubules exist in vertebrate brain, only some of which are sensitive to the action of griseofulvin.

## 1. Introduction

A significant portion of our knowledge concerning microtubule biochemistry and function has been obtained with the use of drugs such as colchicine, the vinca alkaloids, podophyllotoxin, and griseofulvin. Most of these chemical agents, originally called "spindle poisons", can bind to tubulin and by mechanisms yet to be elucidated, prevent the polymerization of tubulin to microtubules (see references 1-5 for reviews). Since most microtubules are in equilibrium with tubulin in soluble pools, the addition of drugs to cells containing them results in dissolution of the microtubules and impairment of the cellular functions which depend upon them. The approach to our current understanding of the biochemical mechanisms of action of chemical agents that disrupt microtubules has involved investigation of the binding characteristics of the agent to tubulin, and more recently, correlating the results with information obtained on the ability of the agents to inhibit microtubule polymerization *in vitro*. These investigations have revealed a number of unexpected results, among them the finding that colchicine, podophyllotoxin, and the vinca alkaloids exert their effects on microtubules *in vivo* through a nonstoichiometric "poisoning" of microtubule assembly.[6,7]

The mechanisms of action by which other antimitotic agents disrupt microtubule function are almost completely unknown. Some progress has been made towards understanding how the drug, griseofulvin, affects microtubules. Earlier studies suggested that griseofulvin affects microtubule function through a mechanism other than inhibition of microtubule polymerization. For example, HeLa cells blocked by griseofulvin at metaphase contain large numbers of long microtubules.[8,9] In addition, microtubules can polymerize *in vitro* in supernatant solutions of embryonic chick brain with high concentrations of griseofulvin present.[8] Nevertheless, the development of viscometric methods for the assay of microtubule polymerization *in vitro*[6] has shown that griseofulvin does affect microtubule assembly. These studies have revealed that different populations of microtubules exist in bovine and chick brain extracts, some of which are blocked from assembling by griseofulvin, others are not.

The purpose of this presentation is to describe the results of two studies currently in progress in our laboratory; one, on the biochemical mechanism of action of the vinca alkaloids, and the second, on the mechanism of action of griseofulvin. Our hope is that the information gained by understanding how these drugs work will permit them to become more useful as experimental probes in the elucidation of microtubule structure and function in cells.

## 2. Correlation of Drug-Binding to Tubulin with the Inhibition of Microtubule Polymerization.

*Colchicine*. The binding of colchicine to tubulins from a variety of sources has been studied carefully, and the results of these studies have been described in a number of reviews.[1-5] Upon incubation of a cell or tissue extract containing tubulin with labeled colchicine, a tight colchicine-tubulin complex forms, which can be measured by several methods including gel filtration or adsorption of the complex on discs of DEAE-Cellulose impregnated filter paper. There is one high affinity colchicine binding site per mole tubulin: The binding constant with tubulins from vertebrate brain (e.g., embryonic chick brain) is $2 \times 10^{6}$ l/mol at 37°. The reciprocal of this value, $5 \times 10^{-7}$ M, is the concentration of colchicine that would saturate exactly half of the colchicine binding sites in a solution of tubulin.

A poisoning mechanism of microtubule assembly by colchicine was first proposed by Olmsted and Borisy,[6] who discovered that the concentration of colchicine required to inhibit microtubule assembly in porcine brain extracts was considerably below that required for half-maximal saturation of the available colchicine binding sites. Attempts in our laboratory to obtain an inhibition constant for colchicine (*i.e.*, the concentration of colchicine that would half-maximally inhibit the initial rate of microtubule polymerization) on the polymerization of embryonic chick brain tubulin have not been successful. The problem is due to the slow rate of colchicine binding to tubulin, which is considerably slower than the initial rate of microtubule assembly. This problem does not exist with podophyllotoxin, which binds at the colchicine binding site, or with the vinca alkaloids, which bind at different sites. Both podophyllotoxin and the vinca alkaloids bind rapidly to tubulin[5], and our attempts to obtain inhibition constants for these drugs have met with much greater success.

Vinca Alkaloids. The characteristics of the binding reaction between the vinca alkaloids and tubulin have not been as extensively studied as those of the colchicine binding reaction. Owellen et al.[10,11] employed a DE81 filter paper disc assay procedure, much like that used for the measurement of colchicine binding to tubulin, to study the binding of tritium-labeled vinblastine and vincristine to porcine brain tubulin. They reported a binding constant of 5-6 x $10^6$ l/mol for the binding of vinblastine to tubulin, and a stoichiometry of 0.5 mol vinblastine per mol tubulin.

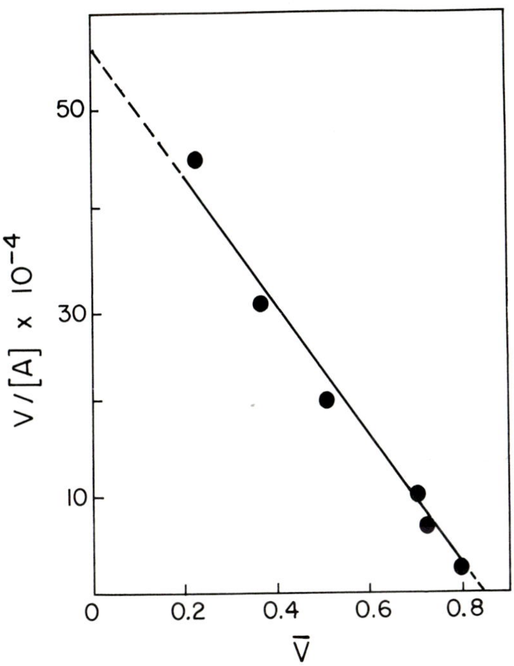


Figure 1. Binding of Vinblastine to Purified Vinblastine-Tubulin Crystals. (Scatchard Plot). Suspensions of purified vinblastine crystals were incubated with different concentrations of tritium-labeled vinblastine for 45 min at 22°. The quantity of bound vinblastine was calculated after measuring the amount of vinblastine in the initial suspensions for each concentration, and the amount of vinblastine remaining in the supernatant after centrifugation of the suspension to pellet the crystals. $\bar{V}$= mol vinblastine bound per mol tubulin. [A]= free vinblastine (mol/l).

Bryan[12] isolated vinblastine crystals from unfertilized sea urchin eggs and found them to contain one mole of vinblastine per mole tubulin. More recently, L. Wilson and J. Bryan have found that purified vinblastine-tubulin crystals will bind exogenously added vinblastine, in a ratio of 1 mole vinblastine per mole tubulin, with a binding constant of 7 x $10^5$ l/mol with freshly purified crystals (Figure 1). It is not yet clear in this work whether the exogenously added vinblastine is exchanging with the previously bound vinblastine, or is binding to a second site.

Other active vinca alkaloid derivatives competitively inhibit the binding of vinblastine to these crystals, as shown with vincristine sulfate in Figure 2.

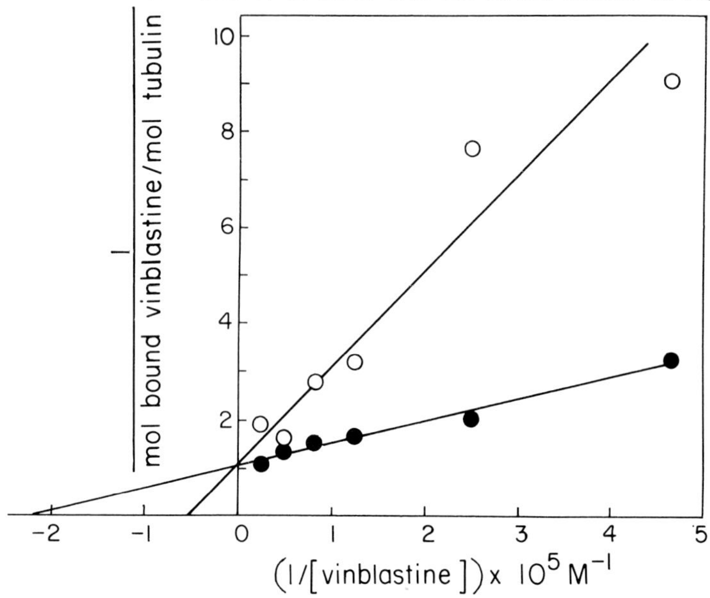


Figure 2. Inhibition of vinblastine binding to purified vinblastine-tubulin crystals by vincristine. Suspensions of crystals were incubated at 21° for 45 min with increasing concentrations of labeled vinblastine in the presence (open circles) and absence (closed circles) of 6 x $10^{-6}$ M vincristine sulfate. The quantity of bound vinblastine was then determined as described in the legend to Figure 1. The affinity for vinblastine decreases with age of the crystals after isolation. In this experiment, the dissociation constant for vinblastine was 2.1 x $10^{-6}$ M. The inhibition constant for vincristine was very similar, 1.4 x $10^{-6}$ M.

We have also studied the binding of vinblastine to purified tubulin from embryonic chick brain by a DE81 paper disc assay procedure[5] and by an equilibrium Sephadex gel procedure, and have found two vinblastine binding sites per mole of tubulin, with binding constants for these sites of approximately 1-3 x $10^5$ l/mol. The differences in vinblastine-tubulin stoichiometry between our results and those of Owellen and his coworkers can be explained by the assay procedures which were employed. With the paper disc assay, the vinblastine-tubulin complex is completely separated from free vinblastine, and the previously bound vinblastine can dissociate from the tubulin without being replaced. Thus, the quantity of vinblastine remaining bound to the tubulin on the paper disc at the end of this assay procedure must be lower than the true equilibrium quantity. However, the differences in binding constants obtained by us and by Owellen et al, cannot be attributed to the non-equilibrium DE81 assay procedure. We have obtained a vinblastine stoichiometry of 0.45 mol vinblastine per mol tubulin with this procedure[5], but still observe the same relatively weak binding constant of 1-3 x $10^5$ liters/mol as with the equilibrium gel procedure. In all of our studies we have obtained binding constants for vinblastine with freshly purified tubulin from chick brain in the range of 1-3 x $10^5$ l/mol, and since one of these procedures is an equilibrium procedure, we feel that our values accurately reflect the true situation for this tubulin. Thus, a vinblastine concentration of the reciprocal of this value, or approximately 5 x $10^{-6}$ M, will half-maximally saturate the vinblastine binding sites.

One way to investigate how the vinca alkaloids disrupt microtubules is to determine whether vinblastine can bind to intact microtubules. If the vinblastine binding sites were not accessible to vinblastine in assembled microtubules, vinblastine could not directly bind to the surface of the microtubules and cause them to depolymerize. This appears to be the situation both with colchicine[13] and with the vinca alkaloids.

Similar to results obtained with colchicine, the vinblastine binding sites are inaccessible to vinblastine in intact sea urchin sperm tail outer doublet microtubules, whereas tubulin solubilized from the outer doublets possesses accessible sites.[14] Vinblastine, like colchicine, does not depolymerize these microtubules. When the microtubules are depolymerized by the method of Stephens[15] at 37°, the addition of vinblastine does not affect the rate of depolymerization. Further evidence that vinblastine does not directly depolymerize microtubules has been obtained with reassembled microtubules from embryonic chick brain. These microtubules are much more stable than those polymerized in vitro in adult chick brain extracts, or in porcine and beef brain extracts. Addition of as much as 5 x $10^{-5}$ M vinblastine to embryonic chick brain extracts containing preassembled microtubules does not result in the depolymerization of the microtubules. The hypothesis is that microtubule assembly in these extracts is 100% inhibited, but the microtubules do not depolymerize in the presence of the drug because of their greater stability coupled with the inability of vinblastine to bind to the microtubules and directly depolymerize them. These microtubules do depolymerize quickly upon addition of 15 mM $CaCl_2$ (in the presence of 10 mM EGTA), and somewhat more slowly upon cooling to 0°. Thus, our present evidence suggests that vinblastine must depolymerize previously existing microtubules by inhibition of the polymerization reaction. With the polymerization reaction blocked, previously existing microtubules, which are in equilibrium with tubulin in soluble pools, will depolymerize at a rate which is determined by the stability inherent in that population of microtubules.

We have employed a viscometric assay for quantitation of the inhibition of in vitro microtubule polymerization by vinblastine and other active drugs. The method, which draws heavily on the viscometric procedures employed in Borisy's laboratory[1,6], consists of preparing 30,000 x g supernatant extracts of embryonic chick brain in 20 mM sodium phosphate, 100 mM sodium glutamate, pH 6.75, and monitoring the rate of specific viscosity increase during the early stages of the polymerization reaction after addition of 10 mM EGTA, 1 mM $MgCl_2$, and 2.5 mM GTP at 30°. The rate of assembly at 30° is sufficiently slow so that a linear increase in specific viscosity is obtained during the first 20-25 min (Figure 3a), and a rate can be calculated from the slope of the line. Inclusion of vinblastine decreases the initial rate. A plot of 1/fraction inhibition vs 1/vinblastine concentration yields a straight line, from which an inhibition constant can be calculated (Figure 3b).

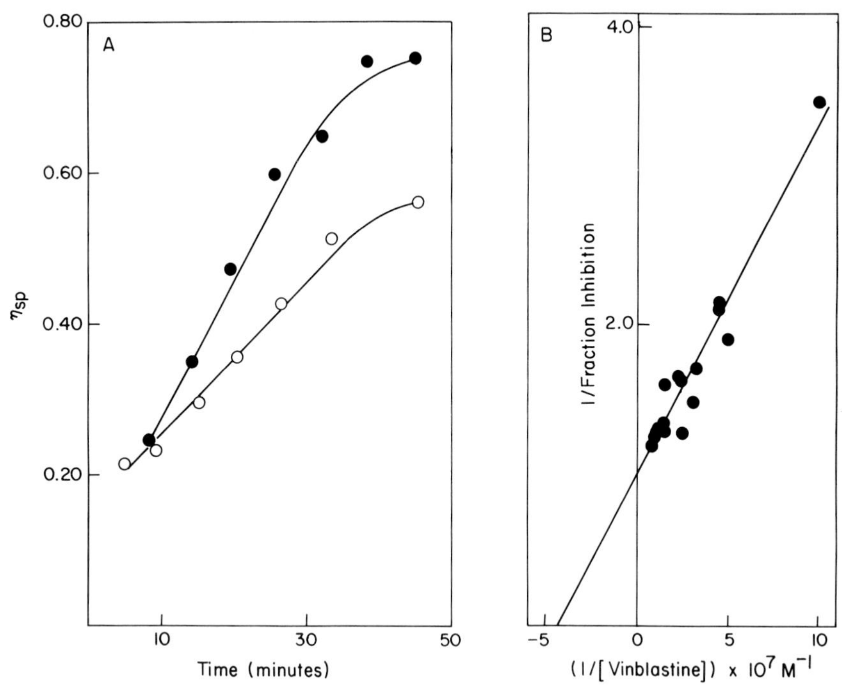


Figure 3a. Increase in specific viscosity with time at 30° in 30,000 x g supernatant fractions of 15-18 day old embryonic chick brain extracts. The presence of vinblastine decreases the initial rate. Closed circles = no vinblastine; open circles = 2.2 x $10^8$ M vinblastine.

Figure 3b. Determination of the inhibition constant for vinblastine. Initial rates of specific viscosity increase at 30° were obtained in the absence and in the presence of different concentrations of vinblastine, as described in the legend to Figure 3a. The inhibition constant, which is considered to be the concentration of vinblastine that inhibits the initial rate of microtubule polymerization by 50%, was 2.3 x $10^{-8}$ M.

The inhibition constant with vinblastine was $2.3 \times 10^{-8}$ M. This value represents the molar concentration of vinblastine necessary to inhibit the initial rate of polymerization by 50%. Thus, the concentration of vinblastine that half-maximally inhibits the initial rate of microtubule polymerization is very similar to the concentration of vinblastine that is required to inhibit mitosis in EHB cèlls by 50%, but is more than 100-fold lower than the measured binding constant to tubulin. But, even more striking, the concentration of tubulin present in the extract (Figure 3b) was $3 \times 10^{-5}$ M. Thus, the total amount of tubulin that was present in solution exceeded the total amount of vinblastine present by a factor of 1000 under conditions of half-maximal inhibition of assembly given a binding constant of vinblastine to tubulin of $9 \times 10^{-6}$ M. The calculated value for the quantity of vinblastine bound per mole tubulin at half-maximal inhibition of assembly is one vinblastine-tubulin complex per 1400 free tubulin molecules in solution.

We have also investigated the ability of vinblastine to inhibit the elongation phase of the polymerization reaction. Ten minutes after initiation of *in vitro* assembly in the absence of vinblastine, different concentrations of vinblastine were added, and as in Figure 3b, an inhibition constant for vinblastine was determined. This value was $9 \times 10^{-8}$ M, or close to 4 times greater than the concentration of vinblastine required to inhibit the rate of assembly half-maximally during the initiation phase of the polymerization reaction. The ratio of vinblastine-bound tubulin to free tubulin during this phase is still very low, at one vinblastine-tubulin complex per approximately 300 free tubulin molecules in solution. These numbers are approximate estimates, but it does seem certain that a very low ratio of vinblastine-bound tubulin to free tubulin is sufficient to inhibit both the initiation and elongation phases of the polymerization reaction. Moreover, the vinca alkaloids have a differential ability to inhibit the initiation and elongation phases.

One of the earlier models explained how drugs such as vinblastine and colchicine inhibited microtubule polymerization in cells by stoichiometrically binding to tubulin in soluble pools. The notion was that the drug-complexed tubulin could not assemble on the growing end of a microtubule. Our results suggest that this is not the way that vinblastine inhibits polymerization.

We feel it is more likely that vinblastine is acting primarily by binding to the growing end of a microtubule during the elongation phase of the assembly reaction, and perhaps to some form of the initiation complex during the initiation phase of polymerization. Our assay procedures for detection of vinblastine binding to intact microtubules are sufficiently sensitive to permit us to conclude that vinblastine is not binding along the surface of outer doublet microtubules, but the binding of a very few molecules of vinblastine at the ends of these microtubules could not have been detected. It is easy to imagine how the binding of just a few vinblastine molecules to the growing ends could prevent any additional tubulin molecules from adding on. However, the question remains as to how such a mechanism should be so highly effective, since vinblastine binding to tubulin is sufficiently weak that the vinblastine bound to the growing ends should rather freely dissociate. One very attractive possibility is that the affinity for vinblastine of the tubulin molecules assembled on the polymer at the growing end have undergone a conformational change such that the affinity for vinblastine is greatly increased. Then, because of a higher affinity for vinblastine, the drug would bind more selectively to these sites at low concentrations, and would remain more strongly attached to the growing end of the polymer.

Podophyllotoxin. That podophyllotoxin competes with colchicine for the colchicine binding site has been shown with tubulins from several different sources[4]. In all cases, inhibition is competitive. Although colchicine and podophyllotoxin share the same binding site on tubulin, the mechanism of the podophyllotoxin binding reaction is different than that of the colchicine binding reaction. In contrast to the binding of colchicine, podophyllotoxin binds rapidly to tubulin, and the binding is not nearly so temperature dependent. The binding constant for podophyllotoxin and tubulin has been estimated by obtaining an inhibition constant for the ability of podophyllotoxin to inhibit the binding of colchicine to tubulin. For chick brain tubulin, the value for half-maximal inhibition of colchicine binding activity is $4\text{-}5 \times 10^{-7}$ M. This is the concentration of podophyllotoxin that would saturate exactly half of the colchicine (and therefore, podophyllotoxin) binding sites on tubulin.

Similar to the situation with vinca alkaloids, the quantity of podophyllotoxin required to inhibit the initial rate of *in vitro* microtubule polymerization in supernatant solutions of chick embryo brain extracts is considerably lower than one would predict if podophyllotoxin were inhibiting polymerization by stoichiometrically binding to tubulin in tubulin pools. The inhibition constant for podophyllotoxin during the initiation phase of assembly at 30° is $3.5 \times 10^{-8}$ M. Similar to experiments carried out with the vinca alkaloids, the ratio of total tubulin to total podophyllotoxin in these extracts was also 1000:1. Thus, it is clear in the case of podophyllotoxin as well as with the vinca alkaloids, that nonstoichiometric poisoning is occurring. In contrast to vinblastine, podophyllotoxin cannot distinguish between initiation and elongation phases of polymerization. The concentration of podophyllotoxin that inhibits polymerization half-maximally 10 minutes after initiation of assembly is $3.9 \times 10^{-8}$ M, statistically identical to that which inhibits polymerization during the initiation phase.

### 3. Griseofulvin's Selective Effect on *In Vitro* Microtubule Polymerization

The site of the antimitotic action of griseofulvin still has not been identified.

Griseofulvin interferes with many microtubule-containing systems in vivo.[16-19] This drug prevents the in vitro assembly of bovine brain microtubules purified[20] by one cycle of assembly and dissassembly; tubulin polymerization in vitro is inhibited half-maximally at a griseofulvin concentration of approximately $5 \times 10^{-6}$ M. Curiously, microtubule formation in crude bovine brain extracts is only partially inhibited by $1 \times 10^{-4}$ M griseofulvin. By electron microscopy, the remaining microtubules are long and appear normal.

The residual griseofulvin insensitivity observed in 30,000 x g brain extracts is due to a subpopulation of "cold stable" microtubules which besides being griseofulvin resistant, are also able to polymerize at 0°. These microtubules polymerize at 0° in 30,000 x g bovine brain extracts on addition of 2.5 mM GTP or GDP ( 1 mM EGTA and 0.5 mM $MgCl_2$ are also added). The structures which form are indistinguishable by electron microscopy from microtubules produced at 30°. $3 \times 10^{-5}$ M podophyllotoxin, colchicine, or vinblastine completely abolish microtubule assembly at 0°, however $1 \times 10^{-4}$ M lumicolchicine or griseofulvin do not.

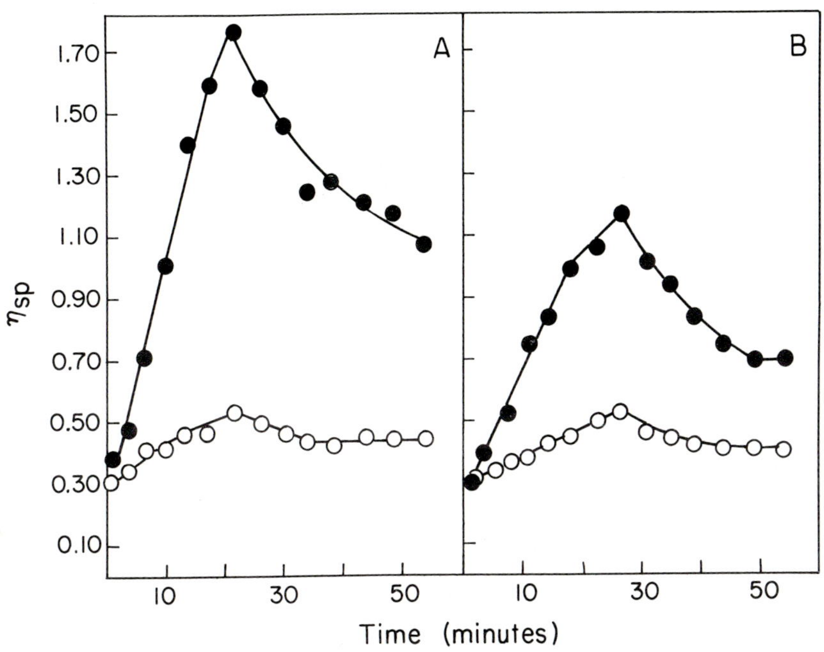


Figure 4. Use of GTP to protect microtubule assembly competence in bovine brain extracts. Bovine brain was homogenized in 0.1 M PIPES pH 6.9 and centrifuged at 2 x 30,000 x g for 45 min, 4°. Supernatant samples were preincubated 2 hours at 0°: a) with 0.5 mM GTP and, b) without GTP. Then, 2 mM GTP, 1 mM EGTA, 0.5 mM $MgCl_2$ and either 0.3% dimethyl sulfoxide (closed circles) or 0.1 mM griseofulvin in 0.3% dimethyl sulfoxide (open circles) was added; Samples were warmed to 30° and polymerization was monitored viscometrically.

Griseofulvin insensitivity can be removed from 30,000 x g brain extracts by centrifugation of the microtubules which form at 0° in homogenates when GTP is added to the homogenizing buffer. Further in vitro tubulin assembly at 30° is completely inhibited by griseofulvin. We conclude that griseofulvin is selectively preventing the polymerization of "cold labile" microtubules.

The specificity of griseofulvin's action has allowed us to identify other differences between the cold stable and cold labile microtubules. For example, GTP added to 30,000 x g brain extracts prevents the loss of the ability to polymerize with time. We find that this loss of assembly competence is due to the preferential inactivation of tubulin from cold labile microtubules. The results presented in Figure 4 demonstrate this observation. Bovine brain supernatant fractions were allowed to age at 0° with and without GTP. After 2 hours GTP was added to samples to initiate assembly, and these samples were immediately warmed to 30°. A significant loss of assembly competence occurred in samples lacking GTP. Note that the griseofulvin-resistant fraction (cold stable microtubules) is the same as in samples protected by GTP. GDP and to a slight degree GMPPCP (0.5 mM) are also able to protect cold labile tubulin, although neither nucleotide promotes cold labile microtubule assembly at 30°. It appears that guanine nucleotides play two roles: 1) preserving assembly competence of soluble tubulin in cold-labile microtubules, and 2) promoting in vitro assembly for both classes of microtubules.

Cold labile and cold stable microtubule populations also exist in chick brain. During chick brain development several changes occur within the microtubule pool. We find that the fraction of cold stable (griseofulvin resistant) microtubules -- approximately 40% of the total -- in embryonic chicks (14-16 days old) is twice that in hatched chicks (1-5 days old). Furthermore, the kinetics of microtubule assembly change markedly during chick brain development. At 30°, a delay of 30-60 minutes is observed during the initiation phase of microtubule assembly in 30,000 x g brain extracts from hatched chicks; no delay occurs with brain extracts from embryonic chicks. The microtubules produced in vitro in hatched chick extracts are depolymerized by $3 \times 10^{-5}$ M podophyllotoxin or vinblastine, whereas the ones formed in embryonic chick extracts remain intact.

How is griseofulvin working? The mechanism by which it affects microtubule assembly is unknown. $^{3}$H-griseofulvin does not bind to DEAE-cellulose purified tubulin[9]. This finding does not rule out the possibility that griseofulvin interacts with tubulin in an activated state or with a non-tubulin component involved in microtubule formation. With regards to the drug's in vivo activity we suggest that the mitotic spindle may contain populations of microtubules which differ in griseofulvin sensitivity. In vivo cytoplasmic microtubule populations heterogeneous with respect to stability to cold temperature[21] or high pressure[22], have been reported. Our results suggest that griseofulvin's antimitotic effects are due to interference with a particular class of microtubules important for spindle integrity.

This work was supported by American Cancer Society Grant CH-4A, and USPHS grant NS09335.

References

1. Olmsted, J.B., and Borisy, G.G., 1973, Ann. Rev. Biochem., 42, 507.

2. Margulis, L., 1973, Int. Dev. Cytol., 34, 333.

3. Wilson, L., Bamburg, J.R., Mizel, S.B., Grisham, L.M., and Creswell, K.M., 1974, Fed. Proc., 33, 158.

4. Wilson, L., and Bryan, J., 1974, in: Advances in Cell and Molecular Biology, vol. 3, ed. E.J. DuPraw (Academic Press, New York) p. 21.

5. Wilson, L., 1975, Ann. N. Y. Acad. Sci., 253, 213.

6. Olmsted, J.B., and Borisy, G.G., 1973, Biochemistry, 12, 4282.

7. Anderson, K., 1975, M.S. Thesis, Stanford University.

8. Grisham, L., Wilson, L., and Bensch, K.G., 1973, Nature (London), 244, 294.

9. Grisham, L.M., Bensch, K.G., and Wilson, L., 1973, J. Cell Biol., 59, 125a.

10. Owellen, R.J, Owens, Jr., A.H., and Donnigian, D.W., 1972, Biochem. Biophys. Res. Commun., 47, 685.

11. Owellen, R.J., Donnigian, D.W., and Kuhar, M.J., 1974, Cancer Res., 34, 3180.

12. Bryan, J., 1972, Biochemistry, 11, 2611.

13. Wilson, L., and Meza, I., 1973, J. Cell Biol., 58, 709.

14. Wilson, L., Anderson, K., and Creswell, K., 1974, J. Cell Biol., 63, 373a.

15. Stephens, R.E., 1970, J. Mol. Biol., 47, 353.

16. Deysson, G., 1964, Ann. Pharm. Fr., 22, 17.

17. Deysson, G., 1964, Ann. Pharm. Fr., 22, 89.

18. Malawista, S.E., 1971, J. Cell Biol., 49, 848.

19. Margulis, L., Neviackas, J.A., and Banerjee, S., 1969, J. Protozool., 16, 660.

20. Borisy, G.G., Olmsted, J.B., Marcum, J.M., and Allen, C., 1974, Fed. Proc., 33, 167.

21. Brinkley, B.R., and Cartwright, Jr., J., 1975, Ann. N. Y. Acad. Sci., 253, 428.

22. O'Connor, T.M., Houston, L.L., and Samson, F., 1974, Proc. Nat. Acad. Sci., 71, 4198.

*Microtubules and Microtubule Inhibitors, eds. M. Borgers and M. de Brabander*
© 1975, North-Holland Publishing Company - Amsterdam


# STUDIES ON MICROTUBULE ASSEMBLY

M.Jacobs, M.Caplow, P.M.Bennett and M.J.Dickens
MRC Cell Biophysics Unit, King's College, London.

The use of polycations has provided information on the initial morphological steps in microtubule assembly. Protamine induces the formation of a microtubule surrounded by an additional wall of tubulin subunits. This structure is a new form of tubulin assembly which we propose calling a duplex microtubule. The duplex is related to a double ring form of tubulin consisting of an 380 Å outer ring and a 250 Å inner ring which contains 13 subunits. A new model for microtubule assembly is postulated in which this double ring acts as a nucleating centre.

The role of nucleotides in microtubule assembly has been further investigated. The dissociation constants for GTP and GDP binding to the E site are 1.3 x $10^{-7}$ and 1.2-2.5 x $10^{-7}$ M and a lower limit of 6.4 x $10^{-4}$M has been determined for ATP. Occupancy of the exchangeable (E) site by GTP is necessary for rapid assembly of tubulin into microtubules. After removal of free and exchangeable nucleotide tubulin does not polymerise in the presence of 0.1 mM ATP. Addition of GDP to the ATP reaction induces microtubule assembly via a pathway involving transphosphorylation in which GTP is synthesised.

## 1. Introduction

There is increasing evidence that microtubule assembly at least 'in vitro' proceeds after a lag period [1,2] during which there appears to be a nucleation step involving an oligomeric ring form of tubulin [1,2]. GTP [1,2,3] is a necessary co-factor in assembly but it is not known what role it has in the nucleation or elongation steps of polymerisation.

In this paper, we describe two approaches to the study of microtubule assembly. In the presence of polycations, a double walled microtubule is formed. This structure is related to a double ring of tubulin subunits which may be involved in the nucleation step. We have also extended our studies on the biochemical mechanism of polymerisation. ATP dependent polymerisation takes place only in the presence of GDP via a transphosphorylation reaction. GDP influences the rate and extent of this reaction and may therefore regulate microtubule formation by adjusting the concentration of tubulin-bound GDP and GTP.

2. A new form of tubulin assembly: duplex microtubule.

A doubled walled microtubule is induced by adding protamine to microtubules reassembled from highly purified tubulin 'in vitro' (fig. 1a). The structure formed is similar to that seen previously by Behnke[4] in brain extracts using a variety of polycations and by Olmsted[1] using DEAE dextran. In the latter case, the extra material observed surrounding the microtubule, was attributed to the DEAE dextran itself.

We propose that the double walled tubule be called a 'duplex microtubule,' since our evidence described below and Behnke's shows that the additional wall consists of tubulin subunits probably complexed with polycations. At protamine concentrations less than the molar concentration of tubulin, only regions of duplex structure are seen (fig. 1a). Initially the microtubules have a diameter of 250 Å and an axial periodicity of 40 Å. After the addition of protamine, the new structure has an diameter of 380 Å. The protofilamentous appearance of the microtubule can still be seen in the centre of the duplex, extending up from the bare regions of microtubule. The predominant feature of the outer wall seen clearly at the edges of the duplex is an 50 Å axial repeat.

---

Fig. 1. (a) Regions of duplex microtubule induced by low levels of protamine. Microtubules (10 μM) were incubated with neutralised protamine (5 μM)

(b) Duplex tubules cooled to 4° .C. for one hour.

(c) Optical diffraction pattern of duplex tubule.

(d) Spermine induced tubule formation. Purified tubulin (1.8 μM) incubated at 4° .C. with neutralised spermine (1.5 μM)

(e) Optical diffraction pattern of spermine tubule.

(f) Double rings and short segments of duplex tubule seen when protamine is added to tubulin at 4° C.

Samples for electron microscopy negatively stained with uranyl 2 % uranyl acetate.

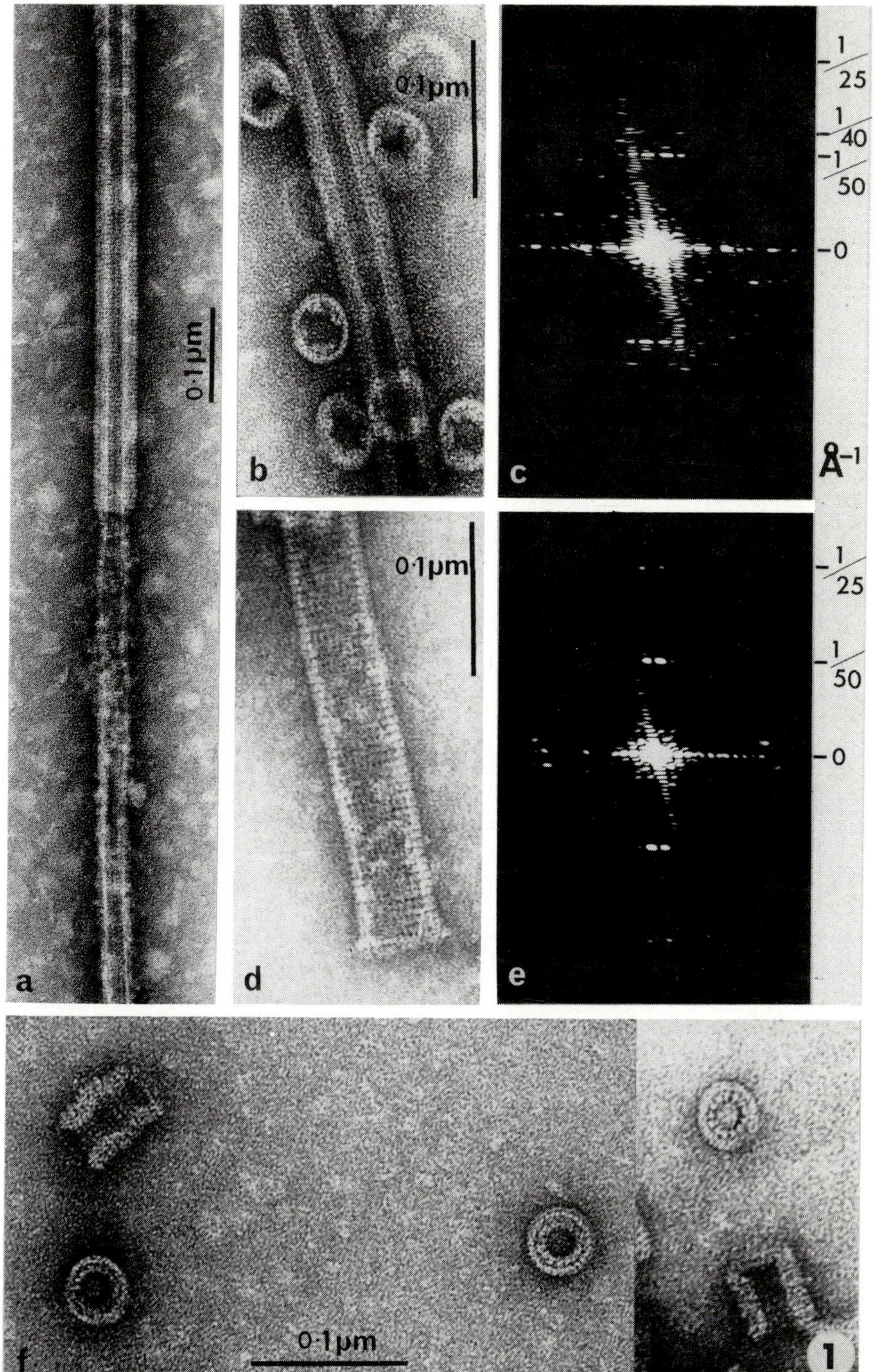
0·1μm
a
0·1μm
b
1/25
1/40
1/50
0
c
Å⁻¹
0·1μm
d
1/25
1/50
0
e
0·1μm
f
1

Microtubules completely converted to the duplex along their entire length can be seen if protamine is added in molar excess over tubulin. These protamine treated microtubules consist of 97 % tubulin and possibly some bound protamine. Since the duplex structure is formed with such a variety of polycations[5], tubulin is the only component present in sufficient quantity to account for the outer wall.

3. Optical diffraction of the duplex.

The diffraction pattern of the duplex (fig. 1c) shows off-meridional reflections with an axial spacing of 40 Å and equatorials which are characteristic of reconstituted[6] and axonemal[7] microtubules. Additional reflections, in particular strong off-meridionals corresponding to a 50 Å axial spacing are identified with the helical arrangement of subunits in the outer wall. A structure which resembles the wall itself though somewhat flattened, is obtained by adding spermine to tubulin at 4°.C. (fig. 1d). Its diameter is similar to that of the duplex tubule and gives rise to off-meridional and equatorial diffraction spots, corresponding to a 50 Å axial spacing and a 40 Å azimuthal spacing respectively (fig. 1e). These spots give the dimensions of the lattice in the spermine tubule namely 50 x 40 Å. This is the same size as the size of the subunits in the microtubule (40 x 50 Å) and implies that the subunits of the inner and outer walls of the duplex tubule are orientated perpendicular to each other in the two layers of the complex.

4. Dissociation of the duplex at 4°.C.

Further information about the duplex microtubule was obtained by incubating them at 4°.C. for one hour. Duplex microtubules still remain though some dissociation has taken place as is shown by the presence of microtubules, short segments of duplex, double and single rings (fig. 1b). Both ring structures have the same outside diameter as the duplex microtubule (table 1) as is seen fortuitously in fig. 1b. where a ring is superimposed on a duplex tubule.

Table 1

Diameters of the ring forms of tubulin

| Type of Ring | Conditions | Diameters Å* | | | |
|---|---|---|---|---|---|
| | | Inner Ring | | Outer Ring | |
| | | Inner | Outer | Inner | Outer |
| Double ring | Tubulin + protamine at 4° | 120 | 250 | 250 | 380 |
| Double ring | Tubules + protamine cooled to 4° | 120 | 250 | 250 | 380 |
| Single rings | Tubules + protamine cooled to 4° | - | - | 250 | 380 |
| Single rings | Tubulin 4° | - | - | 250 | 380 |

* Range of diameter measurements ± 20%.

Double rings together with short segments of helix again with 380 Å diameter were formed by adding protamine to tubulin at 4°.C. (fig. 1f). These rings are so well preserved that often 13 subunits (the number of protofilaments in a microtubule) can be counted in the inner ring. Significantly, the outer diameter of this inner ring is 250 Å, the same as the microtubule, the double ring therefore has the same cross-section as the duplex (table 1). The number of subunits in the outer ring cannot be counted suggesting that the molecules are differently arranged.

Microtubules in the absence of protamine are completely dissociated after 30 minutes at 4°.C. into single rings of outer diameter 380 Å, inner diameter 250 Å (30- 36 S protein) and particles probably dimers (6 S protein). These single rings have therefore the same outer diameter as the duplex tubules and are approximately the same outer diameter as the single rings previously reported under similar conditions[1,6].

5. Model for assembly.

The formation of the duplex 'in vitro' may provide clues to the mechanism of microtubule assembly which is at present unclear. Previous suggestions have involved the unrolling of rings into protofilaments [6,8] or the rearrangement of stacks of rings into a helix[1].

Both these proposals require large subunit rearrangements. Another possibility follows from the observations presented here. Before tubule formation, the dominant form at least in our preparations, is the 380 Å diameter single ring. This ring is able to form a second ring on the inside which is stabilised by polycations. This structure may occur naturally in small amounts and is possibly the disc first described by Borisy et al. (1972).[9]

A plausible sequence of events is the formation of the 380 Å ring from 6 S dimers. This structure then serves as a template for the formation of a second ring on the inside or a short length of duplex. Further dimers are added and the microtubule is propagated on the inner ring or wall. As the first length of tubule grows the outer ring or wall of the duplex comes off. Such a scheme is supported by the observation that 6 S tubulin which normally does not polymerise by itself, will assemble on to the inner wall of short lengths of duplex microtubule[1].

Further studies are in progress to test this model and to determine the relationship between the surface lattices of the various tubule forms.

## 6. Nucleotide binding.

Recently we described the properties of the exchangeable (E) and non-exchangeable (N) nucleotide binding sites [3,10]. During polymerisation GTP binds to the E site, is hydrolysed and the resulting GDP remains bound to the microtubule. Although no significant binding of ATP to the E site was observed, it was reported by others to induce polymerisation[1,2]. GTP, however, may have been synthesised from ATP and GDP by a transphosphorylase[10], an enzyme associated with tubulin.

The Hummel and Dryer column equilibrium method[11] was used to determine the equilibrium constant for GTP binding to the E site. GDP and ATP binding constants were measured by competition, allowing the use of very high ligand concentrations. These constants are summarised in table 2 and their determination will be reported elsewhere in greater detail[12].

Table 2

Dissociation constants ($K_d$) for GTP, GDP, and ATP binding

| Nucleotide | $K_d$ |
|---|---|
| GTP | $1.3 \times 10^{-7}$ M (4°.C.). |
| GDP | $1.2 \times 10^{-7}$ M (4 & 25°.C.). |
| ATP | $6.4 \times 10^{-4}$ M (4°.C.). |

At 4°.C. the E site has approximately equal affinities for GTP and GDP. At the temperature used for polymerisation studies, 25°.C., the dissociation constant for GDP binding is not measurably different from that at 4°.C. A 7000 fold excess of ATP ($5.3 \times 10^{-4}$M) does not influence GTP binding at a concentration of $7.6 \times 10^{-8}$M. Based on the accuracy of the column method a 33 % decrease in GTP binding may go undetected, a lower limit for ATP dissociation constant is then calculated as $6.4 \times 10^{-4}$M. A recent study[13] of nucleotide binding to tubulin using a nitrocellulose filter assay gave dissociation constants equal to $0.5 \times 10^{-6}$, $1.9 \times 10^{-6}$M for GTP and GDP respectively. Very weak ATP binding was also observed which is consistent with our results. The Hummel and Dryer method has also been used to measure GTP binding to bovine tubulin and a dissociation constant equal to $0.4 \times 10^{-6}$M was estimated[14]. However, no variation in the amount of GTP bound was seen over the limited concentration range used.

7. Polymerisation kinetics.

Tubulin prepared by cycles of polymerisation and dissociation contains GDP as a result of the previous polymerisation[10]. Addition of $10^{-4}$M ATP to this protein induced microtubule formation. However removal of free and exchangeable nucleotide by gel filtration through G-25 before polymerisation prevents ATP dependent microtubule assembly unless GDP is added (fig. 2). Polymerisation then proceeds after a time lag. The length of the lag decreases and the initial rate increases with increasing GDP concentration, becoming independent at high concentrations. The initial rate of GTP dependent polymerisation also increases with GTP concentration even above the level calculated to saturate the E site (fig. 3).

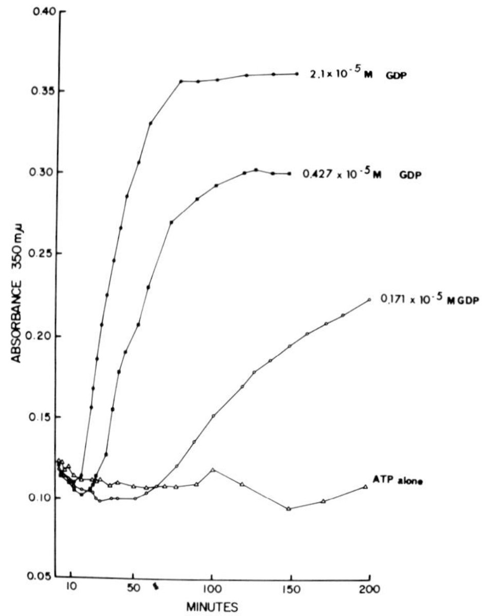


Fig. 2. ATP induced GDP dependent polymerisation. Tubulin 1.7 x $10^{-5}$M. ATP constant 1.1 x $10^{-4}$M. GDP varied, 2.1 x $10^{-5}$M, ● ● 0.427 x $10^{-5}$M, ■ ■ ,0.171 x $10^{-5}$M, ○ ○, no GDP, △ △. Polymerisation was followed by measuring light scattering[2] at 350 mμ .

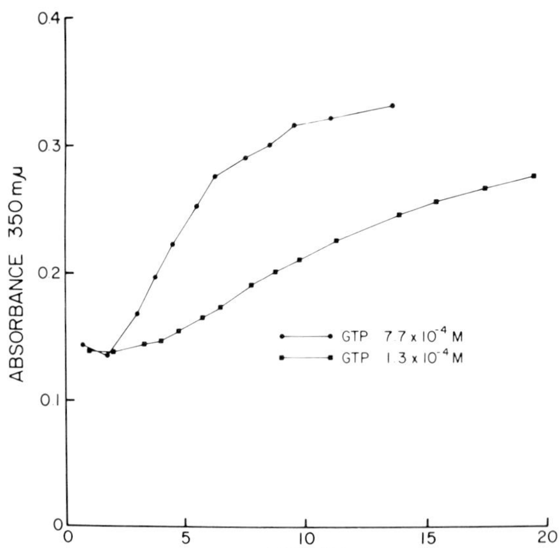


Fig. 3. GTP dependent polymerisation. Tubulin, 1.7 x $10^{-5}$M.

GDP is a potent inhibitor of the GTP dependent polymerisation as indicated by the fact that microtubules are not formed with 1.1 x $10^{-4}$M GTP and 2.7 x $10^{-4}$M GDP. This result is consistent with GDP and GTP having comparable dissociation constants at 25°.C.

The GDP dependence for ATP induced microtubule formation is consistent with a reaction sequence in which GTP is formed by a transphosphorylation reaction and then participates in the polymerisation process. Earlier reports[1,2] found that ATP dependent polymerisation did not require GDP, however, it was not established that the protein was depleted of nucleotide; only extremely low levels of GDP are required in the ATP induced reaction (fig. 2).

The initial rate of the ATP induced reaction becomes independent of the GDP concentration at high levels. This saturation effect is consistent with a mechanism in which the phosphoryl acceptor is either the tubulin bound GDP or the GDP remaining unbound (free GDP). Using the $K_d$ for GDP reported here, the concentration of the species may be calculated at each GDP concentration. The dependence of the rate on these concentrations is given in Fig. 4.

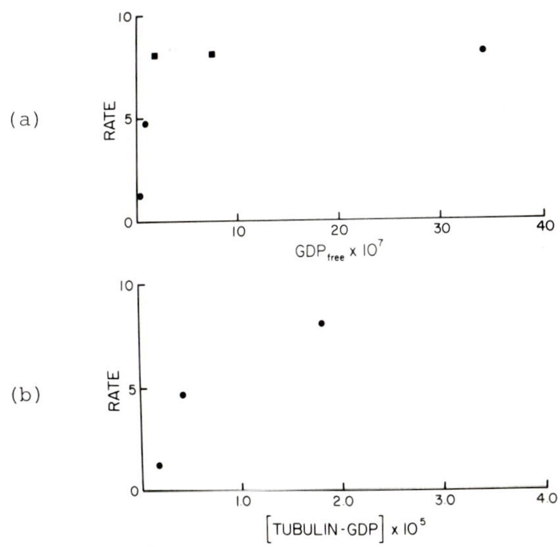


Fig. 4. The effect of GDP concentration on the rate of polymerisation. (a) free GDP using data from 2 experiments, ■ (experiment 1), ● (experiment 2). (b) tubulin bound GDP. The concentrations of the species were calculated using a dissociation constant of 1.3 x $10^{-7}$M.

The results are consistent with a mechanism in which the transphosphorylase substrate is free GDP with an apparent $K_d$ of approximately $7 \times 10^{-8}M$ or the tubulin-GDP complex with apparent $K_d$ of $3.5 \times 10^{-6}M$. Neither of these $K_d$ values are such so as to exclude either mechanism although the very small value calculated for the reaction involving free GDP suggests this is not the mechanism. Studies[14] with a variety of soluble transphosphorylases give $K_d$'s for the phosphate acceptors which are much higher ($5 \times 10^{-5}$- $10^{-3}M$). For this reason it is very unlikely that the transphosphorylation reaction is catalysed by the nucleoside diphosphokinases previously described. A mechanism in which the E and/or the N site GDP are transphosphorylated is favoured.

Our experiments demonstrate that GTP dependent polymerisation does not depend solely on the stoichiometric binding of GTP to the E site. The rate of GTP induced polymerisation and the lag period before it are both dependent on concentration even above that calculated to saturate the E site (fig. 3). Further studies are in progress to see whether a GTP dependent transphosphorylation reaction can explain these results.

Two ways of regulating microtubule assembly are suggested by our studies. Firstly assembly could be switched on and off by the reversible binding of a polycationic molecule which induce ring formation. The correct arrangement of positive charges attached to a membrane or particle could control the initiation sites and spatial arrangement of microtubules. Secondly, since GDP is such a potent inhibitor of polymerisation, a fine control over the rate and extent of assembly could be achieved by modulating the ratio of GTP to GDP. This would be done through the transphosphorylase acting on the bound nucleotide.

## References


1. Olmsted, J.B., Marcum, J.M., Johnson, K.A., Allen, C., and Borisy, G.G., 1974 J. Supramol. Struct. 2, 429.
2. Gaskin, F., Cantor, C.R., and Shelanski, M.L., 1974 J. Mol. Biol. 89, 737.
3. Jacobs, M., Smith, H., and Taylor, E.W., 1974, J. Mol. Biol 89, 455.
4. Behnke, O., 1974. Electron Microscopy Vol. II. Edited by Sanders, J.V., and Goodchild, D.J.
5. Behnke, O. Submitted to Nature.
6. Erickson, H.P., 1974, J. Supramol. Structr. 2, 393.

7. Amos, L.A., and Klug, A.G., 1974, J. Cell. Sci. 14, 523.

8. Kirschner, M.W., and Williams, R.C., 1974, J. Supramol. Struct 2, 412.

9. Borisy, G.G., and Olmsted, J.B., 1972, Science 177, 1196.

10. Jacobs, M., 1975, Ann New York Acad. Sci. 253, 562.

11. Hummel, J.P., and Dreyer, W.J., 1962, Biochim. Biophys Acta 63, 530.

12. Jacobs, M. In preparation. Microtubule Associated Transphosphorylase.

13. Arai, T., Ihara, Arai, K., Kaziro, Y., 1975, J. Biochem. 77, 647.

14. Levi, A., Cimino, M., Mercanti, D., and Calissano, P., 1974, Biochem. Biophys. Acta 365, 450.

*Microtubules and Microtubule Inhibitors, eds. M. Borgers and M. de Brabander*



# STUDIES ON THE PHOSPHORYLATION OF BRAIN MICROTUBULE PROTEIN AND MICROTUBULE-ASSOCIATED PHOSPHOLIPIDS

J.R. Lagnado and E.P. Kirazov

Department of Biochemistry, Bedford College, London, England

Summary. The incorporation of $^{32}$P into rat and chick brain microtubules *in vivo* was investigated. More than half of the total acid-insoluble (bound) radioactivity in microtubules isolated by reassembly *in vitro* was recovered as phospholipid P, most of which was associated with phosphatidylethanolamine and phosphatidic acid. Most of the remaining $^{32}$P could be accounted for as protein-bound phosphoserine P (alkali-labile) associated with tubulin. Attempts to determine the distribution of bound $^{32}$P in the 36S and 6S components of cold- or calcium-depolymerised microtubules from chick brain, after chromatography on Sepharose 6B or high-speed centrifugation, showed that the 36S component contained the bulk of the bound $^{32}$P. However, a significant part of the label found in this fraction was apparently associated with minor high-MW protein components (as phosphoserine P and phospholipid P) which could be separated from 36S tubulin aggregates during re-chromatography in the presence of 0.75 M NaCl. It is proposed that the considerable metabolic activity of microtubule-bound $^{32}$P seen *in vivo* may reflect some aspect of the various membrane-associated phenomena in which microtubules have been implicated, in addition to any more direct role that phosphorylation may have in controlling microtubule assembly.

## 1. Introduction

Tubulin, the main subunit protein of microtubules, constitutes up to one quarter of the total soluble protein of brain homogenates, where it appears to be derived mainly from cytoplasmic microtubules. Recent ultra-structural studies also show that axonal microtubules in cerebral cortex of mammalian brain extend right into nerve endings where, in association with clusters of synaptic vesicles, they are sometimes seen to terminate onto dense projections of the presynaptic membrane [1,2]. This would be in keeping with earlier biochemical evidence for the presence of tubulin in the soluble fraction of isolated nerve-ending particles (synaptosomes) [3,4]. However, a significant proportion of the tubulin present in nervous tissue is also tightly bound to membrane fractions that are derived primarily from isolated nerve-ending particles [3-7]. In fact, tubulin has now been more specifically identified as the major protein component of the subsynaptic dense material characteristic of dendritic spines from type 1 synapses [6,8,9].

Taken together, these observations suggest that microtubular proteins may participate, more directly than was originally anticipated, in mediating or in regulating membrane-dependent physiological processes associated with synaptic function. Clearly, the extent to which microtubules or their subunits are involved in these, or in several other membrane-associated phenomena in non-neural cells (see, e.g., ref. 11) would depend on their ability to form more or less transient functional associations with other soluble or membrane-bound cellular constituents. That such interactions do in fact occur is evident from a number of recent observations (see, e.g., refs. 10, 12, 13), and it was recently proposed that the state of phosphorylation of microtubular protein is an important factor in the control of microtubular assembly and/or function in nervous tissue (see evidence summarised in ref. 10). The aim of the experiments described in this paper was to provide further evidence that brain microtubules are phosphorylated *in vivo* and to determine more precisely the nature of the phosphorylated components labelled *in vivo* that were derived from reassembled microtubules.

## 2. Chemical nature of $^{32}P$ incorporated into brain microtubules *in vivo*

Reassembled microtubules isolated from rat and chick brain labelled *in vivo* with $^{32}P_i$ contained appreciable amounts of acid-insoluble radioactivity (bound $^{32}P$). In general, about 70% of the bound cpm in disassembled microtubular preparations from chick brain, and somewhat less from rat brain, was extractable with 2 : 1 (v/v) $CHCl_3$–MeOH (Tables 1 and 2); similar results were obtained when acidified $CHCl_3$–MeOH was used as lipid solvent (data not shown). Over 80% of the acid-insoluble $^{32}P$ remaining after extraction with lipid solvents was present as alkali-labile P, derived most probably from protein-bound serine residues (see refs 10 and 14 for details of procedures used to identify phosphoserine P in protein): this will be referred to as protein-bound P. Less than 5% of the counts remaining in the alkali-stable fraction could be attributed to nucleic acids.

It was found in preliminary experiments that most of the bound $^{32}P$ present in the crude lipid extracts from unfractionated microtubules or from purified 36S microtubular subfractions could be recovered in the phosphatidylethanolamine and/or phosphatidic acid fractions separated by one- and two-dimensional thin-layer chromatography on silica gel H according to the method of Yagihara *et al*[15]. Further experiments to characterise microtubule-associated phospholipids labelled *in vivo* are being carried out in collaboration with Professor J. N. Hawthorne (Nottingham University, England), but it is apparent from the results obtained so far that none of the label present in the phospholipid fraction could be attributed to phosphoinositides or to phosphatidylserine.

Protein-bound $^{32}P$ in 2× polymerised microtubules or in chromatographically purified 36S tubulin aggregates (see Table 2 and text below) isolated from *in vivo* labelled chick brain appears to be associated with a unique acidic phosphopeptide which was seen in peptide maps of samples digested with trypsin (preliminary observations done in collaboration with P. Dunkley, Institute of Psychiatry, London).

These results indicate that (specific) protein-bound phosphoserine residues and specific classes of phospholipids associated with brain microtubules exhibit considerable metabolic activity *in vivo*, as could be anticipated from earlier studies of microtubule phosphorylation in tissue slice preparations [10,23].

The full significance of these findings cannot be properly assessed, however, until more is known regarding the protein-P and phospholipid-P contents of microtubules *in situ* and of the extent to which this may be regulated by associated enzymes.

3. Electrophoretic analysis of *in vivo* labelled brain microtubules.

It was previously observed that a significant proportion of the radioactivity present in the microtubular fraction of $^{32}P$-labelled slices from guinea pig cerebral cortex was associated with minor high-MW components ('HMW') seen after electrophoresis in SDS-polyacrylamide gels [10]. A similar distribution of $^{32}P$ has now been found for *in vivo* labelled microtubular preparations (1× polymerised) derived from rat and chick brain (Fig.1).

In these preparations, a number of additional minor components migrating between tubulin (TU) and HMW fractions were also labelled; in chick brain, some labelling was also associated with a minor stained component migrating ahead of the $\beta$-tubulin fraction. In addition, a diffuse region of radioactivity of varying intensity was consistently observed to migrate ahead of the dye front (see also refs. 7 and 10), a region which does not stain for protein but contains an excess of SDS.

In an effort to assess the nature of the $^{32}P$ present in the main labelled fractions [i.e., in HMW, ($\alpha$;$\beta$) tubulin, and 'SDS front'], these were eluted electrophoretically from the appropriate slices pooled from 6–7 gels (run in parallel) after fixation and staining of proteins with Coomassie Blue; the eluates were treated with ice-cold 10% TCA (in the presence of carrier albumen) and then as indicated in the legend to Table 1. It was consistently found that most of the $^{32}P$ associated with the HMW and tubulin fractions was recovered as protein-bound P (alkali-labile), while that associated with the SDS front, which accounted for about 10× as much $^{32}P$ as was eluted from the other fractions, was mainly recovered as phospholipid P and acid-soluble P.

The specific activities of protein-bound P in the eluates of the HMW and tubulin fractions were found to give very similar values, an observation that is apparently at variance with the impression gained from the data shown in Fig.1. Furthermore, the observation that most of the $^{32}P$ in the HMW fraction appeared as alkali-labile P does not seem to support an earlier suggestion that $^{32}P$-HMW represents a phosphorylated intermediate of the $Ca^{2+}$ ($Mg^{2+}$)-ATPase activity associated with brain microtubules [10], since this would be expected to behave as acid-stable, alkali-stable P. [14] However, further more direct experiments are needed to clarify these points.

In connection with the above data, it was found that when *in vivo* labelled microtubules were polymerised from a 50% ammonium sulphate precipitate of the initial high-speed supernatant fraction (S) of brain, an additional, highly

Table 1.

The incorporation *in vivo* of $^{32}P$ into the protein and lipid fractions of microtubules isolated from rat brain.

Rat brains were labelled *in vivo* for 2 h after intraventricular injection of $^{32}P_i$ (400 μCi/animal) and MT were isolated by 1× polymerisation in the presence of glycerol and GTP [17] from either the initial high-speed supernatant (S) or from the ammonium sulphate fraction of S (same expt.) precipitating between 0–50% saturation ($S_{0.5}$-ppt.), as indicated. $S_{0.5}$-ppt. was redissolved in reassembly buffer to give protein concn. (ca. 5 mg/ml) similar to that of S. Protein-bound $^{32}P$ was determined as acid-insoluble alkali-labile P (approximately equivalent to bound serine P) in cold 10% TCA-precipitated and 2× washed material, which had been 2× extracted at room temperature with 2:1 (v/v) CHCl–MeOH. [10,14] The combined organic solvent extracts in which about 80% of the $^{32}P$ was present as phospholipids was used without further fractionation to determine $^{32}P$-phospholipid counts. Protein was determined in samples prior to precipitation with TCA by the method of Lowry *et al*. [22] Values shown represent means of triplicate determinations from a single representative expt.

| Starting material | Bound $^{32}P$ | | Distribution of $^{32}P$ | |
|---|---|---|---|---|
| | protein | lipid | protein | lipid |
| | cpm/μg protein | | % | |
| S | 16.7 | 5.9 | 74 | 36 |
| $S_{0.5}$-ppt. | 36.2 | 28.0 | 57 | 43 |

*cpm recovered from acid-insoluble residue

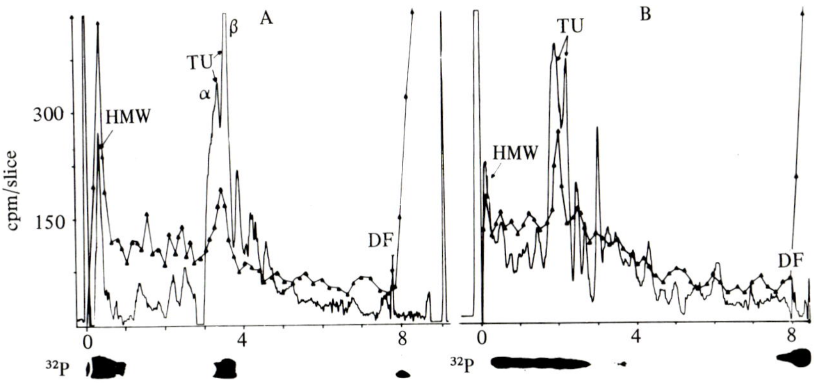


**Fig. 1.** Electrophoretic distribution of $^{32}P$ incorporated into rat (A) and chick (B) brain microtubular components *in vivo*.
Samples of microtubules (150–200 μg protein) isolated by one cycle of assembly-disassembly from rat (A) and chick (B) brain labelled *in vivo* with $^{32}P_i$ (ca. 400 and 600 μCi per animal, respectively) were denatured and separated in SDS-polyacrylamide gels in the presence of 4 M urea as previously described[10]. Graphs show $^{32}P$ cpm counted on 1 mm slices of gels that were fixed and stained for protein with Coomassie Blue; continuous lines are densitometric scans of intact stained gels. Radioautographs of gels run in parallel (prepared as described in ref. 10) are shown below ($^{32}P$). Numbers on abscissa indicate gel length in cm, measured from origin on left; DF = dye front. Smearing on radioautographs is due to overexposure of gels that were overloaded to detect minor $^{32}P$-labelled components. Note high labelling seen in region ahead of dye front.

labelled protein component, MW ca. 48 000, copolymerised with microtubular protein, as illustrated for rat brain ('LMW', Fig.2). This was not seen in microtubular preparations precipitated with ammonium sulphate after reassembly directly from the initial supernatant. In fact, the intense labelling seen in the 'LMW' component and in other, slower-migrating components that are concentrated in microtubular samples polymerised from the $S_{0.5}$-ppt. nearly obscured the labelling of tubulin detected by radioautography (see Fig.2). However, the estimated specific activity of the 'LMW' fraction, as determined after elution (see above), was found to be similar to that of the tubulin fraction.

The data shown in Table 1 also reveal considerable enrichment in $^{32}P$-phospholipids in microtubules reassembled from ammonium sulphate precipitated material. However, the yield of polymerisable protein prepared under these conditions was considerably less (1–2%) than that from the unfractionated supernatant fraction, from which about 10% of the protein was recovered in the microtubular pellet.

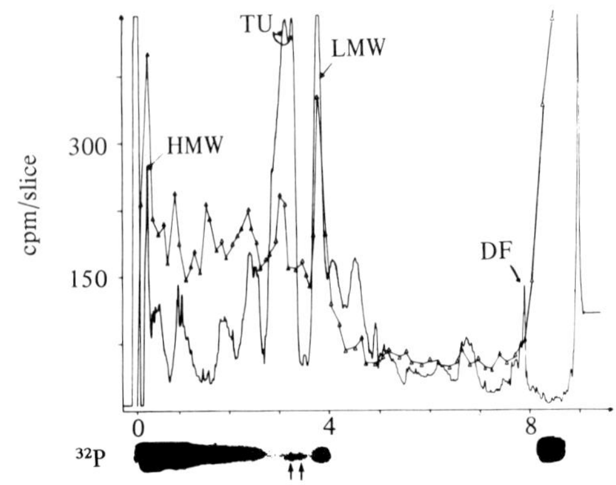


**Fig. 2.** Electrophoretic distribution of $^{32}P$ incorporated *in vivo* into rat brain microtubules.
Experimental details as given in Fig. 1, except that microtubules were isolated from 50% ammonium sulphate precipitate of initial high-speed supernatant (see Table 1 and text). LMW = prominent labelled component not evident in microtubules isolated from unfractionated supernatant (cf. Fig. 1A). Note also concentration of several labelled components between HMW and tubulin (TU) as compared with data shown in Fig. 1A. Arrows on radioautograph indicate relatively weak labelling associated with $\alpha$ and $\beta$ components of tubulin (TU). For further details see text.

The significance of this observation is not apparent, although it is interesting that *in vivo* labelled microtubular protein from guinea pig brain is also enriched in a similar faster-migrating $^{32}P$-protein when it is precipitated directly from the supernatant fraction with high concentrations of Vinblastine or Vincristine ($>10^{-4}$ M) (J.R.L., unpublished observations).

4. The distribution of $^{32}P$ in subfractions of *in vivo* labelled microtubules from chick brain.

An obvious limitation to the experiments described in the preceding section is that it was not possible to distinguish between tubulin derived from the 36S aggregates corresponding to the 400 nm ring-like structures seen by electron microscopy and the 6S form of tubulin ($\alpha;\beta$ dimer) which are both present in cold- or calcium-depolymerised microtubules isolated from *in vitro* reassembled brain microtubules [16-18].

It was recently shown that these two components, which were initially characterised on the basis of their sedimentation properties [18,19], can be separated by chromatography on Sepharose 6B [17-19] to give a first peak, eluted in the void volume, which consisted mainly of 36S 'ring' structures that are readily polymerised at 37°C in the presence of GTP, and a second peak, containing essentially pure tubulin dimer, which polymerises less readily[17] or not at all[18]. Although tubulin was the major protein component found in both peaks (as seen by SDS-gel electrophoresis), the 36S component was shown to be enriched in a number of minor components, the main one corresponding to the HMW fraction discussed above[17].

In addition, it was shown that the 36S component could be reversibly disaggregated into the 6S component in the presence of high salt concentrations[18,19]; under these conditions, a minor heat-stable protein ('tau factor') that is apparently essential for polymerisation can also be released[20].

It is not yet clear to what extent the 36S and 6S forms of tubulin found in depolymerised microtubules are both involved in the assembly process *in vivo*; nor is it apparent how the state of phosphorylation of microtubular protein, or of its associated phospholipids, which may reflect the metabolic turnover of phosphotubulin components[10], could be involved in regulating the assembly and/or functions of microtubules. It has been suggested, however, that the state of aggregation of microtubular protein into microtubular structures or into other forms of aggregates, may determine the extent to which tubulin incorporates $^{32}P$ from [$\gamma$-$^{32}P$]-ATP through its associated intrinsic kinase activity *in vitro*, or in brain slices during incubation with $^{32}P_i$ [10]. More recently, it was shown that the 36S component of disassembled microtubules from pig brain is the preferred P-acceptor during incubation *in vitro* with $^{32}P$-ATP:[21] this has now been confirmed in our laboratory (J.R.L., unpublished observations).

In the present work, we have carried out exploratory experiments to investigate the extent of incorporation of $^{32}P$ into the 36S and 6S components of chick brain microtubules labelled *in vivo*. In addition, we have attempted to separate the minor HMW components, which were found to be nearly exclusively associated with the 36S fraction, in order to facilitate characterisation of the $^{32}P$ incorporated in this fraction. The results of these experiments which are shown in Table 2 and in Figs. 3 and 4, can be summarised and interpreted as follows.

When cold- or $Ca^{2+}$-depolymerised microtubules were fractionated by chromatography on Sepharose 6B (Expts. 1 and 3A), most of the protein-bound $^{32}P$ recovered was found in the peak fraction containing 6S tubulin, while most of the phospholipid-$^{32}P$ recovered was associated with the fraction enriched in the 36S 'ring' fraction (peak I) (see also data for Expt. 4, Table 2). However, only about 20% of the protein was recovered in the 36S fraction (see also Fig. 4A), even though the concentration of protein in samples before separation (MT) was such that over 60% of the protein was expected to be present as 36S aggregates[15]. This suggests that in both experiments, a substantial proportion of the 36S material was disaggregated through dilution during chromatography[17] and/or that only a more stable form of the 36S component can remain intact under these conditions, as was originally suggested by Erickson[17]. Thus, the tubulin present in the 6S (peak II) fraction is probably composed of a mixture of protein originally present as 6S tubulin and of tubulin derived through disaggregation from 36S structures. This could account for the observation that the specific activity of protein-bound $^{32}P$ in the 36S fraction was lower, as compared to that for the 6S fraction, than might have been anticipated on the basis of results obtained *in vitro*[21] (see below).

Table 2.

The distribution of $^{32}P$ in subfractions of *in vivo* labelled microtubules from chick brain.

$^{32}P$ microtubules were isolated from 1-day old chick brain (7–20 animals per expt.), 2 h after intracerebral injection of $^{32}P_i$ (200–600 $\mu$Ci/animal). $^{32}P$ incorporated into protein and lipid fractions was determined as indicated in legend to Table 1 on the TCA-insoluble residues of microtubules that were prepared and fractionated as indicated below. Further explanations of experimental design and of nomenclature of fractions are given in text.

Expts. 1 and 2: 1× polymerised microtubules were disassembled in the cold; the solutions obtained after centrifuging for 60 min at $10^5 g$ (MT, 5–12 mg protein/ml) were fractionated at 4°C on Sepharose 6B essentially as described by Erickson[17], except that in Expt. 2 MT samples were treated with 0.75 M NaCl (MT + NaCl) before chromatography on columns that were equilibrated and eluted with buffers containing 0.75 M NaCl. Elution patterns for total $^{32}P$ and for proteins (measured at 280 nm) in Expt. 1 were similar to those shown for Expt. 3 in Fig. 4A. Protein patterns of peak I and peak II of Expt. 2 are shown in Fig. 3.

Expt. 3: 2× polymerised microtubules were disassembled for 30 min at 4°C in MES-$Ca^{2+}$ buffer (supplemented with 1 mM $CaCl_2$) before chromatography on Sepharose 6B equilibrated in the cold with MES-$Ca^{2+}$ buffer, essentially as described by Weingarten *et al.*[19] In 3B, concentrated peak I ("36S") material of 3A containing 0.75 M NaCl ("36S + NaCl"; 1.8 mg protein/ml) was separated on Sepharose 6B in the presence of 0.75 M NaCl and analysed as described in Fig. 4 and text.

Expt. 4: 2× polymerised microtubule preparations were isolated as in Expt. 3 except that the final disassembly step was carried out at 15°C in presence of MES-$Ca^{2+}$ buffer supplemented with 1 mM GTP and 1 mM $CaCl_2$ (MT; 10.5 mg protein/ml). The pellet fraction (4A, "36S") obtained after centrifuging MT for 190 min × $10^5 g$ (at 8–10°C)[19] was resuspended in the cold in disassembly buffer containing 0.75 M NaCl ("36S + NaCl"; 4.7 mg protein/ml) and centrifuged as indicated above. The resulting pellet contained appreciable amounts of tubulin (denatured aggregates?) in addition to HMW ("HMW + T"), as revealed by SDS-gel electrophoresis (data not shown).

The high specific activity in the phospholipid fraction associated with the 36S component could reflect the preferential association of metabolically active phospholipids with a more stable form of 36S tubulin, or, alternatively, the presence of HMW material, which indeed appears to be exclusively associated with peak I material (see Fig. 4A). The relatively high specific activity of phospholipids associated with the 36S component derived from $Ca^{2+}$-depolymerised material (cf. Expt. 3A and 1) was generally confirmed (Expt. 4 and other data, not shown), but the significance of this observation remains unclear.

The possible significance of the labelling seen in the HMW material that co-chromatographed with 36S tubulin was further investigated in experiments in which the 36S component was disaggregated with NaCl [19] before chromatography.

| Expt. | Fraction analysed | % Protein recovered | $^{32}P$ incorporated | | Distribution of $^{32}P$ | |
|---|---|---|---|---|---|---|
| | | | protein | lipid | protein | lipid |
| | | | cpm/μg protein (%)* | | %** | |
| 1 | MT | | 10.8 | 27.4 | 34 | 66 |
| | '36S' (peak I) | 23 | 4.6 (34) | 39.3 (77) | 19 | 81 |
| | '6S' (peak II) | 77 | 3.4 (66) | 4.4 (23) | 29 | 71 |
| 2 | MT + NaCl | | 12.3 | 37.6 | 29 | 71 |
| | 'HMW' (peak I) | 11 | 28.4 (42) | 31.7 (20) | 49 | 51 |
| | '36S + 6S' (peak II) | 89 | 4.7 (58) | 14.9 (80) | 27 | 73 |
| 3A | MT | | 38.5 | 47.5 | 45 | 55 |
| | '36S' (peak I) | 20 | 11.6 (14) | 128.0 (87) | 13 | 87 |
| | '6S' (peak II) | 80 | 18.6 (86) | 4.7 (13) | 78 | 22 |
| 3B | '36S + NaCl' | | as for '36S' in Part A | | | |
| | 'HMW + T' (peak Ia) | n.d. | n.d. (4) | n.d. (8) | 21 | 79 |
| | 'HMW + T' (peak Ib) | n.d. | n.d. (3) | n.d. (11) | | |
| | '36S − T' (peak Ic) | (60)† | 27.7 (93) | 27.5 (81) | 44 | 56 |
| 4A | MT | | 6.0 | 28.7 | 17 | 83 |
| | '6S' (supnt.) | 10 | 2.9 (8) | 7.1 (3) | 29 | 71 |
| | '36S' (pellet) | 90 | 3.6 (92) | 25.0 (97) | 13 | 87 |
| 4B | '36S + NaCl' | | as for '36S' in Part A | | | |
| | '6S' (supnt.) | 36 | 3.9 (66) | 6.2 (22) | 38 | 62 |
| | 'HMW + T' (pellet) | 64 | 1.1 (34) | 12.8 (78) | 8 | 92 |

*Values in brackets = % of recovered cpm in fraction.
**cpm recovered for protein + lipid in individual fraction = 100%.
†Estimated as a percentage of the protein applied to the column. In most experiments, 50–70% of the protein applied was recovered in combined eluates.

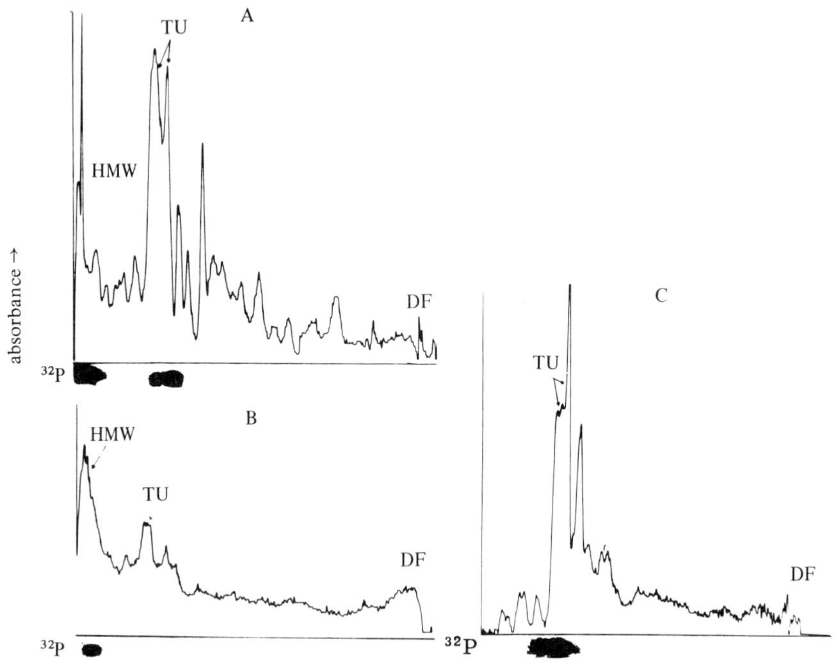


Fig. 3. Electrophoretic distribution in SDS-polyacrylamide gels of proteins and $^{32}P$ in chick brain microtubular components labelled *in vivo*.
Densitometric scans and $^{32}P$ radioautographs (from same gel) shown for unfractionated microtubules (A) and for subfractions obtained after dissociation of 36S microtubular component with 0.75 M NaCl (B and C). For details of preparation see legend for Expt. 3B, Table 2 and text. Apparent heterogeneity of tubulin and radioactivity in C is artefact due to overloading.

In Expt. 3B, the 36S fraction was concentrated by vacuum dialysis in the presence of NaCl and rechromatographed on Sepharose 6B in the presence of NaCl (Fig. 4B). Under these conditions, the material present in the first two peaks of radioactivity eluted (fractions Ia, Ib) gave identical protein patterns after electrophoresis in SDS-gels (data not shown): HMW was present together with some tubulin. A further broad and apparently complex peak of radioactivity (Ic) was eluted in a region coinciding roughly with the elution volume of the 6S component: pooled fractions from this region, which accounted for most of the recovered protein (Table 2, Expt. 3B), were almost exclusively composed of tubulin derived, presumably, from the 36S tubulin aggregates separated in Expt. 3A (see also Fig. 4A). Thus, a partial separation of HMW from 36S tubulin was achieved. It is evident from the limited data shown in Table 2

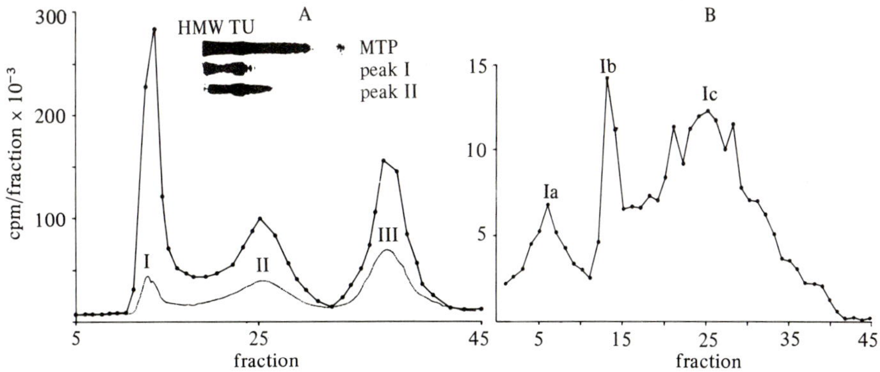


**Fig. 4.** Chromatography on Sepharose 6B of *in vivo* labelled microtubules from chick brain.
(A) $^{32}$P-microtubules were depolymerised in the presence of calcium and fractionated on Sepharose 6B in the absence of NaCl, as indicated for Expt. 2A, Table 2. Total $^{32}$P cpm (solid line) in 1.5 ml eluate fractions is shown together with absorbance recorded at 280 nm (dotted line). Inset shows SDS-gel protein patterns for samples of unfractionated microtubular preparation (MTP) and of material pooled from peak I (fractions 13–15) and peak II (fractions 23–27). HMW = high MW components; TU = tubulin. Note absence of HMW in peak II material. Total protein loaded on column: 14 mg.
(B) Re-chromatography on Sepharose 6B of peak I material of A that was concentrated by vacuum dialysis and chromatographed in the presence of depolymerisation buffer containing 0.75 M NaCl as indicated in legend for Expt. 2B, Table 2. For further explanations, see text.

(Expt. 3A) that most of the bound $^{32}$P recovered as protein-bound P or as phospholipid P was present in the main tubulin fraction (peak Ic). Since, however, too little material was present in peak fractions containing HMW (Ia, Ib, see Fig. 4B) to permit reliable estimations of protein by the Folin–Lowry method, it was not possible to determine specific radioactivities for $^{32}$P in these fractions. Nonetheless. it can be tentatively concluded on the basis of the data giving the percentages of radioactivity recovered that the fractions enriched in HMW are also enriched in $^{32}$P-phospholipid (see discussion of Expt. 3A above).

In Expt. 2 (Table 2, Fig. 3), the possibility of separating HMW from both forms of tubulin was investigated by chromatography of cold-depolymerised microtubules which had been treated with 0.75 M NaCl to dissociate 36S tubulin aggregates, prior to chromatography on columns of Sepharose 6B which were equilibrated and eluted in the presence of 0.75 M NaCl. It was anticipated that this procedure would result in all the tubulin being eluted as 6S tubulin, after elution of the HMW fraction in the void volume, as had been the case when it was present together with the undissociated 36S component (see Fig. 4A). This was found to occur, as shown in the SDS-gel electrophoresis patterns of $^{32}P$-labelled material which behaved similarly during elution to the 36S and 6S components (data not shown): the first peak eluted now contained mainly HMW, while most of the protein present in peak II was identified as tubulin, as shown in Figs. 3A and 3B respectively. Moreover, it is evident from the radioautographs shown (Fig. 3) that virtually all the label detected after electrophoresis of peak I material was associated with several HMW components near the origin, while the label found in peak II material co-migrated almost exclusively with the tubulin fraction. The data given in Table 2 (Expt. 2) show that under these conditions the HMW fraction was considerably enriched in bound $^{32}P$, especially in respect of protein-bound P.

Finally, the results of an experiment in which the 6S and 36S components of tubulin were separated by high-speed centrifugation[19] (Expt. 4A, Table 2) showed that most of the protein was recovered in the pellet fraction which was expected to contain tubulin, present mainly as 36S aggregates. This fraction also contained most of the bound $^{32}P$ and was particularly enriched in $^{32}P$-phospholipids. Although the bulk of the protein-bound $^{32}P$ was also found in the pellet fraction, the specific activities for P-protein were remarkably similar in the supernatant and pellet fractions. The pellet fraction, which contained tubulin and HMW in about the same proportion as in the unfractionated material (MT) (data not shown), was resuspended in buffer containing 0.75 M NaCl and re-centrifuged at high speed. It was anticipated that in this way HMW components, together with some non-specific tubulin aggregates, might be recovered in the pellet, while tubulin derived from dissociated 36S aggregates would remain in solution. This was confirmed by SDS-gel electrophoresis, though an appreciable amount of tubulin was also found in the pellet fraction (denatured tubulin aggregate?). The data shown in Table 2 (Expt. 4) indicate, once again, that the HMW-enriched pellet fraction accounted for the bulk of $^{32}P$-phospholipid (see also Expts. 2 and 3B). On the other hand, the supernatant was considerably enriched in $^{32}P$ protein, despite the absence of any HMW component (cf. Expt. 2). One feature worth noting about this experiment is that the concentration of protein remained high enough throughout to maintain the 36S $\rightleftharpoons$ 6S equilibrium in favour of the 36S species[18,19]; this is in contrast to the situation which obtains during the separation of 6S and 36S components on Sepharose 6B, when dilution of the sample during chromatography tends to favour disaggregation of the 36S species (see above).

5. Conclusions

The evidence presented above indicates that brain microtubules contain protein-bound P and are associated with a fraction of phospholipids, both of which exhibit considerable metabolic activity *in vivo*. This confirms and extends earlier evidence based on studies of microtubule phosphorylation *in vitro* and in tissue slice experiments [10,13,23]. A more precise interpretation of these findings must obviously await further information concerning the P contents of the various components present in microtubules isolated by reassembly *in vitro*, from which labelling due to turnover of bound P and to net phosphorylation might be distinguished. Nevertheless, it was consistently observed that the 36S component of depolymerised microtubule, which consists largely of tubulin polymers stabilised into the ring-like structures, was preferentially labelled *in vivo* under various experimental conditions as compared to the 6S tubulin component. The relative enrichment of the 36S component in metabolically active phospholipid suggests that 36S structures may be loosely associated, *in situ*, with membrane fractions (e.g., plasma membrane) at sites of microtubule–membrane interactions. Indeed, evidence for a close association between enzymes involved in phospholipid metabolism (e.g., diglyceride kinase) and microtubular subunits was recently reported[13]. The possible relationship between microtubules and membrane-associated changes in phospholipid metabolism in functionally active nerve-ending preparations from cerebral cortex should prove a worthwhile area of investigation, especially in view of the recent discovery[1] that microtubules are indeed present in presynaptic terminals *in situ* (see Introduction).

The relatively high incorporation observed of $^{32}P$ into phosphoserine residues of the 36S component may indicate that any regulatory factors controlling the conversion of 6S tubulin to the 36S form, from which microtubules are apparently polymerised [17,18], might be acting, in part at least, as regulators of enzymes concerned with the turnover of protein-bound P. In this connection, our observation that a highly labelled minor protein (HMW), originally seen by SDS-gel electrophoresis, can be separated from salt-dissociated preparations of the 36S component by chromatography may be of interest in the light of recent work[20], showing that a tightly bound, heat-stable protein ('tau factor') apparently required in microtubule assembly can also be released from 36S tubulin preparations by high concentrations of NaCl.

References


1. Gray, E.G., 1975, Proc. Roy. Soc. B, in the press.
2. Gray, E.G., 1975, Prog. Brain Res., in the press.
3. Lagnado, J.R., Lyons, C. and Wickremasinghe, G., 1971, Febs Letters 15, 254.
4. Feit, H., Dutton, G., Barondes, S. and Shelanski, M., 1971, J. Cell Biol. 51, 138.
5. Lagnado, J.R. and Lyons, C., 1971, Biochem. J. 126, 9.
6. Walters, B.B. and Matus, A.I., 1975, Biochem Soc. Trans. 3, 109.
7. Lagnado, J.R., Tan, L.P., Walters, B.B. and Matus, A.I., 1975, submitted for publication.
8. Walters, B.B. and Matus, A.I., 1975, Nature, in the press.
9. Matus, A.I. and Walters, B.B., 1975, J. Neurocytol. 4, 369.
10. Lagnado, J.R., Tan, L.P. and Reddington, M., 1975, Ann. N.Y. Acad. Sci. 253, 577.
11. see 'News and Views' (R. Shields), 1975, Nature 256, 257.
12. Quinn, P.F., 1973, Biochem. J. 133, 273.
13. Daleo, G.R., Piras, M.M. and Piras, R., 1974, Biochem. Biophys. Res. Comm. 61, 1043.
14. Rodnight, R., Reddington, M. and Gordon, M., 1975, in: Research Methods in Neurochemistry, vol. 3, eds. N. Marks and R. Rodnight (Plenum Press, New York), in the press.
15. Yagihara, Y., Bleasdale, J.E. and Hawthorne, J.N., 1973, J. Neurochem. 21, 173.
16. Olmsted, J.B., Marcum, J.M., Johnson, K.A., Allen, C. and Borisy, G.G., 1974, J. Supramol. Struct. 2, 429.
17. Erickson, H.P., 1974, J. Supramol. Struct. 2, 393.
18. Kirschner, M.W., Williams, R., Weingarten, M. and Gerhart, J., 1974, Proc. Nat. Acad. Sci. U.S. 71, 1159.
19. Weingarten, M.D., Suter, M.M., Littman, D. and Kirschner, M.W., 1974, Biochemistry 13, 5529.
20. Weingarten, M.D., Lockwood, A.H., Hwo, S. and Kirschner, M.W., 1975, Proc. Nat. Acad. Sci. U.S. 72, 1858.
21. Kirschner, M.W., Suter, M., Weingarten, M. and Littman, D., 1975, Ann. N.Y. Acad. Sci. 253, .... .
22. Lowry, O.H., Rosebrough, N.J., Farr, A.C. and Randall, R.J., 1951, J. Biol. Chem. 193, 265.
23. Piras, M.M. and Piras, R., 1974, European J. Biochem. 47, 44

# FUNCTIONAL ASPECTS OF MICROTUBULES

# THE ROLE OF THE MICROTUBULAR-MICROFILAMENTOUS SYSTEM IN INSULIN AND GLUCAGON RELEASE BY THE ENDOCRINE PANCREAS

W.J. Malaisse, V. Leclercq-Meyer, E. Van Obberghen, G. Somers, G. Devis, M. Ravazzola, F. Malaisse-Lagae and L. Orci

Laboratory of Experimental Medicine, Brussels University School of Medicine, Brussels, Belgium, and Institute of Histology and Embryology, Geneva University School of Medicine, Geneva, Switzerland

Summary

1. A new antimicrotubular agent, R17934, caused a delayed, not rapidly reversible and partial inhibition of glucose-induced insulin release by isolated islets of Langerhans, this behaviour being superimposable to that previously observed with colchicine or vincristine. 2. Motile events were examined in monolayers culture of neonatal rat endocrine pancreatic cells examined by time-lapse cinematography. Back-and-forth saltatory movements of secretory granules along oriented microtubular pathways were documented. 3. A spontaneous defect in the amount of B-cell microtubular protein concomitant with impaired insulin release was identified in spiny mice. 4. Colchicine, vinblastine, $D_2O$ and cytochalasin B were found to affect glucagon release *in vitro*. 5. These findings, taken together with other recent contributions, afford further evidence in favour of the idea that the microtubular-microfilamentous system in pancreatic endocrine cells is intimately involved in the translocation, functional segregation and eventual release of secretory granules.

## 1. Introduction

The possible role of the pancreatic B-cell microtubular-microfilamentous system in insulin release has been reviewed on several occasions[1-4]. Combined ultrastructural and functional studies on the effect of colchicine, vincristine, $D_2O$ and cytochalasin B in isolated islets and the isolated perfused pancreas of the rat led us to postulate that such a system provides both the structural framework for the functional segregation and oriented translocation of β-granules and the motive force for their intracellular migration and eventual release by exocytosis[4]. We now wish to briefly review more recent contributions relevant to the possible participation of the microtubular-microfilamentous system in hormonal release by the endocrine pancreas.

## 2. Effect of R17934 upon insulin release

We have examined the influence of R17934 upon insulin release evoked by glucose in rat isolated islets, according to a method described in detail elsewhere[5].

The results summarized in Table 1 indicate that R17934 (2 to 20 μg/ml, or 7 to 67 μM) failed to affect glucose-induced insulin release during the first period of incubation (0 to 90 min). However it caused a dose-related inhibition of insulin

Table 1

Effect of R17934 upon insulin release evoked by glucose (16.7 mM) in rat isolated islets

| Expt. | First incubation (0 to 90 min) | | Second incubation (90 to 180 min) | |
|---|---|---|---|---|
| | R17934 | Insulin output | R17934 | Insulin output |
| A | — | 100.0 ± 3.6 (12) | — | 100.0 ± 8.6 (12) |
| | 2 μg/ml | 96.9 ± 2.7 (12) | 2 μg/ml | 76.6 ± 6.7 (12) |
| | 2 μg/ml | 104.7 ± 4.2 (12) | — | 73.1 ± 10.3 (12) |
| B | — | 100.0 ± 4.6 (16) | — | 100.0 ± 10.2 (16) |
| | 20 μg/ml | 104.3 ± 4.4 (16) | 20 μg/ml | 48.2 ± 4.2 (16) |

R17934 was first dissolved in DMSO (5 mg/ml). At the higher concentration here used (20 μg/ml), R17934 was only partially solubilized in the incubation medium. DMSO (0.04 or 0.4 %, v/v) was present at the same final concentration in both control (no R17934) and experimental (with R17934) media ; at this concentration DMSO has no effect upon insulin release[9]. The rates of secretion (mean ± SEM) are expressed in percent of the mean control value found within the same experiment and during the same period of incubation in islets never exposed to R17934. Such control rate of secretion averaged 333 ± 12 and 270 ± 26 μU/islet per 90 min (n = 28, in both cases), during the first (0 to 90 min) and second (90 to 180 min) periods of incubation, respectively. The reduction of the secretory response during the second period (to 79.4 ± 6.0 % of the corresponding mean value found during the first period) confirms prior observations[5,9]. The number of individual observations is shown in parentheses.

release during the second period of incubation (90 to 180 min). The degree of inhibition averaged 24.4 ± 6.7 percent ($P < 0.05$) at a concentration of 2 μg/ml and 51.8 ± 4.2 percent ($P < 0.001$) at a concentration of 20 μg/ml. When the isolated islets were first exposed to R17934 (2 μg/ml) during the first 90 min of incubation, the same degree of inhibition was noticed thereafter whether or not the drug was present during the second period of incubation. This finding suggests that the inhibitory effect of R17934 is not rapidly reversible.

The results obtained in the presence of R17934 are almost identical to those previously collected in the same system with either colchicine or vincristine[5]. The delayed response to antimicrotubular agents is thought to reflect on the time-related progressive disappearance of microtubules[6], which itself may be due to slow binding of mitotic spindle-inhibitors to the islet-cell tubulin[7].

Fig. 1. Isolated islets incubated for 180 min in the absence of R17934 (control).
A. The normal appearance of B-cells is illustrated. Rough endoplasmic reticulum cisternae (RER) appear as flattened profiles. The arrowhead points to a maturing secretory granule in the expanded end of a Golgi cisterna. The arrows indicate microtubules. N = nucleus ; G = Golgi area ; sg = secretory granules.
B. In this field of a B-cell, the typical appearance of microtubules is illustrated at a higher magnification.
Calibration bars equal 1 micron.

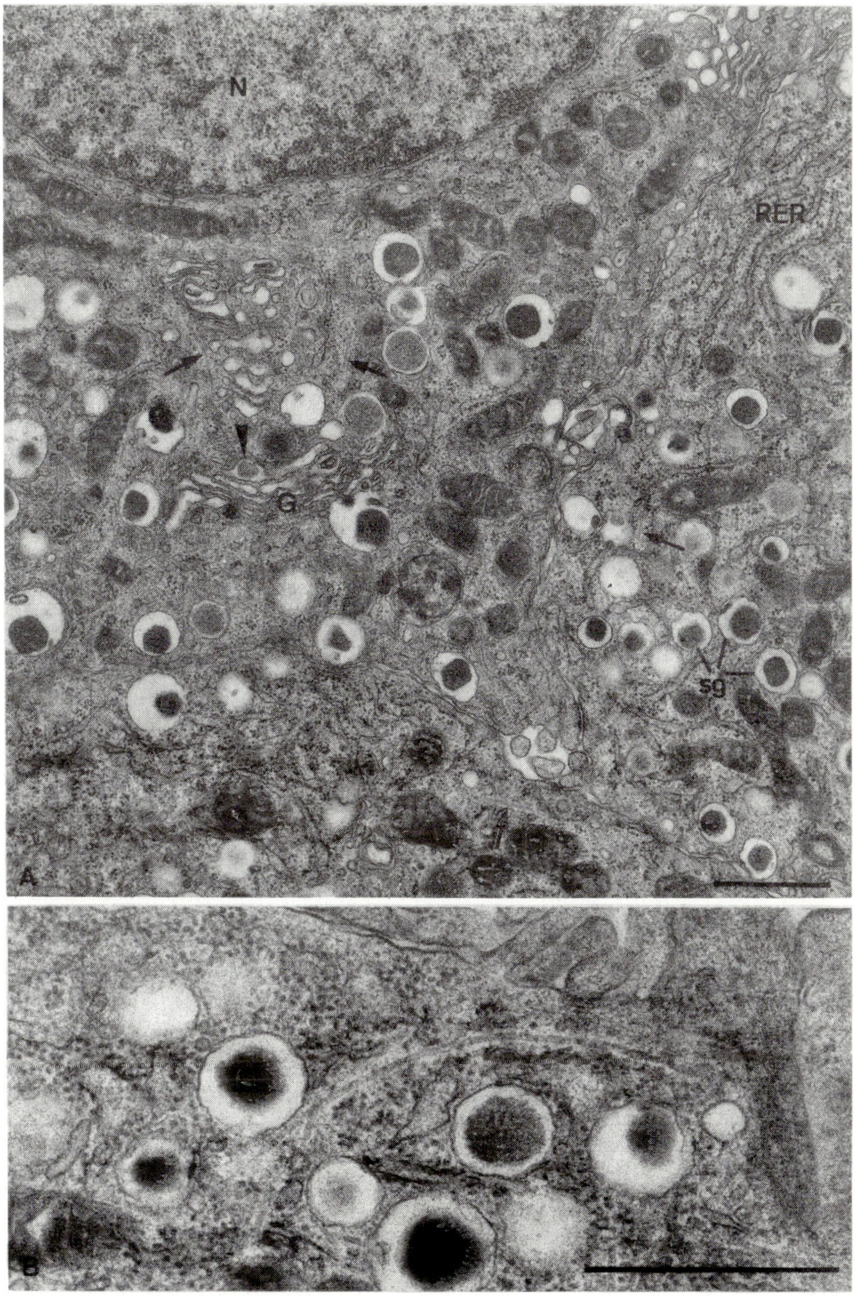
N
RER
G
sg
A
B

At the ultrastructural level (Figs. 1 and 2), in B-cells exposed for 180 min to R17934 (20 μg/ml), the number of identifiable microtubules was apparently reduced. Some residual microtubule-like structures appeared more dense, having lost their normal hollow appearance and were often deprived of their clear surrounding halo. These findings are reminiscent of these found in B-cells exposed to colchicine[8]. Dilatation of both the Golgi-associated vacuoles and cisternae of the rough endoplasmic reticulum were also encountered in B-cells exposed to R17934.

3. Characterization of microtubular and microfilamentous material in islet tissue

Recent work from a number of laboratories has contributed to the characterization of microtubular and microfilamentous material in islet tissue. Montague et al.[7] reported on the presence, purification and colchicine-binding properties of islet-cell tubulin with an apparent molecular weight of 110000. In the isolated islets, about 30 % of the total tubulin content is apparently polymerized as microtubules, as judged by ultracentrifugation separation of soluble and polymerized tubulin from islets homogenized in a 50 % glycerol solution[10]. Actin-like and myosin-like material was chemically identified in islet tissue, which thus contains active ATPase for chemical-to-mechanical energy transduction[11,12]. Using sera containing anti-actin autoantibodies from patients with chronic agressive hepatitis, Gabbiani et al. were able to correlate the immunofluorescent characterization of actin in pancreatic islet cells with the ultrastructural presence of a network of microfilaments (4-7 nm in diameter) known as the cell web[13].

4. Motion analysis in pancreatic endocrine cells

Monolayer cultures of neonatal rat endocrine pancreatic cells[14] were examined by time-lapse cinematography[15]. Detailed motion analysis of these documents is now in progress. Preliminary results so far obtained in our laboratory support the idea that the microtubular-microfilamentous system indeed controls the movement of particles. Fig. 3 indicates that the particles do not move freely in all directions but undergo back-and-forth saltatory movements along oriented pathways. The concept was statistically validated by demonstrating the prevalence of angles close to zero or 180° between successive movements of particles. This is shown in

Fig. 2. Isolated islets incubated in the presence of R17934 (20 μg/ml) for 180 min.

A. Part of a B-cell, showing dilatation of rough endoplasmic reticulum cisternae (RER), the presence of vacuoles (V) and of large-sized secretory granules (arrowheads) in the Golgi area (G). N = nucleus ; sg = ecretory granules. Under this experimental condition, the number of identifiable microtubules is reduced.

B. Filamentous structures reminiscent of microtubules, but lacking the clear halo and the hollow appearance, are sometimes encountered.

Calibration bars equal 1 micron.

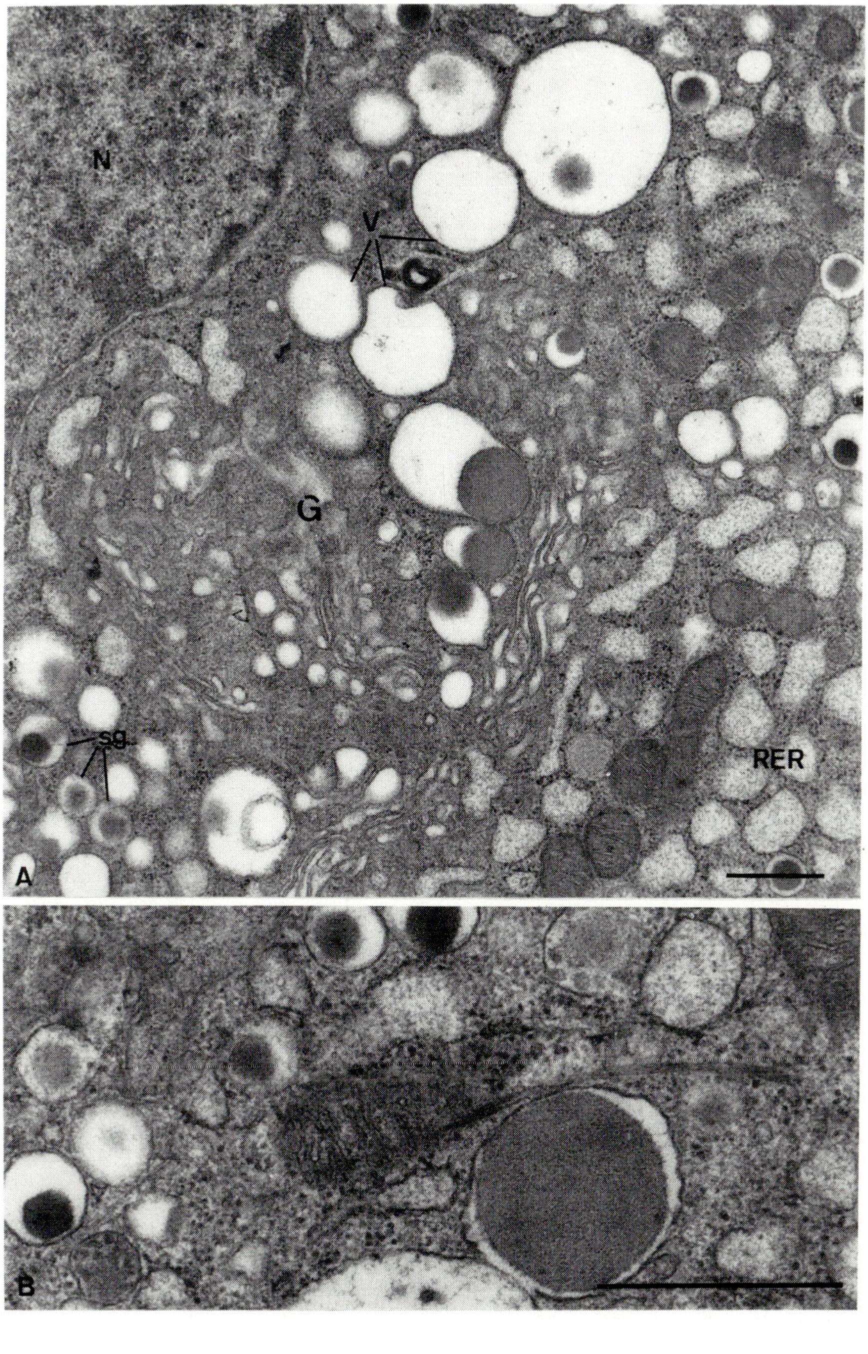
N
V
G
sg
RER
A
B

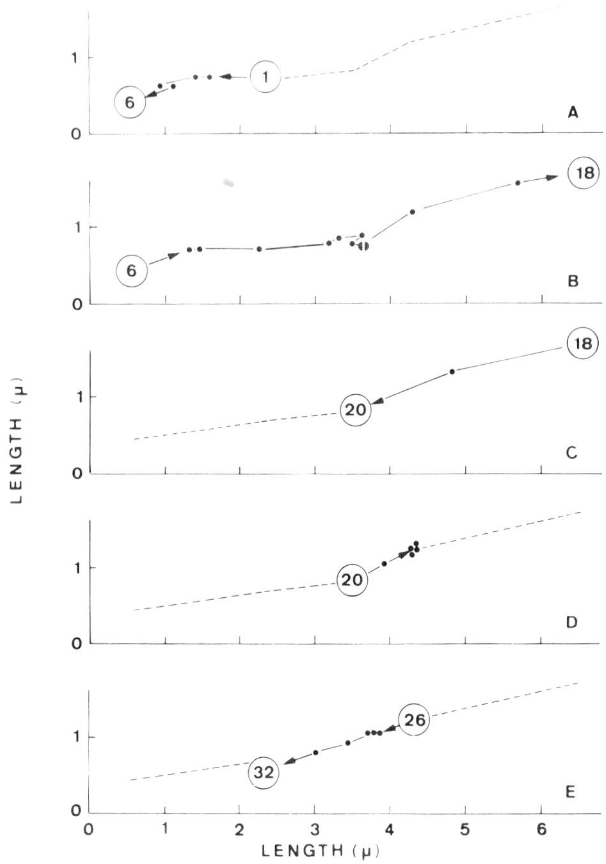


Fig. 3. Time-lapse cinematography of monolayer culture of neonatal rat endocrine pancreatic cells. Particle motion analysis in a secretory cell. The pictures were taken at the rate of one per sec, for 32 successive sec. The size of the particle (approximately 0.3 μ in diameter) is suggestive of a secretory granule, and is shown by the circles, in which each number refers to that of the picture (1 to 32). The points coincide with the center of the particle. Movements in opposite direction are shown on different panels (A to E). The solid lines figure the movements of the particle. The dotted line illustrates the invariability of the overall pathway. The saltatory back and forth movements along this oriented pathway are interrupted by halts which seem to take often place at the same loci.

Fig. 4, which also illustrates the frequency distribution for the amplitude of each movement. More than one fourth of the movements displayed a speed of 1 to 2 μ/sec. In these monolayer cultures, β-granules may be seen arranged in highly ordered rows which interdigitate with microtubules coursing towards and ending in the cell web[4,14]. Taken together, these findings indicate that the microtubular apparatus may indeed serve as a guiding cytoskeleton for the oriented migration of

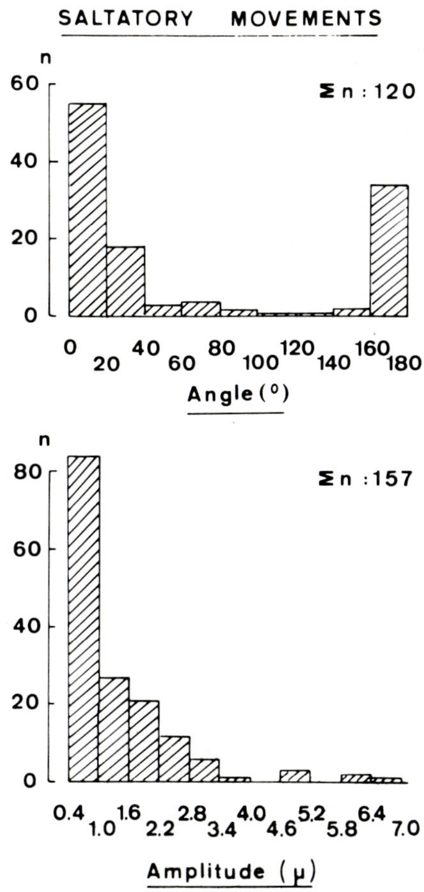


Fig. 4. Frequency distribution for the angulation and amplitude of saltatory movements of particles (presumably secretory granules) in monolayer cultures of neonatal rat endocrine pancreatic cells examined by time-lapse cinematography. The total number of measurements (Σn) is shown in each case. The angles were measured either between 2 successive movements separated by a halt, or within the same movement, when the angle between 2 successive segments of the movement exceeded 45°. A halt is defined by the fact that a given particle (approximately 0.3 μ in diameter) undergoes between 2 successive pictures (taken at one sec interval) a movement of which the amplitude (measured from the center of the particle) does not exceed 0.4 μ. The amplitude of movement was measured between successive halts or between a halt and an angulation of the movement exceeding 45°.

secretory granules towards the cell surface.

5. Spontaneous defect in the B-cell microtubular system of spiny mice

An abnormality of the B-cell microtubular system was recently encountered in spiny mice (Acomys cahirinus)[16]. In this species, an apparent reduction in the number of microtubules in B-cells was first observed. In order to obtain a quantitative assessment of this anomaly, isolated islets of Langerhans were exposed to vincristine, the volume-density of vincristine-induced crystals in B-cell cytoplasm being then determined by morphometry. The volume-density of vincristine-induced deposits was markedly reduced in B-cells of spiny mice, when compared to that found in normal rats and mice and diabetic mutant mice. Simultaneous exposure of the islets to a high glucose concentration (27.8 instead of 5.6 mM) caused a further decrease in volume-density of vincristine-induced crystals in B-cells. This anomaly could account for the paradoxical functional behaviour of the B-cell in spiny mice, i.e. an abnormally low immediate insulin secretory response to various secretagogues coexisting with apparently intact synthesis and storage of the hormone in the B-cell[17].

## 6. Microtubular-microfilamentous system and glucagon release

We have recently examined the effects of cytochalasin B, colchicine, vinblastine and $D_2O$ upon glucagon secretion by pieces of pancreas from duct-ligated rats[18]. Cytochalasin B (10 µg/ml) provoked a rapid and reversible facilitation of arginine-induced glucagon secretion, but failed to affect unstimulated glucagon release. Colchicine (0.1 mM) and vinblastine (0.1 mM) similarly enhanced arginine-induced glucagon secretion (Fig. 5), whilst unstimulated glucagon release was unaffected by colchicine. Edwards and Howell[19] also observed that colchicine and vinblastine facilitate glucagon release by isolated guinea-pig islets of Langerhans. $D_2O$ (100 %, v/v) reversibly inhibited arginine-induced glucagon secretion. These results suggest that a microtubular-microfilamentous system may participate in the process of glucagon secretion. The intimate mechanism(s) by which such a system may apparently play both a restrictive and effector role in glucagon release remains, however, to be elucidated.

## 7. Conclusions

Taken as a whole, the data outlined in the present report indicate that the concept of a microtubular-microfilamentous participation in hormonal release by the endocrine pancreas, first postulated by Lacy et al.[20], is gaining further experimental support. Although we have purposely restricted this presentation to the possible role of the microtubular-microfilamentous system in the process of granules migration and exocytosis, it should be kept in mind that this system may also influence the function of secretory cells by other mechanisms, for instance by affecting either the cell surface topography for carriers, hormonal receptors and other membrane-associated proteins[3] or the pathway for the biosynthesis and

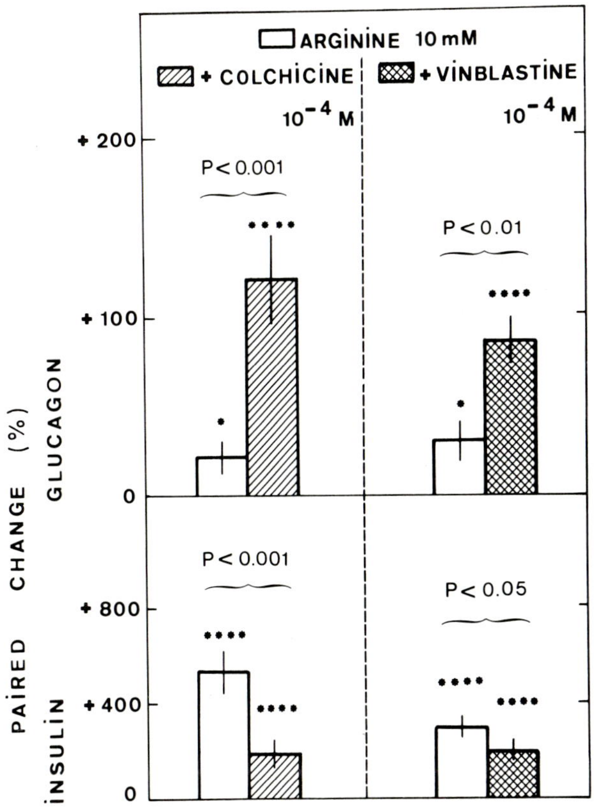


Fig. 5. Effect of colchicine and vinblastine upon arginine-induced glucagon and insulin release by pieces of pancreas from duct-ligated rats. After 120 min (vinblastine) or 150 min (colchicine) preincubation in the presence of glucose (8.3 mM) with or without the mitotic spindle-inhibitor, arginine (10 mM) was added to the media. The arginine-induced increment in secretion rate over the ensuing 30 min is expressed in percent of the paired value found over the preceding 30 min period. Data obtained in the presence of colchicine or vinblastine (each 0.1 mM ; shaded columns) are compared to those found in the control experiments performed in the absence of mitotic spindle-inhibitor (open columns). Also shown is the statistical significance for the arginine-induced increment in secretion rate (* $P < 0.05$ ; **** $P < 0.001$) and for the influence of colchicine and vinblastine upon such an increment. Note the facilitation of glucagon release and the concomitant inhibition of insulin release provoked by the mitotic spindle-inhibitors.

conversion of hormonal precursors (unpublished observations).

Acknowledgement. This work was supported in part by grants 20001, 20193, and 3.4527.75 from the Fonds de la Recherche Scientifique Médicale (Brussels), grant

20353 from the Fonds voor Geneeskundig Wetenschappelijke Onderzoek (Brussels), grants 3.8081.72, 3.0310.73 and 3.0553.75 from the Fonds National de la Recherche Scientifique (Bern), and a contract of the Ministère de la Politique Scientifique (Belgium) within the framework of the Association Euratom - Universities of Brussels and Pisa. We wish to thank B. Noël for excellent secretarial help. We also like to thank Dr. M. De Brabander (Janssen Pharmaceutica ; Beerse, Belgium) for gift of R17934.

References

1. Malaisse, W.J., Malaisse-Lagae, F., Van Obberghen, E., Somers, G., Devis, G. and Orci, L., 1973, Excerpta Medica I.C.S. 273, 282.
2. Lacy, P.E. and Malaisse, W.J., 1973, Recent Progr. Hormone Res. 29, 199.
3. Malaisse, W.J., Devis, G., Somers, G., Van Obberghen, E., Malaisse-Lagae, F., Ravazzola, M., Blondel, B. and Orci, L., 1975, in press (MTP, Lancaster).
4. Malaisse, W.J., Malaisse-Lagae, F., Van Obberghen, E., Somers, G., Devis, G., Ravazzola, M. and Orci, L., 1975, Ann. N.Y. Acad. Sci. 253, 630.
5. Malaisse, W.J., Malaisse-Lagae, F., Walker, M.O. and Lacy, P.E., 1971, Diabetes 20, 257.
6. Devis, G., Van Obberghen, E., Somers, G., Malaisse-Lagae, F., Orci, L. and Malaisse, W.J., 1974, Diabetologia 10, 53.
7. Montague, W., Howell, S.L. and Green, I.C., 1975, Biochem. J. 148, 237.
8. Somers, G., Van Obberghen, E., Devis, G., Ravazzola, M., Malaisse-Lagae, F. and Malaisse, W.J., 1974, Europ. J. clin. Invest. 4, 299.
9. Malaisse, W.J., Hager, D.L. and Orci, L., 1972, Diabetes 21, 594.
10. Pipeleers, D.G., Pipeleers-Marichal, M.A. and Kipnis, D.M., 1975, Endocrinology (suppl. to vol. 96), 88 (abstract).
11. Ostlund, R.E. and Kipnis, D.M., 1975, Endocrinology (suppl. to vol. 96), 88 (abstract).
12. Charles, M.A. and Clarke, M., 1975, Endocrinology (suppl. to vol. 96), 90 (abstract).
13. Gabbiani, G., Malaisse-Lagae, F., Blondel, B. and Orci, L., 1974, Endocrinology 95, 1630.
14. Orci, L., Like, A.A., Amherdt, M., Blondel, B., Kanazawa, Y., Marliss, E.B., Lambert, A.E., Wollheim, C.B. and Renold, A.E., 1973, J. Ultrastruct. Res. 43, 270.
15. Orci, L., Blondel, B., Malaisse-Lagae, F., Ravazzola, M., Wollheim, C., Malaisse, W.J. and Renold, A.E., 1974, Diabetologia 10, 382 (abstract).
16. Malaisse-Lagae, F., Ravazzola, M., Amherdt, M., Gutzeit, A., Stauffacher, W., Malaisse, W.J. and Orci, L., 1975, Diabetologia 11, 71.
17. Stauffacher, W., Orci, L., Amherdt, M., Burr, I.M., Balant, L., Froesch, E.R. and Renold, A.E., 1970, Diabetologia 6, 330.
18. Leclercq-Meyer, V., Marchand, J. and Malaisse, W.J., 1974, Diabetologia 10, 215.
19. Edwards, J.C. and Howell, S.L., 1973, FEBS Letters 30, 89.
20. Lacy, P.E., Howell, S.L., Young, D.A. and Fink, C.J., 1968, Nature 219, 1177.

*Microtubules and Microtubule Inhibitors, eds. M. Borgers and M. de Brabander*



# IN VIVO AND IN VITRO EVIDENCES FOR A ROLE OF MICROTUBULES IN THE SECRETORY PROCESSES OF LIVER

Y. Le Marchand, A. Singh, C. Patzelt, L. Orci and B. Jeanrenaud
Laboratoires de Recherches Médicales and
Institut d'Histologie et d'Embryologie
Geneva University, Medical School, Geneva

## INTRODUCTION

It is well established that the liver is a major site of biosynthesis and release of very low density lipoproteins (VLDL). Until recently, however, the mechanism by which these lipoproteins are secreted into the blood stream was ill-defined. When we initiated a series of studies on this particular problem we were helped by previous observations that these lipoproteins could be identified with the electron microscope as opaque particles, 300 to 1'000 Å in diameter[1-3]. Since several hormones stored as vesicle-enclosed granules are secreted, at least in part, via a functioning microtubular system[4-8], we hypothetized that the vesicle-enclosed VLDL might also be transported intracellularly, and therefore secreted out, via the hepatic microtubular system which has been shown to be scattered throughout the parenchyma[9].

The study of the role of the microtubular system in the secretory processes of mouse liver has been carried out using a in situ technique previously described[10-11].

## EFFECT OF COLCHICINE OR VINCRISTINE ON TRIGLYCERIDE SECRETION

These drugs are known to interfere with the microtubular system of several different cell types[12-17]. Colchicine has been shown to be highly specific for liver tubulin to which it binds, thus presumably preventing the polymerization of tubulin to microtubules[18]. The binding of vincristine to tubulin appears to be somewhat less specific than that of colchicine, the drug binding apparently to some other proteins[18]. When used at high enough concentrations ($2 \times 10^{-3}$ M) vincristine not only binds to tubulin but the resulting complexes precipitate as paracrystals that are easily recognizable at the electron microscope[13]. Colchicine and vincristine thus provide us with a means of assessing the role of the microtubular system in VLDL release.

In our experiments, the net triglyceride release by perfused mouse livers has been taken as an index of liver VLDL secretion since VLDL triglycerides could be shown to account for more than 80 percent of total secreted triglycerides[11]. We could observe that the output of triglyceride was markedly curtailed by the presence of colchicine or vincristine in the perfusate, the lowest effective doses tested being 0.4 µM and 1 µM for vincristine and colchicine, respectively (see Figure 1).

Figure 1: Production of triglycerides by perfused livers of normal fed mice : effect of various concentrations of colchicine.

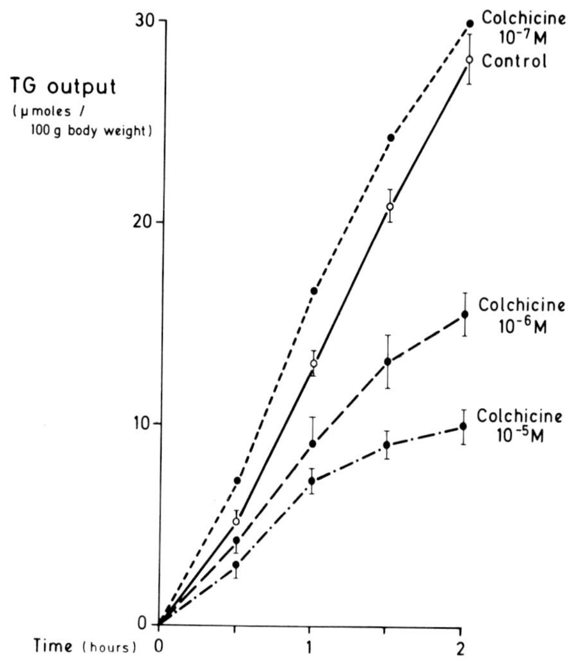


Krebs-Ringer bicarbonate buffer with bovine serum (25 %), albumin (1.5 g/100 ml) and RBC (20 %). 1.5 mM oleate.
Each point, mean of 4 values ± S.E. (Colchicine $10^{-7}$M, mean of 2 values)

Clearly, such inhibitory effect of colchicine or vincristine could be the result of either a specific interference of these drugs with the hepatic microtubular system or that of nonspecific effect(s) upon various metabolic pathways. To rule out the possibility that colchicine or vincristine would have mainly unspecific, microtubule-unrelated effects, several basic functions of the liver were measured. It was observed that, at least for a 10 µM colchicine or a 2.5 µM vincristine concentration, triglyceride secretion was markedly decreased at a time when cellular ATP, glycogenolysis, urea production, $O_2$ consumption, fatty acid uptake and oxidation were unmodified by these drugs. Despite such results, the observed decrease in triglyceride-rich lipoproteins produced by vincristine or colchicine could still be caused by decreased lipogenesis, decreased esterification of fatty acid to triglycerides or decreased protein synthesis. However, lipogenesis (estimated by labeled precursors incorporation into hepatic saponifiable lipids) was found not to be decreased by these drugs. Furthermore, and as illustrated in Figure 2, it was observed that the incorporation of labeled oleate into hepatic triglycerides was actually increased by vincristine or colchicine, the secretion of labeled triglycerides being, as expected, markedly decreased.

Figure 2: Incorporation of labeled oleate into liver and perfusate triglycerides: effect of vincristine.

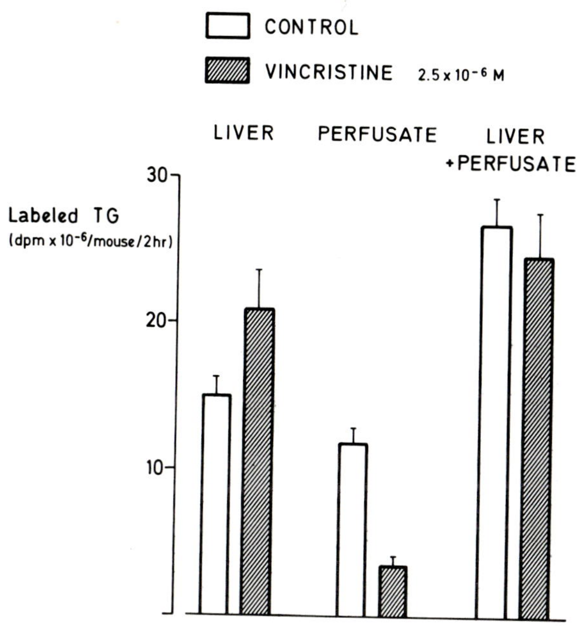


Krebs-Ringer bicarbonate buffer with bovine serum (25%), albumin (1.5 g/100 ml) and RBC (20 %). 1.5 mM oleate. Each point, mean of 4 values ± S.E.

More importantly (see Figure 2), total oleate incorporation into liver plus perfusate triglycerides was similar in control and "mitotic-spindle inhibitor"-treated livers. These results strongly suggested that although total oleate esterification was not modified by either colchicine or vincristine, the newly formed (i.e. labeled) triglycerides were retained within the cytoplasm and not secreted outside the cells. The concept of a blockade in the intracellular movement of VLDL was corroborated by the observation that, after a 2 h perfusion in the presence of either one of these drugs, unlabeled triglycerides actually accumulated within the liver, an accumulation that was represented, at the electron microscope, by numerous clusters of vesicles containing VLDL-like particles. Concomitantly, microtubules had either virtually disappeared (colchicine treatment) or were apparently replaced by amorphous material that could represent precipitated microtubular protein (vincristine treatment)[11,19]. The decrease in VLDL secretion brought about by colchicine or vincristine was unlikely to be related to a marked alteration of protein synthesis (with secondary impairment of the apoprotein synthesis) as is apparent by the following series of experiments:

EFFECT OF COLCHICINE OR VINCRISTINE ON PROTEIN SECRETION

Thus, in other experiments, schematized by Figure 3, we observed that although the overall protein synthesis (estimated by labeled amino acid incorporation into liver plus perfusate proteins) was not modified by either

Figure 3: Incorporation of labeled amino acids into liver and perfusate albumin: effect of vincristine.

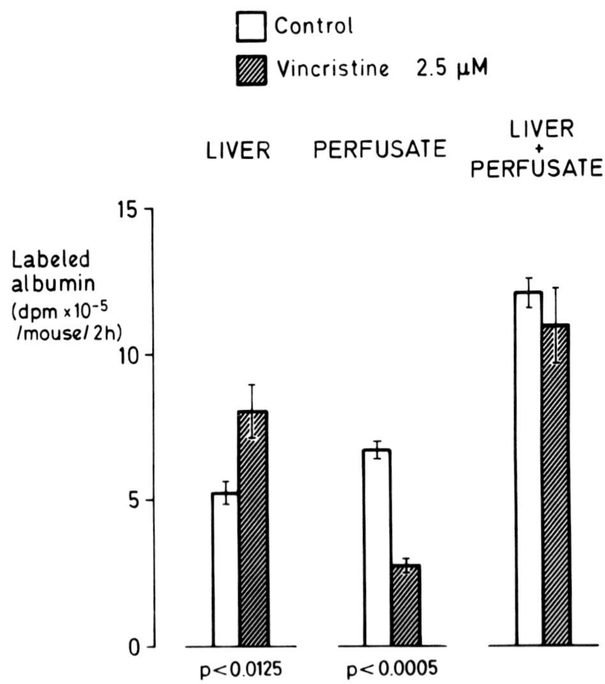


Krebs-Ringer bicarbonate buffer with albumin (3g/100 ml) and RBC (20 %). 20 mg and 40 μCi amino acids / 100 ml. Each point, mean of 4 - 7 values ± S.E

colchicine or vincristine, the secretion of proteins was markedly inhibited by these drugs. As illustrated by Figure 3, we were struck by the observation that at a time when, for instance, albumin secretion was decreased by colchicine, albumin accumulated within the liver while total albumin synthesis (from labeled amino acids) was apparently unaltered. This suggested again, in a way that was similar to that previously observed for labeled oleate incorporation into triglycerides, that the translocation of albumin had been hampered out not its total synthesis. To determine whether the inhibition of protein secretion observed in livers perfused with vincristine or colchicine was specific for one particular group of secreted proteins, chromatographic separation of the various proteins released by perfused livers was carried out, as shown in Figure 4.

Figure 4:

1 globulins

2 albumin

3 small polypeptides

4 amino acids

(reprinted from the J. Clin. Invest.[20])

**G 100 sephadex chromatographic pattern of labeled proteins secreted by perfused mouse livers : Effect of vincristine or colchicine.**

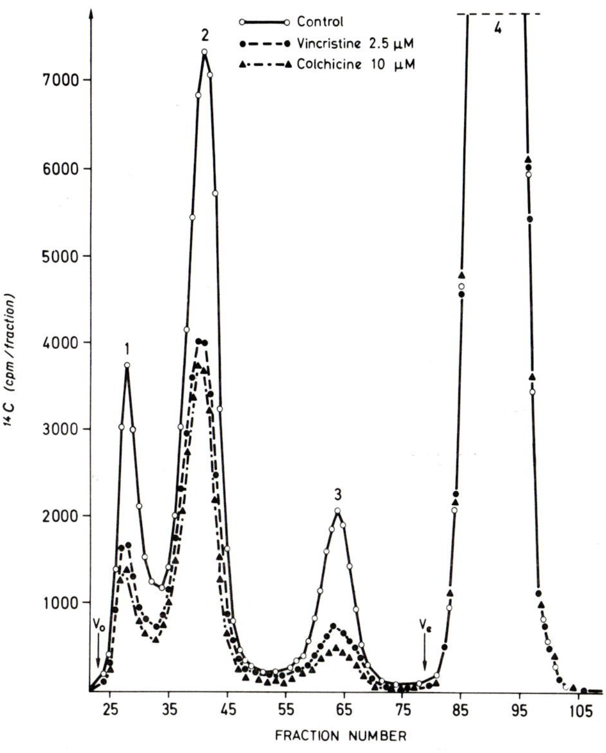


It was observed that the output of a whole spectrum of proteins was inhibited by colchicine or vincristine, namely the output not only of albumin but also of various globulins and even of small polypeptides. We were surprised to observe that the secretion of such small polypeptides appeared to require a functioning microtubular system since it was thought that these products would simply diffuse out. Further experiments indicated, however, that these small polypeptides were not degradation products derived from either albumin or globulins and that their secretion *per se* involved an intact function of microtubules[20].

## LACK OF EFFECT OF LUMICOLCHICINE

It has been shown that lumicolchicine, an analog of colchicine obtained by UV exposure of the latter does not compete with colchicine when using a labeled colchicine binding assay in a high speed liver supernatant. This indicates that lumicolchicine does not bind to tubulin[18], although it has been reported to bind to plasma membrane[21]. Of interest was therefore the observation, summarized by Figure 5, that lumicolchicine ($10^{-5}$M) did not

Figure 5: **Production of triglycerides by perfused livers of normal fed mice: effect of lumicolchicine and colchicine.**

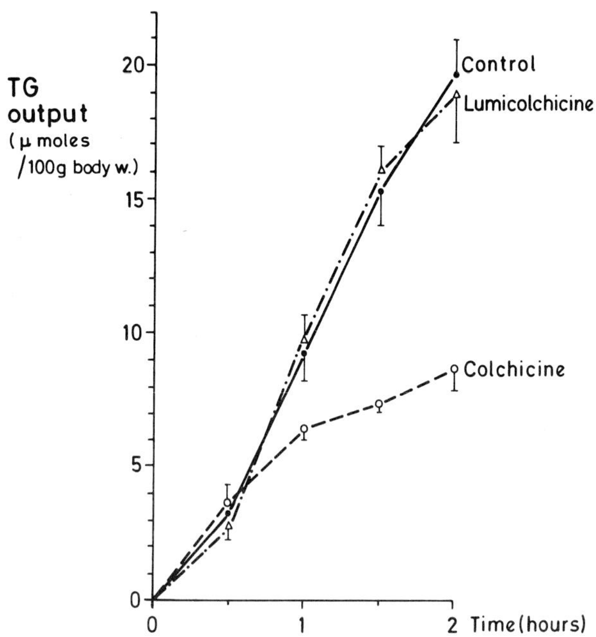


Krebs-Ringer bicarbonate buffer with bovine serum (25 %), albumin (1.5 g / 100 ml) and RBC (20 %). 1.7 mM oleate.
Lumicolchicine and colchicine: $10^{-5}$ M.
Each point, mean of 5-9 values ± S.E.

alter triglyceride secretion by perfused mouse livers while equimolar concentration of colchicine did. These results further substantiated the concept, without directly proving it, that the colchicine effects were likely to be primarily related to an alteration of the function of the microtubular system, not to other metabolic alterations.

## EFFECT OF $D_2O$ ON TRIGLYCERIDE SECRETION

$D_2O$ has been claimed to "stabilize" the microtubular system, and due to this, to inhibit secretions such as insulin secretion[5]. We were therefore interested to see whether $D_2O$ would also inhibit the output of triglyceride by perfused mouse livers. $D_2O$ did not appear to be very effective or specific in altering the hepatic microtubular system. Thus, when water was replaced by $D_2O$ to the extent of 33%, a slight inhibition only of triglyceride secretion was observed. Increasing $D_2O$ to 45% of total water in the perfusion medium resulted in a greater inhibition of triglyceride output, as shown in Table 1.

Table 1

Effect of various concentrations of $D_2O$ on triglyceride secretion by perfused mouse livers

| | Nb. of exp. | Triglyceride output after (µmoles/100g body weight) 60 min | 120 min |
|---|---|---|---|
| Control | 3 | 13.9 ± 0.8 | 22.4 ± 0.7 |
| $D_2O$ 33% | 3 | *10.6 ± 0.7 | *18.3 ± 0.2 |
| Control | 7 | 7.0 ± 1 | 16.7 ± 1.3 |
| $D_2O$ 45% | 6 | *3.8 ± 0.6 | *10.3 ± 0.7 |

*statistically different from controls at $p<0.01$
Livers were perfused as described in Figure 1 but water was replaced by $D_2O$ to an extent of 33% or 45%.

This inhibitory effect could not, however, be ascribed exclusively to an alteration of the microtubular system since, in contrast to what was observed with colchicine or vincristine, fatty acid oxidation to ketone bodies was also found to be decreased by $D_2O$ thus casting doubts about the specificity of its action upon the microtubular system.

## EFFECT OF COLCHICINE *IN VIVO*

Having observed striking and seemingly specific effects of colchicine

proteins. These experiments do not prove that the effects of colchicine or vincristine are the result of a "disruption" of microtubular function. Such a mechanism is, however, strongly suggested by the fact that, at the concentrations used, no deleterious effects of these drugs could be detected. Biochemically and ultrastructurally, the state of preservation of vincristine- or colchicine-treated livers was satisfactory. In colchicine-treated livers virtually no microtubules could be observed, an observation that is consistent with the concept that the polymerization of microtubules had probably been interfered with by the binding of colchicine to tubulin[18]. In vincristine-treated livers, masses of amorphous material were seen within the cytoplasm, masses that are tentatively interpreted as being the result of precipitation of microtubular proteins by vincristine, as described elsewhere[18]. In both instances, i.e. in the presence of either colchicine or vincristine, the most striking observation was the presence of an accumulation of either triglyceride (chemically or radioactively measurable, recognizeable at the electron microscope as VLDL-like particles) or of proteins (radioactively measurable) within the parenchyma at a time when their appearance in the perfusate was markedly decreased. This observation was consistent with the concept of a "paralysis" of the intracellular movement of these products due to altered microtubular function. Unspecific effects of vincristine or colchicine appear unlikely since, in the presence of these drugs, basic functions of the liver, as well as fatty acid uptake, overall esterification of oleate, lipogenesis, overall protein synthesis, were not changed significantly. The likely specificity of colchicine upon the microtubular system was also strengthened by the observation that lumicolchicine, a colchicine analog that does not bind to tubulin but apparently binds to plasma membrane had no effect upon triglyceride secretion when added to perfused livers.

Although circumstantial evidence indicates that microtubules are likely to be of importance in the secretion of various products by the liver, the actual function of the microtubular system is not clarified by the present experiments. It is not clear yet how microtubules or microtubule precursors interact with vesicles containing the products to be secreted, and how these structures favor the intracellular, directional movement of the vesicles. It is not yet established with certainty whether vesicles are specific for one particular product of secretion or whether several different products are transported within the same vesicles. It is not known whether vesicles are all alike or, on the contrary, represent an heterogenous population. It is not known whether some soluble proteins could interact with the microtubular system without being enclosed in vesicles (Figure 7).

Figure 7: POSSIBLE MEANS OF INTRACELLULAR TRANSLOCATION OF SECRETORY PRODUCTS IN LIVER.

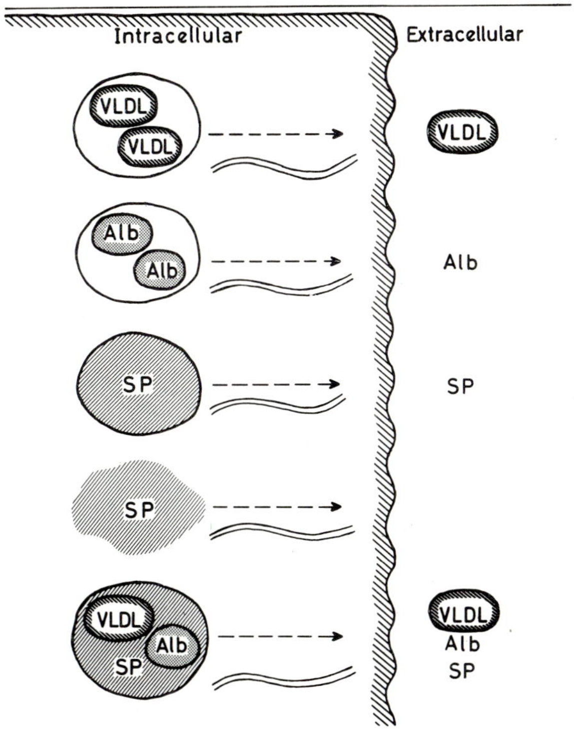


Finally, it is not known yet whether the microtubular system of the liver might be a site of regulation (by hormones, substrates, ions, etc.) or whether the regulation of the different secretion processes of the liver is located at some other points, e.g. at the step of actual synthesis of packaging of the products to be secreted.

Although we have given special emphasis in this summary to work from our laboratories, we would like to point out that similar work leading to analogous conclusions has also been carried out by Stein et al[23] and Redman et al[24].

ACKNOWLEDGMENTS

This work has been supported by Grants n° 3.552.71, 3.8080.72 SR and 3.2180.73 of the "Fonds National Suisse de la Recherche Scientifique" Berne, Switzerland; and by Grants-in-aid of Nestlé Alimentana, Vevey, Switzerland; Zyma S.A., Nyon, Switzerland; Stiftung zur Förderung der Ernährungsforschung in der Schweiz.

## REFERENCES


1. Jones, A.L., Ruderman, N.B. and Herrera, M.G., 1967, J.Lipid Res. 8, 429.
2. Stein, O. and Stein, Y., 1967, J.Cell Biol. 33, 319.
3. Hamilton, R.L., Regen, D.M., Gray, M.E. and LeQuire, V.S., 1967, Lab. Invest. 16, 305.
4. Lacy, P.E., Howell, S.L., Young, D.A. and Fink, C.J., 1968, Nature 219, 1177.
5. Malaisse, W.J., Malaisse-Lagae, F., Walker, M.O. and Lacy, P.E., 1971, Diabetes 20, 257.
6. Nève, P., Willems, C. and Dumont, J.E., 1970, Exp.Cell Res. 63, 457.
7. Williams, J.A. and Wolff, J., 1972, J.Cell Biol. 54, 157.
8. Poisner, A.M. and Bernstein, J., 1971, J.Pharmacol.Exp.Ther. 177, 102.
9. Bruni, C. and Porter, K.R., 1965, Amer.J.Pathol. 46, 691.
10. Assimacopoulos-Jeannet, F., Exton, J.H. and Jeanrenaud, B., 1973, Amer. J.Physiol. 225, 25.
11. Le Marchand, Y., Singh, A., Assimacopoulos-Jeannet, F., Orci, L., Rouiller, Ch. and Jeanrenaud, B., 1973, J.Biol.Chem. 248, 6862.
12. Bensch, K.G. and Malawista, S.E., 1969, J.Cell Biol. 40, 95.
13. Malawista, S.E. and Sato, H., 1969, J.Cell Biol. 42, 596.
14. Borisy, G.G. and Taylor, E.W., 1967, J.Cell Biol. 34, 535.
15. Adelman, M.R., Borisy, G.G., Shelanski, M.L., Weisenberg, R.C. and Taylor, E.W., 1968, Fed.Proc. 27, 1186.
16. Malawista, S.E., 1965, J.Exp.Med. 122, 361.
17. Temple, R., Williams, J.A., Wilber, J.F. and Wolff, J., 1972, Biochem. Biophys.Res.Commun. 46, 1454.
18. Patzelt, C., Singh, A., Le Marchand, Y., Orci, L., and Jeanrenaud, B., 1975, J.Cell Biol., in press.
19. Orci, L., Le Marchand, Y., Singh, A., Assimacopoulos-Jeannet, F., Rouiller, Ch. and Jeanrenaud, B., 1973, Nature 244, 30.
20. Le Marchand, Y., Patzelt, C., Assimacopoulos-Jeannet, F., Loten, E.G. and Jeanrenaud, B., 1974, J.Clin.Invest. 53, 1512.
21. Stadler, J., Franke, W.W., 1974, J.Cell Biol. 60, 297.
22. Singh, A., Le Marchand, Y., Orci, L. and Jeanrenaud, B., Europ.J.Clin. Invest., in press.
23. Stein, O., Sanger, L., Stein, Y., 1974, J.Cell Biol. 62, 90.
24. Redman, C.M., Banerjee, D., Howell, K. and Palade, G.E., 1975, J.Cell Biol. 66, 42.

*Microtubules and Microtubule Inhibitors, eds. M. Borgers and M. de Brabander*



# CHARACTERIZATION OF HEPATIC MICROTUBULES: POSSIBLE FUNCTIONAL MODELS

C. Patzelt, A. Singh, Y. Le Marchand, and B. Jeanrenaud

Laboratoires de Recherches Médicales and
Institut d'Histologie et d'Embryologie
Geneva University, Medical School, Geneva

The kinetics and the specificity of colchicine binding have been studied in high speed (100,000 xg) supernatant obtained from mouse livers. It was found that about 80% of colchicine binding activity of a total homogenate was present in this soluble fraction. A colchicine binding protein was isolated subsequently and identified as being tubulin by its many similarities to isolated mouse brain tubulin. The dose-response curves of colchicine binding of high speed supernatant as well as of tubulin from liver were found to be in good agreement with that of the inhibitory effect of colchicine upon triglyceride secretion by perfused livers. In a similar fashion the specificity and sensitivity of liver tubulin for vinblastine was demonstrated.

A method has been developed to measure, in one liver, the respective amounts of free tubulin and of tubulin assembled as microtubules. It was observed that, under basal conditions, only about 40 percent of total tubulin was in the form of microtubules. For microtubules isolated from brain the establishment of a similar equilibrium was observed. This equilibrium appears, in both systems, to be altered in a analogous manner by exposure to low temperature or by the addition of vinblastine, whereas the sensitivity for colchicine was completely different. These observations are discussed in terms of a functional model in which the assembly-disassembly cycle of microtubules may be of relevance.

## 1. INTRODUCTION

In the liver, as in many other tissues, the microtubular system appears to be implicated in secretory processes[1-7]. In addition, the liver seems to us to present a suitable model for characterizing cytoplasmic microtubules as well as for studying their function, due to the considerable size of this organ and the possibility of its perfusion <u>in vitro</u>. Following the electron microscopic demonstration of the presence of microtubules in hepatocytes[1,2,8] and the studies on the metabolic consequences brought about in the liver by microtubule inhibitors[1-7] the identification of the hepatic microtubular structure protein, i.e. tubulin, was the next logical step in characterizing the system.

It has often been claimed, though not proven, that microtubules and a precursor pool of free tubulin exist in the tissue in a dynamic equilibrium. To substantiate such a concept, we have employed a method that allows for the measurement of these two states of the microtubular system, i.e. the structured microtubular one and that of unstructured free tubulin, in the intact liver. This procedure offers the possibility of a more direct study of the effect of agents known to interfere with microtubular function. Furthermore, we have taken advantage of the well investigated model of isolated brain microtubules[9-12], to facilitate the interpretation of these studies, which should eventually help to develop a functional model of microtubule action.

## 2. METHODOLOGY

Six weeks old male Swiss albino mice, which were fed _ad libitum_, were used throughout these studies.

The binding assay for $^3H$-colchicine described by Weisenberg[13] has been used after minor modifications that have been described previously[14]. The procedure for isolation of brain microtubules was that of Shelanski et al[10]. The technique of liver perfusion has been described in detail elsewhere[1,15].

## 3. INTRACELLULAR DISTRIBUTION OF THE COLCHICINE BINDING ACTIVITY OF THE LIVER

Studies on the distribution of the colchicine binding activity in the liver between soluble and particulate phases showed that the bulk of this binding activity was found in the high speed supernatant (100,000 xg) of a liver homogenate. This finding is consistent with the cytoplasmic localization of microtubules (Table 1).

Table 1

Distribution of the colchicine binding activity in the liver as measured by fixation to DEAE-Sephadex

| | Bound [$^3H$] colchicine | | |
|---|---|---|---|
| | pmol/mg protein | pmol/ml | % of total |
| Total homogenate | 6.22 | 1258 | 100 |
| 1st supernatant | 32.07 | 955 | 79 |
| 2nd supernatant | 21.32 | 117 | 9 |
| Pellet | 5.35 | 297 | 23 |

Livers were homogenized with two volumes of $10^{-2}$mM phosphate buffer, pH 6.5, containing $10^{-2}$M $MgCl_2$ and $10^{-3}$MGTP. Homogenates were centrifuged (100,000 xg, 4°C, 60 min), the pellet of the 1st centrifugation was resuspended and centrifuged once more resulting in the 2nd supernatant and the final pellet. Supernatants and pellet were made up to the original volume to be comparable at a volume basis. Each value represents the mean of 3 experiments.

The binding activity was found to show a linear dependency on the protein concentration, and a temperature sensitivity similar to that reported for brain tubulin. Furthermore, it did not compete with lumicolchicine.

One should emphasize that a binding activity of about 20% of that of total homogenate remained in the final pellet (Table 1). Such binding activity seems to be tightly particle bound since it could not be solubilized by detergents such as lubrol (1%).

4. PURIFICATION OF LIVER MICROTUBULAR PROTEIN

When comparing the colchicine binding activities of liver and brain high speed (100,000 xg) supernatant (Table 2), it was found that this activity, when related to protein concentration, was about 15 times higher in brain than in liver.

Table 2

Comparison of colchicine activity of high speed supernatants and purified microtubular protein from liver and brain

| | Bound $[^3H]$ colchicine (pmol/mg protein) | |
|---|---|---|
| | liver | brain |
| 100,000 xg supernatant | 38.7 | 577.1 |
| Purified microtubular protein | 3859.2 | 4135.8 |

If the colchicine binding activity from liver high speed supernatant was purified by a method described for brain tubulin[16], an electrophoretically pure protein was obtained that had, when compared to brain tubulin, a similar specific binding activity for colchicine (Table 2), an identical electrophoretical behavior, and a very similar electron microscopic appearance (paracrystals) upon precipitation with vinblastine[14].

5. COLCHICINE BINDING KINETICS OF LIVER HIGH SPEED SUPERNATANT AND PURIFIED TUBULIN FROM LIVER AND BRAIN

The colchicine dose-dependency of the binding of tritiated colchicine to either liver high speed supernatant or purified tubulin preparations from liver and brain is plotted in Figure 1. For all these samples a plateau of binding was reached at a colchicine concentration of about $10^{-4}$M with a half maximal binding at about $10^{-5}$M colchicine. At colchicine concentrations higher than $2 \times 10^{-4}$M a further increase of the binding could be observed for liver high speed supernatant (Figure 1A). This increase was absent in the case of purified tubulins (Figure 1B). Furthermore, this low affinity binding activity could not be precipitated by $2 \times 10^{-3}$M vinblastine, whereas the binding activity with the higher affinity was completely precipitated upon addition of this alkaloid (Figure 1A, Figure 2).

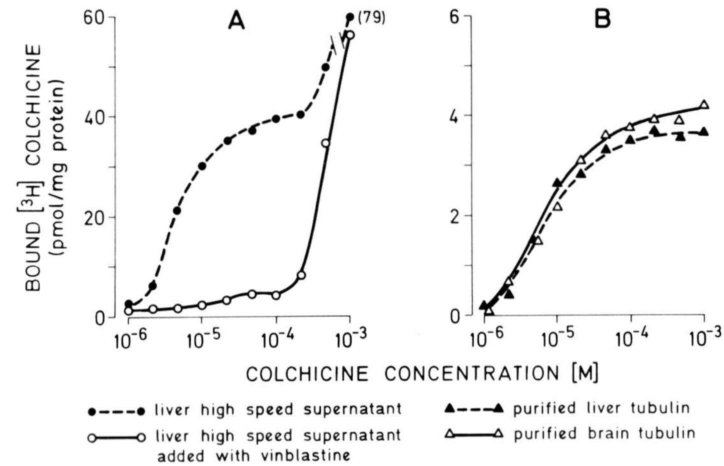


Fig. 1. Colchicine dose-dependency of the colchicine binding activity of (A) liver high speed supernatant and (B) purified liver and brain tubulin. Vinblastine precipitation was performed at a concentration of 2 x $10^{-3}$M. After 30 minutes incubation at 37°C the samples were centrifuged (20 minutes, 10,000 xg) and the colchicine binding activities measured in the supernates.

6. EFFECT OF VINBLASTINE ON PURIFIED TUBULIN FROM LIVER AND BRAIN

Vinblastine is known to precipitate tubulin as typical paracrystalline structures[14,17,18]. The precipitation of purified liver or brain tubulin and its relationship to the vinblastine concentration is shown in Figure 2. Although there seems to be no competition between the binding of colchicine and that of the vinca alkaloids to tubulin[19], vinblastine has been reported to stabilize to some extent the colchicine binding site, which exhibits temperature-dependent lability[20]. As also shown in Figure 2, this stabilizing property of vinblastine upon colchicine binding activity can be demonstrated for either one of these proteins. Stabilization and precipitation of tubulin from either liver or brain occured at similar concentrations of vinblastine. While incubation of mouse brain 100,000 xg supernatant with this alkaloid resulted in a nearly selective precipitation of tubulin, that of liver high speed supernatant lead to precipitation of many other proteins, indicating that vinblastine was of rather low specificity for the liver microtubular system.

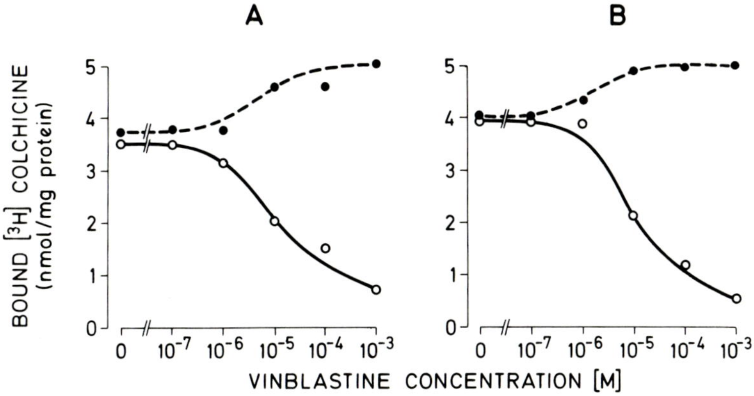


Fig. 2. Effect of vinblastine of purified tubulin from (A) liver and (B) brain. Tubulin solutions were incubated with the indicated concentrations of vinblastine for 60 minutes at 37°C. The colchicine binding activity was then measured either directly or in the supernates after centrifugation for 20 minutes at 10,000 xg.

## 7. DETERMINATION OF THE PORTIONS OF MICROTUBULES AND FREE TUBULIN IN THE LIVER

Microtubules obtained from various tissues have been reported to be rapidly destroyed by exposure to low temperature. Moreover, they can be stabilized, under these conditions, by various agents, one of them being glycerol[10]. Thus, the rapid preparation of tissue homogenates either in the presence or in the absence of glycerol (4 M), and the consecutive incubation of these two different homogenates at low temperature will result in a complete destruction of microtubules in the sample containing no glycerol, whereas the homogenate with glycerol will contain stabilized microtubules. After ultracentrifugation of the two samples, tubulin concentration in the respective supernatants can be measured by the binding assay using a colchicine concentration of $10^{-4}$M. The supernatant of the glycerol-free homogenate will contain the total amount of tubulin, since no microtubules are left, whereas that of the glycerol-containing homogenate will contain only that portion of tubulin which was not originally present as formed microtubules, since stabilized microtubules of this sample are pelleted by centrifugation. The difference in the colchicine-binding activities of these two supernatants is therefore a measure of that portion of tubulin that is assembled as microtubules in the tissue. This method is also applicable to isolated microtubules, the homogenization step being omitted in this case. The procedure just described is summarized in Figure 3, and is explained in more detail elsewhere[14]. It should be mentioned that the glycerol concentration of the two homogenates has been equalized after exposure to the cold to avoid artifacts during the further procedure. It could be shown that the colchicine-binding assay *per se* was not influenced by the presence of 4 M glycerol.

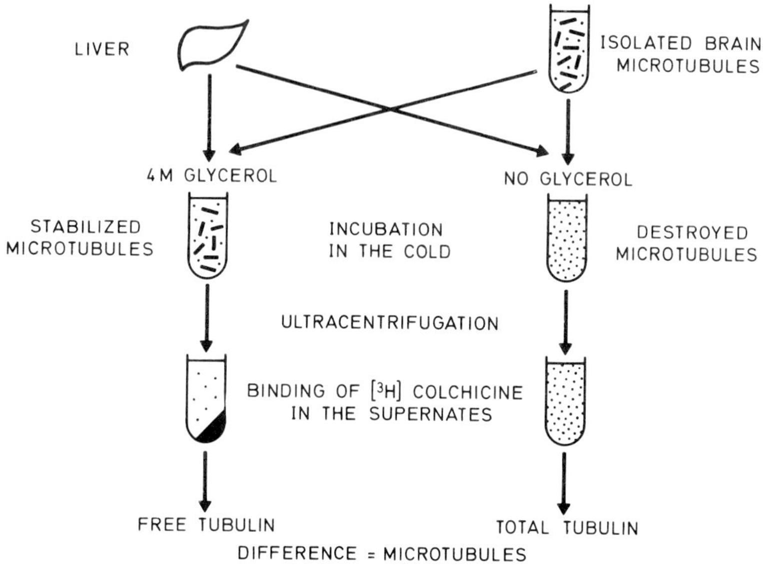


Fig. 3. Schematic representation of the procedure to determine the portions of free tubulin and microtubules in the liver and in a sample of isolated brain microtubules.

## 8. EQUILIBRIUM BETWEEN MICROTUBULES AND FREE TUBULIN UNDER BASAL CONDITIONS

When applying the above-mentioned procedure to liver taken from normal fed mice, it was found that liver cells contained about 40% of their tubulin in the form of microtubules (Figure 4A, time 0). This percentage did not change in livers perfused for 80 minutes under basal conditions[1] (Figure 5A, and 6A, controls).

When tubulin from mouse brain was purified by assembly-disassembly cycling (a procedure which, for unknown reasons, is not possible to apply to liver), and finally incubated in assembly buffer (MES-MEG)[10] at 37°C, the formation of microtubules reached a plateau after 30 minutes. Upon measuring, by the method described before, the amount of tubulin assembled as microtubules at this time, it was also found to be about 40% of total tubulin. This proportion remained essentially constant during a further incubation period of 60 minutes and was found, above the critical value needed for microtubule formation[11,12], to be independent on the concentration of tubulin. These two preparations, i.e. the liver after a short perfusion of 20 minutes, and isolated brain microtubules, formed during 30 minutes from a tubulin solution (2.5 mg protein/ml) were then further studied and compared, as summarized below.

9. INFLUENCE OF LOW TEMPERATURE, VINBLASTINE AND COLCHICINE ON THE STATE OF LIVER MICROTUBULES AND ON ISOLATED BRAIN MICROTUBULES

When livers from anesthetized mice were removed following a rapid washout perfusion with warm MES-MEG buffer, to be thereafter incubated in cold (0°C) buffer, a complete disappearance of microtubules could be observed within 5 minutes while the total amount of tubulin remained stable (Figure 4A). When measured by an analogous method, isolated brain microtubules were destroyed in a similar time-dependent manner (Figure 4B). Such rapidity of microtubule destruction is consistent with results obtained on isolated microtubules, using the technique of changes in turbidity[11,14].

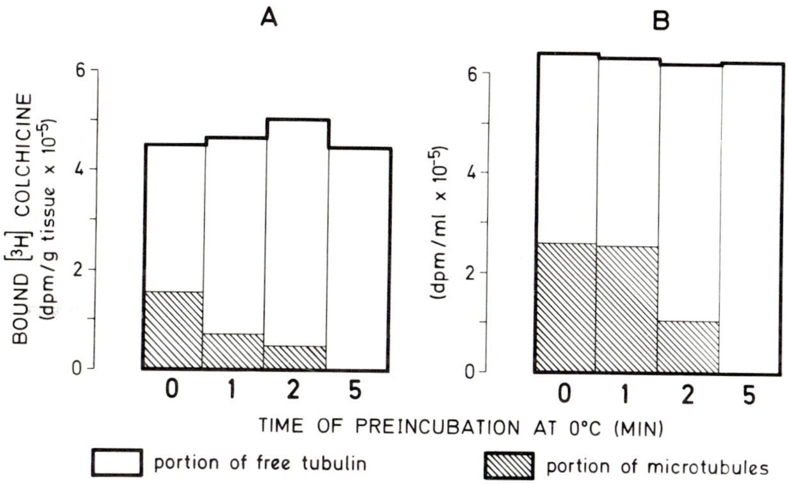


Fig. 4. Effect of exposure to low temperature on the state of the microtubular system in (A) liver and (B) on isolated brain microtubules. Each bar represents the mean of at least 3 experiments.

The effect of vinblastine, which has been extensively investigated upon isolated tubulin, is not yet understood with respect to its inhibitory action on microtubular function. It was therefore of interest to determine how this compound could influence the equilibrium between free and structured tubulin. In Figure 5A, the results obtained for mouse livers perfused for 60 minutes with various concentrations of vinblastine added as a single dose to the perfusate after 20 minutes of preperfusion are summarized. By analogy, isolated brain microtubules were incubated for 60 minutes with identical concentrations of vinblastine (Figure 5B). As can be seen, in both systems the presence of vinblastine resulted in a concentration-dependent decrease of both the total amount of tubulin and of the microtubule portion.

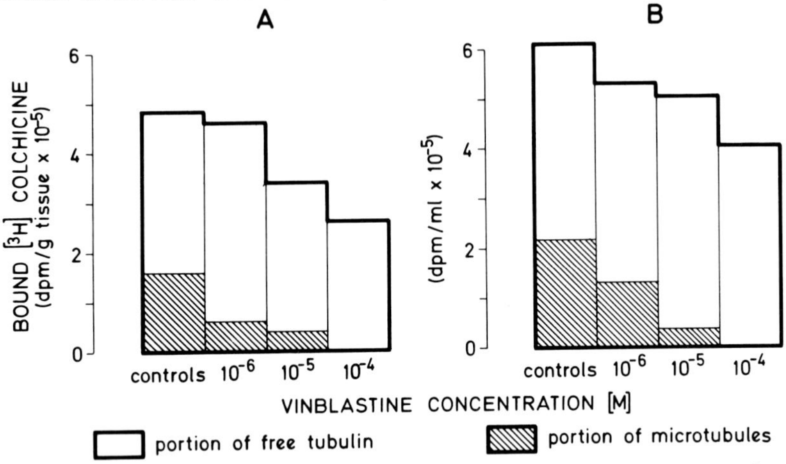


Fig. 5. Effect of vinblastine on the state of the microtubular system in (A) the perfused liver and (B) on isolated brain microtubules. Each bar represents the mean of at least 3 experiments.

Using a similar experimental design, the effect of colchicine was then studied. No interference between the initial exposure step to colchicine and the final $^{3}$H-colchicine-binding assay was observed, provided the perfusion (liver) or incubation (brain) was performed with labeled colchicine of

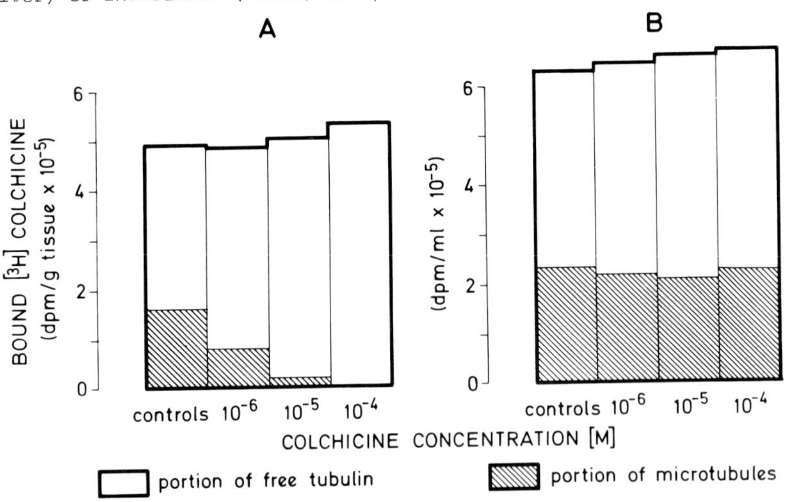


Fig. 6. Effect of colchicine on the state of the microtubular system in (A) the perfused liver and (B) on isolated brain microtubules. Each bar represents the mean of at least 3 experiments.

specific activity identical to that used later for the binding assay. Liver perfusion (Figure 6A) and incubation of isolated brain microtubules (Figure 6B) were performed for 60 minutes in the presence of various amount of $^{3}H$-colchicine. While the total amount of tubulin remained essentially constant in both systems, there was a surprising discrepancy in the behavior of respective microtubules: Liver microtubules disappeared in a dose-dependent manner in the presence of colchicine, whereas the brain microtubule portion remained almost unaltered, even in the presence of $10^{-4}$M colchicine, a concentration which is able to completely inhibit the formation of these microtubules.

## 10. DISCUSSION

From the results summarized here it can be concluded that the liver contains a colchicine-binding protein which, considering its intracellular distribution, its specificity for colchicine and its several similarities to purified brain tubulin, can be identified as being the hepatic microtubular protein, i.e. tubulin. In addition, there is good agreement between the dose-dependent binding kinetics of colchicine of either liver high speed supernatant or purified liver tubulin, and the actual inhibitory effect of colchicine upon hepatic secretory processes[1,4]. Thus, in both cases maximal effect of colchicine is obtained at $10^{-4}$M, the lowest effective concentration being about $10^{-6}$M. The further increase in colchicine-binding activity of liver supernatant at colchicine concentrations greater than $2 \times 10^{-4}$M can hardly be related to microtubular protein since this activity was found to be not precipitable by vinblastine nor was it present in purified microtubular protein. The present results dealing with vinblastine cannot be compared directly to vinca alkaloid-induced inhibition of perfused liver secretion[1,4] since such effect was mainly studied with vincristine. With this reservation in mind, it should be noted that the range of vinblastine concentrations needed to obtain precipitation of tubulin corresponds rather well with the inhibitory action of vincristine upon microtubular function. Taken together, these results support the contention that colchicine and vinca alkaloids display an inhibitory action on intrahepatic transport processes because of their interference with the microtubular protein.

To investigate further the actual state of the liver microtubular system, a method has been developed to measure the respective amounts of unassembled (free) tubulin and microtubules. Based upon characteristics previously determined for isolated brain microtubules[10], this method offers several advantages: 1) determinations can be done in one single liver; 2) handling of samples serving for measurement of either free or total tubulin differs only in that one of them contains glycerol during homogenization and exposure to cold; 3) exposure to cold can be kept very short due to rapid destruction of microtubules at 0°C; 4) it is applicable to isolated microtubules, thus offering an additional tool for the study of these structures beside the described measurements of changes in turbidity[10,11], viscosity[12] or flow birefringence[21], methods which have the drawback of not allowing determinations of the remaining free tubulin.

When using this method, the remarkable observation was that the relationship between free and polymerized tubulin was very similar (i.e. 40% as microtubules) in both intact liver and isolated brain microtubules, under conditions, in which the systems were seemingly in equilibrium. This equilibrium may be defined as the situation in which rates of assembly and disassembly are equal. Based principally upon previous experiments carried out with isolated microtubules[9,10,12,22] the factors that appear to influence the assembly-disassembly cycle are schematized in Figure 7. It may be assumed that the outlined situation is, to a certain extent, comparable to the liver, when taking in consideration the actual concentrations of trinucleotides and the low cytosolic calcium concentrations ( $10^{-6}$M) in the hepatocyte.

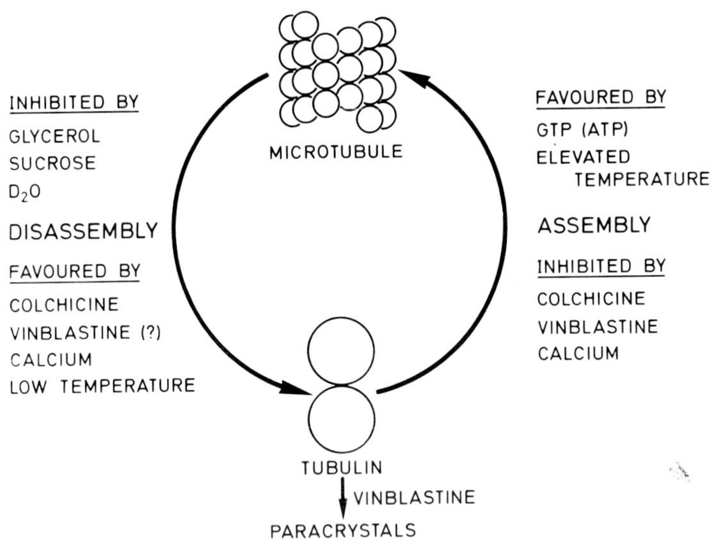


Fig. 7. Schematic view of the assembly-disassembly cycle of isolated microtubules.

Colchicine is assumed to interfere with the assembly-disassembly cycle by its binding to free tubulin thus preventing its incorporation into microtubules[22,23]. The fact that liver microtubules are broken down by colchicine whereas isolated brain microtubules are rather insensitive to this drug may indicate, that the turnover rate of the cycle is very high in the liver and low in the case of isolated microtubules. Varying sensitivities of microtubules for colchicine have been reported for many biological systems, this sensitivity ranging from almost nil[23,24] to rather high rates[25,26,27] which are comparable to the liver.

More difficult yet to understand are the effects of vinblastine on either liver or isolated microtubules. The reduction of the total amount of tubulin, presumably by precipitation, may influence the equilibrium thus resulting in a disappearance of microtubules. However, the assumption of a low assembly-disassembly rate of isolated microtubules mentioned above does not appear to fit this explanation. Alternatively, vinblastine may induce precipitation of microtubules per se thereby producing unsoluble material no longer measurable by the present method.

Extrapolating the present data, which are restricted to cytoplasmic microtubules, to the very complex area of microtubular function, one could develop the following hypotheses: a) the equilibrium between free tubulin and microtubules could be altered by activation or inactivation of the microtubular system, the rates of assembly and disassembly being regulated by trinucleotides availability, calcium concentration or some other unknown triggers; b) since the position of this equilibrium appears to be similar for the two systems used, i.e. liver and isolated brain microtubules, while the assembly-disassembly rate may be very different, therefore the turnover rate of microtubules may be a direct index of their function. One may imagine that microtubules could grow in one direction while depolymarizing at their tail end. Such a directional growth has been already established for isolated brain microtubules[28,29]. The realization that the second possibility might not necessarily be reflected in different ratios of free tubulin to microtubules does indicate that several combined approaches will be needed to eventually understand the existing relationship between the microtubular system and the vesicle-enclosed products which they are presumed to transport intracellularly.

ACKNOWLEDGEMENTS

This work has been supported by grants n° 3.522.71, n° 3.8080.72 SR and 3.2180.73 of the "Fonds National Suisse de la Recherche Scientifique", Berne, Switzerland and by grants-in-aid from the Zyma SA, Nyon, Switzerland, of the "Stiftung zur Förderung der Ernährungsforschung in der Schweiz" and from "Farbwerke Hoechst", Frankfurt, West-Germany.

REFERENCES

1. Le Marchand, Y., Singh, A., Assimacopoulos-Jeannet, F., Orci, L., Rouiller, Ch. and Jeanrenaud, B., 1973, J. Biol. Chem. 248, 6862.
2. Orci, L., Le Marchand, Y., Singh, A., Assimacopoulos-Jeannet, F., Rouiller, Ch. and Jeanrenaud, B., 1973, Nature 244, 30.
3. Stein, O. and Stein, Y., 1973, Biochim. Biophys. Acta 306, 142.
4. Le Marchand, Y., Patzelt, C., Assimacopoulos-Jeannet, F., Loten, E.G. and Jeanrenaud, B., 1974, J. Clin. Invest. 53, 1512.
5. Stein, O., Sanger, L. and Stein, Y., 1974, J. Cell Biol. 62, 90.
6. Singh, A., Le Marchand, Y.,Orci, L. and Jeanrenaud, B., 1975, Europ. J. Clin. Invest., in press.
7. Redman, C.M., Banerjee, D., Howell, K. and Palade, G.E., 1975, J. Cell Biol. 66, 42.
8. Bruni, C. and Porter, K.R., 1965, Amer. J. Path. 46, 691.
9. Weisenberg, R.C., 1972, Science 177, 1104.
10. Shelanski, M.L., Gaskin, F. and Cantor, R.C., 1973, Proc. Nat. Acad. Sci. USA 70, 765.
11. Gaskin, F., Cantor, C.R. and Shelanski, M.L., 1974, J. Mol. Biol. 89, 737.
12. Olmsted, J.B. and Borisy, G.G., 1973, Biochemistry 12, 4282.
13. Weisenberg, R.C., 1972, J. Cell Biol. 54, 266.
14. Patzelt, C., Singh, A., Le Marchand, Y., Orci, L. and Jeanrenaud, B., 1975, J. Cell Biol., in press.
15. Assimacopoulos-Jeannet, F., Exton, J.H. and Jeanrenaud, B., 1973, Amer. J. Physiol. 225, 25.
16. Weisenberg, R.C. and Timasheff, S.N., 1970, Biochemistry 9, 4110.
17. Bensch, K.G., Marantz, R., Wisniewski, H. and Shelanski, M., 1969, Science 165, 495.
18. Marantz, R. and Shelanski, M.L., 1970, J. Cell Biol. 44, 234.
19. Bryan, J., 1972, Biochemistry 11, 2611.
20. Wilson, L., 1970, Biochemistry 9, 4999.
21. Haga, T., Abe, T. and Kurokawa, M., 1974, FEBS Letters 39, 291.
22. Filner, P. and Behnke, O., 1973, J. Cell Biol. 59, 99a.
23. Wilson, L. and Meza, J., 1973, J. Cell Biol. 58, 709.
24. Murphy, D.B. and Tilney, L.G., 1974, J. Cell Biol. 61, 757.
25. Lacy, P.E., Howell, S.L., Young, D.A. and Fink, C.J., 1968, Nature 219, 1177.
26. Williams, J.A. and Wolff, J., 1970, Proc. Nat. Acad. Sci. USA 67, 1901.
27. Weissman, G., Dukor, P. and Zurier, R.P., 1971, Nature New Biol. 231, 131.
28. Borisy, G.G., Olmsted, J.B., Marcum, J.M. and Allen, C., 1974, Fed. Proc. 33, 167.
29. Binder, L.I., Dentler, W. and Rosenbaum, J., 1973, J. Cell Biol. 59, 24a.

*Microtubules and Microtubule Inhibitors, eds. M. Borgers and M. de Brabander* 


# MICROTUBULES AND MICROTUBULE INHIBITORS AND CYTOCHALASIN IN THE THYROID GLAND

Pierre Nêve, Pierre Rocmans, Paule Ketelbant-Balasse

Laboratories of Experimental Medicine, Electron Microscopy and Pathology
and Institut de Recherche Interdisciplinaire en Biologie Humaine et Nucléaire,
Université Libre de Bruxelles, Brussels, Belgium

## Summary

As in many other systems, microtubules are also found in the thyroid cells of any species. They are associated so closely to microfilaments that one speaks about "microtubule-microfilament system". Microtubules seem to play a role of cytoskeleton in the cream hamster thyroid. As suggested by *in vitro* experiments in presence of vinka alkaloids, the microtubules seem to be responsible for the stratification of organelles in the thyroid cells of the cream hamster.

The microtubule-microfilament system seems to be involved in the formation of pseudopods engulfing colloid droplets : a mechanism that represents the first step of thyroid secretion after acute administration of thyrotropin.

By using vincristine or cytochalasin B in incubated dog thyroid slices after six daily *in vivo* injections of thyrotropin, it is possible to demonstrate the existence of a second mechanism of thyroid secretion different from colloid phagocytosis by newly formed pseudopods. This mechanism is unsensitive to microtubule inhibitors and to cytochalasin B.

## Introduction

Microtubules are ubiquitous cellular organelles and the thyroid cell does not break from this rule. As already observed by Seljelid[1] in the rat thyroid, microtubules have been described in the thyroid cells of many species. They are very often associated with microfilaments in such way that authors very often speak about "microtubular-microfilamentous system" when they try to determine the roles played by these organelles in the thyroid[2,3]. Drugs which affect microtubules and also perhaps the microfilaments have enabled to suggest some microtubular functions in the thyroid and to advance the knowledge of thyroid physiology.

---

Work realized thanks to grant n° 20472 and 3.9001.75 of the Fonds de la Recherche Scientifique Médicale, Brussels, and under contract of the Ministère de la Politique Scientifique within the framework of the Association Euratom - University of Pisa - University of Brussels.

Description

As in other organs and systems[4,5,6,7], the microtubules in the thyroid appear as hollow, usually straight cylinders 250-280 Å wide ; their length is difficult to determine as it depends on the incidence of sectioning (up to 4 μm). No obvious termination to any organelle or membrane can be detected. The microtubules can be easily recognized in cross sections thanks to the clear area about 100 Å wide, free of any organelle surrounding the body of the microtubule (Fig. 1). The microtubules occupy preferential localizations in the cytoplasm : along the lateral membrane, perpendicular to the cellular base and apex, around the nucleus and also parallel to the apical border[8,9]. Microtubules also take part to the constitution of the mitotic spindle. At least in the dog thyroid two different populations of microfilaments have been identified[3]. The first, most easily visible variety is made up of 80-100 Å wide microfilaments often associated in bundles and widely dispersed throughout the cytoplasm. Microfilaments of the second type are less prominent : they are principally located in the subcortical area of the cells : they are thinner (50-70 Å wide) and seem shorter : this would explain why their recognition requires some special incidences of sectioning. They frequently appear as a fine filamentous network inserting upon the inner surface of the plasma membrane (Fig. 2).

Cytoskeletal role of the microtubules

Microtubules are particularly abundant in the cream hamster thyroid cell that is characterized by a stratification of their organelles[8]. The nucleus and rough endoplasmic cisternae occupy the lower half of the cytoplasm whereas the upper half is formed by superposed sheaths made up of Golgi cisternae, lysosomes and apical vesicles the latter being just beneath the microvilli at the apex. Microtubules are so closely associated with these organelles that it has been suggested that the microtubules might play a role of intracellular scaffolding responsible for this stratification of organelles.

In order to test this hypothesis whole cream hamster thyroids have been incubated in presence of vincristine that disrupts the microtubules. Control glands even after 7 hours of incubation do not show any ultrastructural difference with glands directly fixed after removal. On the other hand, after incubation in presence of vincristine at a concentration higher than $5.10^{-4}$ M major morphological changes appear : microtubules vanish, paracrystalline structures are formed and the stratification of the organelles becomes less obvious. Lysosomes are found at the base

---

Fig. 1. Transmission electron microscopy. Apical part of a dog thyroid follicular cell. Microtubules in longitudinal and in cross sections (arrows) are easily recognizable. A subcortical filamentous network is connected with microfilaments founds inside the microvilli. (× 54,000).

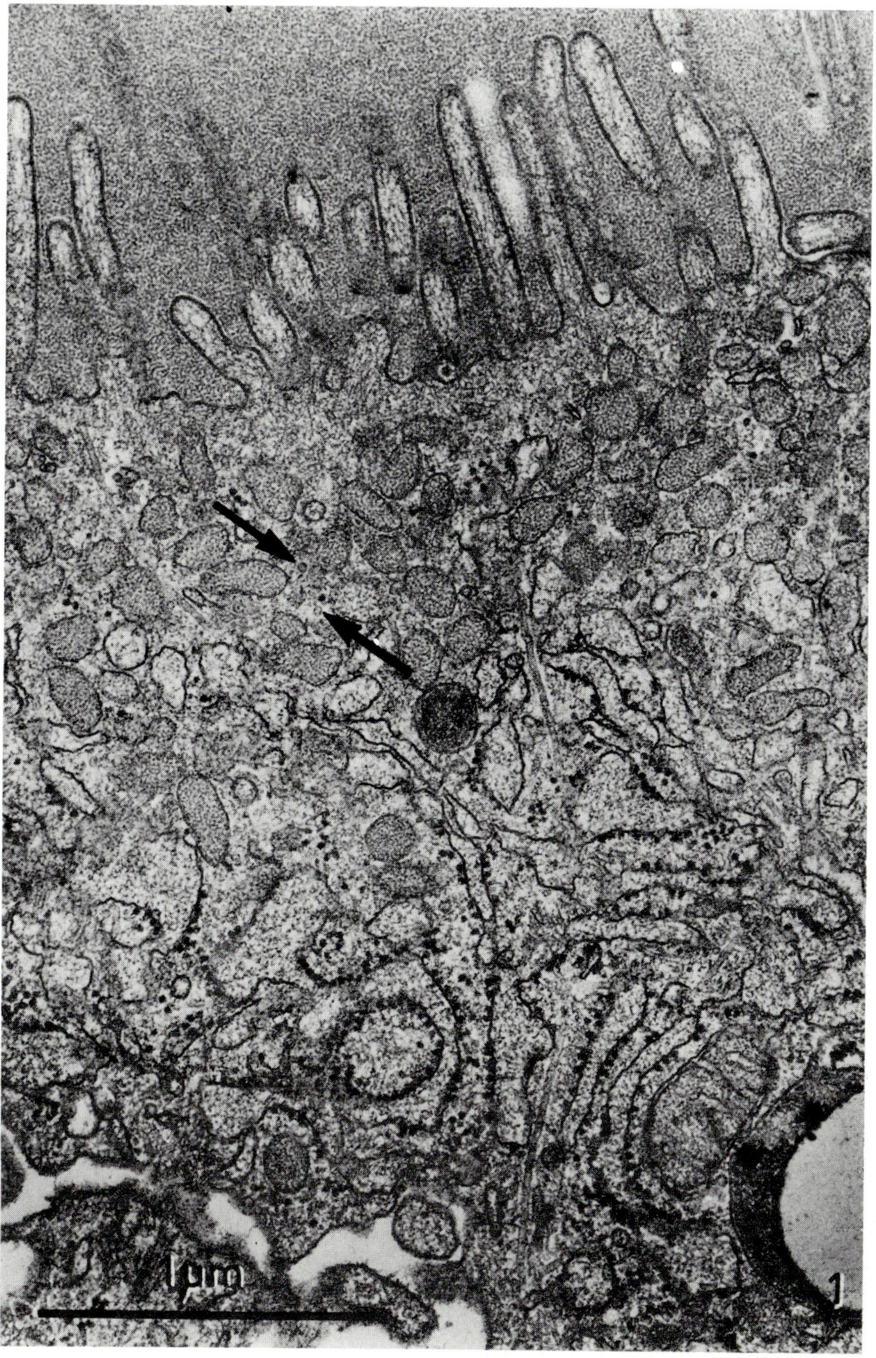
1µm
1

of the cells and Golgi cisternae are dispersed everywhere in the cytoplasm. The more the vincristine concentration is increased, the more these changes are evident. At the vincristine concentration of $9.10^{-4}$ M ergastoplasmic cisternae become rounded, are dispersed throughout the cytoplasm ; stratification of the organelles is completely lost but no change involves the centriole, the zonula occludens, mitochondria and desmosomes.

In conclusion disruption of the microtubules by vincristine in incubated cream hamster thyroids is correlated with loss of organelle stratification as though the microtubules would play some scaffolding role maintaining the organelles in specific locations[10]. However, these effects are only obtained when high doses of the drug are used in such manner that an unspecific toxic effect of vincristine cannot be excluded.

## Thyroid secretion and the microtubular-microfilamentous system

Acute addition of thyrotropin to resting thyroids either *in vivo*[11,12] or *in vitro*[13] induces the formation of pseudopods engulfing parts of colloid. These phagocytised colloid droplets thereafter fuse with lysosomes. The lysosomal enzymes hydrolyse thyroglobulin and release thyroid hormones which leave the cells probably by diffusion[14,15]. Formation of pseudopods engulfing colloid represents the first step of thyroid secretion after acute stimulation with thyrotropin (Fig. 3). As shown farther the secretory events are different after repeated stimulations with thyrotropin.

## Acute thyrotropin stimulation

In order to test the role of microtubules and microfilaments in thyroid secretion after acute TSH stimulation[3,16], agents interfering with microtubules (colchicine, ethanol, deuterated water, vinblastine and vincristine) and with microfilaments (cytochalasin) are added with TSH to the incubation medium of dog thyroid slices. The effects of these agents are similar to those observed in other systems : microtubules vanish after treatment with colchicine, vincristine and vinblastine : they are maintained with ethanol and deuterated water ; no morphological change involves the microfilaments ; paracrystalline structures are formed after incubation with vinblastine or vincristine. The TSH-stimulated secretion of

---

Fig. 2. Transmission electron microscopy. Tangential section of a dog thyroid follicle through connection between three cells near to the apex. Bundles of microfilaments (100 Å thick) are particularly abundant close to the junctional complexes and desmosomes. Microtubules (arrows) are present. (× 36,000).

Fig. 3. Scanning electron microscopy. Newly formed pseudopods emerge above the microvilli at the apex of follicular cells from a dog thyroid slice incubated in presence of thyrotropin. (× 6,000).

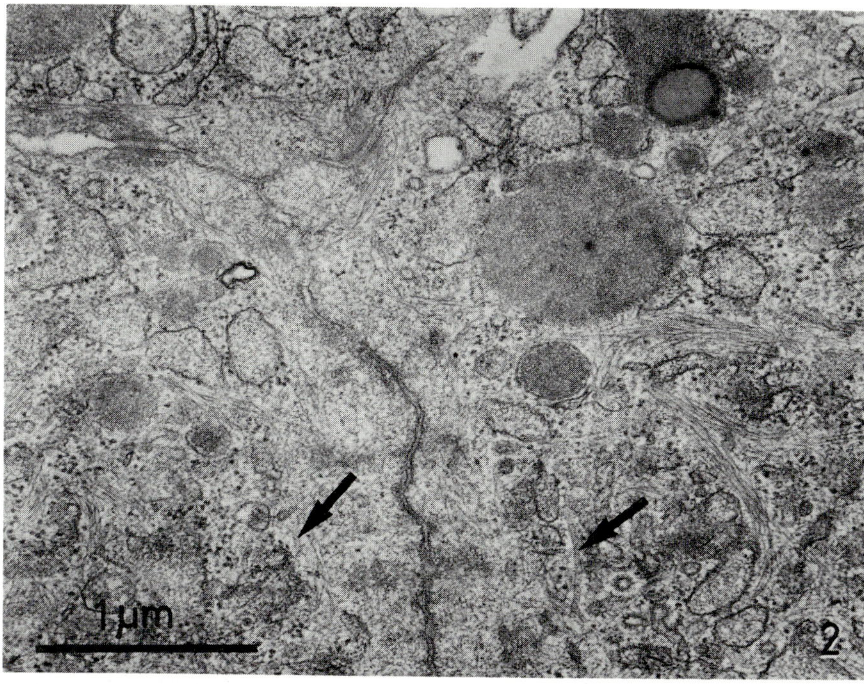
1 μm
2

5 μm
3

hormones by $^{131}$I-prelabelled thyroid slices is measured by the release of butanol extractible radioiodine ($BE^{131}I$) in the medium[17] : this TSH effect is inhibited both by microtubule inhibitors and cytochalasin[16]. Such observations have been confirmed by other authors *in vitro*[2,18,19] and *in vivo*[20]. Transmission electron microscopy and scanning electron microscopy of the thyroid slices demonstrate that the inhibitors completely abolished pseudopod formation and intracellular colloid droplet formation in the stimulated cells. However in slices incubated in presence of cytochalasin B, herniations of the apical membrane different from pseudopods appear : they might be induced by cytochalasin-disruption of some subcortical filamentous network (Figs. 4 and 5). This suggests that the inhibition of thyroid secretion by these agents bears on the first step of secretion : the phagocytosis of colloid[3,9,21,22].

The inhibitory effects induced by $D_2O$ and cytochalasin are reversible, which rejects the possibility of a non-specific toxic effect. As these inhibiting agents blocked phagocytosis of colloid and secretion in both TSH and in dibutyryl cyclic 3',5'-AMP stimulated slices, it is concluded that they affect cyclic 3',5'-AMP action rather than cyclic 3',5'-AMP formation. However, this *in vitro* effect was not observed *in vivo* : vinblastine had no inhibitory influence of the dbc-AMP-induced formation of pseudopods and colloid droplets in mice[20]. Instead the vinblastine-induced crystalloid structures disappeared upon treatment with dbc-AMP, and reappearance of microtubules occurred. These authors suggest that the integrity of microtubules *in vivo* seems to depend on an adequate supply of cAMP. It has also been reported that thyroid generation of $H_2O_2$, formation of hormone and oxidation of glucose-1-C require microtubular integrity[23].

Although cytochalasin does not act only[24,25,26,27] on microfilaments, it has been suggested that both microfilaments and microtubules are required for the acute TSH-stimulated thyroid secretion process and principally for the formation of pseudopods. The mechanism through which these organelles are involved in the formation of these pseudopods remains hypothetical so much the more that plasma membranes of thyroid tissue contain an intrinsic or firmly bound tubulin-like protein[28] and that in other systems it has been shown that colchicine would act directly on membranes[29].

Fig. 4. Scanning electron microscopy. Cellular herniations at the apex of follicular cells from dog thyroid slice incubated for short time in presence of cytochalasin (2 µg/ml). Ruffling of the plasma membranes differenciates the herniations from pseudopods. (× 13,500).

Fig. 5. Scanning electron microscopy. Part of follicle in a dog thyroid slice incubated in presence of thyrotropin added with vincristine : no pseudopod is formed. (× 3,900).

1 µm
4

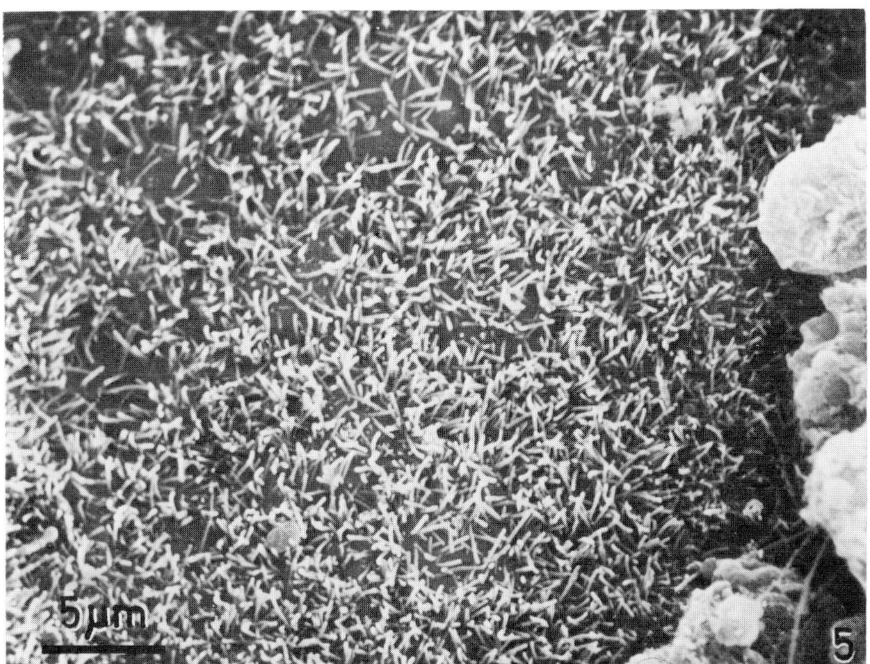
5 µm
5

Repeated stimulation by thyrotropin

The thyroid secretion process seems to occur in a different way when the glands are repeatedly stimulated by thyrotropin *in vivo*[12]. In these conditions hypertrophy and hyperplasia occur and nearly no pseudopods and/or colloid droplets are encountered except when thyrotropin is administered a few hours before fixation. *In vitro* experiments are performed on incubated dog thyroid slices from animals repeatedly stimulated for six days by thyrotropin and prelabelled *in vivo* with $^{125}I$. Ultrastructural observations (TEM and SEM) are permanently coupled with thyroid secretion determinations measured by $BE^{125}I$ release. The level of thyroid secretion is 2 to 9 times higher in repeatedly stimulated state than in basal state. TSH added *in vitro* (5 mU/ml) further enhances the release of $BE^{125}I$ like in non-stimulated slices. However, cytochalasin B or vincristine added *in vitro* with TSH inhibit partially or completely the acute stimulation of the $BE^{125}I$ release induced by TSH. A parallel inhibition of the acute stimulation effect of TSH is observed for the $BE^{125}I$ release, the number of apical pseudopods and the number of colloid droplets. The mechanism of this "cytochalasin-" or "vincristine-sensitive" secretion is obviously phagocytosis[30].

When complete inhibition of colloid phagocytosis has been obtained with thyrotropin and cytochalasin or vincristine, the level of secretion in the repeatedly stimulated slices remains higher than in reference slices. This suggests that the remaining $BE^{125}I$ release corresponds to a "cytochalasin-" or "vincristine-unsensitive mechanism" of thyroid secretion different from colloid phagocytosis. This "cytochalasin-" or "vincristine-unsensitive mechanism" must correspond to an active phenomenon. Indeed when the repeatedly stimulated dog thyroid slices are incubated at 21°C of temperature instead of 37°C (and consequently when the physical state of the cell membrane is modified), the $BE^{125}I$ release is depressed to a lower level, close to what is observed in reference thyroid slices without any thyrotropin stimulation. Therefore, in repeatedly stimulated thyroids two different mechanisms of secretion seem to occur. The first one is sensitive to cytochalasin or vincristine and probably corresponds to phagocytosis of colloid and can be always increased by direct addition of thyrotropin. The second one is cytochalasin or vincristine unsensitive and seems prominent in case of cellular hyperplasia : it might correspond to micropinocytosis[31].

In conclusion the use of microtubular inhibitors and of cytochalasin enables to dissociate two different mechanisms of thyroid secretion.

References

1 Seljelid, R., 1967, *J. Ultrastr. Res.* 17, 401.

2 Williams, J.A. and Wolff, J., 1971, *Biophys. Biochim. Res. Comm.* 44, 422.

[3] Nève, P., Ketelbant-Balasse, P., Willems, C. and Dumont, J.E., 1972, *Exp. Cell Res.* 74, 227.

[4] Rebhun, L.I., 1972, *Int. Rev. Cytol.* 32, 93.

[5] Dustin, P., Jr., 1972, *Arch. Biol.* 83, 419.

[6] Dustin, P., 1974, *Arch. Biol.* 85, 263.

[7] Nève, P., 1975, *in:* Eukaryotic cell function and growth : regulation by intracellular cyclic nucleotides, eds. J.E. Dumont, B. Brown and N. Marshall (Plenum Publishing Corporation, New York) to be published 1975.

[8] Nève, P. and Wollman, S.H., 1971, *Anat. Rec.* 168, 23.

[9] Nève, P., Ketelbant-Balasse, P., Rodesch, F. and Pasteels, J.M., 1972, *Bull. Acad. Méd. Belg.* 12, 647.

[10] Ketelbant-Balasse, P. and Nève, P., 1975, *Cell and Tissue Research*, to be published 1975.

[11] Wetzel, B.K., Spicer, S.S. and Wollman, S.H., 1965, *J. Cell Biol.* 25, 593.

[12] Nève, P. and Dumont, J.E., 1970, *Z. Zellforsch.* 103, 61.

[13] Nève, P. and Dumont, J.E., 1970, *Exp. Cell Res.* 63, 285.

[14] Wollman, S.H., 1969, *in:* Lysosomes in biology and pathology, vol. 2, eds. J.T. Dingle and H.B. Fell (North-Holland, Amsterdam) p. 483.

[15] Pantic, V.R., 1974, *Int. Rev. Cytol.* 38, 153.

[16] Nève, P., Willems, C. and Dumont, J.E., 1970, *Exp. Cell Res.* 63, 457.

[17] Willems, C., Rocmans, P. and Dumont, J.E., 1970, *Biochim. Biophys. Acta* 222, 474.

[18] Williams, J.A. and Wolff, J., 1970, *Proc. Nat. Acad. Sci. USA* 67, 1901.

[19] Berens, S.C., Williams, J.A. and Wolff, J., 1971, *Biochim. Biophys. Acta* 252, 314.

[20] Ekholm, R., Ericson, L.E., Josefsson, J.O. and Melander, A., 1974, *Endocrinology* 94, 641.

[21] Ketelbant-Balasse, P., Rodesch, F., Nève, P. and Pasteels, J.M., 1973, *Exp. Cell Res.* 79, 111.

[22] Ketelbant-Balasse, P. and Nève, P., 1974, *in:* Scanning electron microscopy, Proceedings of the Workshop on Advances in Biomedical Applications of the SEM, IIT Research Institute, Chicago, p. 761.

[23] Chiraseveenuprapund, P. and Rosenberg, I.N., 1974, *Endocrinology* 94, 1086.

[24] Carter, S.B., 1972, *Endeavour* 31, 77.

[25] Mizel, S.B. and Wilson, L., 1972, *J. Biol. Chem.* 247, 4102.

[26] Estensen, R.D. and Plagemann, P.G.W., 1972, *Proc. Nat. Acad. Sci. USA* 69, 1430.

[27] Kletzien, R.F., Perdue, J.F. and Springer, A., 1972, *J. Biol. Chem.* 247, 2964.

[28] Bhattacharyya, B. and Wolff, J., 1975, Excerpta Medica, International Medical Abstracting Service, Seventh International Thyroid Conference, Boston, p. 21.

[29] Wunderlich, F., Müller, R. and Speth, V., 1973, *Science* 182, 1136.

[30] Ketelbant-Balasse, P., Rocmans, P., Frederic, N., Golstein, J. and Nève, P., 1975, Excerpta Medica, International Medical Abstracting Service, Seventh International Thyroid Conference, Boston, p. 24.

[31] Seljelid, R., Reith, A. and Nakken, K.F., 1970, *Lab. Invest.* 23, 595.

*Microtubules and Microtubule Inhibitors, eds. M. Borgers and M. de Brabander*



# THE EFFECT OF COLCHICINE ON THE DISTRIBUTION OF APICAL VESICLES AND TUBULES IN ABSORPTIVE CELLS OF CULTURED HUMAN SMALL INTESTINE

L.A. Ginsel, W.F. Debets and W.Th. Daems

Laboratory for Electron Microscopy, University of Leiden,
Rijnsburgerweg 10, Leiden, The Netherlands

## SUMMARY

The effect of colchicine on the absorptive cells of cultured human small intestine was investigated, with special attention to the influence on the apical vesicles and tubules. These structures occurring mainly in the terminal web area of the absorptive cells, are thought to have a function in the transport of cell-coat material from the Golgi apparatus to the cell surface. The apical vesicles and tubules in the absorptive cells cultured in the presence of colchicine showed time- and concentration-dependent changes. Firstly, the number of such vesicles and tubules in the area of the terminal web decreased strongly, but increased deeper in the cytoplasm. This change in the distribution of these structures in the apical cytoplasm is explained as a colchicine-induced disturbance of the intracellular transport of cell-coat material, and supports the assumed transport role of apical vesicles and tubules. Secondly, the accumulation of the apical vesicles and tubules, together with the presence of a relatively large number of lysosome-like bodies, suggests a - crinophagic - function of the lysosomal apparatus in the disposal of the contents of the apical vesicles and tubules.

## INTRODUCTION

Numerous studies on the small intestine of mammals have shown that the absorptive cells synthesize their own cell coat[13,14,17], and the important role of the Golgi apparatus in the production of cell-coat material has been described[2,3,16]. However, little information is available about the intracellular mechanism of transport from the Golgi to the apex of the cell and the final excretion of this cell-coat material at the top of the absorptive cell.

Studies done in the rat with radioactive cell-coat precursors have provided evidence for a role of smooth-surfaced vesicles, which occur in the Golgi region as well as in the apical cytoplasm, in the transport of cell-coat material to the surface of the absorptive cells[2,3], and a number of studies suggest the involvement of lysosomes in this process[4,7,11,16]. The absorptive cells of the small intestine of children show similar small vesicles both in the Golgi region and directly below the microvillar border, often fused with the lysosome-like bodies. A possible crinophagic regulatory function of these lysosomes in the secretion of cell-coat material has been suggested[7,11].

With respect to further studies to investigate this hypothesis, the following points are of importance. Colchicine and other agents causing microtubular disruption showed an inhibitory effect on the secretory process in cells of various types[8,15,21]. Administration of these drugs to cultured bone cells often resulted in the accumulation of secretory vacuoles[9,19]. These findings prompted us to study the effect of colchicine on the transport and secretion of cell-coat material in the absorptive cells of cultured suction--biopsy specimens of adult human small intestine.

## MATERIAL AND METHODS

### Culturing

Suction-biopsy specimens were obtained from the jejunum of five individuals. Directly after collection, one specimen was prepared for routine electron microscopy and the others were cut in two and placed in the organ culture system. Organ cultures were prepared according to Browning and Trier[6] with the following modifications[12]. The specimens were cultured in NCTC-135 medium (Flow Laboratories) supplemented with 10% fetal calf serum (Difco Laboratories) as well

---

Fig. 1. Partial view of three absorptive cells of a villus in a 6-hr control culture. The cylindrical absorptive cells have a microvillous brush border bearing a distinct cell coat. In the apical part of the cytoplasm the rootlets of the microvilli (mw) are inserted in the terminal web (tw), which also contains apical vesicles and tubules (small arrows). Directly under the terminal web, multivesicular bodies (mvb) and dense bodies (edb) are present. The Golgi apparatus (Ga) in the vicinity of the nucleus consists of a numberof cisternae and vesicles. At the lower left, two nuclei (N) can be seen.

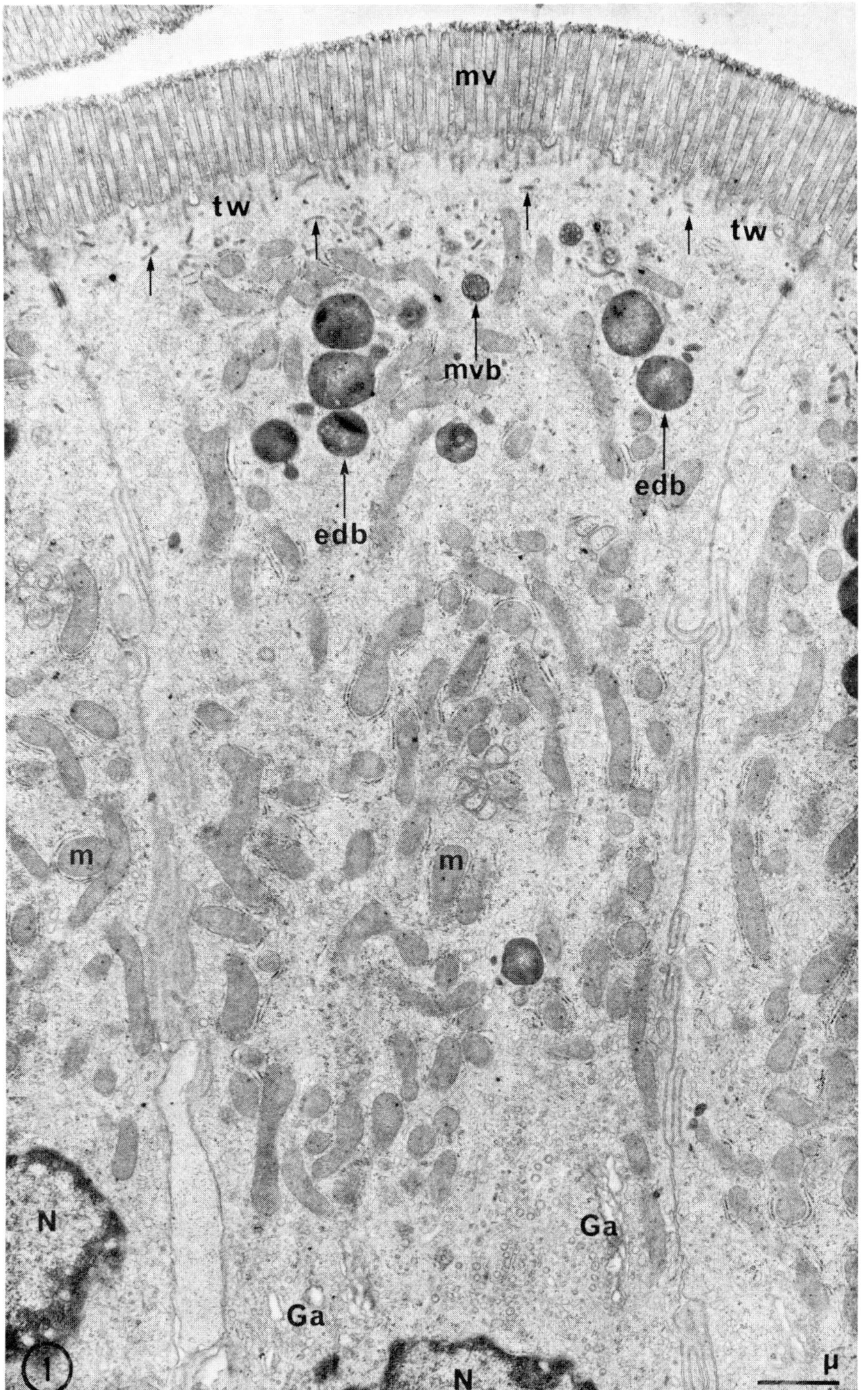
mv
tw
tw
mvb
edb
edb
m
m
N
Ga
Ga
N
1
μ

as 0.6 U insulin and 50 μg gentamicin per ml. Control specimens were cultured in this medium for 1, 3, and 6 hr. Test specimens were cultured for the same period in medium to which 10 or 100 μg colchicine (Merck, A.G.) was added per ml.

Electron microscopy

The specimens were then divided into small blocks and fixed at room temperature for 60 min in 1.5% glutaraldehyde in cacodylate buffer (pH 7.2), followed by 60 min at 4°C in 1% osmium tetroxide buffered with phosphate to pH 7.2. The fixatives had a final osmolarity of 320 mOsm. Dehydration, embedding, and sectioning were performed according to standard techniques. The sections were stained first with uranyl acetate and then with lead hydroxide, and were studied in a Philips EM 200 or a Siemens Elmiskop I.

Morphometry of the apical vesicles and tubules

The number and location of the apical vesicles and tubules in the apical 5-μ zone of the cytoplasm of the absorptive cells in control and test specimens were determined. Per specimen, photographs were made of the apical cytoplasm of 10 absorptive cells in which both the microvilli and the nucleus were visible. A test-line preparation was used to count the vacuoles and tubules in cell compartments situated at given distances below the base of the striated brush border in each photograph.

---

Fig. 2. The apical cytoplasm of an absorptive cell of a normal intestinal villus cultured in control medium for 6 hr. Directly below the microvilli the terminal web (tw) contains numerous dense apical vesicles and tubules (small arrows). Between the mitochondria (m) there are multivesicular bodies (mvb) and dense bodies (edb). Note the continuity between the multivesicular body and an apical vesicle (double arrows).

Fig. 3. The apical cytoplasm of an absorptive cell cultured in the presence of colchicine (10 μg/ml) for 6 hr. The terminal web (tw) is enlarged and contains a few apical vesicles and tubules. Deeper in the cytoplasm there are accumulations of apical vesicles and tubules (small arrows) together with multivesicular bodies (mvb) and numerous small and normal-sized dense bodies (edb).

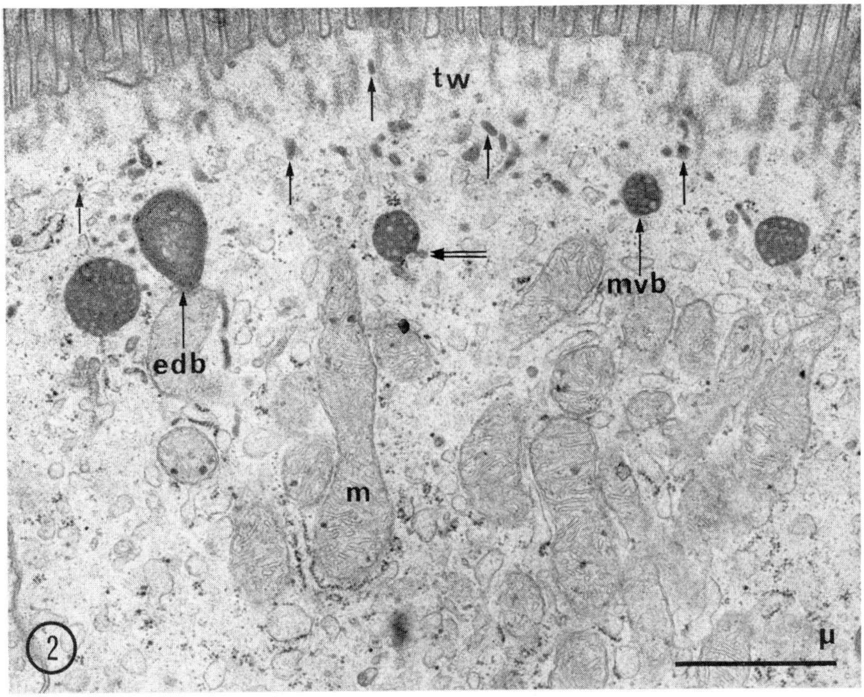
tw
mvb
edb
m
2
μ

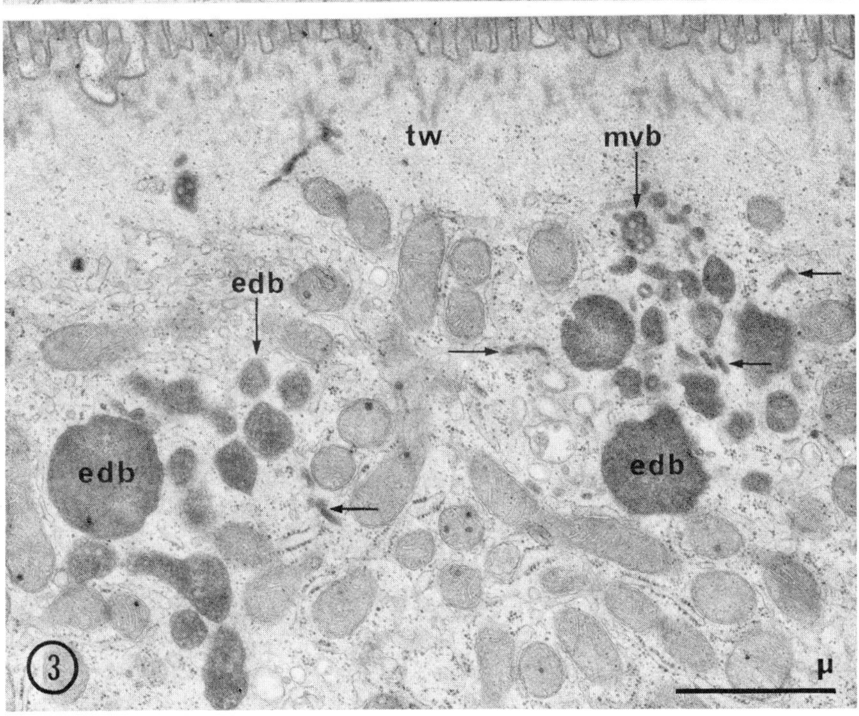
tw
mvb
edb
edb
edb
3
μ

RESULTS

The following description of the absorptive cells is restricted to the number and location of the apical vesicles and tubules (avt) and to their relation to the lysosome-like bodies. Microtubules were only occasionally observed in the cytoplasm of the absorptive cells (see ref.1), and will not be described here. For detailed information on the fine structure of the lysosome-like bodies, the Golgi apparatus, the apical vesicles and tubules, and the cell coat in normal absorptive cells and in absorptive cells cultured for periods of up to 48 hours, the reader is referred to the studies on juvenile and adult human intestine[7,11,12,18,20].

Apical vesicles and tubules in uncultured absorptive cells

The apical cytoplasm of normal absorptive cells generally contains, in addition to mitochondria and the elements of the endoplasmic reticulum, two types of lysosome-like bodies: multivesicular

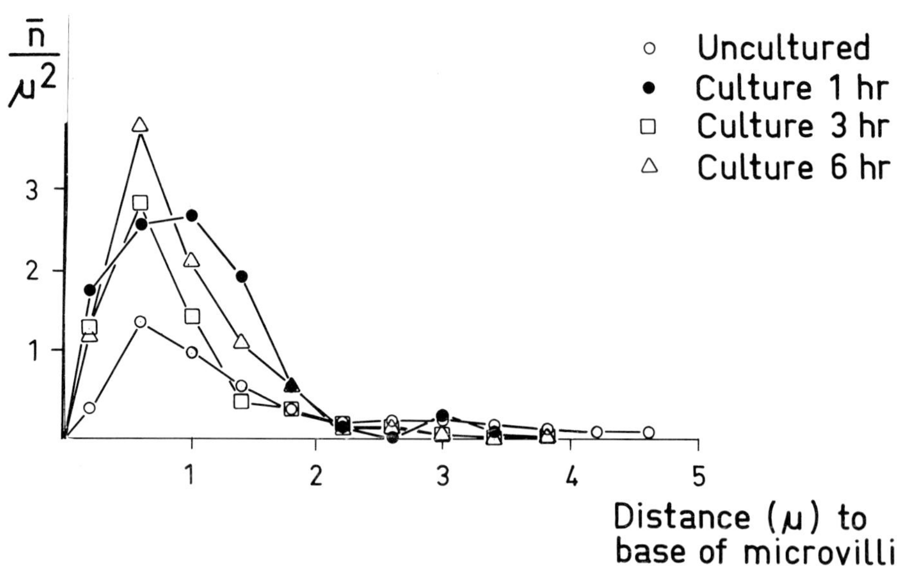


Fig. 4. Distribution of apical vesicles and tubules (avt) in the apical cytoplasm of uncultured absorptive cells and cells cultured in control medium for 1, 3, and 6 hr. Note in the apical cytoplasm of the cultured absorptive cells the increased number of avt in a zone situated about 0.6 μ below the base of the microvilli.

bodies and dense bodies (cf. fig.1). Both were found in small numbers in a relatively narrow zone directly below the terminal web[11].

The terminal web area is a cytoplasmatic zone varying in width up to about 0.7 μ and lying below the base of the microvilli. This zone is normally free of cell organelles (cf. fig.2), but contains a varying number of small vesicles and tubules whose usually electron-dense matrix is bounded by a single membrane. In the present study these apical vesicles and tubules were found mainly in the apical cytoplasm about 0.6 μ below the base of the microvilli (fig.4). The avt show a preference for the vicinity of the multivesicular bodies, and continuities between both structures were frequently observed. Less frequently continuities were observed between avt and the dense bodies[11].

Effect of culturing on the apical vesicles and tubules

In 1-, 3-, and 6-hr cultures (figs.2 and 6) the number of avt in the apical cytoplasm was markedly increased (fig.4). They were mainly located - as in the uncultured controls - in the terminal web area, between or directly beneath the rootlets of the microvilli, but fusions of the avt with the apical-cell membrane were not observed. Many avt showed continuities with the multivesicular bodies (fig.2) and, to a lesser extent, with the dense bodies.

Effect of colchicine on the apical vesicles and tubules in cultured absorptive cells

After 1 hour of culture: Compared with the 1-hr control specimen, there were no obvious changes in the number and location of the avt in the absorptive cells cultured for 1 hr in the presence of colchicine (10 and 100 μg/ml) (fig.5). The avt were located predominantly in the terminal web, and their interrelationship with the lysosome-like bodies appeared unchanged.

After 3 hours of culture: In comparison with the 3-hr controls, absorptive cells cultured with colchicine (10 and 100 μg/ml) showed concentration-dependent changes. The terminal web often looked rather empty, due partly to an enlargement of the terminal web area but mainly to a decrease in the number of avt in this area (fig.5). Deeper in the cytoplasm, the number of avt increased and accumulations of these structures were frequently seen, often grouped together with multivesicular bodies and a larger number of small and normal-sized dense bodies (cf. fig.3). Fusions between avt and the lysosome-like bodies were regularly observed. The multivesicular bodies had a normal appearance; within the single trilaminar membrane, the electron-dense matrix contained many electron-lucid vesicles. The population of dense bodies differed from the normal picture by an increased number of small edb.

After 6 hours of culture: The changes in the location of the avt in the apical cytoplasm of the colchicine-treated cells showed further progression at 6 hr (fig.5). Absorptive cells cultured with 10 µg/ml colchicine showed an increase in the size of the terminal web area (fig.7) which was even more pronounced after culture with 100 µg/ml (fig.8). This phenomenon was accompanied by a strong decrease in the number of avt in the same area (fig.5). Deeper in the cytoplasm the avt increased in number. Clusters of avt with multivesicular bodies and a large number of small and normal-sized dense bodies were even more conspicuous than in the 3-hr cultures (fig.3).

## DISCUSSION

In the present study colchicine induced time- and concentration--dependent changes in the absorptive cells of cultured human intestine. The number of apical vesicles and tubules decreased strongly in the terminal web but increased deeper in the cytoplasm, and this area showed large accumulations of avt together with multivesicular bodies and dense bodies.

Reports on ultrastructural studies on the secretory activity of intestinal cells refer to the role of small vesicles in the transport of cell-coat material[2,3,5,7,11]. The results of electron--microscopical autoradiographic studies with labelled cell-coat precursors in the absorptive cells of the rat small intestine suggest that the smooth-surfaced vesicles occurring in the Golgi region as well as immediately below the microvillar border, have a function

Fig. 5. Distribution of apical vesicles and tubules in the apical cytoplasm of absorptive cells cultured in the presence of colchicine (10 and 100 µg/ml) for 1, 3, and 6 hr. The apical vesicles and tubules cultured in the presence of colchicine show time- and concentration-dependent changes. There are now fewer avt in the terminal web area directly below the base of the microvilli, but deeper in the cytoplasm the number has increased.

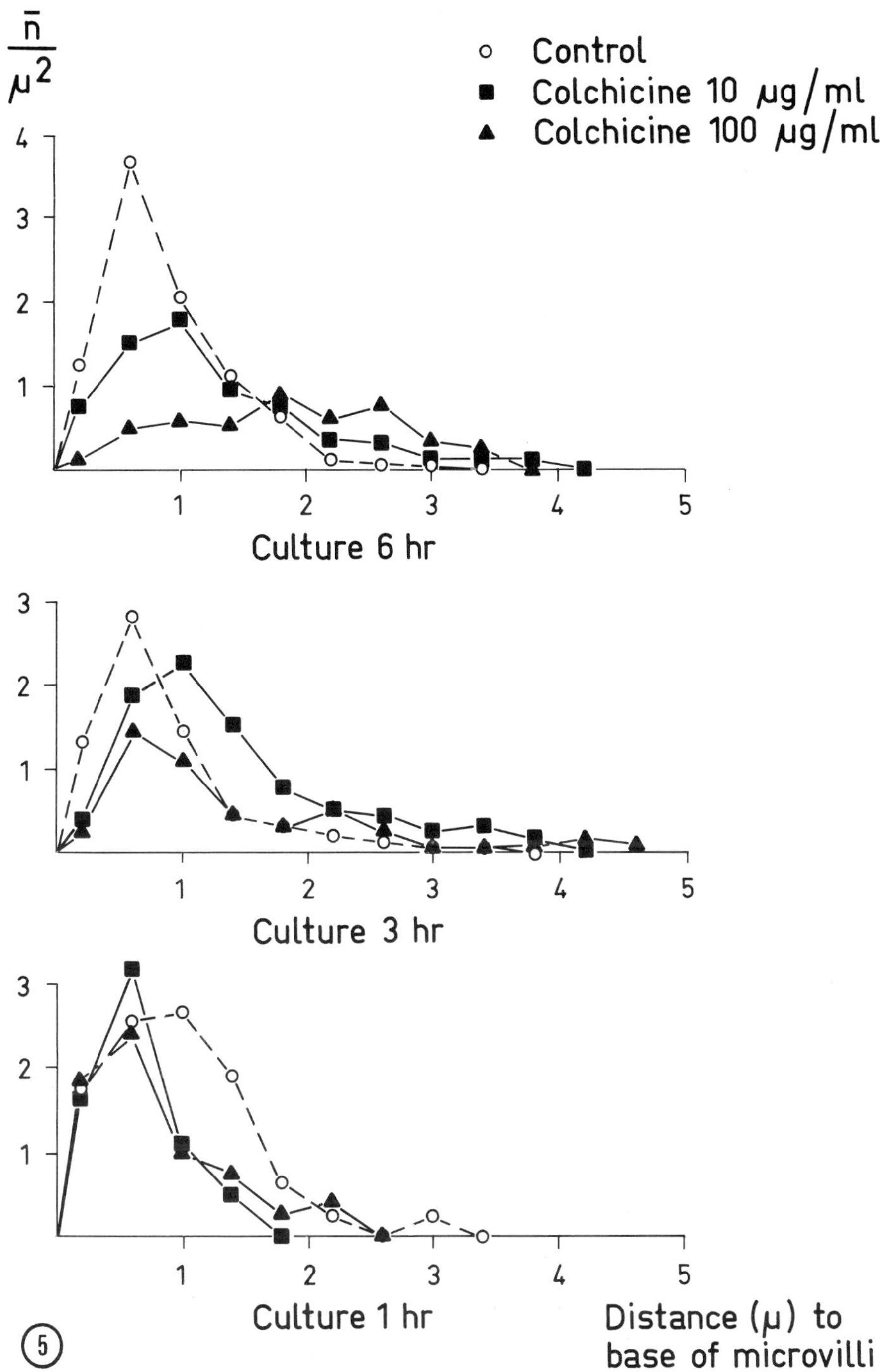

n̄/μ²
Control
Colchicine 10 μg/ml
Colchicine 100 μg/ml
4
3
2
1
1
2
3
4
5
Culture 6 hr
3
2
1
1
2
3
4
5
Culture 3 hr
3
2
1
1
2
3
4
5
Culture 1 hr
Distance (μ) to
base of microvilli
5

in the transport of cell-coat material to the cell surface[2,3,4]. In our studies on the absorptive cells of children, small vesicles and tubules located in the area of the terminal web showed glycoprotein staining properties similar to those of the cell coat, the cisternae and vesicles of the Golgi apparatus, and the lysosome-like bodies [7,11]. These observations support a carrier function of avt in the transport of cell-coat material from the Golgi apparatus to the cell surface.

In a number of different cell types, i.e., the insulin-producing beta cells of the islets of Langerhans[15], thyroglobulin-producing cells[21], and collagen-producing osteoblasts[19], colchicine and other agents causing microtubular disruption strongly inhibit secretion. Our results show that as compared with normal uncultured absorptive cells of the small intestine, there is an increase in the number of apically situated vesicles and tubules in the absorptive cells of control specimens cultured for 1, 3, and 6 hours. In both cultured and uncultured controls these structures were mainly located in the area of the terminal web. The absorptive cells cultured for 1, 3, and 6 hours in the presence of colchicine showed a time- and concentration-dependent effect. The number of apical vesicles and tubules in the terminal web area strongly diminished, whereas an increase of these structures was noticed deeper in the cytoplasm. If all these vesicles and tubules indeed derive from the Golgi apparatus, this change in their distribution over the apical cytoplasm can be interpreted as evidence of a disturbance of the intracellular transport.

These dislocated apical vesicles and tubules were very often found together with multivesicular bodies and with a large number of small and normal-sized dense bodies. Such clusters of vesicles and dense bodies did not occur in the absorptive cells cultured in the absence of colchicine. Fusions between the vesicles and tubules and

Fig. 6. Apical part of an absorptive cell cultured in control medium for 6 hr. The terminal web area (tw) is relatively small and contains a large number of apical vesicles and tubules (small arrows).

Fig. 7. Apical part of an absorptive cell cultured in the presence of colchicine (10 μg/ml) for 6 hr. Compared with fig. 6, the terminal web area (tw) is enlarged and fewer apical vesicles and tubules are visible. Note the clusters of avt (small arrows) and the dense bodies (edb).

Fig. 8. Apical part of an absorptive cell cultured in the presence of colchicine (100 μg/ml) for 6 hr. The terminal web area (tw) is extremely enlarged in some places. Below the rootlets of the microvilli the clear apical cytoplasm contains very few apical vesicles or tubules (small arrow).

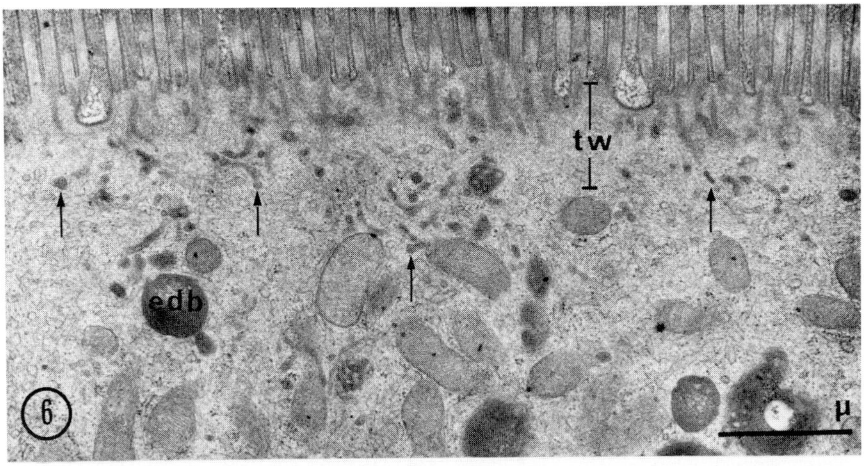
tw
edb
6
μ

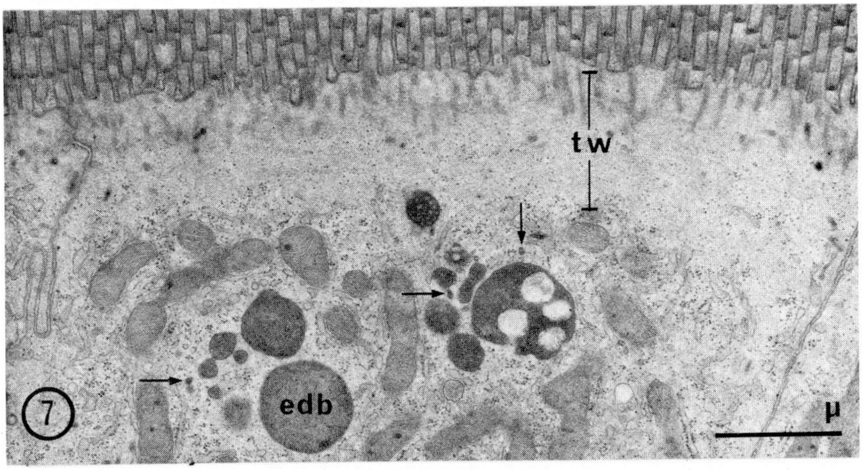
tw
edb
7
μ

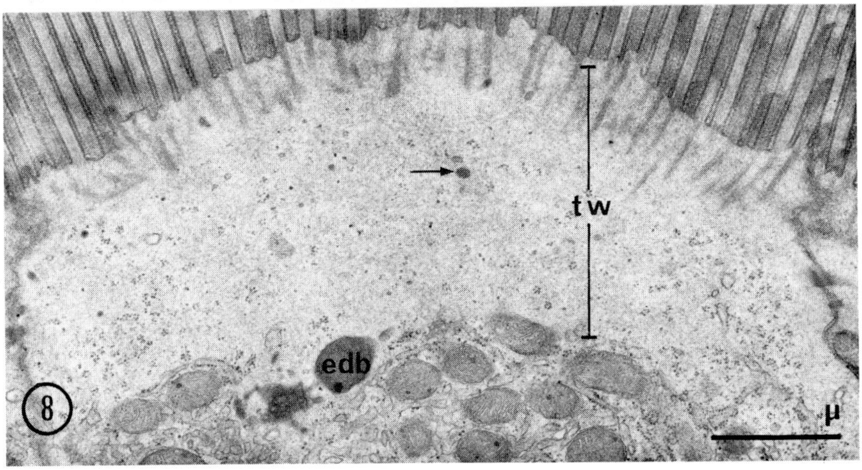
tw
edb
8
μ

the multivesicular bodies were more frequently seen than between avt and the dense bodies. These observations are suggestive of a crinophagic function of the lysosome-like bodies in intestinal absorptive cells, a mechanism by which the membrane of a secretion granule fuses with the membrane of a lysosome[7,10,11]. In the material studied, this mechanism is probably operative in the disposal of excess secretory products[10], accumulating due to a transport impairment, induced by the action of colchicine.

ACKNOWLEDGEMENTS

The authors greatfully acknowledge the help of Drs. A.S. Peña and P.M.A. Vio in providing the biopsy material, and the technical assistance of L.D.C. Verschragen, J.J. Beentjes, J.J. Magdelijns, and M. de Gruil-van Egmond. They are also indebted to Drs. E. Wisse, P. Brederoo, W.C. de Bruijn, and H.J. van der Rhee for their valuable criticism.

The investigations were supported in part by the Foundation for Medical Research (FUNGO), which is subsidized by the Netherlands Organisation for the Advancement of Pure Research (ZWO).

REFERENCES

1. Behnke,O.: J.Ultrastruct.Res. 11, 139 (1964).
2. Bennett,G.: J.Cell Biol. 45, 668 (1970).
3. Bennett,G. and Leblond,C.P.: J.Cell Biol. 46, 409 (1970).
4. Bennett,G. and Leblond,C.P.: J.Cell Biol. 51, 875 (1971).
5. Borgers,M. and De Nollin,S.: J.Parasitol. 60, 953 (1974).
6. Browning,Th.H. and Trier,J.S.: J.Clin.Invest. 48, 1423 (1969).
7. Daems,W.Th., Van Gemund,J.J., Vio,P.M.A., Willighagen,R.G.J., and Den Tandt, W.R. In: Lysosomal Storage Diseases (Hers,H.G. and Van Hooff, H. eds.) Academic Press 1973, p. 575.
8. Dehm,P., Prockop,D.J.: Biochim. Biophys.Acta 264, 375 (1972).
9. Ehrlich,H.P., Ross,R., and Bornstein,P.: J.Cell Biol. 62, 390 (1974).
10. Farquhar,M.G. In: Lysosomes in Biology and Pathology (Dingle,J.T. and Fell,H.B. eds). Vol. II. North-Holland Publ. Comp. 1969, p. 462.
11. Ginsel,L.A., Daems,W.Th., Emeis,J.J., Vio,P.M.A., Van Gemund, J.J.: Virchows Arch. Abt.B.Zellpathol. 13, 119 (1973).
12. Ginsel,L.A., Van der Want,J.J.L., and Daems,W.Th.: in preparation (1975).
13. Ito,S. and Revel,J.P.: J.Cell Biol. 23, 44a (1964).
14. Ito,S.: Fed.Proc. 28,12 (1969).
15. Lacy,P.E., Howell,S.L., Young,D.A., and Fink,C.J.: Nature 219, 1177 (1968).
16. Rambourg,A., Bennett,G., Kopriwa,B., and Leblond,C.P.: J.Microscopie 11, 163 (1971).
17. Revel,J.P., and Ito,S.: The Specifity of Cell Surfaces. Englewood Cliffs, N.Y., Prentice Hall 1967, p. 211.
18. Rubin,W.: Amer.J. Clin. Nutr. 24, 45 (1971).
19. Scherft,J.P., and Heersche,J.N.M.: Cell Tiss.Res. 157, 353 (1975).
20. Toner,P.G.: Int.Rev.Cytol. 24, 233 (1968).
21. Williams,J.A., and Wolff,J.: Proc.Natl.Acad.Sci.U.S.A. 67, 1901 (1970).

*Microtubules and Microtubule Inhibitors, eds. M. Borgers and M. de Brabander* 


# EFFECTS OF COLCHICINE AND VINBLASTINE ON THE MOBILIZATION OF LYSOSOMES IN PHAGOCYTIZING HUMAN LEUKOCYTES

Stephen E. Malawista

Department of Internal Medicine
Yale University School of Medicine
New Haven, Connecticut

## SUMMARY

The evidence presented here supports the hypothesis that microtubules facilitate the getting-together of lysosome and phagosome in human polymorphonuclear leukocytes (PMN), and that anti-inflammatory properties of colchicine and vinblastine may depend in part upon inhibition of this process and of the associated specific extracellular release of granular (lysosomal) enzymes.

## 1. Introduction

Some time ago we suggested that anti-inflammatory effects of colchicine, seen most prominently in the treatment of acute gouty arthritis, are related to its anti-mitotic effects through a common mechanism: the interference in cells with the organization of labile fibrillar systems concerned with structure and movement, of which the mitotic spindle of dividing cells is only one example.[1,2] The model for this hypothesis was the frog dermal melanocyte, a cell in which the rapid, reversible dispersion of melanin granules, thought to be accompanied by reversible cytoplasmic structural alterations (gel-to-sol transformations), could be easily controlled and measured.[1,3] Subsequently, the demonstration by many workers of the near-ubiquity of an appropriate structure (the cytoplasmic microtubule), the isolation and ongoing characterization of microtubule and associated proteins from different sources, and the interaction of these proteins with anti-mitotic agents such as colchicine and vinblastine, have produced an explosion of information about apparently microtubule-associated functions in a wide variety of cells, and have made both necessary and useful the synthetic efforts of symposia such as this one. It is a particular honor to contribute to a meeting under the chairmanship of Professor Dustin, whose exceptional book on colchicine[4] was so helpful to my early acquaintance with that unusual alkaloid.

The work to be reviewed here concerns effects of colchicine and vinblastine on the mobilization of granular (lysosomal) enzymes that occurs in human PMN during phagocytosis, and the possible role of microtubules in this process. Since much of this work has recently been summarized in some detail,[5] the present review will be presented in outline form.

2. Mobilization of granular enzymes

a. Morphologic findings. After the phagocytosis of bacteria, colchicine-treated PMN showed evidence in stained specimens, of less lysosomal degranulation and vacuolization (formation of phagolysosomes) than in controls, even though similar numbers of bacteria appeared to have been ingested.[6,7] Similar results were obtained with vinblastine.[7]

b. Enzymic findings. To obtain an independent measure of the degree of degranulation, cells were fractionated after standard incubations, and the intracellular distribution of the classical lysosomal enzyme, acid (β-glycero-) phosphatase, was determined. In cells not given bacteria to ingest, a mean of about 80% of enzyme activity was recovered from the granule fraction. With phagocytosis, and the concomitant dissolution of granular enzymes as lysosomal granules fused with phagosomes, the granule-associated enzyme activity decreased by roughly one-half. In cells pretreated with colchicine, significantly more enzyme activity remained in the granule fraction after phagocytosis than in controls.[6] Once again, similar results were obtained with vinblastine.[7] These measurements support the morphologic findings that drug-treated leukocytes do not degranulate after phagocytosis as much as controls.

Moreover, it was first noted in these studies that the <u>total</u> cellular activity of this rather stable enzyme also decreased during phagocytosis, again significantly less so in drug-treated cells.[6] This finding raised the interesting possibility that granular enzymes were being released from the cells during phagocytosis, where they might contribute to the inflammatory response, and that colchicine and vinblastine might modify that release.

The release of granular enzymes during phagocytosis has subsequently been demonstrated by several investigators.[8-17] The release appears to be specific, in that it occurs from apparently intact, functioning cells.[8,12,15-17] For example, we demonstrated the loss of the granular enzyme, lysozyme, from cells during phagocytosis, and its accumulation in the media, with no concomitant loss of the cytoplasmic enzyme, lactic dehydrogenase.[8] Release of lysozyme and ingestion of bacteria occurred <u>pari passu</u>, and release was a function of both incubation time and dose of bacteria (with significant release after as small a phagocytic challenge as two bacteria per PMN). These findings suggest that the concomitant release of granular enzymes during phagocytosis may be a normal secretory function of PMN that is important for the inflammatory response. As could be expected with the use of an agent that inhibits the mobilization of lysosomes in phagocytizing cells, colchicine pretreatment resulted in the appearance of significantly less enzyme activity in the media after phagocytosis than in controls, and more activity in the cells.[18] These findings may explain in part the anti-inflammatory activity of colchicine in acute gouty arthritis.[2,19]

3. Colchicine-sensitive microtubules.

Our hypothesis has been that microtubules facilitate the getting-together of lysosome and phagosome, and that colchicine and vinblastine disrupt that process by interfering with the organization of microtubules.[2] The evidence for that hypothesis is as follows:

a. The effect of these drugs on the mobilization of granular enzymes during phagocytosis, appeared to be indirect; colchicine had no stabilizing effect on isolated lysosomes from several sources.[20] Subsequently, in human leukocytes treated with radioactive colchicine or vinblastine and then fractionated, most of the bound radioactivity was indeed associated with cytoplasmic protein and not with granules.[21]

b. PMN contain microtubules that are sensitive to colchicine.[22] This point is no longer surprising, but at the time it was made, microtubules had been looked for in PMN and not found. Similar studies with vinblastine also revealed the dissolution of normal microtubules in leukocytes, and led to the chance discovery of vinblastine-induced paracrystalline aggregates of microtubular protein.[23-25] In fractionated leukocytes, the colchicine-binding protein is precipitable by vinblastine.[21]

c. In other cell types, both lysosomes and ingested material are among the structures that undergo long saltatory movements, which are (like microtubules) both generally radially directed and colchicine-sensitive.[26] Such movements might normally facilitate the fusion of these two structures (i.e., lysosomal degranulation).[2] Dissolution of microtubules, and cessation of such movements, would then impede the process of degranulation.

d. Vinblastine has the same metabolic,[6,7] morphologic and enzymic effects on phagocytizing leukocytes as does colchicine (see above). We began using vinblastine on leukocytes after finding, in studies of the mitotic spindle, that this drug shares with colchicine the ability to produce a decrease in the number and length of microtubules and eventual complete dissolution of these organelles.[27] These findings were complemented by the accumulating biochemical work on the cytoplasmic protein that binds (at different sites) colchicine and vinblastine,[28-35] and which polymerizes to form microtubules.

e. While the standard concentration of colchicine or vinblastine ($2.5 \times 10^{-5}$M) that was used to produce functional effects in human leukocytes was far above that required to bind microtubule protein, several other biochemical effects studied in human leukocytes occurred only at much higher concentrations.[21] Thus, although vinblastine inhibited the incorporation of [$^{14}$C] valine into leukocyte protein at concentrations down to 6 X $10^{-6}$M, colchicine did not affect this incorporation until levels of 3 X $10^{-4}$M were attained. Concentrations of these agents as high as 1 X $10^{-4}$M did not inhibit the incorporation of [$^{3}$H] uridine into RNA or of [$^{14}$C] acetate into lipid.

4. The phagocytic stimulus.

Since the extent of mobilization of lysosomes during phagocytosis depends upon the dosage of particles ingested,[8] it is important to show that the suppressive effects of drugs on mobilization do not simply result from a decrease in the uptake of particles. We have determined in three ways that ingestion is similar in drug-treated leukocytes and in controls.

a. The first way was by direct inspection of leukocytes after phagocytosis, which failed to reveal significant differences in the uptake of particles between colchicine- or vinblastine-treated and control groups.[6,7] To produce the observed differences in enzyme release seen after treatment with colchicine, phagocytosis would have had to have been decreased by 50% or more.[8,19]

b. The second method involved the recovery of live bacteria from media and cells after various intervals of incubation.[6,7] Cells treated with either colchicine or vinblastine were able to take up staphylococci from the media as well as controls, and, despite the evidence of significantly less degranulation than controls, were also able to kill equally well the bacteria that were taken up (as measured by the small recovery of live, cell-associated bacteria). Since killing of staphylococci is thought to be primarily an intracellular event, and since drug-treated bacteria, incubated without cells, grew as well as controls, we concluded that ingestion was indeed equal with or without drug.

c. Finally, we have employed a method designed to distinguish particles adherent to cells, but not ingested, from those that are intracellular; it involves the concomitant uptake from the media of radiolabelled ($^{131}$I) human serum albumin (HSA).[36,37] The $^{131}$I-HSA does not accumulate in resting cells, and cell-associated radioactivity provides a very sensitive time- and dosage-dependent measure of ingestion.[37]

When we used colchicine in this system, we found,[5] as others have,[38,39] that with a given concentration and time of incubation, colchicine may inhibit the uptake of one particle (e.g., starch) but not of another (e.g., live staphylococci). In addition, cells and sera from different donors appear to contribute to variability in results. Therefore, we

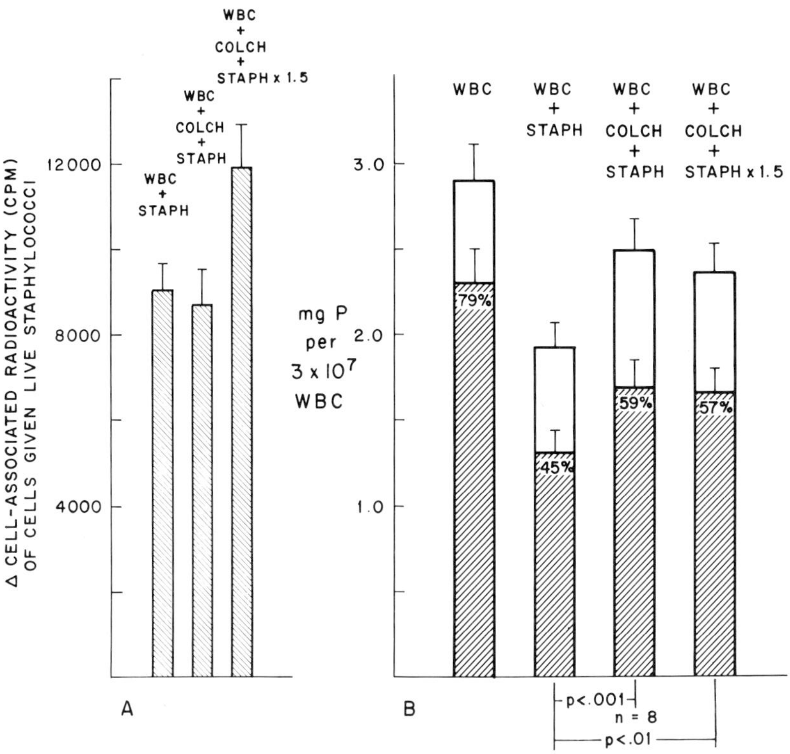


Figure 1. The effects of colchicine on phagocytosis and on lysosomal mobilization under conditions employing comparable cells, sera, particles, and incubations. (A) In each of the 8 experiments shown, there was no significant difference in cell-associated radioactivity between cells preincubated for 30 minutes with or without colchicine, and then given the same number of live staphylococci for an additional 40 minutes ($n=4$, $p>.05$). Ratios of bacteria:PMN varied from 9 to 23:1 (median 14:1). However, with colchicine-treated cells given half again as many live staphylococci ("STAPH X 1.5"; ratios 14 to 35:1, median 21:1), each had significantly higher counts than the same controls ($n=4$, $p<.05$). Standard errors of the means are indicated ($n=32$). (B) In each of the same 8 experiments, 4 additional sets of triplicate flasks were incubated in a similar fashion, and the triplicates were combined, assayed in duplicate for total and for granule-associated acid ($\beta$-glycero-) phosphatase activity, and the duplicate values averaged. ("Granule-associated" [hatched bars] in this series includes all elements sedimented at 8,200 X g for 20 minutes, after the cells, suspended in 0.34 M sucrose, had been disrupted by Teflon homogenization in ice. The activity in the cell supernatant fraction was also determined [clear bars]. Their sum is the total cellular activity.) The means of the 4 sets of 8 values for total and for granule-associated activities are shown, with their standard errors. The percentages are of the total activity of resting cells (i.e., of "WBC"). Colchicine-treated cells degranulated significantly less than controls during phagocytosis, even when they were stimulated by 50% more bacteria ("STAPH X 1.5") than the controls. Colchicine had no effect on resting cells (not shown). The p values (calculated from the t-test for paired variables) apply to differences both in granule-associated activity and in total activity. (From Ann. N.Y. Acad. Sci.[5] By permission of the publisher.)

measured total and granule-associated acid phosphatase activity in experiments in which the measurements of ingestion were comparable with and without colchicine (Figure 1A).[5] The overall uptakes in the two groups were within 3% of each other. Moreover, we also studied super-stimulation, by giving half again as many bacteria to some samples of colchicine-treated cells. They showed an increase in uptake of 32% (Figure 1A) (and highly significant increases in cell-associated staphylococci in microscopic counts). As before, colchicine-treated cells degranulated less than controls during phagocytosis. Moreover, even the super-stimulated colchicine-treated cells degranulated less than the normally stimulated controls (Figure 1B). Super-stimulated controls (not shown in Figure 1) degranulate more than normally stimulated controls.

Thus, in our system the drug effect on the mobilization of lysosomes does not appear to be attributable to an altered phagocytic stimulus.

This work was supported in part by grants from the United States Public Health Service (AM-10493, AM-5639, AI-271) and the Arthritis Foundation. I am grateful to Mrs. Wanda D. Prinz for expert preparation of the manuscript.

References


1. Malawista, S.E., 1965. On the action of colchicine: the melanocyte model. J. Exp. Med. 122, 361.
2. Malawista, S.E., 1968. Colchicine: a common mechanism for its anti-inflammatory and anti-mitotic effects. Arthritis Rheum. 11, 191.
3. Malawista, S.E., 1975. Microtubules and the movement of melanin granules in frog dermal melanocytes. Ann. N.Y. Acad. Sci. 253, 702.
4. Eigsti, O.J. and Dustin, P., Jr., 1955. Colchicine (Iowa State College Press, Ames, Iowa).
5. Malawista, S.E., 1975. Microtubules and the mobilization of lysosomes in phagocytizing human leukocytes. Ann. N.Y. Acad. Sci. 253, 738.
6. Malawista, S.E. and Bodel, P.T., 1967. The dissociation by colchicine of phagocytosis from increased oxygen consumption in human leukocytes. J. Clin. Invest. 46, 786.
7. Malawista, S.E., 1971. Vinblastine: colchicine-like effects on human blood leukocytes during phagocytosis. Blood 37, 519.
8. Wright, D.G. and Malawista, S.E., 1972. The mobilization and extracellular release of granular enzymes from human leukocytes during phagocytosis. J. Cell Biol. 53, 788.
9. Martin, R.R. et al., 1967. Human reactions to staphylococcal antigens: a possible role for leukocyte lysosomal enzymes. J. Immunol. 99, 269.
10. Pruzanski, J.J. and Patterson, R., 1967. Subcellular distribution of histamine in human leukocytes. Proc. Soc. Exp. Biol. Med. 124, 56.

11. Baehner, R.I. et al., 1969. Degranulation of leukocytes in chronic granulomatous disease. J. Clin. Invest. 48, 187.

12. Crowder, J.G. et al., 1969. Release of histamine and lysosomal enzymes by human leukocytes during phagocytosis of staphylococci. J. Lab. Clin. Med. 74, 436.

13. Tew, J.G. et al., 1969. Lysozyme and β-lysin release stimulated by antigen antibody complexes and bacteria. J. Immunol. 102, 743.

14. May, C.D. et al., 1970. The effects of compounds that inhibit antigenic release of histamine and phagocytic release of lysosomal enzymes on glucose utilization by leukocytes in humans. Proc. Soc. Exp. Biol. Med. 133, 758.

15. Weissmann, G. et al., 1971. Effect of cyclic AMP on release of lysosomal enzymes from phagocytes. Nature New Biol. 231, 131.

16. Henson, P.M., 1971. Interactions of cells with immune complexes: adherence, release of constituents, and tissue injury. J. Exp. Med. (Suppl.) 134, 114.

17. Hawkins, D. and Peeters, S., 1971. The response of polymorphonuclear leukocytes to immune complexes in vitro. Lab. Invest. 24, 483.

18. Wright, D.G. and Malawista, S.E., 1973. Mobilization and extracellular release of granular enzymes from human leukocytes during phagocytosis. Inhibition by colchicine and cortisol but not by salicylate. Arthritis Rheum. 16, 749.

19. Malawista, S.E., 1975. The mechanism of action of colchicine in acute gouty arthritis. In: The 1974 Conference on Gout and Purine Metabolism. J.R. Klinenberg, ed. Arthritis Rheum. (in press).

20. Weissmann, G., 1966. Lysosomes and joint disease. Arthritis Rheum. 9, 834.

21. Creasey, W.A., Bensch, K.G. and Malawista, S.E., 1971. Colchicine, vinblastine and griseofulvin: pharmacological studies with human leukocytes. Biochem. Pharmacol. 20, 1579.

22. Malawista, S.E. and Bensch, K.G., 1967. Human polymorphonuclear leukocytes: demonstration of microtubules and effect of colchicine. Science 156, 521.

23. Bensch, K.G. and Malawista, S.E., 1968. Microtubule crystals: a new biophysical phenomenon induced by Vinca alkaloids. Nature 218, 1176.

24. Bensch, K.G. and Malawista, S.E., 1969. Microtubular crystals in mammalian cells. J. Cell Biol. 40, 95.

25. Malawista, S.E. and Sato, H., 1969. Vinblastine produces uniaxial, birefringent crystals in starfish oöcytes. J. Cell Biol. 42, 596.

26. Freed, J.J. and Lebowitz, N.M., 1970. The association of a class of saltatory movements with microtubules in cultured cells. J. Cell Biol. 45, 334.

27. Malawista, S.E. et al., 1968. Vinblastine and griseofulvin reversibly disrupt the living mitotic spindle. Science 160, 770.

28. Borisy, G.G. and Taylor, E.W., 1967. The mechanism of action of colchicine. Binding of colchicine-$H^3$ to cellular protein. J. Cell Biol. 34, 525.

29. Borisy, G.G. and Taylor, E.W., 1967. The mechanism of action of colchicine. Colchicine binding to sea urchin eggs and the mitotic apparatus. J. Cell Biol. 34, 535.

30. Shelanski, M.L. and Taylor, E.W., 1967. Isolation of a protein subunit from microtubules. J. Cell Biol. 34, 549.

31. Wilson, L. and Friedkin, M., 1967. The biochemical events of mitosis. I. The in vivo and in vitro binding of colchicine in grasshopper embryos and its possible relation to inhibition of mitosis. Biochemistry 6, 3126.

32. Creasey, W.A., 1967. The binding of antimitotic agents by cell-free extracts of tumor cells. Pharmacologist 9, 192.

33. Weisenberg, R.C. et al., 1968. The colchicine-binding protein of mammalian brain and its relation to microtubules. Biochemistry 7, 4466.

34. Bensch, K.G. et al., 1969. Induction in vitro of microtubular crystals by Vinca alkaloids. Science 165, 495.

35. Marantz, H. et al., 1969. Vinblastine-induced precipitation of microtubule protein. Science 165, 498.

36. Chang, Y., 1969. Studies on phagocytosis I. Uptake of radio-iodinated ($^{131}I$) human serum albumin as a measure of the degree of phagocytosis in vitro. Exp. Cell Res. 54, 42.

37. Cannarozzi, N.A. and Malawista, S.E., 1973. Phagocytosis by human blood leukocytes measured by the uptake of $^{131}I$-labeled human serum albumin: inhibitory and stimulatory effects of cytochalasin B. Yale J. Biol. Med. 46, 177.

38. Chang, Y., 1968. The effect of anti-inflammatory drugs on phagocytosis in vitro. Arthritis Rheum. 11, 473.

39. Lehrer, R.I., 1973. Effects of colchicine and chloramphenicol on the oxidative metabolism and phagocytic activity of human neutrophils. J. Infect. Dis. 127, 40.

*Microtubules and Microtubule Inhibitors, eds. M. Borgers and M. de Brabander*



# THE EFFECT OF MICROTUBULE INHIBITORS ON DEGRADATION OF SERUM LIPOPROTEINS

T. Chajek, O. Stein, and Y. Stein

Lipid Research Laboratory, Department of Medicine B, Hadassah University Hospital, and Department of Experimental Medicine and Cancer Research, Hebrew University-Hadassah Medical School, Jerusalem, Israel

Summary

The release of plasma lipoprotein lipase by heparin was studied in fed and food-deprived rats pretreated with colchicine and vinblastine. Four hours after the administration of either drug the lipoprotein lipase activity released by heparin was only half of that found in controls. Colchicine affected the release of both protamine-sensitive and protamine-resistant lipoprotein lipase. The effect of the antimicrotubular agents on the release of the protamine sensitive lipoprotein lipase was studied in the perfused rat heart. Administration of colchicine or vinblastine 4 h prior to perfusion of the heart caused a very marked reduction in lipoprotein lipase activity released into the perfusate within 1 min of heparin perfusion. At the same time an increase in residual heart lipase occurred so that the total lipoprotein lipase content of the heart (heparin releasable plus residual) did not change. The colchicine effect was dose and time dependent; no decrease in heparin-releasable enzyme activity occurred after only 30 min of pretreatment or upon addition of colchicine into the perfusate. These results indicate that colchicine did not impede enzyme synthesis or its release from the cell surface, but may have interfered with the transport of lipoprotein lipase from the site of its synthesis to the endothelial cell surface.

## Introduction

Very low density lipoproteins and chylomicrons, which are the main carrier of serum triglyceride are catabolized in two main areas, in the capillary bed and in the liver. The main enzyme responsible for the catabolism of the triglyceride is a lipase, which requires a certain serum cofactor for its activity and is known as lipoprotein lipase. The first stage of catabolism occurs in the vascular system and the evidence, which has accumulated to date, indicates that the hydrolysis of the triglyceride occurs on the surface of the endothelial cell. This evidence is based on the finding that chylomicrons can be shown to adhere to endothelial surfaces. Moreover, degradation of adhering particles has been demonstrated after incubation of various organs under suitable conditions (1, 2). Further evidence for the surface location of lipoprotein lipase is its extremely rapid availability for release by heparin, as the enzyme activity reaches its maximal activity within several minutes of heparin injection (3). However, a second injection of heparin administered within an hour of the first will elicit a much lower release. If the interval between the two injections is increased to 4 hours, the response to the second injection will be the same as to the first (4). Thus it seems that the readily releasable enzyme can be easily depleted and that time is needed for its repletion. Such a time lag could indicate that the enzyme has to be transported from the site of its synthesis to the site of its action. Since the release of lipoprotein lipase is reminiscent of an endocrine secretory process it seemed of interest to determine whether it will be affected by agents such as colchicine and vinblastine, which during the recent years have been shown to affect release of hormones (5), glycoproteins (6), lipoproteins and proteins into the circulation (7, 8).

## Results

The lipoprotein lipase activity was determined 10 min after injection of heparin. When colchicine was injected 4 h prior to the injection of heparin the lipoprotein lipase activity was 32% of the control values. The lower activity of lipoprotein lipase was apparently not due to the presence of colchicine in the circulation as no difference in enzymic activity was detected if colchicine was administered 40 min prior to the termination of the experiment. Addition of 30 and 60 $\mu$g colchicine to 1.0 ml of the assay medium did not interfere with the detection of enzymic activity. Injection of vinblastine at an equimolar dose to that of colchicine produced the same result. Since colchicine treatment could have affected the food intake of the rats and hence the activity of lipoprotein lipase, the determinations were repeated in rats deprived of food for the duration of the experiment, between time 0 and 270 min. As seen in Table II, the effect of colchicine was the same in food deprived rats as in rats fed ad libitum (Table I). Since post heparin lipoprotein lipase of plasma is derived most probably from both the liver and from extrahepatic tissues (9), an attempt was made to determine whether colchicine does affect enzyme activity derived from the different sources. One way to distinguish between the hepatic and extrahepatic enzyme is by their different resistance to protamine inhibition (9) and therefore the enzyme assay was run both in the presence and absence of protamine. The lipoprotein lipase in plasma of colchicine-treated rats was significantly lower than in controls, both in the absence and in the presence of protamine and the decrease was more pronounced when the enzymic activity was measured in the absence of protamine. The hepatic and extrahepatic enzyme activity was computed according to Krauss et al. (9) and the results indicated that the depression of the extrahepatic enzyme by colchicine was somewhat more marked than that of the hepatic enzyme.

TABLE I

Effect of colchicine and vinblastine on the release of plasma lipoprotein lipase by heparin

Rats fed ad libitum were injected intraperitoneally with colchicine, 0.5 mg/100 g by wt or with vinblastine 1 mg/100 g by wt. Control animals were injected with 0.9% NaCl. Heparin 10 u/100 g by wt was injected intravenously. Post heparin plasma was obtained 10 min after the injection of heparin. Values are means ±S.E. Figures in parentheses represent number of rats. a vs c, d, $p < 0.01$.

| Treatment | Fatty acid released ($\mu$moles/ml plasma/h) | Percent of control activity |
|---|---|---|
| Heparin at 0 min | 10.3 ± 1.2 (19)[a] | 100 |
| Colchicine at 0 min<br>Heparin at 30 min | 11.1 ± 0.2 (3)[b] | 100 |
| Colchicine at 0 min<br>Heparin at 270 min | 3.4 ± 0.5 (3)[c] | 32 |
| Vinblastine at 0 min<br>Heparin at 270 min | 5.3 ± 0.7 (3)[d] | 51 |

Adapted from Chajek et al. (10).

TABLE II

Effect of colchicine on the release of plasma lipoprotein lipase by heparin in fasted rats

Rats were deprived of food for the duration of the experiment, 0-270 min (between 8 am and 12.30 pm). Colchicine, 0.5 mg/100 g by wt was administered intraperitoneally. Control animals were injected with 0.9% NaCl. Heparin 10 units/100 g by wt was injected intravenously. Post heparin plasma was obtained 10 min after the injection of heparin. Values are means ±S.E. Figures in parentheses represent number of rats. a vs. b. $p < 0.001$.

| Treatment | Fatty acid released ($\mu$moles/ml plasma/h) | Percent of control activity |
|---|---|---|
| Heparin at 270 min | 10.9 ± 0.3 (4)[a] | 100 |
| Colchicine at 0 min<br>Heparin at 270 min | 4.0 ± 0.35 (4)[b] | 37 |

Adapted from Chajek et al. (10).

The heart is known as a rich source of extrahepatic lipoprotein lipase (11) which can be readily released into the perfusate by heparin in an in vitro system (4, 11). Hence the effect of colchicine on peripheral lipoprotein lipase was investigated in an isolated perfused rat heart. This system enables one to determine whether the fall in heparin-releasable lipase caused by colchicine treatment is due only to the inhibition of its transport to the endothelial cell surface, or might have also been affected by inhibition of enzyme synthesis.

The results presented in Table III show that the release of lipoprotein lipase occurs almost immediately after introduction of heparin and that between 0 - 1 min a sharp peak of enzyme activity appears in the perfusion medium. The rate of lipoprotein lipase release decreases sharply during the subsequent time intervals studied. Thus, between 1 - 3 min of perfusion, the activity releasable per min decreases by more than 80% and during the next 2 min less than 10% of the initially releasable activity appeared in the perfusate per min. Pretreatment of the rats with 0.5 mg/100 g body weight of colchicine 4 h prior to perfusion resulted in a very sharp decrease ($p < 0.001$) in heparin-releasable activity during the first min of perfusion. The enzyme activity which appeared in the perfusate during the subsequent two periods of perfusion (1 - 3 min and 3 - 5 min) did not differ significantly from that in the control group (Table III). As in the colchicine-treated rat more lipoprotein lipase was found in the heart than in the saline injected rat ($p < 0.001$), it is evident that the total enzyme activity (which is the sum total of perfusate and homogenate activities) was the same in the experimental and control groups. A similar degree of decrease of the rapidly releasable enzyme activity was found also after pretreatment of the rats with vinblastine, given at an equimolar concentration to that of colchicine. In order to learn more about the effect of colchicine in this system, the drug was added to the perfusate at a concentration which might have been reached after injection of colchicine in vivo, assuming a homogeneous distribution in body water. As seen in Table IV the presence of $2.5 \cdot 10^{-5}$ M colchicine in the perfusate did not lower the enzyme release by heparin and did not affect the enzyme activity recovered in the heart homogenate. The time course of colchicine action is shown also in Table IV. No effect was seen after 30 min of pretreatment and a smaller, but statistically significant fall ($p < 0.02$) in the rapidly releasable enzyme peak was observed 2 h after pretreatment. Pretreatment with colchicine for 4 h resulted in a very marked decrease in the lipoprotein lipase activity releasable during the first min of perfusion. The comparison of the activity recovered in the heart at the end of perfusion shows that in rats pretreated with colchicine for 2 or 4 h more enzyme activity ($p < 0.001$) was recovered in the heart homogenate than in the hearts of the 0.9% NaCl-injected rats.

TABLE III

Effect of colchicine and vinblastine on the heparin-releasable lipoprotein lipase in perfusates of rat heart

All rats were deprived of food at zero time and were injected with the indicated materials; hearts were removed and perfused 4 h after the injection. Figures in parentheses represent number of experiments; values are means ±S. E., fatty acid released, $\mu$mol/g tissue/h

| Treatment | Perfusate* (min after heparin injection) | | | Heart at end of perfusion | Total enzyme activity |
|---|---|---|---|---|---|
| | 0 - 1 | 1 - 3 | 3 - 5 | | |
| 0.9% NaCl (15) | 28.7 ± 1.9 | 10.1 ± 1.4 | 5.1 ± 0.9 | 56.7 ± 5.5 | 100.6 ± 5.1 |
| Colchicine 0.5 mg** (5) | 5.0 ± 0.6 | 9.1 ± 1.5 | 5.5 ± 1.6 | 85.7 ± 6.3 | 105.3 ± 6.4 |
| Vinblastine 1.0 mg** (4) | 9.5 ± 1.7 | 11.5 ± 2.0 | 4.3 ± 0.9 | 85.1 ± 6.0 | 110.6 ± 4.2 |

* The activity in the perfusate was expressed per g heart weight.

** Per 100 g body weight.

Adapted from Chajek et al. (12)

TABLE IV

Heparin-releasable lipoprotein lipase in perfusates of rat heart at various time intervals after pretreatment with colchicine

All rats were deprived of food throughout the experimental period. Colchicine in perfusate was $2.5 \cdot 10^{-5}$ M, the dose of colchicine in vivo was 0.5 mg/100 g body weight. Figures in parentheses represent number of experiments; values are means ±S. E., fatty acid released, $\mu$mole/g tissue/h.

| Treatment | Perfusate* (min after heparin injection) | | | Heart at end of perfusion | Total enzyme activity |
|---|---|---|---|---|---|
| | 0 - 1 | 1 - 3 | 3 - 5 | | |
| 0.9% NaCl (6) | 31.7 ± 0.5 | 14.3 ± 3.1 | 4.6 ± 0.9 | 52.6 ± 3.8 | 102.5 ± 6.7 |
| Colchicine added to perfusate (4) | 37.9 ± 2.5 | 14.7 ± 1.5 | 3.7 ± 0.2 | 56.5 ± 6.7 | 111.7 ± 8.2 |
| Colchicine in vivo | | | | | |
| 30 min (3) | 33.4 ± 2.7 | 11.5 ± 2.7 | 4.0 ± 1.3 | 72.1 ± 4.8 | 120.1 ± 3.3 |
| 120 min (3) | 21.9 ± 2.9 | 11.3 ± 0.3 | 7.0 ± 1.5 | 88.3 ± 3.3 | 128.6 ± 4.7 |
| 240 min (3) | 4.8 ± 0.6 | 9.7 ± 3 | - | 99.3 ± 6.7 | 113.9 ± 5.6 |

* The activity in the perfusate was expressed per g heart weight.

From Chajek et al. (12).

To define more closely the lipolytic activity measured, assays were performed also on samples which had been preincubated with protamine or 1 M NaCl. Protamine and high NaCl concentration inhibited the activity of the enzyme released into the perfusate, as well as that determined in homogenates of hearts at the end of perfusion. No enzyme activity in either perfusate or heart could be measured in the absence of serum in the assay system.

## Discussion

The most pronounced effect of colchicine was on the enzymic activity released into the perfusate during the first min of heparin perfusion of the rat heart. Antimicrotubular agents, colchicine and vinblastine have been shown to arrest processes which involve cellular transport of various products in secretory vesicles or vacuoles (5 - 8). Since impedance with secretion of proteins could also be due to the interference with their synthesis, it became important to determine whether under our experimental conditions (estimated colchicine concentration $2 \cdot 10^{-5}$ M) colchicine does affect the synthesis of the lipoprotein lipase. Since the total lipoprotein lipase activity in hearts of colchicine-treated rats was not lower than that of the controls, it seems that in the intact rat, colchicine did not affect enzyme production, in analogy to the synthesis of serum lipoproteins and serum albumin (13, 14).

Direct interference with the release of the enzyme by heparin was ruled out by the experiments in which colchicine was present in the perfusate. These results as well as the lack of effect of colchicine after 30 min of pretreatment do not support the possibility that binding of colchicine to plasma membrane is involved in the fall of heparin-releasable enzyme activity. The increase of the colchicine effect with the time of pretreatment indicates that the drug has influenced the renewal of the heparin-releasable enzyme, which is apparently localized on the endothelial cell surface.

The mechanism of lipoprotein lipase renewal on the cell surface is not known, but if the enzyme might be considered as a "secretory" enzyme its transport would involve Golgi derived secretory vesicles. These vesicles would also participate in the transport of the enzyme, if the latter is a part of the so-called cell coat or membrane glycoproteins. Colchicine could affect the transport of secretory vesicles by its known action on the microtubular system, but it could also interfere with the fusion process between the vesicles and the plasma membrane.

Acknowledgments

The excellent technical help of Mrs. Y. Dabach and Mr. G. Hollander is gratefully acknowledged. This study was supported in part by a grant from the Joint Research Foundation of the Hebrew University and Hadassah for Dr. T. Chajek, and by a grant from the Delegation Generale a la Recherche Scientifique et Technique of the French Government. Dr. Y. Stein is an Established Investigator of the Ministry of Health.

References

1. Schoefl, G.I. and French, J.E. 1968. Proc. Roy. Soc. Ser. B, Biol. Sci. 169, 153.
2. Blanchette-Mackie, E.J. and Scow, R.O. 1971. J. Cell Biol. 51, 1.
3. Robinson, D.S. 1970, in: Comprehensive Biochemistry, vol. 18, Lipid Metabolism eds. M. Florkin and E.H. Stotz (Elsevier Publishing Co.). pp. 51 - 105.
4. Aktin, E. and Meng, H.C. 1972. Diabetes, 21, 149.
5. Lacy, P.E., Howell, S.L., Young, D.A. and Fink, C.J. 1968. Nature 219, 1177.
6. Rossignol, B., Herman, G. and Keryer, G. 1972. FEBS Lett. 21, 189.
7. Stein, O. and Stein, Y. 1973. Biochim. Biophys. Acta 306, 142.
8. Orci, L., LeMarchand, Y., Singh, A., Assimacopollos-Jeannet, F., Rouiller, Ch. and Jeanrenaud, B. 1973. Nature 244, 30.
9. Krauss, R.M., Windmueller, H.G., Levy, R.I. and Fredrickson, D.S. 1973. J. Lipid Res. 14, 286.
10. Chajek, T., Stein, O. and Stein, Y. 1975. Biochim. Biophys. Acta 380, 127.
11. Robinson, D.S. and Jennings, M.A. 1965. J. Lipid Res. 6, 222.
12. Chajek, T., Stein, O. and Stein, Y. 1975. Biochim. Biophys. Acta 388, 260.
13. Stein, O., Sanger, L. and Stein, Y. 1974. J. Cell Biol. 62, 90.
14. Redman, C.M. 1973. Colchicine and vinblastine inhibit the secretion of albumin by rat liver slices. Ninth International Congress of Biochemistry, Stockholm. 259.

*Microtubules and Microtubule Inhibitors, eds. M. Borgers and M. de Brabander*



# MICROTUBULE DISTRIBUTION AND MELANOSOME MOVEMENTS IN FISH MELANOPHORES

M. Schliwa

Arbeitsgruppe Kinematische Zellforschung,
Universität Frankfurt am Main

Summary. Recent results in the study of pigment movements within melanophores of the angelfish, Pterophyllum scalare, are presented. Special emphasis is laid upon changes in cell shape and the distribution of microtubules. Melanophores with pigment dispersed are disc-shaped cells with a prominent microtubular system emanating from the cytocentrum, where they are associated with dense cytoplasmic material. In the cell processes, a cortical arrangement parallel to the long axis of the cell processes prevails. During pigment aggregation, the cell shape changes from a disc to a sphere with collapsed peripheral margins. Serial sections spaced at regular intervals of partly dispersed melanophores indicate an abrupt decrease in the number of microtubules beyond the margins of the pigment. It is concluded that melanosome movements depend on the assembly of microtubules during dispersion and their disassembly during aggregation. According to the regularity of paraxial microtubular associations in the cortex of cell processes with dispersed pigment, it is possible that lateral interactions between these microtubules contribute to the stabilization of the tubular shape of the cell processes as well as the flat, discoidal shape of the whole cell. As to the mechanism of melanosome transport, it is proposed that movement of granules is a result of the interaction between assembly/disassembly of microtubules and a contractile component of the cytoplasm.

In recent years, microtubules have been recognized as elements that play an important role in the development and maintenance of cell shape[1,2] as well as in the movement of cytoplasmic particulates[3-5]. Despite of considerable efforts which have been made to elucidate their participation in the organization of cytoplasm, it is still not clear whether microtubules only provide pathways for particle movement or whether they assist in the movement as well.

Since the demonstration of microtubules within fish melanophores[6], both these possibilities have been considered to be essential for the movement of pigment granules. In melanophores of Fundulus heteroclitus, microtubules were observed to form channels through which the movement of melanosomes is more or less directed[6,7]. Experiments with colchicine, vinblastine, low temperature and high hydrostatic pressure confirmed that microtubules are required for both centrifugal and centripetal migrations[8,9]. However, the precise nature of the involvement of microtubules in pigment movements is still open to further discussion.

## Cell shape changes

Melanophores of the angelfish, Pterophyllum scalare, which were used for the investigations described here, are commonly found just under the epidermis and are usually surrounded by collagenous tissue. In the black body regions of the fish, melanophores lie close together in a plane parallel to the body surface. Cell processes radiate from the cell center in such a way that the melanophore has a circular outline (Fig. 1). Their overall diameter varies between 100 and 140 µm.

Vertical sections of melanophores with completely dispersed pigment demonstrate that they are of relatively constant thickness (Figs. 2 and 4). Apart from the narrow intercellular spaces separating neighbouring processes (Fig. 1), melanophores with pigment dispersed are of discoidal shape with a vertical diameter ranging from 1 µm near the periphery to about 3 µm in the central region.

---

All scale bars represent 0,1 µm unless otherwise noted.

Fig. 1. Horizontal section of a melanophore with dispersed pigment. Cell processes radiate from the cell center. All processes are of approximately the same length; as a consequence, the outline of the cell is roughly circular. Scale bar = 1 µm. Inset: Intercellular space between neighbouring cell processes near the basis.

Fig. 2 . Survey electron micrograph of a vertical section through a melanophore cut somewhat off the center. Thickness of the cell varies between 2 and 3 µm. Indented nucleus (N) in the center of this micrograph. Scale bar = 1 µm.

Fig. 3. Vertical section through the central region of a melanophore with pigment nearly completely aggregated. Together with a nucleus (N), the melanosomes are compactly gathered in the cell center, while the cell processes appear collapsed (arrows). Thickness of the central region about 10 µm. Scale bar = 1 µm.

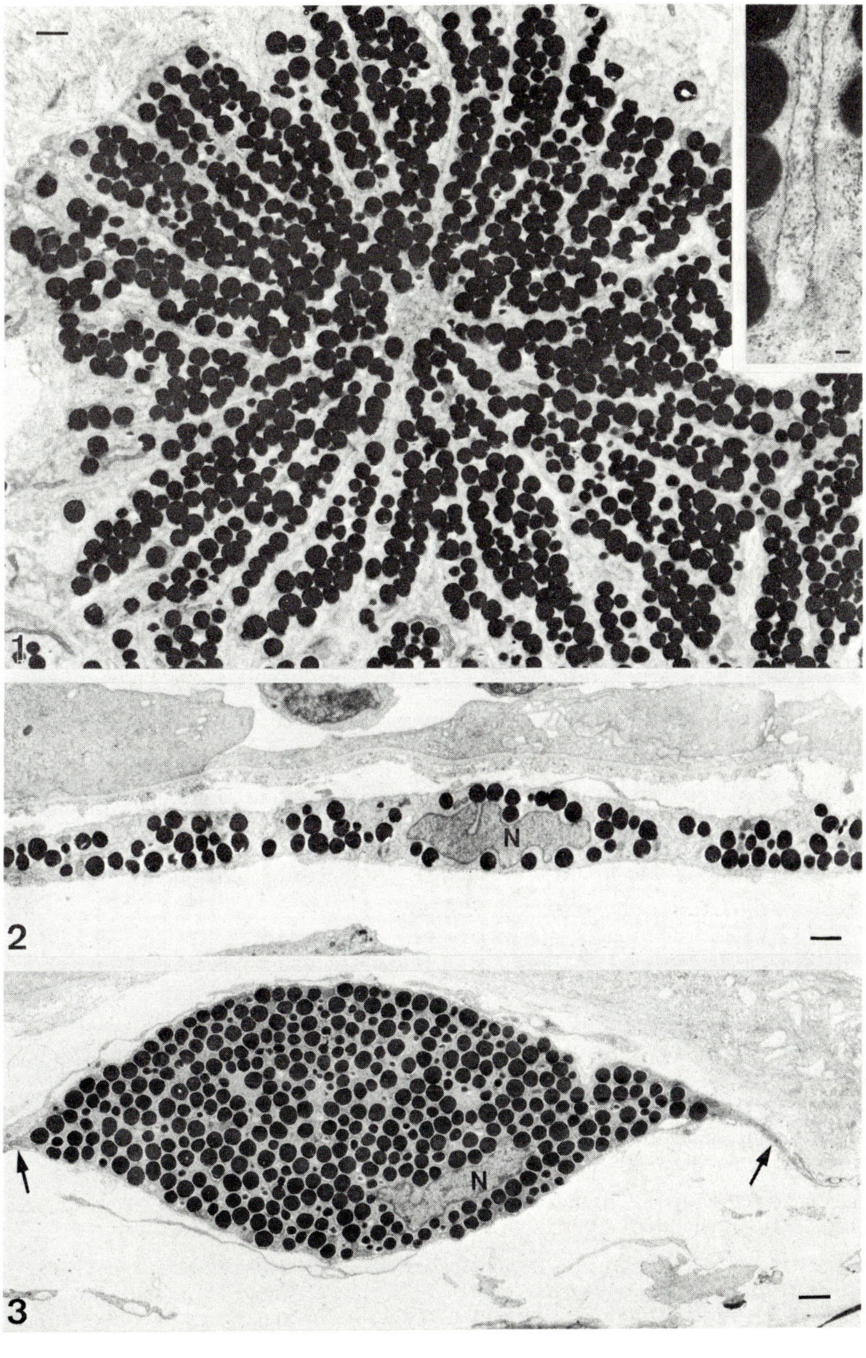
1
2
N
3
N

When melanophores have been stimulated to aggregate their pigment, the appearance of these cells contrasts with the structure of dispersed melanophores (Fig. 3). Melanosomes fill the centrosphere, whose thickness thereby increases up to 14 μm, while the cell processes collapse and flatten. Nevertheless, the overall cellular volume does not appear to change, as judged from cell profiles in vertical section. There is no apparent indication of retrograde cytoplasmic flow during melanosome aggregation. Since cell shape changes were frequently noted in melanophores from different species of fish[10-14] as well as amphibians[15,16], it is reasonable to conclude that they are a characteristic accompaniment of pigment movements.

## Microtubule distribution

The striking changes in melanophore morphology during pigment movements are of great importance for an understanding of the role of microtubules. Exact knowledge of the arrangement of microtubules is, therefore, a prerequisite for a consideration of their function.

Microtubules are oriented radially relative to the cytocentrum. Within this central portion of the melanophore, however, microtubules are randomly distributed (Fig. 5) and show a close association with diffuse electron-opaque cytoplasmic densities, from which they appear to arise (Figs. 5 and 6). In the aggregated as well as the dispersed state, there is a conspicuous lack of structural relationship to centrioles (Fig. 7), which may be separated

---

Fig. 4. Vertical section of a cell process cut parallel to its long axis. The cell process is outlined to show the relatively constant vertical diameter (about 1,5 μm).

Fig. 5. Horizontal section of part of the central region near the nucleus. Electron-opaque cytoplasmic densities (arrows) are associated with the microtubules.

Fig. 6. Vertical section of a central region similar to that seen in Fig. 5. Microtubules appear to converge upon dense material (arrow). Apparent cross-bridging between microtubules is indicated by an arrowhead.

Fig. 7. Centrioles (C) of melanophores. a. Dispersed state. b. Aggregated state. No microtubules appear in association with centrioles.

Fig. 8. Vertical section of a melanophore with pigment dispersed cut near the basis of a cell process (periphery of the cell center). Circular cross sections of microtubules are evenly distributed throughout the cross section area.

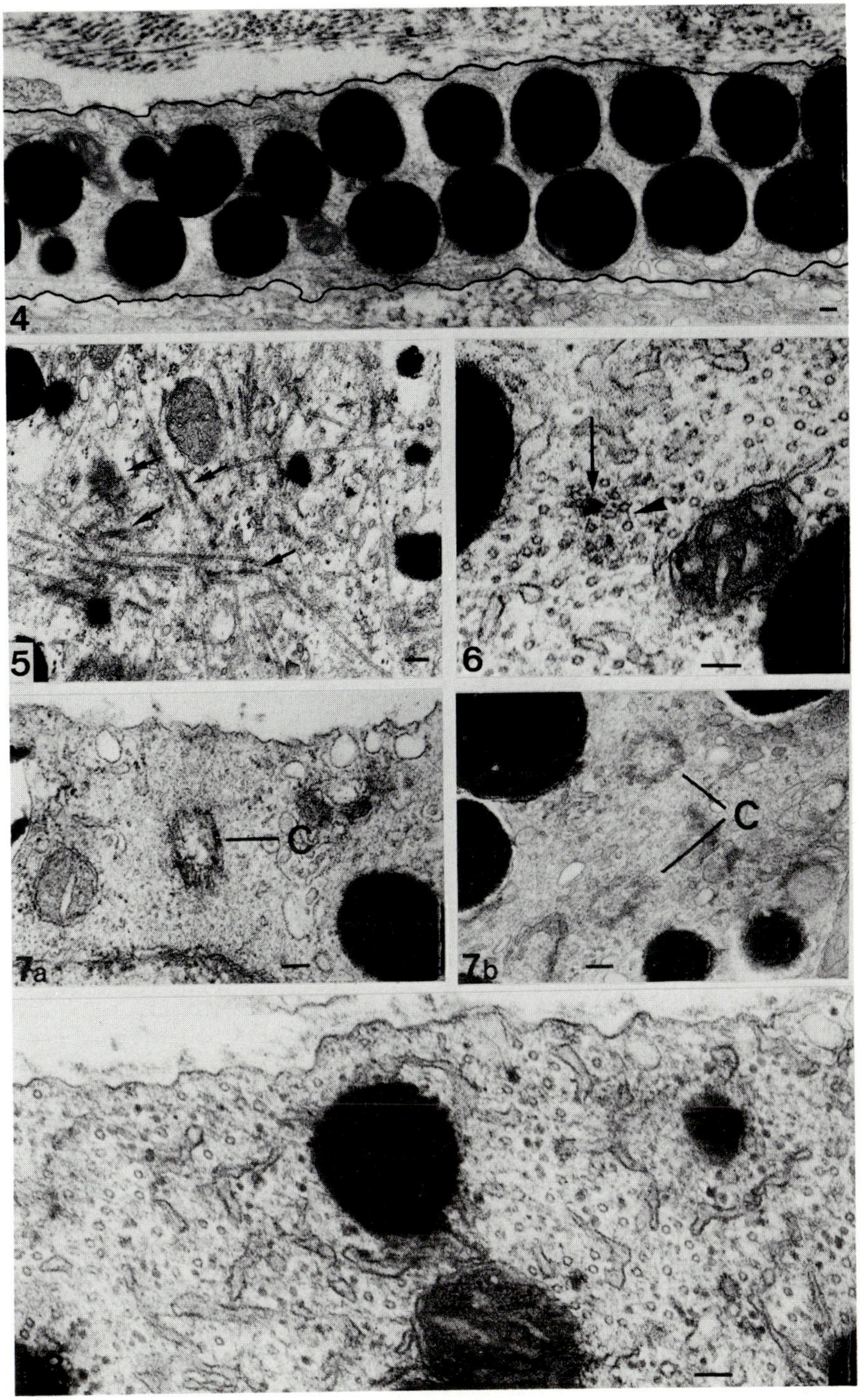
4
5
6
C
7a
C
7b

from the cytocentrum by more than 5 µm. Microtubules cluster around the cytoplasmic densities and in their vicinity reach a locally high density.(Fig. 6).In vertical sections cut somewhat off this center near the bases of the cell processes, those microtubules arising from the cytocentrum have apparently spread out and are distributed evenly throughout the cross section area (Fig. 8). In the cell processes, microtubules are longitudinally aligned (Fig. 9) and are arranged predominantly in the cortical zone of cytoplasm (Figs. 10 and 11). In these regions, relatively constant surface-to-surface distances are observed (about 30 nm). This arrangement commonly prevails along the entire length of a cell process. Near its tapering end, microtubules are still found around the periphery, but their paraxial association may be less extensive. In any case, there is an apparant lack of association between microtubules and melanosomes.

Upon removal of melanosomes from the cell processes and their accumulation in the cell center during aggregation, the cell processes collapse and flatten, and fibrous cytoplasmic material can be observed distal to the migrating pigment (Fig. 12).At the same time, the vertical diameter of the cell center increases essentially (Fig. 3). Fig. 13 shows a series of electron micrographs of a cell process cut paralell to its long axis. At the basis of this process, several mitochondria have accumulated near the pigment mass (Fig. 13 a). Microtubules are seen near the plasma membrane. In the mid-region (Fig. 13 b), microtubules are evenly distributed throughout the cell process, while at the end (Fig. 13 c), no microtubules can be recognized. In this most peripheral zone the cell process is extremely collapsed (vertical diameter about 0, 1 - 0, 2 µm) and exhibits a condensed cytoplasmic matrix of moderate electron density. Cross sections of cell processes corresponding to the

---

Fig. 9. A microtubule (arrows) following the course of the cell membrane of a longitudinally cut cell process.

Fig. 10. Cross section of part of a cell process with dispersed pigment. Microtubules are almost exclusively found near the periphery of the process.

Fig. 11. Higher magnification of a cell process similar to that of Fig. 10, showing the relatively regular spacing of microtubule profiles (arrowheads).

Fig. 12. Vertical section of the distal part of a cell process cut parallel to its long axis. Melanosomes had moved part way towards the cell center. The cell process distal to the outermost melanosome (arrows) is collapsed and shows increased electron density of the fibrous cytoplasmic matrix.

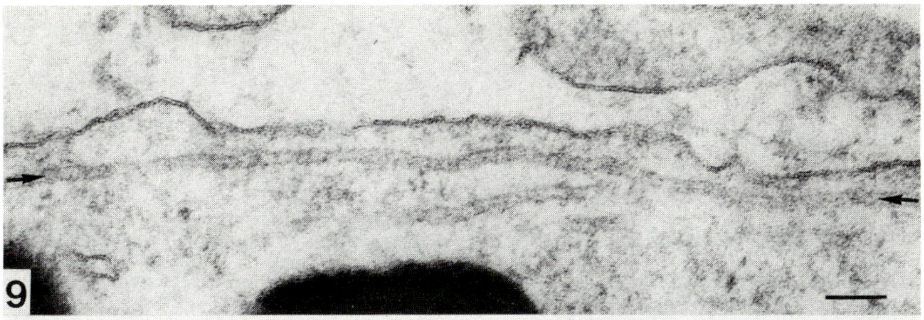
9

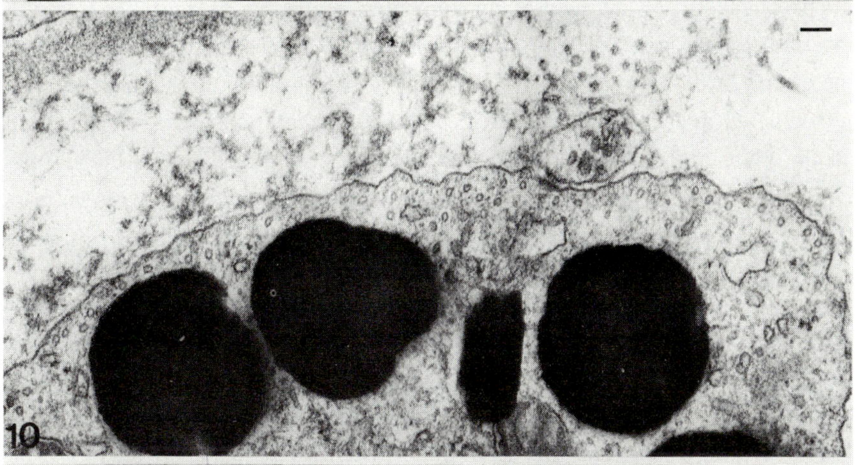
10

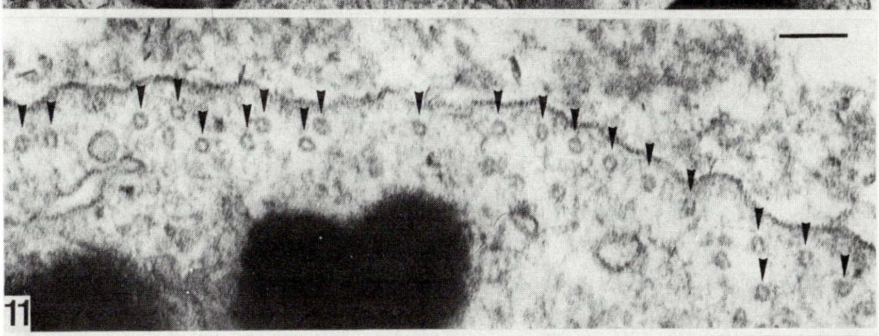
11

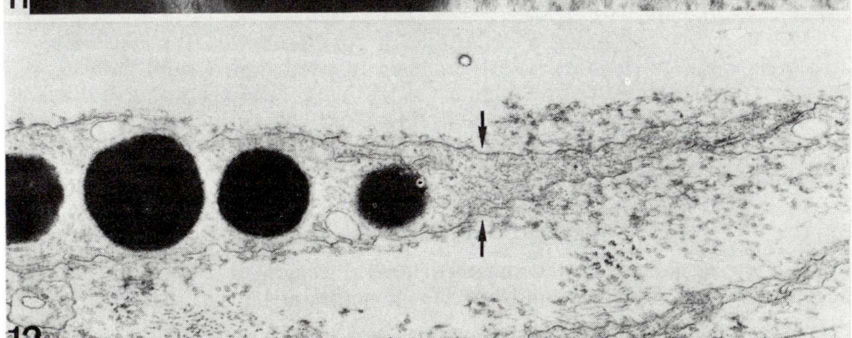
12

mid-region and tapered end shown in Figs. 13 b and c, respectively, are depicted in Figs. 14 and 15. In the former, a limited number of microtubules is recognizable, while in the latter, no microtubules can be seen. Because the circumferences of cell processes with aggregated pigment are comparable to those with dispersed pigment, there is a remarkable overlap of neighbouring processes (Fig. 15). From these observations it is apparent that during pigment aggregation, a considerable amount of microtubules (about 60%) disassembles, leaving a fibrous cytoplasmic component in the collapsed cell processes. The number of microtubules seems to decrease from the basal part towards the periphery of a cell process. In the basal zone, some microtubules are still associated with the plasma membrane. In the mid-region, microtubules are evenly distributed; intimate associations between microtubules and the plasma membrane (which are commonly observed in the dispersed state) are rarely to be seen in the aggregated state. As a rule, no microtubules are present in the most peripheral portions of the cell processes.

Microtubule assembly and disassembly during pigment movements

In order to examine whether melanosome movements and microtubule assembly or disassembly occur simultaneously, an attempt was made to study the distribution of microtubules in intermediate stages of dispersion. For these investigations, pulsating melanophores showing rythmic inward and outward motions of their pigment were fixed and prepared for electron microscopy. Starting from the cytocentrum, vertical sections spaced at regular intervals of about 2 µm were cut and the number of microtubules as well as the cross section area was evaluated. The results of one of these experiments are given in Fig. 16. As can be seen in this graph, microtubule density ($mt/\mu m^2$)

---

Fig. 13. Vertical section of a cell process cut parallel to its long axis. Completely aggregated state. Vertical bars in each micrograph indicate the distance from the margin of the pigment mass (µm). Note progressive thinning of the cell process. a. Funnel-shaped basis of the cell process with mitochondria (M). b. Mid-region; some microtubules are present (arrowheads). c. Tapering end with condensed cytoplasm. No microtubules.

Fig. 14. Cross section of a cell process cut in a region similar to that depicted in Fig. 13 b. Microtubule profiles are indicated by arrowheads.

Fig. 15. Cross section of the most peripheral region of a cell process. No microtubules are present. Note overlap of neighbouring cell processes.

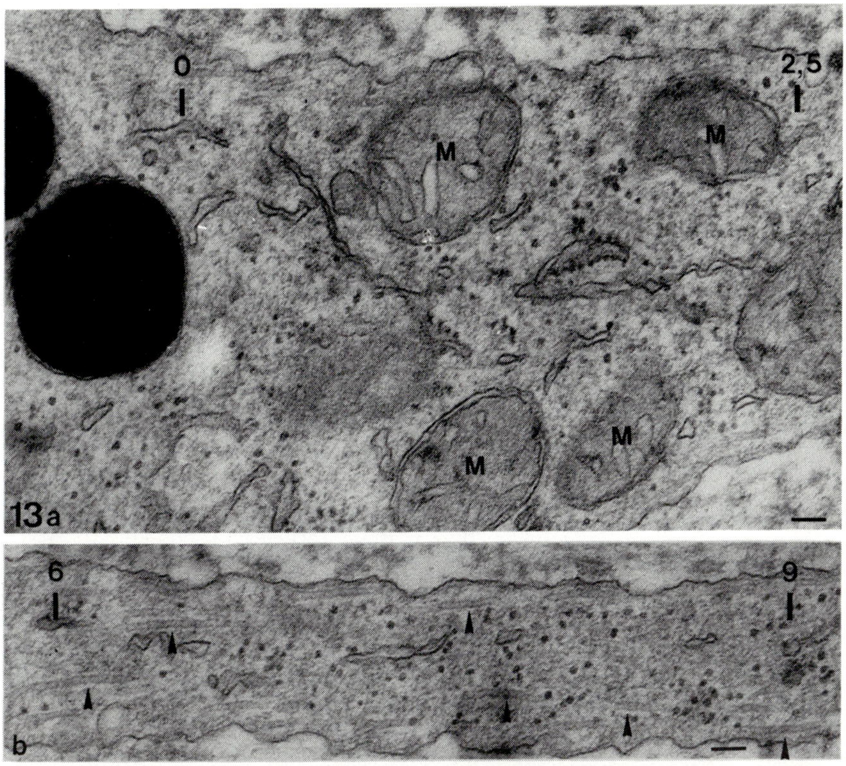
0
2,5
M
M
M
M
13a
6
9
b
14
17
c

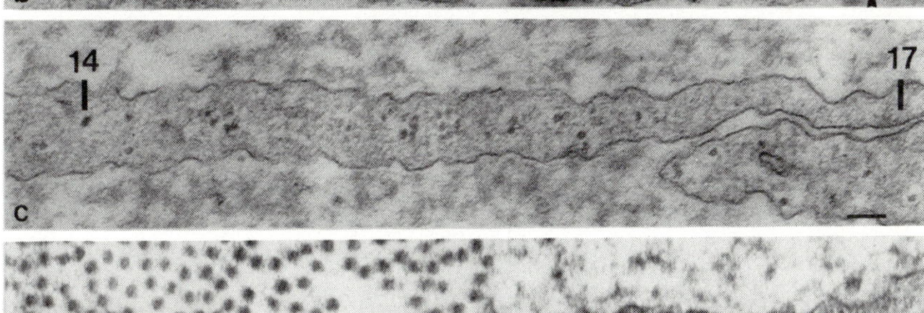
14

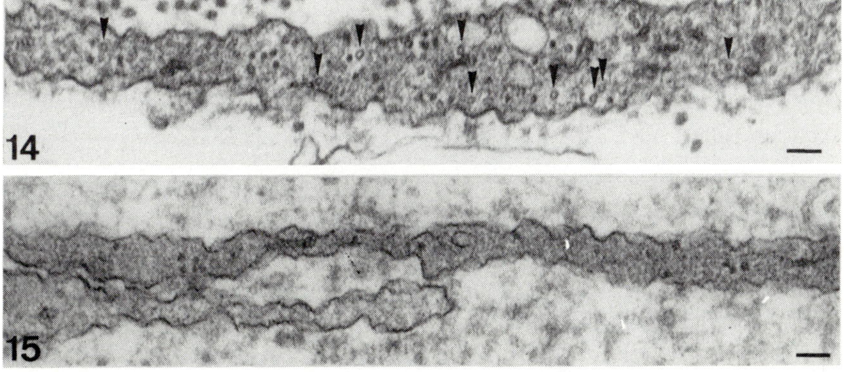
15

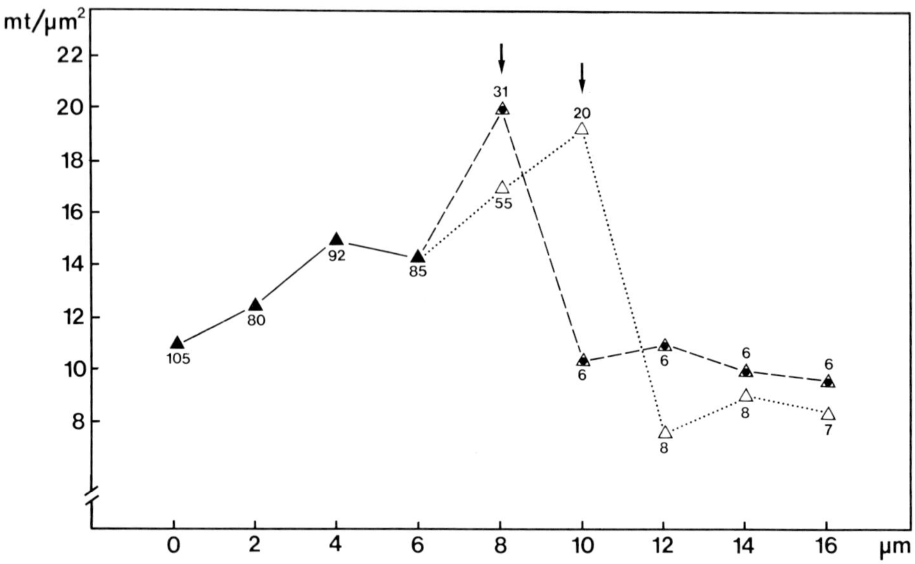


Fig. 16. Microtubule density (mt/µm²) and total number of microtubules counted (indicated in the graph) within a cross-sectioned cell process. Vertical sections spaced at about 2 µm intervals along the length of a cell process were obtained from the basis (0 µm) towards the periphery (16 µm). The cell process branches at a distance of 6-8 µm from the basis. Arrows indicate the last appearance of melanosomes within the cell processes.

increases with increasing distance from the basis of the cell process and reaches a maximum shortly after the cell process has branched. Beyond the margins of the pigment (at 10 and 12 µm, respectively), there is a striking decrease in microtubule density. The total number of microtubules is highest near the basis of the cell process, remains fairly constant between 2 and 8 µm (80-90 microtubules) and is remarkably low after pigment has vanished from both the branches (6-8 microtubules). Of particular interest is the observation that the number of microtubules is relatively constant within the unbranched process (2-8 µm) as well as the melanosome-free portions of both the branches (10-16 µm). Similar results were obtained in other experiments where transverse sections spaced at stated intervals have been made. From these observations it is not clear, however, whether single microtubules extend a considerable portion or even the entire length of the cell processes. These microtubule tracings have to be carried out by means of serial thin sectioning.

The results outlined above coincide with the observations of Porter[17] on erythrophores of Holocentrus ascensionis. This author observed disassembly of the bulk of the microtubules comprising the central population during pigment aggregation and reassembly during pigment dispersion.

It is not possible at present to state undoubtedly which microtubules are disassembled during aggregation and which survive the aggregation process. It should be mentioned, however, that intimate associations between microtubules and the plasma membrane (which are commonly observed in the dispersed state) are rarely to be seen in the aggregated state. In addition, the number of microtubules within the central portions of the cell processes of melanophores with pigment dispersed is roughly comparable to that within cell processes of melanophores with pigment aggregated. These observations suggest a disassembly of the cortical set of microtubules. Porter[17], in his study on erythrophores of Holocentrus ascensionis, suggests that in this particular chromatophore, the cortical populations of microtubules persist during pigment aggregation.

## The role of microtubules

From experiments with antimicrotubular agents it is apparent that normal distribution of microtubules is necessary for the processes of pigment aggregation and dispersion[8,18-20]. After treatment with vinblastine in appropriate concentrations, microtubules disappeared and vinblastine-induced microtubular crystals are observed throughout the cell. Their appearance coincides with the inhibition of melanosome movements. Application of this alkaloid results in an especially rapid and profound inhibition of melanosome dispersion and thus seems to interfere predominantly with the process of microtubule repolymerization[8]. During the initial phase of vinblastine action, cyclic AMP is capable of counteracting the inhibitory effects of vinblastine on pigment dispersion[21].

The action of colchicine is somewhat unexpected in that this drug only partially inhibits granule movement[8,18] and apparently does not alter the number or distribution of microtubules in the aggregated as well as the dispersed state[8]. Only after prolonged exposure, microtubule number seems to decrease[19]. When melanophores are treated with low temperature or high hydrostatic pressure, microtubules are disrupted and melanosome movements cease[9]. These observations confirm that microtubules are an essential component of the mechanism for melanosome aggregation and dispersion.

What is the precise role of microtubules during pigment movements? The apparent lack of association between microtubules and melanosomes seems to rule out a sliding mechanism via thin bridges or arms[22,23] that might generate the motive force for sliding of pigment granules over microtubules[9]. Instead, microtubules are arranged predominantly around the periphery of cell processes in the dispersed state and show close paraxial association. Their surface-to-surface distance of about 30 nm is similar to that of the so-called long link microtubule bridge observed within highly organized microtubular patterns[24,25]. Although microtubule bridges are usually not adequately preserved with the methods used in this study, it may be concluded that the cortical array of microtubules provides a structural stabilization to the cell processes. As the microtubule frame disassembles during aggregation, the cell processes collapse. Conversely, as microtubules are laid down in the cortex during dispersion, the tubular shape of the cell processes is reestablished. During the process of melanosome dispersion, direction and extent of the restructuring of the microtubular apparatus are predetermined by the position and length of the cell processes. This seems to set pigment movements apart from other microtubule-dependent processes of cell elongation, including neurulation[26], spermiogenesis[25], and myogenesis[27]. As a consequence, rapid redistribution of cytoplasm with resulting velocities of pigment displacements ranging from 2 µm/sec during aggregation to about 0,4 - 0,8 µm/sec during dispersion is possible.

The processes of breakdown and reestablishment of the microtubular apparatus seem to be accompanied by changes in the cytoplasmic matrix. Work recently completed provides indirect evidence, based on studies of the internal cellular pressure, for a contractile component which contracts during aggregation and relaxes during dispersion[28]. At the ultrastructural level, unarranged fibrous meshworks can be observed in cell processes of melanophores with aggregated pigment. Their significance is difficult to deduce on the basis of morphology alone, but they are likely candidates for a contractile component. The movement of pigment granules may, therefore, be the result of an interaction between assembly/disassembly of microtubules and a contractile component. The precise nature of this interaction will be the subject of future investigations.

Supported in part by a grant from the Deutsche Forschungsgemeinschaft.

References

1. Porter, K.R.: Cytoplasmic microtubules and their functions. In: G.E.W. Wolstenholme and M. O'Conner (eds.), Ciba Foundation Symp. on Principles of Biomolecular Organization, p. 308-345. Churchill, London 1966.

2. Tilney, L.G.: Ordering of subcellular units. The assembly of microtubules and their role in the development of cell form. Develop. Biol. 2, 63-102 (1968)

3. Schmitt, F.O.: Fibrous proteins and neuronal dynamics. In: S.H. Barondes (ed.), Cellular dynamics of the neuron, Symp. Int. Soc. Cell Biol. 8, 95-111 (1969)

4. Freed, J.J., Lebowitz, M.M.: The association of a class of saltatory movements with microtubules in cultured cells. J. Cell Biol. 45, 334-354 (1970)

5. Wagner, R.C., Rosenberg, M.D.: Endocytosis in Chang liver cells: the role of microtubules in vacuole orientation and movement. Cytobiologie 7, 20-27 (1973)

6. Bikle, D., Tilney, L.G., Porter, K.R.: Microtubules and pigment migration in the melanophores of Fundulus heteroclitus L. Protoplasma 61, 322-345 (1966)

7. Green, L.: Mechanism of movement of granules in melanocytes of Fundulus heteroclitus. Proc. Nat. Acad. Sci. 59, 1179-1186 (1968)

8. Schliwa, M., Bereiter-Hahn, J.: Pigment movements in fish melanophores: morphological and physiological studies. III. The effects of colchicine and vinblastine. Z. Zellforsch. 147, 127-148 (1973)

9. Murphy, D.B., Tilney, L.G.: The role of microtubules in the movement of pigment granules in teleost melanophores. J. Cell Biol. 61, 757-779 (1974)

10. Kamada, T., Kinosita, H.: Movement of granules in fish melanophores. Proc. Japan Acad. 20, 484-492 (1944)

11. Fujii, R.: Correlation between fine structure and activity in fish melanophores. In: G. Della Porta and O. Mühlbeck (eds.), Structure and control of the melanocyte, p. 114-123. Springer, Berlin 1966

12. Fujii, R.: The physiology of fish melanophores. In: T. Kawamura, T.B.

Fitzpatrick and M. Seiji (eds.), Biology of normal and abnormal melanocytes, p. 31-46. Univ. Tokyo Press, Tokyo 1971

13. Schliwa, M., Bereiter-Hahn, J.: Pigment movements in fish melanophores: morphological and physiological studies. II. Cell shape and microtubules. Z. Zellforsch. 147, 107-125 (1973)

14. Obika, M.: The changes in cell shape during pigment migration in melanophores of a teleost, Oryzias latipes. J. exp. Zool. 191, 427-432 (1975)

15. Wise, G.E.: Ultrastructure of amphibian melanophores after light-dark adaption and hormonal treatment. J. Ultrastruct. Res. 27, 472-485 (1969)

16. Gartz, R.: Adaptationsmorphologie der Melanophoren von Krallenfrosch-Larven. Cytobiologie 2, 220-234 (1970)

17. Porter, K.R.: Microtubules in intracellular locomotion. In: Locomotion of tissue cells, Ciba Foundation Symp. 14, 149-166 (1973)

18. Wikswo, M., Novales, R.R.: The effect of colchicine on migration of pigment granules in the melanophores of Fundulus heteroclitus. Biol. Bull. 137, 228-237 (1969)

19. Wikswo, M., Novales, R.R.: The effect of colchicine on microtubules in the melanophores of Fundulus heteroclitus. J. Ultrastruct. Res. 41, 189-201 (1972)

20. Junqueira, L.C., Raker, E., Porter, K.R.: Studies on pigment migration in the melanophores of the teleost Fundulus heteroclitus (L). Arch. Histol. Japon. 36, 339-366 (1974)

21. Schliwa, M., Bereiter-Hahn, J.: Pigment movements in fish melanophores: morphological and physiolgical studies. IV. The effect of cyclic adenosine monophosphate on normal and vinblastine treated melanophores. Cell Tiss. Res. 151, 423-432 (1974)

22. McIntosh, J.R., Hepler, P.K., Van Wie, D.G.: Model for mitosis. Nature 224, 659-663 (1969)

23. Smith, D.S.: On the significance of cross-bridges between microtubules and synaptic vesicles. Phil. Trans. Soc. London Ser. B 261, 395-405 (1971)

24. Roth, L.E., Pihlaja, D.J., Shigenaka, Y.: Microtubules in the heliozoan axopodium. I. The gradion hypothesis of allosterism in structural proteins. J. Ultrastruct. Res. 30, 7-37 (1970)

25. MacKinnon, E.A., Abraham, P.J., Svatek, A.: Long link induction between the microtubules of the manchette in intermediate stages of spermiogenesis. Z. Zellforsch. 136, 447-460 (1973)

26. Burnside, B.: Microtubules and microfilaments in amphibian neurulation. Am. Zool. 13, 989-1006 (1973)

27. Warren, R.H.: Microtubular organization in elongating myogenic cells. J. Cell Biol. 63, 550-566 (1974)

28. Schliwa, M., Bereiter-Hahn, J.: Pigment movements in fish melanophores: morphological and physiological studies. V. Evidence for a microtubule-independent contractile system. Cell Tiss. Res. 158, 61-73 (1975)

*Microtubules and Microtubule Inhibitors, eds. M. Borgers and M. de Brabander*



# MICROTUBULE AND CONTRACTILE SUBUNIT PROTEINS OF BLOOD PLATELETS ROLE IN HAEMOSTATIC ACTIVITIES

N. CRAWFORD and A.G. CASTLE

Department of Biochemistry, University of Birmingham
P.O. Box 363, Birmingham B15 2TT, England

SUMMARY

The cytoplasm of the platelets contains a variety of polymeric proteins associated as fibrillar elements. One system, consisting of actinoid and myosinoid proteins constitutes the microfilament structures and another, whose subunit composition is less well defined, represents the microtubules. The microtubules appear in electron micrographs as concentric rings lying beneath the inner face of the plasma membrane but the microfilaments are more randomly distributed, often intercalated between the microtubule structures and the surface membrane and elsewhere in the cytoplasm associated with the boundary membranes of granular organelles.

In this paper the properties and subcellular localisation of the platelet actin- and myosin-like proteins are reviewed and compared with those of the analogous components of skeletal muscle myofibrils. The myosin appears to be predominantly located in the cytosol of the cell whilst much of the actin is firmly associated with membrane elements.

Recently we have isolated and purified a microtubule subunit protein "tubulin" from platelets, which has properties similar to tubulins from cilia, flagella and mammalian brains. It binds $^{3}$H-colchicine, has a sedimentation constant of 6S and the monomer (MW $\pm$ 55,000), coelectrophoreses with authentic tubulins, running in gels as a closely spaced doublet similar to the $\alpha$ and $\beta$ components of brain tubulins. We also present some views on how the microtubule and microfilament proteins may operate with interplay in platelet motile activities and on their possible importance in the cells haemostatic activity.

## 1. Introduction

Following the introduction of glutaraldehyde fixation in electron microscopy, microtubules and microfilaments have become well recognised features of the cytoplasm of blood platelets. In thin sections, despite some size heterogeneity, these two fibrillar complexes can generally be readily distinguished, the microtubules being around 250-300 Å in diameter and the actinoid and myosinoid filaments, falling in the range 30-100 Å. Both types of structure can often be seen lying adjacent to and appear to be closely associated in some way with, the inner face of the plasma membrane. During the last few years considerable attention has been directed towards their structure and function because of their possible involvement in, not only the platelet's normal haemostatic activities which are concerned with sealing injury sites in blood vessel walls, but also in the pathological events preceding the formation of intravascular thrombi. These two important 'in vivo' processes have common features in that they both are initiated by certain alterations in the platelet's normal morphology and behaviour which may include shape changes (from discs to spheres), spike and pseudopodia formation, aggregation and the specific release of haemostatically active, intragranular constituents from the cell. All these latter phenomena are examples of the platelets wide range of motile activities some of which are believed to be triggered via surface receptors which are responsive to one or more of the various haemostatic stimuli such as thrombin, ADP, collagen or biogenic amines.

Circulating platelets are normally discoid in shape (Fig. 1) and in well prepared sections, processed at room temperature, the microtubules are clearly seen in the apical regions of the lentiform cell (also Fig. 1). In sections taken through an equatorial plane the microtubules then appear as a concentric bundle partially or fully encompassing the cell and lying close to the inner wall of the surface membrane. Fig. 2 shows a diagrammatic representation of the distribution of microtubules in the normal discoid cell. Intercalated between the microtubules and the plasma membrane and also scattered elsewhere in the platelet cytoplasm are the microfilament structures. These fibrillar elements are much smaller than the microtubules and can often be seen to interconnect with each other and with the plasma membrane and boundary membranes of the granular organelles (for excellent reviews of the general morphology and distribution of these platelet contractile filaments and microtubules see the following papers and reviews. Behnke [1,2,3] White [4] Hovig [5] Bettex-Galland and Lüscher [6-7] and Zucker-Franklin [8-9]. When the platelet becomes activated as part of a haemostatic or thrombotic event, pseudopodia form and a change in the distribution of the contractile microfilament structures then occurs. They appear to concentrate into a central zone of the cytoplasm where they surround and interconnect clumps of granular organelles [4]. It is not known whether they are instrumental in the centralisation of the organelles or are fortuitously located there due to their organelle membrane association. In addition the organised concentric array of microtubules seems to disassemble and then reform in a more randomised fashion often extending from the centre of the cell into, and throughout, the length of the pseudopodia (Fig. 2).

---

Fig. 1. Electron micrographs of platelets showing:- Top - normal discoid shape as seen in the ciruculating blood. Lower left - enlargement showing various intracellular features (M - mitochondria, G - lysosome-like granules, OCS - open canalicular system, DTS - dense tubular system and Mt - microtubules). Lower right - enlargement of apical region of platelet showing microtubule bundles in cross section (Hovig T. Ref. 5 ).

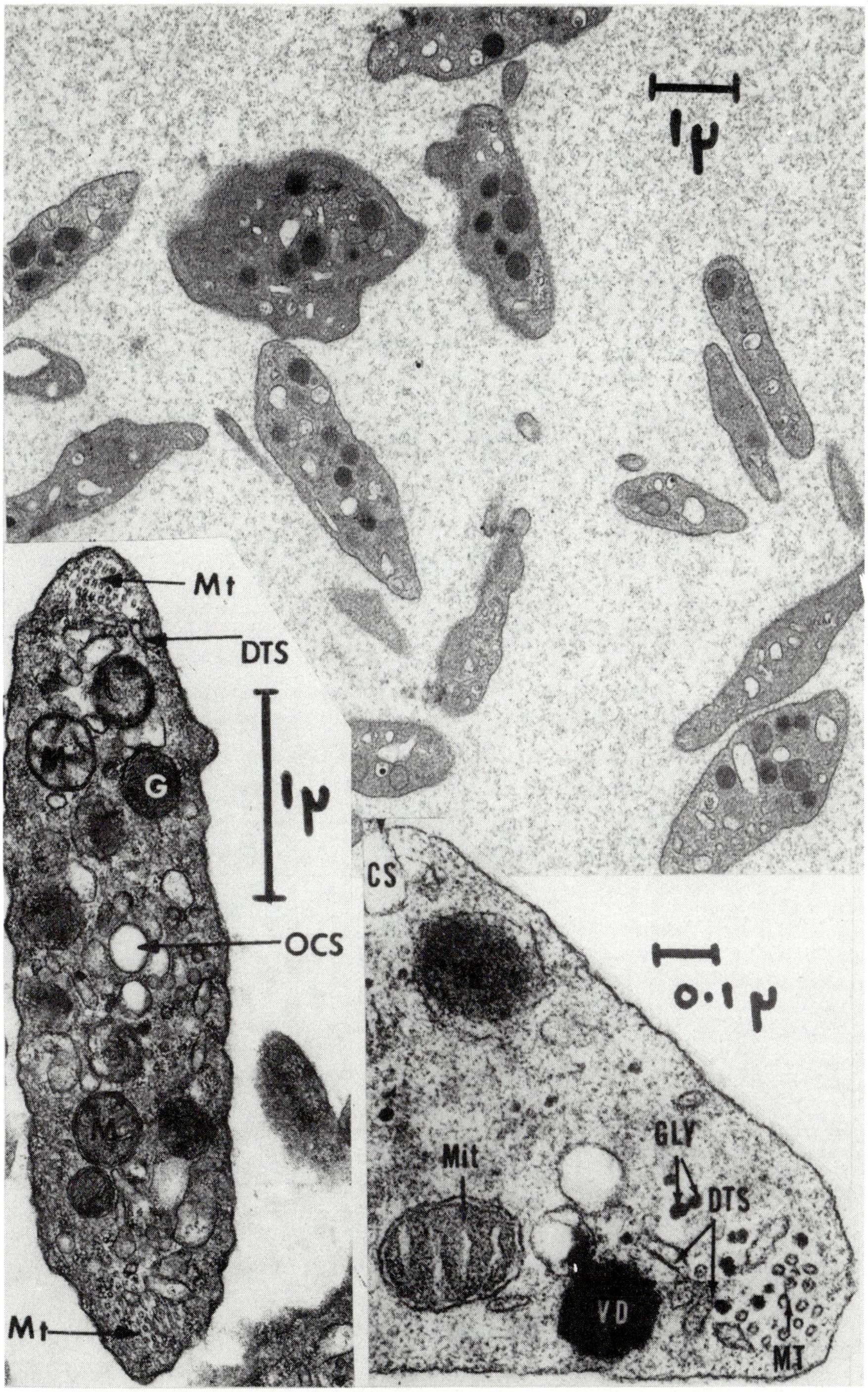
1μ
Mt
DTS
1μ
G
OCS
Mt
CS
0·1μ
GLY
DTS
Mit
VD
MT

During the last few years we have been actively studying the subunit protein composition of both the platelet contractile and microtubule systems and have particularly focussed our attention upon the physical and chemical properties of the organised polymers and the subunit pools from which they may be derived. In this paper we will attempt to summarise our findings and also present some views on how these two important structural and contractile polymeric complexes may be formed 'in vivo' and how they may interact and operate in certain motile phenomena.

## 2. The subunit proteins of the platelet microfilaments

Although the biochemical evidence for the presence of actin- and myosin-like proteins in the platelet is now extensive [6,7,10-15], the differential identification of cytoplasmic structures corresponding to the thick and thin filaments of muscle myofibrils has been much more difficult to establish. Using the heavy meromyosin (HMM) decorating procedure [16] the small diameter (30-50 Å) microfilaments present in glycerinated platelets can be made to display arrow-head structures and these have therefore been classified as "actinoid" in character [3]. The larger (70-100 Å) non-decorated filaments have been assumed to be the myosinoid polymeric assemblies but since their presence in the cytoplasm is more fugitive and they exhibit considerable size heterogeneity within the same cell their definitive identification is more difficult. Some two or three years ago we considered that biochemical information on the subcellular localisation of the individual components of the platelet contractile system would be a necessary adjunct to the morphological evidence if one is to consider the actomyosin complex as a force generating candidate for at least some of the cells motile activities, and then begin to explore the factors affecting its assembly, integrity and regulation. Our first finding was that a large proportion of the myosin-like protein of pig platelets located in the soluble phase compartment of the cell and could be readily isolated and purified from a soluble phase fraction of the cell prepared by 6 million g min centrifugation of platelet homogenates [15,17]. This protein, which represented around 10% of the total soluble phase protein, showed in SDS-polyacrylamide gel electrophoresis a typical myosin-like

---

Fig. 2. Upper Diagrammatic representation of platelet sections seen left, as sectioned in equatorial plane and right in transverse section (PM - plasma membrane. MITO - mitochondria, VAC - vacuoles, LYS - lysosomes VD - very dense granules or 5 HT storage bodies. Mts - microtubules, Mfs - microfilaments). Modified from diagram published by Behnke [1]. Lower left, Electron micrograph of activated platelet showing microtubules extending into pseudopodia. Lower right, SDS - Polyacrylamide gel electrophoretic separation of isolated soluble phase protein (myosin) showing heavy chain (ca. 200,000) and light chain components (13-22000).

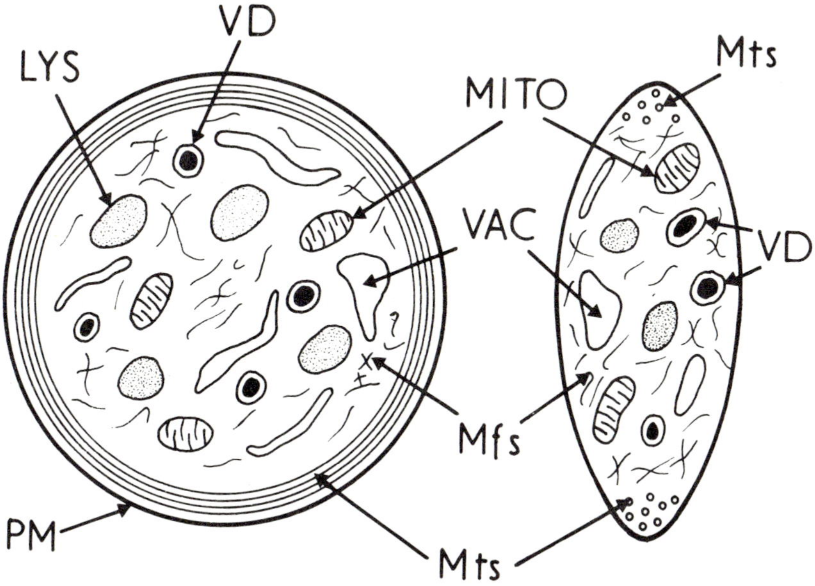
VD
LYS
Mts
MITO
VAC
VD
Mfs
PM
Mts

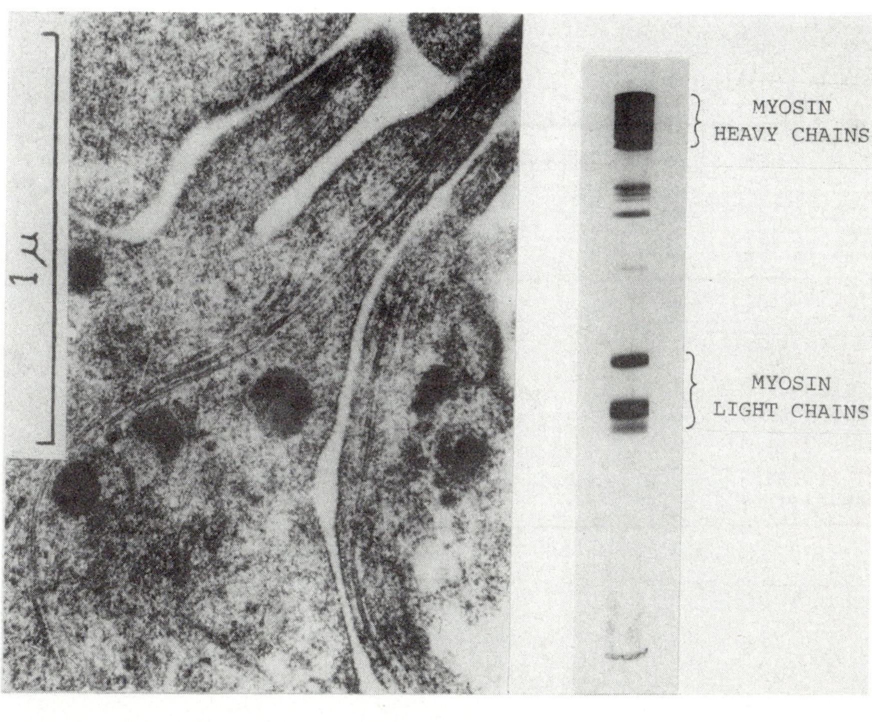
1μ
MYOSIN
HEAVY CHAINS
MYOSIN
LIGHT CHAINS

heavy chain component (200,000 daltons) and 2 or 3 low molecular polypeptides believed to be light chain species (Fig. 2). The ATPase activity of the isolated presumptive myosin was $Ca^{2+}$ stimulated, inhibited by low concentrations of $Mg^{2+}$ and gave a pH profile similar to that of skeletal muscle myosin with two major peaks of activity in the regions pH 6.0 and pH 9.0 [15]. That the soluble phase of the platelet was probably a valid compartment 'in vivo' for this platelet myosin was indicated by the simultaneous release into the soluble phase of both the myosin protein and a soluble phase marker enzyme, lactate dehydrogenase, when the soluble phase was prepared from platelets homogenised by a variety of cell disruption procedures (pestle and blender homogenisers and ultrasonication) for timed periods between a few seconds and 5 minutes (Fig. 3). Despite an increasing yield of homogenate and soluble phase total protein with the increased time of exposure to the homogenisation procedure the ratio of lactate dehydrogenase (total activity units), to the amount of myosin-like protein which could be extracted from the soluble phase by cycles of high ionic strength extraction and low ionic strength precipitation remained fairly constant. The following tables show the figures for two such experiments using pestle homogenisation (Table 1) and ultrasonication (Table 2).

Table 1

Blender homogenisation of platelets and measurement of lactate dehydrogenase (LDH) and the isolated myosin-like protein in the soluble phase fraction prepared from the homogenates by 6 million-g-min centrifugation (for full experimental details see ref. 15).

| | PROTEIN | | | | | RATIO |
|---|---|---|---|---|---|---|
| Homog time sec | Homog mg | Sol. phase mg | Sol. phase % of homog | Sol. phase protein pptd at low $\mu$ mg | Sol. Phase LDH total units of activity | TOTAL LDH / low $\mu$ protein |
| 15 | 68.3 | 54.8 | 80.2 | 6.8 | 52.6 | 7.4 |
| 30 | 78.1 | 62.3 | 79.8 | 7.3 | 65.6 | 8.9 |
| 60 | 78.9 | 64.4 | 81.5 | 8.4 | 69.8 | 8.3 |
| 100 | 82.5 | 66.5 | 80.6 | 9.5 | 83.8 | 8.8 |
| 200 | 86.9 | 70.9 | 81.6 | 12.7 | 91.6 | 7.3 |
| 300 | 90.2 | 72.7 | 81.7 | 13.6 | 103.4 | 7.6 |

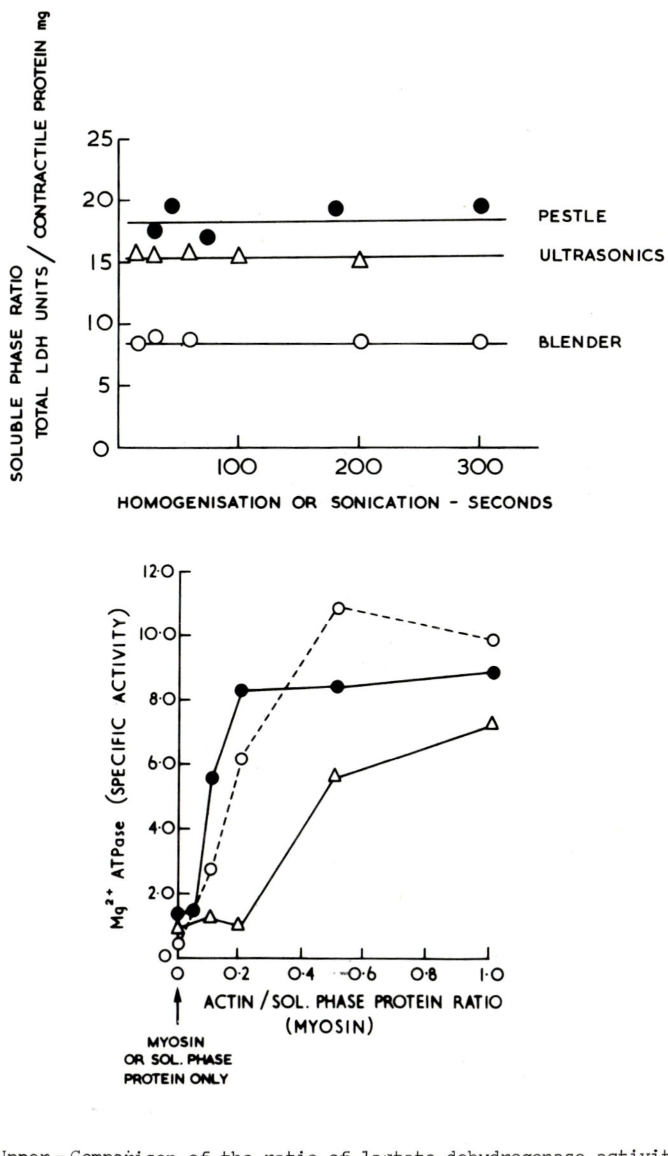


Fig. 3 Upper - Comparison of the ratio of lactate dehydrogenase activity, LDH, (total activity units) to the yield of contractile protein (mg), from the soluble phases prepared by 3 different homogenisation procedures. Lower $Mg^{2+}$ ATPase activities of synthetic actomyosins prepared by combining soluble phase myosin-like proteins (isolated by dilution precipitation from three different platelet preparations, ●, 0, △ ), with skeletal muscle actin. Specific activities calculated with respect to the concentration of soluble phase protein in each mixtures.

Table 2

Ultrasonication of platelets and measurement of lactate dehydrogenase (LDH) myosin ratios in soluble phases prepared by 6 million-g-min centrifugation. For full experimental details see Ref. 15.

| | | | PROTEIN | | RATIO |
|---|---|---|---|---|---|
| Sonication time sec | Sonicate mg | Sol. phase mg | Sol. phase protein pptd at low $\mu$ mg | Sol. phase LDH total units of activity | Total LDH / Low $\mu$ protein |
| 15 | 124 | 98 | 18.5 | 291 | 15.7 |
| 30 | 134 | 105 | 14.2 | 223 | 15.6 |
| 60 | 138 | 108 | 18.1 | 284 | 15.7 |
| 100 | 136 | 104 | 15.7 | 243 | 15.5 |
| 200 | 135 | 102 | 18.9 | 283 | 15.0 |

In the former experiments there is a progressive increase in homogenate and soluble phase protein yield with increasing time whereas in the latter the protein yield varies little over the range 15-200 seconds implying maximum cell rupture at low sonication times. In both experiments however the LDH/myosin ratios remain constant over the time scales studied suggesting the simultaneous liberation of the two proteins from the same intracellular compartment. In addition, evidence that the platelet soluble phase myosin was contractile-competent was gained from combination studies in which synthetic homologous and heterologous actomyosins were produced from mixtures of the myosin at various wt/wt ratios with skeletal muscle actin or purified pig platelet actin. All such mixtures showed ATP induced viscosity changes and enhancement of the low ionic strength $Mg^{2+}$ ATPase activity, often as high as 10 fold. Fig. 3 shows some typical experiments from these studies. One other property of platelet myosin, first noticed by us in analytical ultracentrifugation studies of whole cell actomyosin preparations (thrombosthenin), containing both unassociated myosin and actomyosin, and in studies of the soluble phase proteins was that both monomeric ($S^{o}_{20w}$ - 5.4) and dimeric ($S^{o}_{20w}$ - 8.0) forms may coexist in solution. These investigations also revealed that the propensity to form higher mol. wt. aggregates could be promoted by the addition of muscle F-actin and $Mg^{2+}$-ATP. The situation can be represented by the following equation derived from extensive analytical ultra-centrifugation studies of isolated thrombosthenin and soluble phase fractions published elsewhere [14,17].

$$\text{MYOSIN MONOMER}\quad \underset{\text{Absence of F-actin}}{\overset{\text{Presence of F-actin and ATP}}{\rightleftharpoons}}\quad \text{MYOSIN DIMER}$$

$$S^{o}_{20w} = 5.4 \qquad\qquad S^{o}_{20w} = 8.0$$

The importance of such an equilibrium shift in the formation and contractile function of the platelet actomyosin system will be considered later.

Turning our attention next to the location of platelet actin it was found that when the various particulate elements of platelet homogenates (membranes, lysosomes, mitochondria and amine storage granules, subfractionated by zonal rotor centrifugation [18],were solubilised in SDS/Urea/mercaptoethanol and subjected to SDS-polyacrylamide gel electrophoresis, all showed the presence of a prominent 43-45000 dalton component which comigrated with authentic skeletal muscle actin [19]. Fig. 4 shows a typical densitometric scan of a platelet membrane fraction prepared by zonal gradient centrifugation and electrophoresed after solubilisation in SDS/Urea/mercaptoethanol, with and without added muscle actin. Other particulate fractions gave essentially similar profiles indicating that the putative actin component was ubiquitously distributed throughout all the subcellular fractions and associated with a variety of the membrane and granular organelles. Additional evidence for identifying the 43-45000 polypeptide as actin was provided by hybridisation experiments with skeletal muscle myosin. In these experiments, despite extensive dialysis of the particulate fractions against low salt solutions (favouring F-actin depolymerisation), all showed significant enhancement of the $Mg^{2+}$ATPase activity of muscle myosin indicating that the particle-associated actin was present in a polymerised and combinable form (Fig. 4). Platelet particulate fractions also contained the unusual amino acid 3-methylhistidine, a feature of the primary sequence of muscle actins and myosins and absent in other proteins. More recently it has been established, [19], using detergent solubilisation procedures applied to isolated membrane fractions and $^{125}$-iodide-labelled purified platelet actin added to membrane fractions in sucrose density gradients, that the presence of actin in platelet membranes probably reflects a genuine association by firm protein-protein linkages rather than hydrophobic interactions with the lipid moeities. A proportion of the actin was also found to be present in the platelet soluble phase but we have consistently failed to demonstrate myosin $Mg^{2+}$ATPase increases in hybridisation experiments with the soluble phase and added myosins. This suggests that the soluble phase actin is either in the G-form or may represent denatured molecular species lacking the property of myosin combination.

To summarise therefore platelet myosin may exist in an unassociated form in the soluble compartment of the cell as either monomeric or dimeric units. Its potential to form larger polymeric aggregates may be influenced by the presence of F-actin, $Mg^{2+}$ and ATP, although morphological criteria have not yet revealed true thick filament forms analogous to those seen in skeletal muscle myofibrils. The

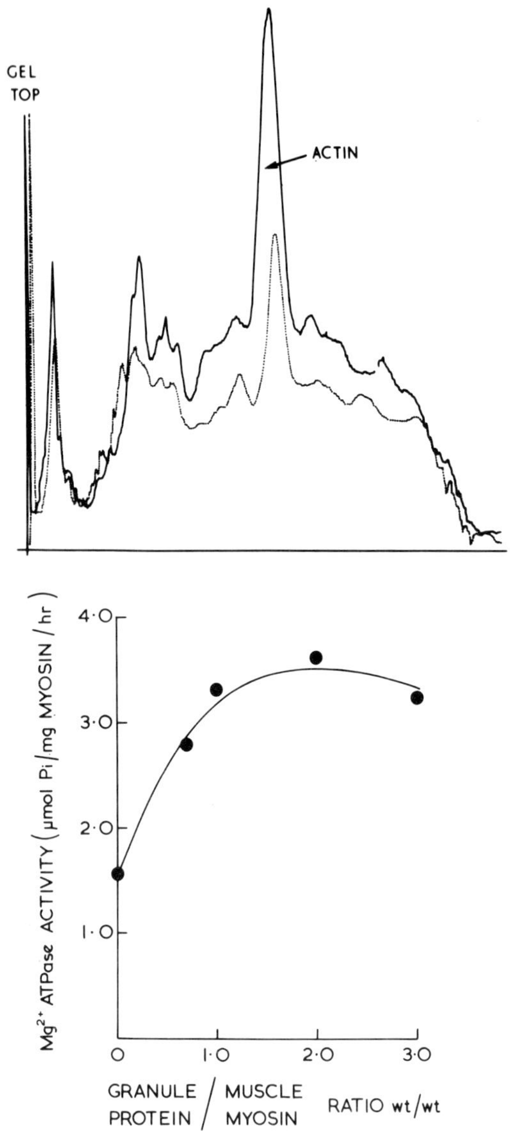


Fig. 4. Upper Densitometric scan of SDS-Polyacrylamide gel electrophoretic separation of the membrane fraction polypeptides soluble in SDS/Urea/mercapto-ethanol. Broken trace-without added muscle actin. Continuous trace-with added muscle actin. Lower Activation of the $Mg^{2+}$ATPase of muscle myosin by additions of a platelet granule fraction prepared by zonal centrifugation of homogenate in sucrose density gradients.

actin-like protein of the platelet though present to some extent in the soluble phase is predominantly located with and tightly bound to platelet membranes (surface and intracellular), and to the boundary membranes of the various intracellular granular organelles. Moreover this actin exists in a form in which the length and polarity of the polymer, as attached to the membrane, is favourable towards the formation of synthetic actomyosin complexes with added muscle myosin. This differential distribution of the two major components of the platelet actomyosin-like contractile system confirms the morphological evidence and may be of considerable importance for the consideration of the various cytoplasmic factors which may affect its performance as a biochemical mediator for platelet motile activity.

3. The platelet microtubule subunits.

Microtubule structures have also been implicated in a wide range of motile activities in eukaryotic cells viz. cytoplasmic streaming [20], membrane movement [21], cell division [22], ciliary motion and flagella motility [23,24], nerve cell elongation [25] and the migration of intracellular particles and organelles [26,27]. Although their presence in the blood platelet is well established from the elegant electron microscopic studies of Behnke [1,2 and 3], White [4] and Hovig [5] and Bettex-Galland and Lüscher [6,7] there have been few reported attempts to isolate and identify their subunit components and to compare and contrast them with the components of microtubule systems in other mammalian cells and tissues.

According to the views of Wilson & Bryan [28] microtubules can be broadly classified into two major groups, the "stable" and the labile". Stable microtubules are typified by those seen in cilia and flagella and are organised in a chracteristic 9 + 2 formation when seen in cross section. These microtubules can be isolated intact from the organism, do not depolymerise at low temperatures and are relatively resistant to colchicine and other anti-mitotic agents. By contrast "labile" microtubules, examples of which are the spindle fibres of dividing cells and the cytoplasmic microtubules seen in large numbers in brain and nerve tissues, are sensitive to temperature, bind colchicine and interact with agents such as vinblastine and vincristine. Platelet microtubules appear to fall into this latter category, tend to disassemble at low temperatures, are able to bind to $^{3}$H-colchicine and form crystalline lattices within the cell when treated with vinblastine [29]. Perhaps the best studied microtubules of mammalian systems however, are those of brain and much is now known about their architecture at the molecular level. The major subunit protein of brain microtubules is a dimer protein of molecular weight 110-120000 with a sedimentation constant of around 6S and containing bound GTP. This is known as "tubulin" and whilst as the dimer it binds colchicine, this property is lost in the presence of denaturing agents which split the protein into two non-identical monomers of molecular weights about 53000 and 56000.

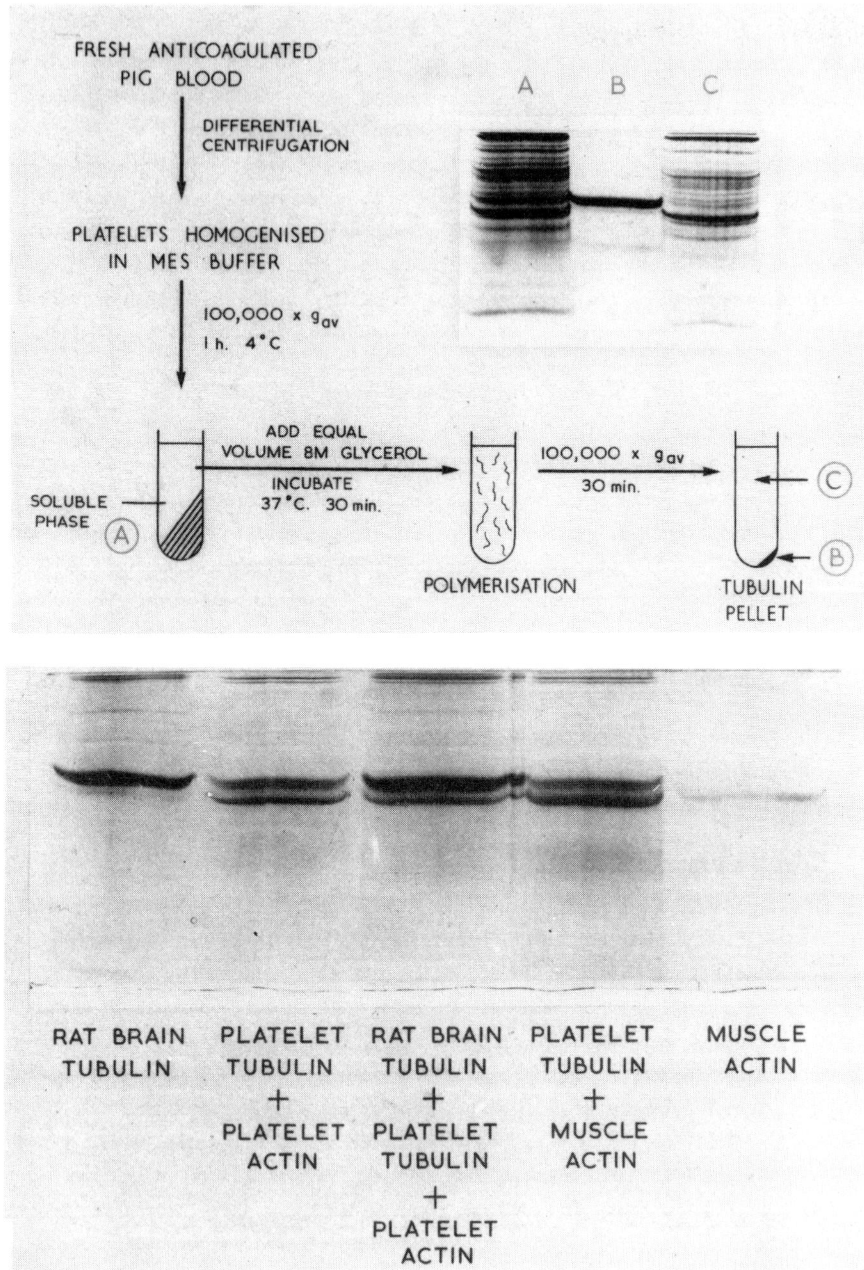


Fig. 5. Upper Procedure for the isolation of platelet tubulin by a temperature dependent polymerisation. Electrophoretic separations A, B and C illustrate the degree of purification as monitored at the stages indicated in the flow diagram. Lower Co-electrophoresis studies of platelet tubulin, rat brain tubulin and mixtures of the tubulins with purified muscle actin. All proteins were treated with SDS/Urea/mercaptoethanol before electrophoresis in 7% SDS-Polyacrylamide slab gels.

In our earlier attempts to isolate the microtubule subunit proteins of blood platelets, some of the established procedures which have been successful with brain [30,31] proved to be inapplicable to this cell because of the larger proportion of actin present. A procedure was developed based upon the temperature-dependent polymerisation technique of Shelanski et al [32] with some modifications made to avoid co-polymerisation of the actin. The procedure has been described in detail [33] and is summarised diagrammatically in Fig. 5. Included also in this figure are SDS-polyacrylamide gel electrophoretic separations of the platelet soluble phase and isolated tubulin to indicate the degree of purification achieved. Routinely, colchicine-binding activity was determined on the fractions during the isolation procedure and Table 3 shows the results for a typical experiment, showing that the colchicine-binding activity of the isolated tubulin-like protein was 35 times greater than that of the supernatant remaining after tubulin removal.

Table 3

[$^3$H]-Colchicine-binding activity of tubulin isolated from the platelet soluble phase

| Fraction | Protein mg | Specific activity DPM/µg |
|---|---|---|
| Platelet tubulin | 14.2 | 411.7 |
| Supernatant after tubulin removal | 238.0 | 11.7 |

The molecular weight of the platelet tubulin was determined using SDS-polyacrylamide gel electrophoresis which gave a value of 55,000 daltons (S.D. $\pm$ 1500,n = 8). Further electrophoresis experiments with purified tubulins, isolated from rat, rabbit and guinea pig and bovine brains by the unmodified Shelanski procedure, showed co-migration of these proteins with the 55,000 tubulin band in the purified platelet preparations. An example of these co-electrophoresis experiments is shown in Fig. 5. The platelet tubulin migrated in discontinuous SDS gels as a closely-spaced doublet similar to that seen in some preparations of brain tubulin. Also present in both the brain and platelet tubulin preparations were 2 or 3 high molecular weight polypeptides ( $>$250,000 daltons). These have been referred to, in brain preparations, as dynein-like proteins [34,35] because of their similarity to the high molecular weight components seen in preparations of tubulin from cilia and flagellae. Analytical ultracentrifuge and sucrose density gradient centrifugation studies of the platelet tubulin revealed that the colchicine-binding dimer has a sedimentation constant (6S) similar to that of brain tubulin (Fig. 6). Amino acid analysis gave results essentially the same as those determined for guinea pig and rat brain tubulins with a high content of acidic amino acids (Fig. 6).

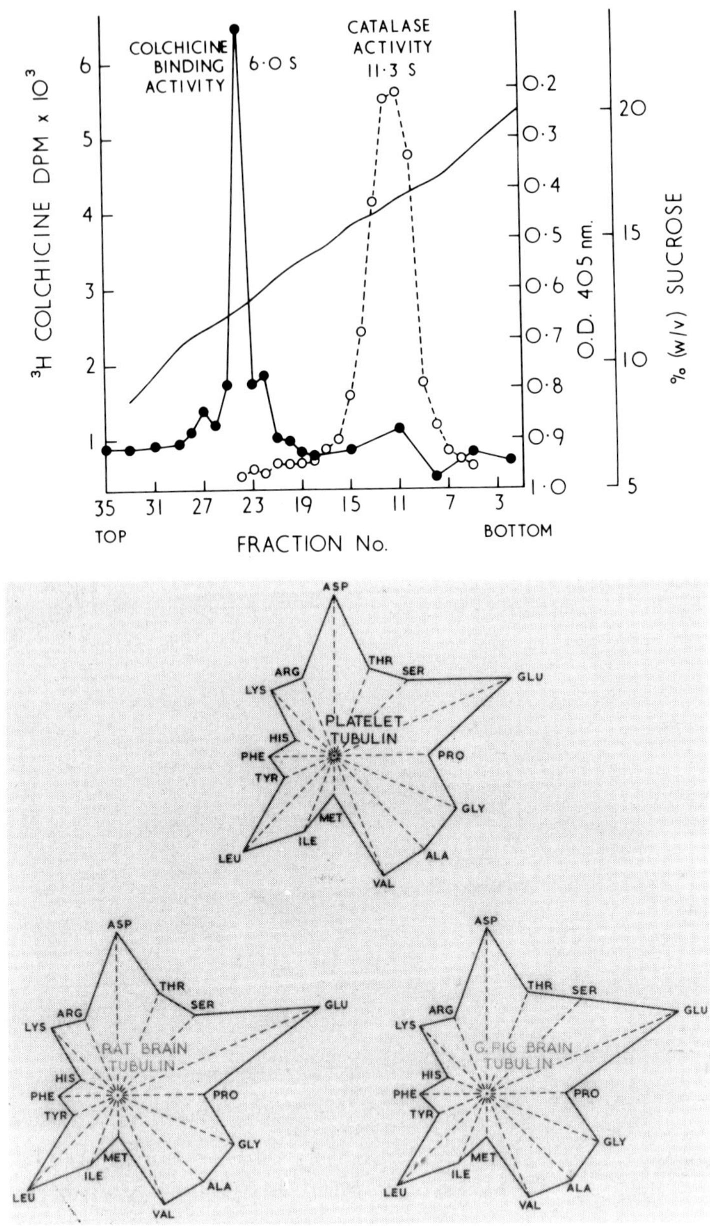


Fig. 6. Upper Ultracentrifugation of depolymerised platelet tubulin in sucrose density gradient (7-20% w/v sucrose). Samples were taken by aspiration from the bottom of the tube and measured for catalase activity (O-O) and colchicine binding activity (●-●). Lower Amino acid analyses of purified samples of platelet (top) rat brain (bottom left) and guinea pig brain (bottom right) tubulins. Radial lines are proportional to amino acid concentration in moles %.

Electron micrographs of the platelet tubulin polymer showed linear aggregation giving filaments of varying length and width but none as wide as the 200-300 Å microtubule structures seen in whole platelet electron micrographs. Since a microtubular subfilamentous structure with centre to centre spacing of around 50-60 Å, has been observed in whole platelets [38], the linear polymer formed in our procedure may represent an intermediate stage in the microtubule assembly process.

4. General discussion and comments on the two platelet polymeric systems.

In earlier studies of the blood platelet, the morphological similarity of the microtubule subfilaments and the contractile microfilaments led to some speculation that they may be formed "in vivo" from the same basic subunit. Our studies of the subunit proteins present in platelet cytoplasm, with a clear identification of a tubulin protein coexisting with actin, suggest that they have entirely separate identities and each are involved in the formation of two quite different polymeric complexes. In addition to containing a subunit pool for microtubule assembly, the soluble phase compartment of the platelet is also the major source of myosin. This myosin shows a tendency to form higher molecular weight aggregates under certain conditions, which in the platelet, though producing less well organised structures than in muscle, may be the analogous components of the thick filaments of the myofibrils.

The identification of actin tightly associated with both intracellular membranes and surface membrane components may indicate a possible function as loci for polymeric extension of actin fibres on which aggregated myosin molecules may then assemble for actomyosin-like action. The formation and function of such a contractile complex will, of course, be greatly influenced by other soluble phase constituents (ATP, $Mg^{2+}$ and $Ca^{2+}$), as also by the local ionic strength conditions prevailing in the cytosol. Although a tropomyosin-like protein has also been identified in platelets [37] and $Ca^{2+}$ sensitivity of the $Mg^{2+}$-ATPase activity of platelet actomyosin has also been clearly established [38], to date no definitive evidence for the presence of troponin-like regulatory subunits, analogous to those found in skeletal muscle, has been presented. This, of course, raises the question of whether platelet contractility is simply a "one off event" or can operate through a finely regulated contraction/relaxation cycle. All these questions require answers before we can begin to define a role for the microfilament structures in platelet behaviour. The platelet is a secreting cell and in the "Release Reaction", a motile event which culminates in the specific extrusion from the platelet of intragranule stored components such as ATP and 5HT, extracellular $Ca^{2+}$ ions are stimulatory [39]. The phenomenon however will proceed in the presence of EDTA [40,41] suggesting that if an actomyosin-like contractile complex is involved in this process, changes in the location of intracellular $Ca^{2+}$ may be the important initiator.

The microtubular system of the platelet is even more perlexing in a number of respects. Whilst on the one hand these structures do appear to be of cytoskeletal importance to the cell, and shape change phenomena seem to some extent to be correlated with their integrity, in other cell systems they are certainly not contractile but disassemble readily in the presence of $Ca^{2+}$ and obviously have some role in intracellular motility. Moreover, there is some evidence too that platelets devoid of organised microtuble structures, after treatment with Vinca alkaloids, still operate quite well in "in vitro" clot retraction experiments. It is our view that the key to further understanding of the action and interplay of these two polymeric systems may well lie in the nature of the association of the microfilament structures with surface and intracellular membrane structures on the one hand and the subadjacent concentric bundle of microtubules on the other. Such an association would, of course, add some constraints on the lateral diffusability of certain membrane proteins within the lipid layer but this fact may be important too,in the topographical integrity of surface-orientated transport and receptor complexes for shape change and motile activity. Our results, which are in partial agreement with the views of Behnke [2], suggest that, at least, two if not three, quite separate equilibrium systems may co-exist within the cytoplasm of the cell. The first,is that between soluble monomeric G-actin and the polymerised F-actin which is clearly bound in some way with the membrane elements (both the inner face of the plasma membrane and the boundary membranes of the granular organelles). The cytoplasm of the platelet also contains myosin monomers and dimers which may be in equilibrium "in vivo" with larger myosin aggregates. This myosin equilibrium, however, in the non-activated cell, is pushed predominantly in the direction of monomeric species. In the activated cell given favourable ionic strength conditions these equilibria may shift readily and, perhaps, irreversibly in the direction of polymer formation and the components for actomyosin formation and contractile action then become available. In contrast, however, although the lability of the platelet microtubules suggests a similar equilibrium existing between organised microtubules and a subunit pool, in the normal cell the energetics for this seem to be well in the direction of the organised polymeric structures which are protected in some way from the action of $Ca^{2+}$. This organisation is lost during shape change phenomena and a more random distribution of assembled microtubules then takes place.

Whilst it is realised that these foregoing ideas may be somewhat speculative we have presented them here because they may provide a stimulus for new experimental approaches to the problems and we also feel that our results indicate the value of the platelet as a useful experimental model for the study of a wide variety of microtubule and microfilament-dependent motile and locomotory events.

5. Acknowledgements

We gratefully acknowledge the financial support of the Medical Research Council and the British Heart Foundation for certain aspects of the work and our thanks are also extended to our colleagures Drs. D.G. Taylor and P. Jackson for allowing us to draw freely upon some of their more recent experimental findings.

6. References

1. Behnke, O., 1967, Anat. Rec. 158, 121.

2. Behnke, O., 1970, Int. Rev. Exp. Pathol. 9, 1.

3. Behnke, O., Kristenson, B.I. and Engdahl Nielson L., 1971 in: Platelet Aggregation ed. J. Caen (Masson & Co. Paris) p.3.

4. White, J.G., 1971 in: Platelet Aggregation ed. J. Caen (Masson & Co. Paris) p.15.

5. Hovig, T. 1968, Ser. Haematol. 1, 3.

6. Bettex-Galland, M. and Lüscher, E.F., 1960, Thromb. et Diath. Haemorrh. 4, 178.

7. Bettex-Galland, M. and Lüscher, E.F., 1965, Adv. in Protein Chem. 20, 1.

8. Zucker-Franklin, D., 1969, J. Clin. Invest. 48, 165.

9. Zucker-Franklin, D., 1970, J. Cell Biol. 47, 293.

10. Grette, K., 1962, Acta Physiol. Scand. 56, Suppl. 195, 1.

11. Nachman, R.L., Marcus, A.J. and Safier, L.B., 1967, J.Clin.Invest. 46, 1380.

12. Cohen, I., Bohak, Z., de Vries, A. and Katchalski, E. 1969, Europ.J.Biochem. 10, 388.

13. Booyse, F.M., Hoveke, J.P., Zschocke, D. and Rafelson, M.E., 1971. J.Biol.Chem. 246, 429.

14. Cove, D.H. and Crawford, N., 1975, J. Mechanochem. & Cell Motility 3, in press.

15. Harris, G.L.A. and Crawford, N., 1975, J. Mechanochem. & Cell Motility 3, in press.

16. Huxley, H.E., 1963, J. Mol. Biol. 7, 281.

17. Crawford, N., 1975, in: Contractile Proteins, FEBS Meeting Proceedings Budapest 1974, pub. Hungarian Acad. of Sciences, 1974, p.205.

18. Taylor, D.G. and Crawford, N., 1975, in: Methodological Developments in Biochemistry, Vol. 4 Subcellular Studies, ed. E. Reid. (Longmans) p.319

19. Taylor, D.G. and Crawford, N., 1975, Biochem. Soc. Trans. (553rd meeting) 161.

20. Porter, K.R., 1966, Principles of Biomol.Organ. Ciba Foundation Symp. 1965, p.308.

21. Tilney, L.G., 1968, J. Cell Sci. 3, 549.

22. Taylor, E.W., 1965, J. Cell Biol. 25, 154.

23. Shelanski, M.L. and Taylor, E.W., 1968, J. Cell Biol. 38, 304.

24. Rosenbaum, J.L. and Carlson, K., 1969, J. Cell Biol. 40, 415.

25. Daniels, M.P., 1972, J. Cell Biol. 53, 164.

26. Rebhum, L.I., 1967, in: The Contractile Process ed. A. Stracher (Little Brown Boston) p.223.

27. Dupraw, E.J., 1965, Develop. Biol. 12, 53.

28. Wilson, L. and Bryan, J., 1974 in: Adv. in Cell & Molec. Biol. ed. E.J. Dupraw (Academic Press) p.21.

29. Behnke, O., 1969, J. Ultrastruct. Res. 26, 111.

30. Weisenburg, R.C., Borisy, G.G. and Taylor, E.W., 1968, Biochemistry 7, 4466.

31. Bryan, J. and Wilson, L., 1971, Proc. Natl. Acad. Sci. US 68, 1762.

32. Shelanski, M.L., Gaskin, F. and Cantor, C.R., 1973, Proc. Natl. Acad. Sci. US. 70, 765.

33. Castle, A.G. and Crawford, N., 1975, FEBS Letters, 51, 195.

34. Burns, R.G. and Pollard, T.D., 1974, FEBS Letters, 40, 274.

35. Gaskin, F., Kramer, S.B., Cantòr, C.R., Adelstein, R. and Shelanski, M.L. 1974, FEBS Letters 40, 281.

36. Behnke, O., 1967, J. Ultrastruct. Res. 19, 147.

37. Cohen, I. and Cohen, C., 1972, J. Mol. Biol. 68, 383.

38. Thorens, S., Schaub, M.C. and Lüscher, E.F., 1973, Experientia 29, 349.

39. Holmsen, H., Day, H.J. and Stormorken, H., 1969, Scand. J. Haematol. Suppl. 8, p.1.

40. Holmsen, H. and Day, H.J., 1968, Nature (Lond) 219, 760.

41. Mürer, E.H., 1968, Biochim. Biophys. Acta 162, 320.

*Microtubules and Microtubule Inhibitors, eds. M. Borgers and M. de Brabander*



# STIMULUS TRANSMISSION ALONG MICROTUBULES IN SENSORY CELLS: AN HYPOTHESIS

J. Atema

Boston University Marine Program

Marine Biological Laboratory

Woods Hole, MA 02543

U.S.A.

Microtubules may have at least four distinct functions: structural including morphogenetic, motile, transport, and signal transmission. It is the purpose of this paper to present evidence for the latter function, which has not yet been demonstrated experimentally.

The arguments are that microtubules are characteristically present in modified ciliary form in most metazoan sensory cells; that their chemical structural properties make them possible candidates for intracellular signal transmission; and that complex protist behaviour may require a fast internal communication system.

A hypothesis for the origin of neurosensory cells of higher animals is presented, based on principles of evolution and genetics, which assume that more complex forms of life evolved from subsequent utilization of preadapted units.

Directed and adaptive behaviour must involve perception of external stimuli, their transduction, and their transformation into a motor output. It can thus be postulated that the earliest and most primitive forms of life were capable of perceiving environmental stimuli to which they responded with appropriate actions (i.e. changes in behaviour).

Some of the simplest forms of life studied are bacteria; and indeed it is well known that they show adaptive behaviour in chemical gradients and to other gradient fields. From studies on genetic mutants of *Escherichia coli* it has become clear that in bacteria a separation exists between reception of chemical stimuli and the nutritional value of the substances[1]. This points to the existence of a primitive "reflex arc" consisting of reception, transduction, and adaptive behaviour. However, for this paper prokaryote life will not be further considered since true ( 25 nm diameter, 13 protofilament) microtubules have not been described here, although similar mechanisms of stimulus transmission may be involved in prokaryotes and eukaryotes.

Work similar to that on *E. coli* has been done with the ciliate protozoan, *Paramecium aurelia*[2]. Generalizing from these and other studies it is clear that reception of sensory stimuli and the resultant - often complex and integrated - orientation behaviour are a normal property of protists. Not only do protists show orientation behaviour to external stimuli, they also display precisely timed and carefully regulated internal behaviour (e.g. feeding behaviour in *Nassula's* pharyngeal microtubular basket [3] ) and organelles of intriguing complexity ("eyes" in a dinoflagellate consisting of four modified organelles and connected with a complex network of microtubules to motile organelles)[4]. Behaviour and organelle development of this complexity seem to indicate a fast intracellular communication system which ties receptor sites to distant motor units. Directed protein fibers, such as microtubules, could provide the basis for such a system. While all proteins that form cells may exhibit a general excitability to a great number of high energy stimuli, it appears that stimulus specific low energy receptors and even specialized organelles developed early in evolution - chemotaxis by specific, genetically fixed receptor molecules in E. coli and P. aurelia; photoreceptor organelles in Euglena and dinoflagellates, etc.

Metazoan forms of life can be seen as having originated from protozoan forms through a process of either symbiosis (cells of different genetic and functional background getting together) or clone differentiation (a cluster of identical cells formed by mitosis followed by functional, anatomical and genetic specialization). It can be hypothesized that in this process all differentiated cells retained in their organelles some level of general responsiveness to environmental stimuli, but that natural selection led to anatomical and functional specializations of one or another organelle depending on the cell's function in the whole organism. Some cells focused on chemical perception, others on visual or tactile perception, while most others did not specialize in perception at all. In this ongoing process of specialization it is likely that new structures evolved based on existing potential. Two of the organellar pools of potential for distance transmission may be polarized membranes and polymerized protein fibers.

Today's metazoans all have clearly distinguishable nervous systems and sensory receptors. Some receptors, do not reach the ultimate levels of specialization with which we are most familiar. Dual receptor function appears to exist next to specialized receptors: e.g., Limulus: visual and mechanical at physiological stimulus levels[5]; Balanus larvae: chemical and mechanical[6]; and others[7].

However, most well-studied receptors show great specialization and impressive sensitivites, which all lie near the level of absolute thresholds: 1 photon for visual receptors; 1 molecule for chemoreceptors; 10 nm displacement for vertebrate hair cells and insect mechanoreceptors.[15] Despite this great specialization some generalities in the blueprints of structure and function of sensory cells can be discovered. Of these, perhaps the only possible organelles with properties for distance transmission are membranes and microtubules.

The membrane, which obviously forms the interface between cell and environment has been extensively studied and shows many attractive features for reception and transmission of sensory stimuli such as its potential for stimulus amplification through internal transmitters (perhaps $Ca^{2+}$), its cable properties in common with well studied nerve axons, its receptor features in common with post-synaptic transmitter reception, its presence in all cells, and its physical exaggerations in the form of microvilli and cilia (see Table 1). Membrane resistance changes in general have been documented extensively and only the most pertinent example of interest to sensory physiology will be mentioned here. In photoreceptors a resting current between distal and proximal segment is modified by photon triggered changes in rhodopsin structure, which open up $Ca^{2+}$ gates. This so-called dark current spans the distance between distal and proximal segment and its modification represents stimulus information to the receptor cell. The process appears similar in ciliary and rhabdomeric photoreceptors[8] and serves as a clear example of distance transmission through a membrane regulated process.

Table 1

Major external sensory systems

| Ciliary | Microvillar |
|---|---|
| Vertebrate hair cells (mechanoreceptor): vestibular, hearing, lateral line | Chemoreception Vertebrate taste (microtubules in microvilli) |
| Invertebrate mechanoreception | Vertebrate vomeronasal (cilia occasionally on receptor cells, always on supporting cells) |
| Chemoreception vertebrate and invertebrate olfaction, invertebrate taste, | Vision, rhabdomeric type |
| Vision, ciliary type | Electroreception, teleost fishes |
| Electroreception, elasmobranch and chondrost fishes | |

Microtubules, however, are also present in all receptor cells and have attractive features for stimulus transmission, as described earlier.[14] Their most obvious presence in sensory cells is in ciliary structures. The presence of cilia in most receptor cells has provided a focus of interest for many years and various theories have been advanced to explain their sensory function. An almost trivial explanation is that cilia merely serve to extend the membrane outwards, but it is not clear why cilia are required, since microvilli apparently serve this function equally well without an elaborate 'internal skeleton'. The rest of this paper will argue how ciliary and perhaps non-ciliary microtubules may be involved actively in sensory transduction.

An hypothesis for active involvement of sensory cilia in the transduction process was clearly expressed by Lowenstein _et al_.[9] who stated that kinocilia in vertebrate vestibular hair cells may act as motile cilia in reverse. They pictured a process of chemical-mechanical transduction, which could act (1) to drive a motile cilium with chemical 'stimuli' (supplied near the basal bodies), and (2) to deliver chemical stimuli to the cell in response to mechanical bending at the sensory cilium tip. In other words, in these mechanoreceptor cells they pictured a flagellar beat following the initial mechanical stimulus. This beat would provide the cell body with an energy pulse to be converted into a generator potential, which in turn modulates the afferent spike trains one can record. Moreover, directional sensitivity in hair cells is correlated with the orientation of the central pair of ciliary microtubules and with the position of the basal foot, a ciliary structure often observed in both mechanoreceptors and motile cilia.

A similar proposal has been advanced by Moran et al. who explicitly consider the mechanoreceptor cilia of grasshopper chordotonal organs "motile cilia that respond to the mechanical stimulus with an 'active stroke' which excites a transducer membrane at the dendrite tip".[10] Earlier experiments had shown that structural disassembly of chordotonal microtubules corresponded to cessation of mechanoreceptor function.[11]

While a flagellar beat explanation seems effective for those mechanoreceptor cilia, which show the presence of dynein arms and maintain a 9+2 or 9+0 configuration, it does not explain what happens in mechanoreceptors where a 9+0 structure dissolves distally[12], or why cilia are found in the majority of chemoreceptors and in ciliary electroreceptors, or the frequent cases in which receptor cilia are split distally into single microtubules. It seems likely that such microtubular specializations (as found in careful structural analysis of some insect chemo- and mechanoreceptors[12]) indicate direct and specific participation of microtubules in the sensory transduction process.

I will thus propose the following hypothesis for a sensory function of microtubules.

Microtubules can act singly or in cilia as conduction fibers through the propagation of a wave of conformational changes in protein sub-units. While the primary stimulus in some receptors (chemoreceptors) may interact first with the membrane, where local amplification can take place, perhaps through the local influx of $Na^+$, in other receptors (mechanoreceptors) it may act directly on the microtubules. The stimulus energy - amplified or not - forces one (or more) sub-units into a changed conformation. This sub-unit then no longer fits in its tubule lattice which causes a subsequent conformational change in neighboring sub-units, Fig.1. The resulting wave of conformational changes can thus be propagated to one or both ends of a microtubule, depending on bonding of sub-units. The energy for this process is stored in the microtubule structure, and the microtubule "potential" is restored after each "depolarization" wave by cell metabolic processes. Arrival of this wave at the ciliary base results in local membrane permeability changes perhaps at the ciliary 'necklace' or at the ciliary 'granule plaques'. The latter are membrane bound $Ca^{2+}$ binding sites just opposite

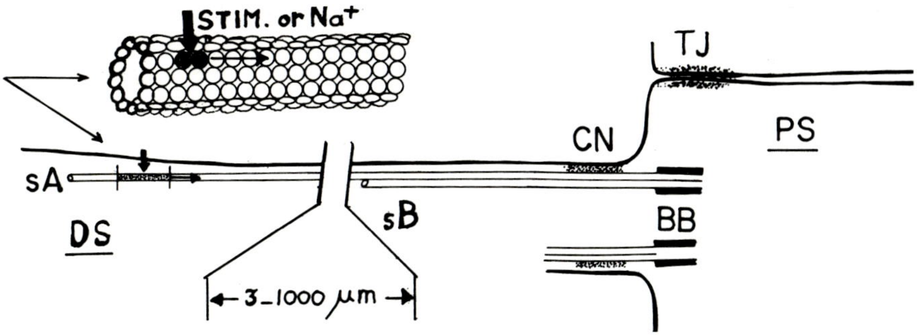


Fig.1. Stimulus induced change in tubulin sub-unit in microtubule lattice (hypothetical). Abbreviations for all figures: DS distal segment; PS proximal segment or dendrite; A axon; MTcil ciliary microtubule doublet; MT microtubule; sA,sB sub-fibers A,B; C cilium; CN ciliary necklace; BB basal body or centriole; R rootlet; TJ tight membrane junction; MI mitochondrium; MV microvilli; AC accessory or supporting cell; F parallel fibers

the ciliary microtubule doublets discovered in cilia of Paramecium.[13]

Through the use of a local internal transmitter (e.g. $Ca^{2+}$) the membrane permeability (for $Na^{+}$?) of the sensory dendrite (inner segment) tip is changed and a local depolarization (or current change) follows. This depolarization may spread electrotonically to the basal membrane to cause transmitter release at the first synapse (in epithelial sense cells) or to spike generation sites for further axonal conduction in bipolar neurosensory cells. When - as is most common - receptor cells are found in epithelia the connecting circuit of receptor cells and supporting cells through low resistance membrane junctions may further amplify the generator current. Since biochemical and structural support for this hypothesis has been given elsewhere[14] I will only mention some experiments which tend to support the hypothesis as well as point the direction in which more conclusive evidence is likely to be found.

Cockroach campaniform sensilla are mechanoreceptors which contain up to 1000 microtubules in their distal segment as well as a cilium which connects the distal segment to the dendrite (proximal segment) where one finds a basal body with rootlets, Fig. 2. Electrical responses to mechanical stimuli applied to the distal tip can be recorded from afferent axons. When colchicine or vinblastine was applied in dosages high enough to disassemble the distal segment microtubules, the electrical responses ceased. However, when tetrodotoxin was applied no disassembly or functional impairment occured. Apparently microtubules played an essential role in the stimulus transduction process, while membrane bound sodium gates - which are specifically blocked by tetrodotoxin - were not involved.[11]

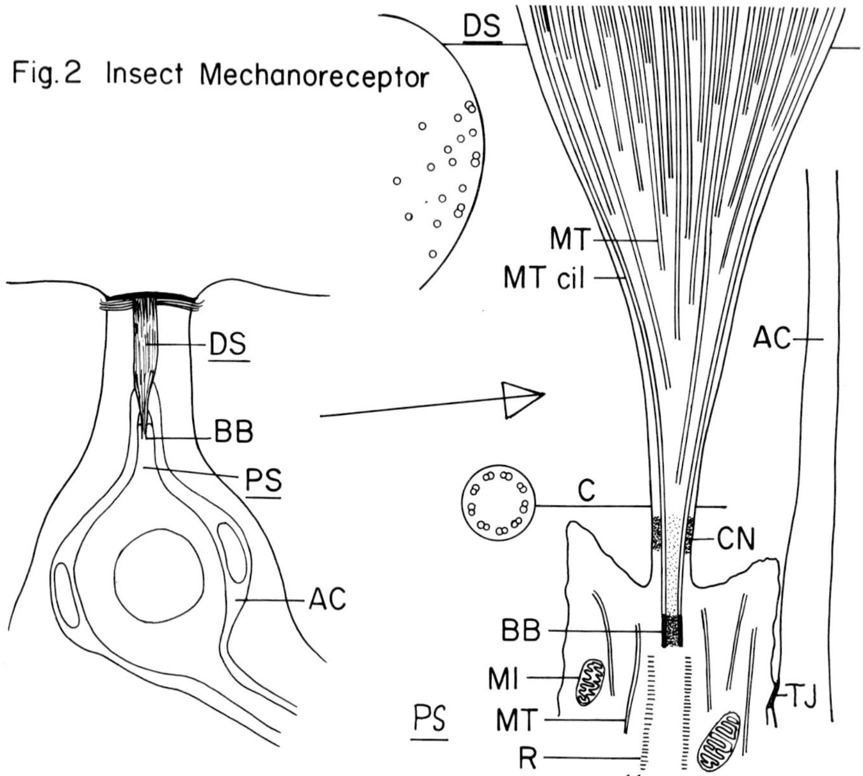


Fig.2. Mechanoreceptor cell of insect campaniform sensillum.[11] A 9 + 0 cilium divides shortly above the apical dendrite into separate doublet tubules. A great number (300-1,000) of non-ciliary microtubules appear in the distal region. Rootlets descend from a basal body into the proximal segment, which is filled with numerous cytoplasmic microtubules and mitochondria. Special junctions connect the cell with its supporting cells in a place a little below the centriole. According to the hypothesis presented above distal microtubules will receive and transmit a periferal cuticle deformation, which has been estimated at 10-50 nm.[15] Special membrane areas near the ciliary base may in turn be stimulated by the ciliary microtubules and turn on an ionic transmitter process. Abbreviations as in Fig. 1.

---

Fig.3. Olfactory cells. a. vertebrate,[22] b. crustacean,[19] c. insect.[12] Cilia in olfactory receptor cells maintain their 9 + 2 or 9 + 0 structure only for a short distance after which the doublet microtubules separate; further distally B subfibers begin to open up and drop out while non-ciliary microtubules begin to occur. In Crustacea, single microtubules within a membrane, which shows regular swellings, are the only organelles present in the distal segment which may reach a length of almost 1000 µm. The cilia always have basal bodies and rootlets. Cytoplasmic microtubules and mitochondria characterize the apical dendrite region, which is connected with special tight junctions to neighboring cells. According to the hypothesis the many arms of the divided cilia will provide extended interaction surfaces for odor molecules. Local interactions are thought to be propagated along single microtubules to the ciliary base, perhaps after initial local amplification by ionic processes involving the membrane. The proximal segment may process the stimuli as in other types of receptor cells. Abbreviations as in Fig. 1.

Fig. 3 Olfactory Receptor Cells

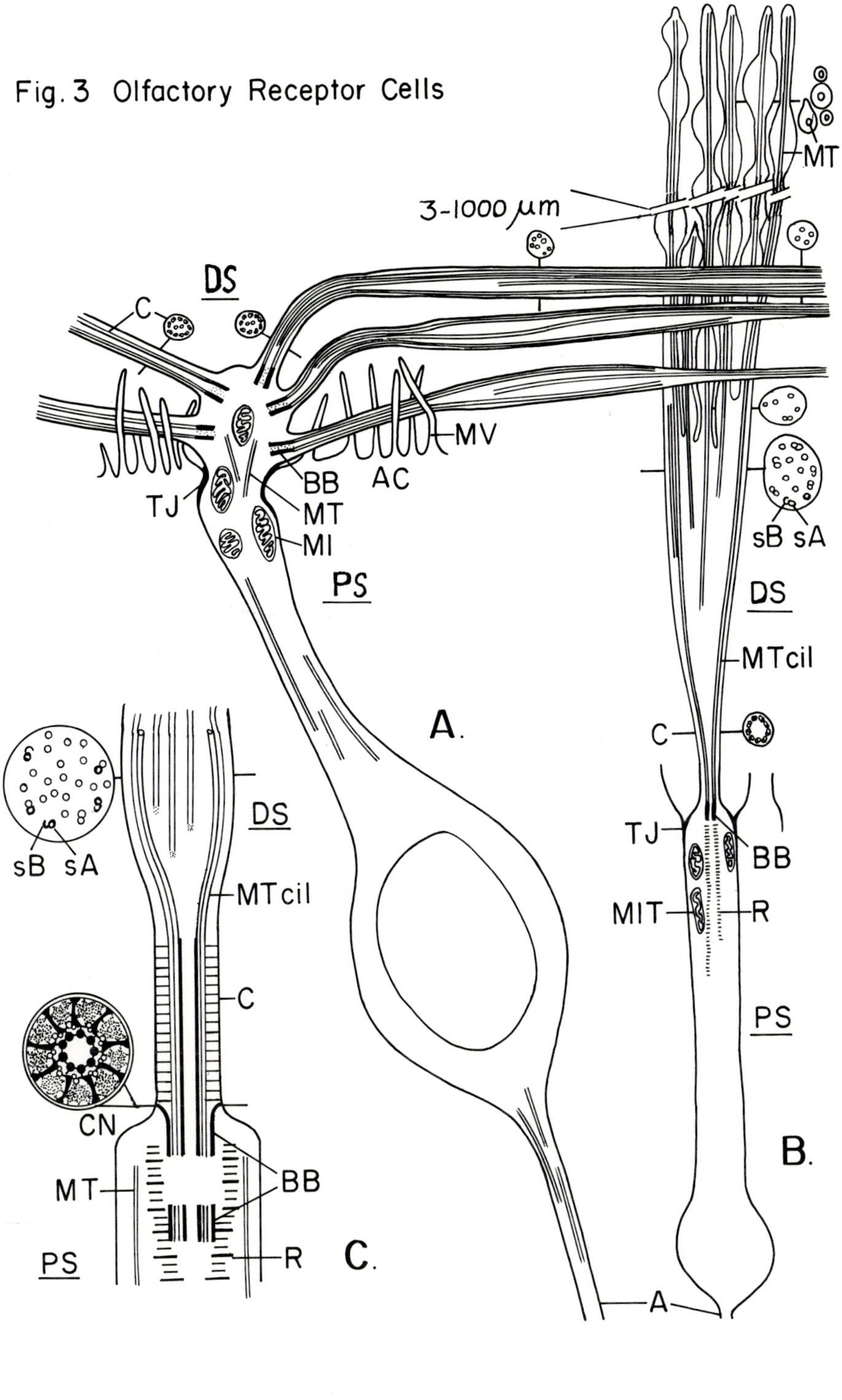

The same laboratory investigated a similar structure, the chordotonal organ. Their results led to the hypothesis that its cilium transduces the miniscule mechanical distortions (>10 nm) of the distal tip[15] by an active stroke, which in turn stimulates the apical portion of the dendrite.[10]

Since an argument can be made that in a mechanoreceptor passive translation takes place, whereby microtubules and cilia act as stiff rods, perhaps stronger evidence can be generated from chemoreceptors. When cockroach antennae were tested for their mediation of two behavioral responses, one to sex pheromone and one to aggregation pheromone, it was found that colchicine, and especially vinblastine and vincristine, applied to the antennae, would greatly reduce the responses to sex pheromone, but not to aggregation pheromone. Formaldehyde, however, had the opposite effect.[16] Although these results do not provide conclusive evidence, they do imply a receptor/transducer role for microtubules, since colchicine, vinblastine and vincristine are considered specific microtubule blocking agents, whereas formaldehyde may have affected a great number of proteins. More extensive experiments must be done to elucidate these results.

Further suggestive chemoreceptor evidence comes from experiments of protozoan chemotaxis and settling. Crypthecodinium cohnii (dinoflagellates) showed preferential settlement on DOPA and avoided settling on choline impregnated agar plates. They were then exposed to levels of colchicine, vincristine, vinblastine, griseofulvin or trifluralin, which did not appear to affect normal swimming behaviour or ability to attach to substrate. The chemically mediated choline avoidance response was reduced by all antitubulins except cochicine, while preferential settling on DOPA was not affected.[17] Thus, chemoreception appears affected by microtubule blocking agents.

An hypothesis of receptor function of mechanoreceptor cilia with 9+2 configurations and with dynein cross bridges as in all motile cilia could well be conceived as a 'motile cilium in reverse', producing an active beat upon stimulation[9,10]. But one would perhaps not expect to see this true ciliary motility work as a stimulus transducer when the typically motile structures are absent as in mechanoreceptor cilia without dynein arms or with a 9+0 configuration, as found in the cockroach subgenual organ. Moreover, here the 9 doublet microtubules separate widely and run each along the inside of membrane folds.[18] It is even more difficult to see how this transduction mechanism would take place in olfactory cilia, where 9+0 configuration is normal[+] and where often the cilia splay into single membrane wrapped microtubules sometimes only shortly distal from the basal bodies.[19] Additional non-ciliary microtubules appear in the distal area both in mechanoreceptors[10] and in chemoreceptors[12, 19]. The farthest removed from this situation is the vertebrate taste cell where fine 5-7 nm diameter fibers and single cytoplasmic microtubules descend from microvilli.[20]

+ except in frog olfactory cilia which are slightly motile and preserve a short proximal section of 9+2 configuration before separating.

If the taste cell requires protein conduction, it can be speculated that it is done here through non-ciliary microtubules, or perhaps even bundles of fine protein filaments. See Figs. 3 and 4.

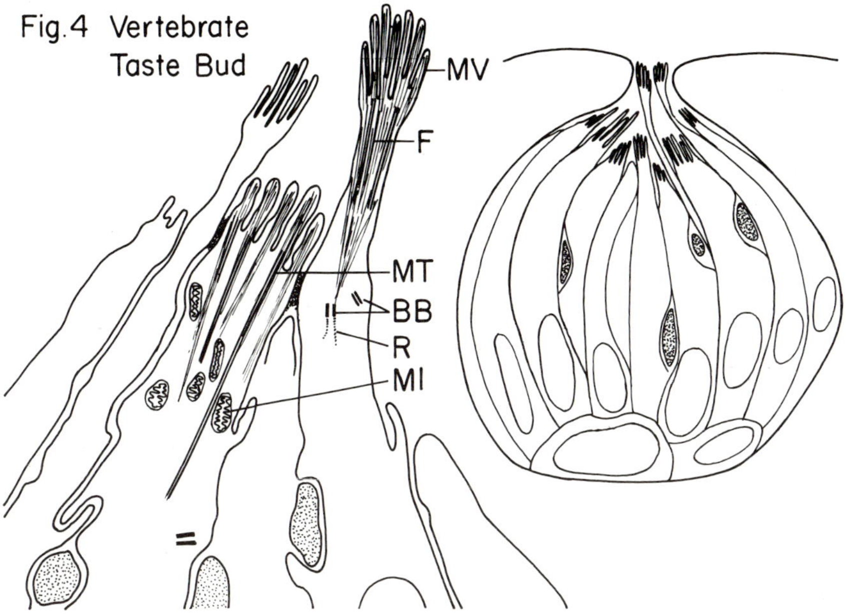


Fig.4. Vertebrate taste cell.[20] Left: whole taste bud; right: detail of apical cell portions. Microvilli extend distally into the taste pit. In the long neck which connects the distal microvilli with the proximal cell body, regular parallel fibers are found next to cytoplasmic microtubules Centrioles can be observed in the region where these fibers end, as well as much more proximally, as far as near the nucleus. The rather compact shape of this cell type perhaps does not require distance transmission via protein fibers as suggested for other sensory cell types. However, the arrangement of large numbers of small parallel fibers and cytoplasmic microtubules running from the inside of the microvilli through the neck of the cell to the supranuclear areas, is intriguing in the context of this hypothesis. Abbreviations as in Fig. 1.

It thus seems plausible that the microtubules themselves, with or without ciliary configuration, must be considered for a transduction function. The strongest case for their active function in sensory transduction must necessarily come from olfactory or other ciliary chemoreceptors. It is here that special conduction mechanisms have been postulated[21], that single periferal microtubles occur, and must stretch a length of up to 1000 μm[19] (pers.obs.) and that high sensitivity for specific stimuli (pheromones) are known.

The possible roles of microtubules can thus be summarized as:

1) structure, e.g. cytoskeleton, morphogenesis, mitosis,
2) transport, e.g. neuronal transport, pigment migration, perhaps with actin as force generator,
3) motility, e.g. motile cilia, and
4) intracellular communication.

It is thought to be likely that more than one role can be performed simultaneously. If proven true this concept will have implications for such diverse fields as neurophysiology (dendrite conduction), protist evolution and behaviour, and clinical medicine.

Acknowledgements

I want to express my gratitude to Dr. Lynn Margulis for many stimulating discussions, continuous support and proof reading of the manuscript; to Ilyse Rosenthal Atema for the illustrations; and to Boston University Graduate School for financial assistance.

References

1. Adler, J., 1973, in: Behaviour of Micro-organisms, ed. A. Perez-Miravete (Plenum, New York).

2. Kung, C., 1973, in: Behaviour of Micro-organisms, ed. A. Perez-Miravete (Plenum, New York).
Kung, C., et.al., 1975, Science 188, 898.

3. Tucker, J.B., 1971, J. Cell Sci. 9. 539.

4. Mornin, L. and D. Francis, 1967, J. Micr. 6: 759.
Greuet, C. 1968, Protistologica 4 : 209.
Greuet, C. 1969, Protistologica 5 : 481.
Greuet, C. 1975, pers. comm. via M. Steinert.

5. Barlow, R.E. Jr., 1975. pers. comm.

6. Nott, J.A. and B.A. Foster, 1969, Phil. Trans.R. Soc. Lond. B 256,115.

7 Von Bekesy, G. 1964, Science 145, 834.
Anderson, H.T. 1970, in : Taste and Smell in Vertebrates, eds. G.E.W. Wolstenholme and J. Knight (Churchill, London).

8. Hagins, W.A., 1972, Ann. Rev. Biophys. Bioeng. 1, 131.

9. Lowenstein, O., et.al., 1964, Proc. R. Soc. Lond. B 160, 1.

10. Moran, D.T., et al., 1975, Cell Tiss. Res. (in press).

11. Moran, D.T. and F.G. Varela, 1971, Proc. Nat. Acad. Sci. U.S.A. 68, 757.

12. Gaffal, K.P. and U. Bassemir, 1974, Protoplasma 82, 177.

13. Plattner, H., 1975, J. Cell Sci. 18. 257.

14. Atema, J., 1973, J. Theor. Biol. 38, 181.

15. Spinola, S.M. and K.M. Chapman, 1975, J. comp. Physiol. 96. 257.

16. Block, E.F. and W.J. Bell, 1974, J. Insect Physiol. 20, 993.

17. Levandowsky, M., *et. al.*, 1975, J. Bact. (in press).

18. Moran, D.T. and J.C. Rowley, 1975, Tiss. Cell 7, 91.

19. Ghiradella, H., *et.al.*, 1968, Protoplasma 66, 1.

20. Murray, R.G. and A. Murray, 1967, J. Ultrastruct. Res. 19, 327.

21. Ottoson, D. and G.M. Shepherd, 1967, Prog. Brain Res. 23, 83.
Beidler, L.M., 1970, in: Taste and Smell in Vertebrates, eds. G.E.W. Wolstenholme and J. Knight (Churchill, London).

22. Graziadei, P. 1972, in: Handbook of Sensory Physiology IV, 1, ed. L.M. Beidler, (Springer Verlag, New York).

*Microtubules and Microtubule Inhibitors, eds. M. Borgers and M. de Brabander*



# EFFECTS OF COLCEMID AND LITHIUM IONS ON PROCESSES OF CULTURED CELLS DERIVED FROM THE NERVOUS SYSTEM.

G. Reiser, E. Lautenschlager and B. Hamprecht.

Max-Planck-Institut für Biochemie, 8033 Martinsried,
Federal Republic of Germany

## 1. Summary

Glioma cells and neuroblastoma x glioma hybrid cells treated with dibutyryl cyclic AMP extend long processes. Colcemid added to such cultures causes the retraction of the processes within 2 to 3 hours. Lithium ions (10 mM), when added together with colcemid, strongly delay this effect of colcemid. In contrast, the effect of vinblastine, which also causes disappearance of processes of the hybrid cells is not prevented by $Li^+$. The usual cessation of cell proliferation in the presence of colcemid, also observed with the hybrids and their parental cells, is delayed by 10 mM $Li^+$. The results may be explained by a stabilizing action of $Li^+$ on the microtubules exposed to colcemid. It is not yet known whether $Li^+$ ions interfere with the binding of colcemid to tubulin.

## 2. Introduction

Microtubules serve a variety of functions in eukaryotic cells, the common denominator of which is the maintenance and the motility of cellular structures. The microtubules are involved in cell movement, intracellular organization and shaping of cells, chromosome movement during cell division and exocytotic processes [1-3]. They are essential for the outgrowth of fibers from nerve cells and cultured neuroblastoma cells [1]. Colchicine and colcemid disrupt the microtubular structure, and thus interfere with most of the processes in which microtubules play a central role [1]. These agents prevent the extension of neurites from neurons [4-9] and neuroblastoma cells [10,11], or cause the retraction of such cellular extensions. In addition, disintegration of the microtubules by colchicine inhibits axoplasmatic transport in nerve fibers [12-17]. Rat glioma [18] and mouse neuroblastoma x rat glioma hybrid cells [19,20] generate long processes, if treated with $N^6,O^{2'}$-dibutyryl adenosine 3':5'- cyclic monophosphate (dibutyryl cyclic AMP). Here we report that such processes are retracted, if colcemid is added to the culture medium, and that this effect is strongly retarded in the presence of lithium ions.

## 3. Methods

Cell culture. Mutant rat glioma line C6-BU-1 [21] and mutant mouse neuroblastoma line N18TG2 [22] and the mouse neuroblastoma x rat glioma line 108CC5 were regularly cultured in plastic dishes, as earlier described [23]. For extension of processes the glioma [18] and the hybrid cells [19,20] were incubated in culture medium supplemented with 1 mM dibutyryl cyclic AMP. For study of the effects of colcemid (Calbiochem), vinblastine sulfate (Eli Lilly) and LiCl, these compounds were used at final concentrations of 0.13, 10 and 10000 μM respectively, in 60 mm plastic dishes (Greiner, Nürtingen FRG) containing 5 ml of culture medium (37° C). Stock solutions of these substances were added to the culture dishes without renewal of the culture medium. For establishing the growth curves, the cells were seeded in 85 mm plastic dishes and cultured and counted as described [23].

Photography. The change in the morphology of the cells treated with various agents was recorded by serial photographs on an inverted microscope with phase contrast optics either by hand (Leitz "Diavert") or by a time-lapse cinematographic equipment (Reichert "Biovert" plus "Leicina" camera).

## 4. Results

On prolonged treatment with dibutyryl cyclic AMP, the neuroblastoma x glioma hybrid cells differentiate by extending long processes (fig. 1 A). After addition of colcemid to the culture medium, the processes are retracted and the cells detach from the surface of the culture dish within 2 to 3 hours (fig. 1 B-D). No processes are seen in cultures that have been treated with colcemid for 10 hours (fig. 1 E). However, if 10 mM LiCl was present, processes are abundant even after 10 hours of exposure to colcemid (fig. 1 F).

Fig. 1. Influence of colcemid and $Li^+$ on cell processes of neuroblastoma x glioma hybrid cells. 4 x $10^4$ viable hybrid cells were treated with dibutyryl cyclic AMP for 14 days. Subsequently the cells were treated with colcemid and $Li^+$. A. cells before exposure to colcemid. B,C,D,E cells after exposure to colcemid for 30, 60, 145 and 600 min. F. Cells treated for 600 min with colcemid + $Li^+$. At all times except at 600 min, the same area of the dish was photographed. Cells not treated with colcemid or colcemid + $Li^+$ are indistinguishable from those in A. Analogously, this also applies to figures 2 and 3. The bar in this figure and figure 2 and 3 represents 100 μm.

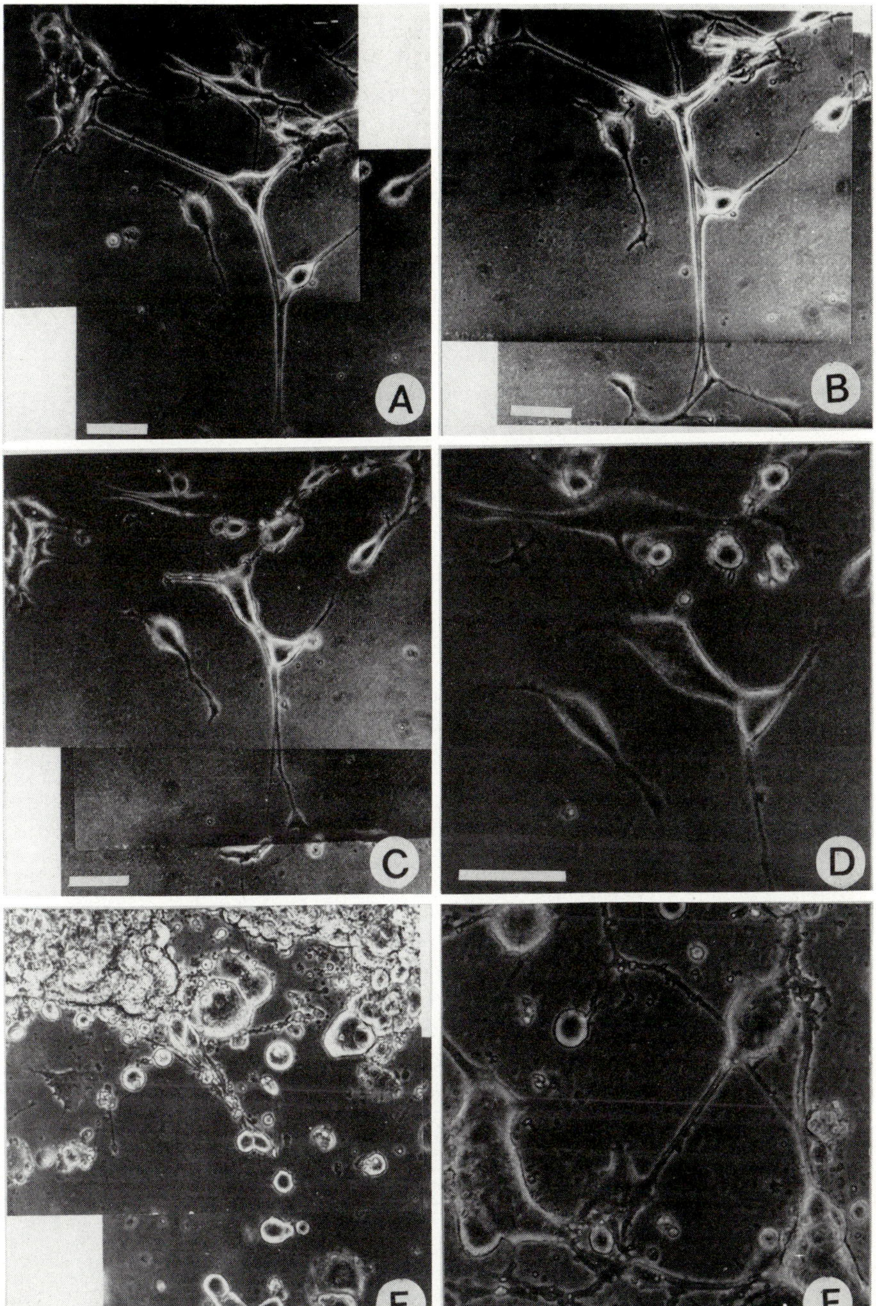
A
B
C
D
E
F

In the presence of LiCl, a reduction in the number of processes is observed only after 24 hours of incubation with colcemid. In the absence of colcemid, control cultures (with or without LiCl) do not obviously change their number of processes. An appreciable effect of $Li^+$ was only observed in concentrations above lmM. It was of interest to see whether or not the hybrid cells (as model neurons) were protected by $Li^+$ against the effect of colcemid. Experiments with rat glioma cells demonstrated that this antagonism between colcemid and $Li^+$ existed also in other cell types (fig. 2 A-F). In the presence of colcemid the glioma cells differentiated by treatment with dibutyryl cyclic AMP lose their processes and detach from the plates even more rapidly than the hybrid cells (within 1 to 2 hours). Again $Li^+$ strongly delays the loss of processes.

Although vinblastine, like colcemid, acts antimitotically by disrupting the microtubules, it binds to sites on tubulin that are different from the binding sites for colchicine [1]. As seen from fig. 4 A - F, vinblastine causes a strong reduction in the number of processes within a few hours. In the case of vinblastine the processes are not retracted but rather dwindle into small vesicles. Lithium ions only slightly retard this process, if at all. After 24 hours of treatment with the vinca alcaloid, all processes have disappeared, irrespective of the presence of $Li^+$.

Colchicine and its analogue colcemid, at the low concentrations used here, appear to specifically disrupt microtubules engaged in a wide variety of cellular events [1]. If $Li^+$ specifically antagonized this action of colcemid on microtubules, it was to be expected to do this generally. It was, e.g., expected that $Li^+$ would partially prevent the antimitotic action of colcemid. In fig. 3 growth curves of hybrid cells (fig. 3 A), glioma cells (fig. 3 B) and neuroblastoma cells (fig. 3 C) are shown. On addition of colcemid to the cultures, cell proliferation ceases rapidly. However, the number of cells does not remain constant. Rather, due to cell rupture, it decreases with the time of exposure to colcemid. This was demonstrated by time lapse cinematography. In cultures exposed to colcemid and $Li^+$,

Fig. 2. Influence of colcemid and $Li^+$ on cell processes of rat glioma cells. $2 \times 10^4$ viable glioma cells were treated with dibutyryl cyclic AMP for 14 days. Subsequently, the cells were treated with colcemid and $Li^+$. A, cells before exposure to colcemid. B, F cells after exposure to colcemid for 1.5 h and 10 h. C,D,E cells were incubated with colcemid and $Li^+$ for 1.5, 7, 10 h.

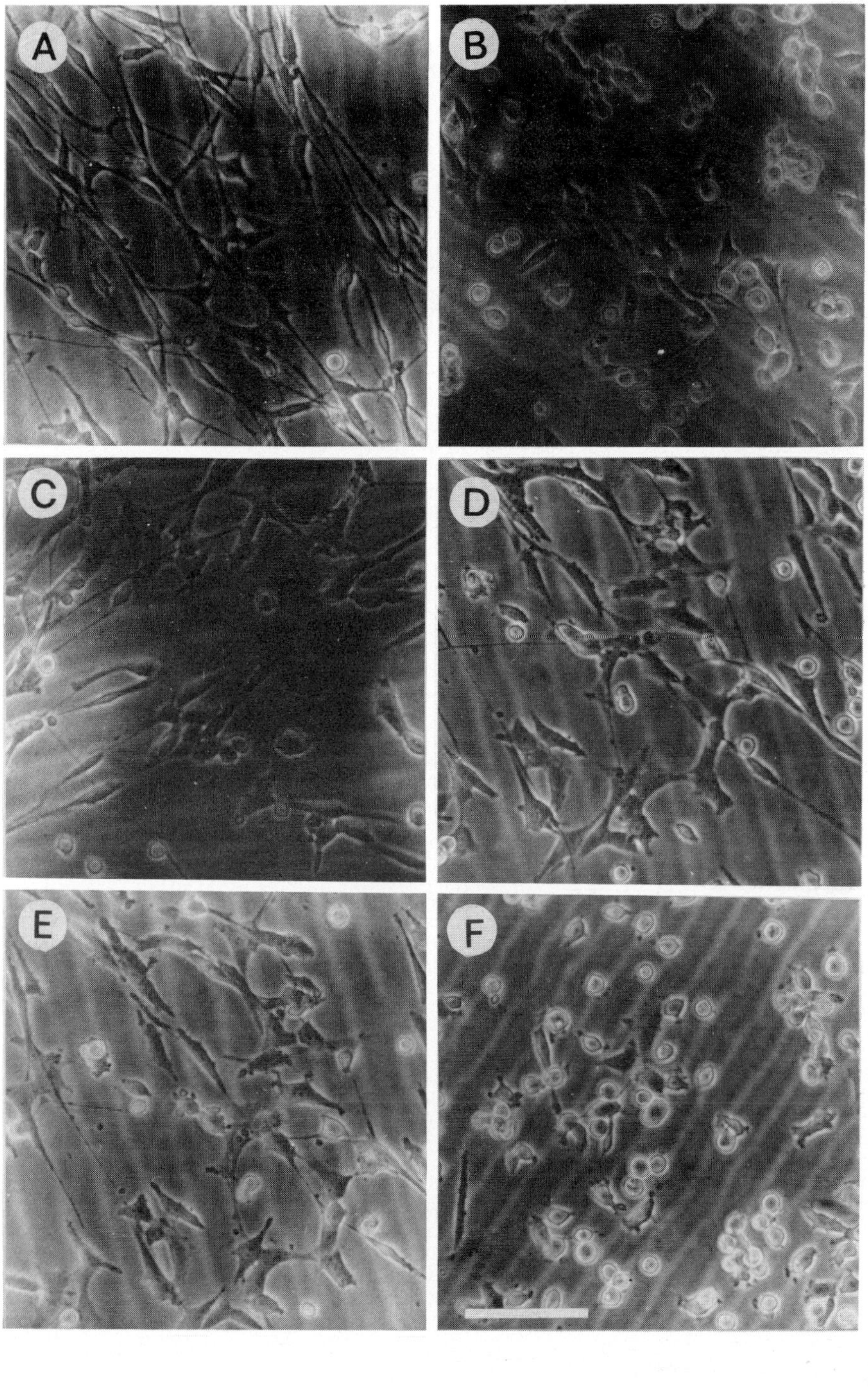
A
B
C
D
E
F

cell proliferation ceases less rapidly than in those treated with colcemid alone (fig. 3 A-C).

Treatment of the hybrid cells and their parental lines with colcemid, loosened the attachment of the cells to the surface of the culture plates. Hardly any trypsinization was required for detachment of the cells. In the presence of $Li^+$ the effect of colcemid on the adhesion of the cells to the plates was strongly attenuated. It is important to note that even in the absence of colcemid a marked increase in cell adhesion to the plates is observed after addition of $Li^+$. In this case prolonged trypsinization and much more vigorous rocking of the plates is necessary for detaching the cells.

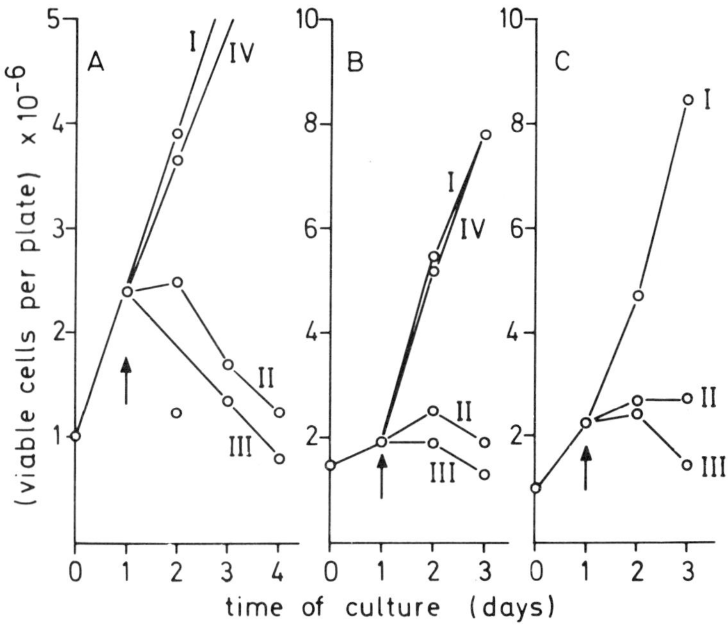


Fig. 3. Effect of colcemid and $Li^+$ on the growth of hybrids (A) glioma (B) and neuroblastoma (C) cells. I, control (absence of colcemid and $Li^+$). II, colcemid and $Li^+$; III, colcemid; IV $Li^+$. The additions were made at the times indicated by the arrows.

---

Fig. 4. Influence of vinblastine sulfate on cell processes of hybrid cells. 4 x $10^4$ viable cells were treated with dibutyryl cyclic AMP for 14 days. Subsequently, the cells were treated with vinblastine. A, cell before exposure to vinblastine. B,C,D,E cells after treatment with vinblastine for 0.5, 1.5, 2.5, 19 h. The arrow points at the same cell soma, as observed at different times.

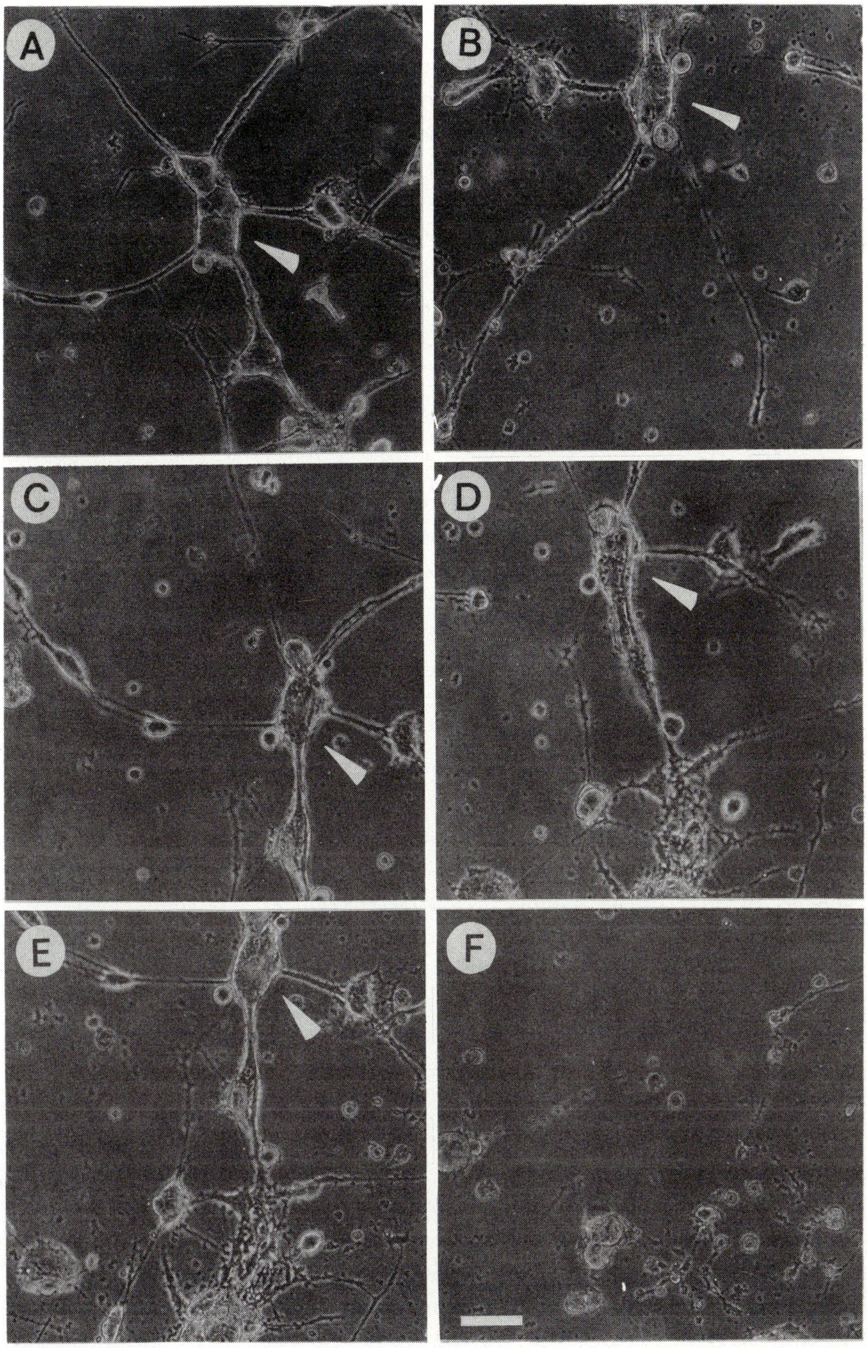
A
B
C
D
E
F

5. Discussion

Rat glioma cells[24] and rat glioma x mouse neuroblastoma cells[25] have been used as model systems for studying glia and nerve cell properties, respectively, since they possess many qualities characteristic of these cell types. In the present study they were used to obtain some insight into the effects of $Li^+$ ions on cellular functions requiring the participation of microtubules. In these glioma and glioma x neuroblastoma hybrid cells, colcemid causes the retraction of cellular extensions. In both cell lines, this effect is antagonized by $Li^+$ ions in millimolar concentrations. The colcemid concentration used (0.13 μM) was probably low enough to assure exclusive action on microtubules [3]. Thus $Li^+$ might have stabilized the microtubules or might have prevented the binding of colcemid to tubulin. Experiments are in progress to elucidate the mechanism of the interference of $Li^+$ with the action of colcemid. From the hybrid cells processes also disappear on treatment with vinblastine; this effect is not influenced by $Li^+$ ions. Thus, the effect of $Li^+$ appears to be directed rather specifically towards the action of colcemid, but not of vinblastine. Further evidence for an involvement of $Li^+$ with microtubular structure comes from the fact that the alkali ion also partially obviates the action of colcemid on cell proliferation.

$Li^+$ by itself has obvious effects on the undifferentiated and differentiated glioma and hybrid cells. It enhances the adhesion of the cells to their substrate. This is not necessarily in disagreement with the finding that neuroblastoma microtubules are not affected by exposure of neuroblastoma cells in culture to $Li^+$ ions, [26] since no further details were given by the author. Our observation of increased cell adhesion in the presence of $Li^+$ points again to the involvement of microtubules in the action of $Li^+$. It is known that disruption of microtubules by colchicine causes loss of cell adhesion [27]. Furthermore, it has been reported that $Li^+$ inhibits the depolymerization of at least some parts of the aggregated material in the mitotic apparatus [28].

$Li^+$ salts have gained increasing importance in the therapy of mania and probably also depression [29-36]. There are indications that $Li^+$ affects the rate of metabolism of monoamines in the brain [37-43]. The processes involved in monoamine action, axoplasmic transport [12-17] and release [44], are dependent on intact microtubules. Consequently, in view of our findings that $Li^+$ protects microtubules from disruption by colcemid, it cannot be excluded that an important action of $Li^+$ in therapy of mania and depression is on the metabolism of neurotransmitters. At the moment, this is mere speculation.

## 6. Acknowledgement

We are grateful to Drs. N. Matussek and M. Schou for helpful discussions and to the Sonderforschungsbereich 51 of the Deutsche Forschungsgemeinschaft for support.

## References

1. Olmsted, J.B. and Borisy, G.G., 1973, Ann. Rev. Biochem. 42, 507-540.
2. Margulis, L., 1973, Internatl. Rev. Cytol. 34, 333-361.
3. Du Praw, E.J., 1974, Adv. Cell. Mol. Biol. 3, 21-72.
4. Lyser, K., 1968, Devel. Biol. 17, 117-142.
5. Tennyson, V., 1970, J. Cell. Biol. 44, 62-79.
6. Yamada, K., Spooner, B. and Wessells, N., 1970, Proc. Natl. Acad. Sci. U.S. 66, 1206-1212.
7. Roisen, F. and Rebhuhn, L., 1971, Physiologist 14, A 220.
8. Daniels, M., 1972, J. Cell Biol. 53, 164-176.
9. Hier, D., Arnason, B. and Young, M., 1972, Proc. Natl. Acad. Sci. U.S. 69, 2268-2272.
10. Seeds, N., Gilman, A., Amano, T. and Nirenberg, M., 1970, Proc. Natl. Acad. Sci. U.S. 66, 160-167.
11. Schubert, D., Humphreys, S., de Vitry, F. and Jacob, F., 1971, Devel. Biol. 25, 514-546.
12. Dahlström, A., 1968, Eur. J. Pharmacol. 5, 111-112.
13. Kreutzberg, G., 1969, Proc. Natl. Acad. Sci. U.S. 62, 722-728.
14. Davison, P., 1970, Adv. Biochem. Psychopharm. 2, 289-302.
15. Sjöstrand, J., Frizell,M. and Hasselgren, P., 1970, J. Neurochem. 17, 1563-1570.
16. Fernandez, H., Burton, P. and Samson, F., 1971, J. Cell. Biol. 51, 176-192.
17. Ochs, S., 1972, Science 176, 252-260.
18. Hamprecht, B., Jaffe, B.M. and Philpott, G.W., 1973, FEBS Lett. 36, 193-198.
19. Hamprecht, B., Traber, J. and Lamprecht, F., 1974, FEBS Lett. 42, 221-226.
20. Hamprecht, B., 1974, Mosbacher Kolloquium 25, 391-423.
21. Amano, T., Hamprecht, B. and Kemper, W., 1974, Exp. Cell Res. 85, 399-408.
22. Minna, J., Glazer, D. and Nirenberg, M., 1972, Nature New Biol. 235, 225-231.

23. Hamprecht, B. and Schultz, J., 1973, Hoppe-Seyler's Z. Physiol. Chem. 354, 1633-1641.

24. Benda , P., Lightbody, J., Sato, G., Levine, L. and Sweet, W., 1968, Science, 161, 370-371.

25. Traber, J., Fischer, K., Buchen, C. and Hamprecht, B., 1975, Nature 255, 558-560.

26. Wolff, J., Proc. 9th Congr. Collegium Internationale Neuropsychopharmacologicum, in press.

27. Asnes, C.F. and Smith, E.E., 1974, Fed. Proc. 33, 1392.

28. Petzelt, C., 1972, Exp. Cell Res. 70, 333-339.

29. Cade, J.F.J., 1949, Med. J. Austr. 36, 349-352.

30. Schou, M., Juel-Nielsen, N., Stromgren, E. and Voldby, H., 1954, J. Neurol. Neurosurg. Psychiat. 17, 250-260.

31. Schou, M., 1959, Psychopharmacologia 1, 65-78.

32. Gershon, S. and Yuwiler, A., 1960, J. Neuropsychiat. 1, 229-241.

33. Maggs, R., 1963, Brit.J. Psychiat. 109, 56-65.

34. Schou, M., 1968, J. Psychiat. Res. 6, 67-95.

35. Davis, J.M. and Fann, W.E., 1971, Ann. Rev. Pharmacol. 11, 285-302.

36. Murphy, D.L. and Beigel, A., 1974, Arch. Gen. Psychiat. 31, 643-648.

37. Corrodi, H., Fuxe, K., Hökfelt, T. and Schou, M., 1967, 11, 345-353.

38. Schanberg, S.M., Schildkraut, J.J. and Kopin, I.S., 1967, Biochem. Pharm. 16, 393-399.

39. Stern, D.N., Fieve, R.R., Neff, N.H. and Costa, E., 1969, Psychopharm. 14, 315-322.

40. Greenspan, K., Aronoff, M.S. and Bogdanski, D.F., 1970, Pharmacology, 3, 129-136.

41. Ho, A.K.S., Loh, H.H., Craves, F., Hitzemann, R.J. and Gershon, S., 1970, Eur. J. Pharmacol. 10, 72-78.

42. Grahame-Smith, D.G. and Green, A.R., 1974, Brit.J. Pharmacol. 52, 19-26.

43. Peterson, N.A.,and Raghupathy, E., 1974, Biochem. Pharm. 23, 2491-2494.

44. Gillespie, E., Levine, R. and Malawista, S., 1968, J. Pharmacol. Exp. Ther. 164, 158-165.

*Microtubules and Microtubule Inhibitors, eds. M. Borgers and M. de Brabander*



# NEUROTUBULIN AND BRAIN DEVELOPMENT

J. Nunez, A. Fellous, J. Francon and A.M. Lennon

Unité de Recherche sur la Glande Thyroide et la Régulation Hormonale
I.N.S.E.R.M., Bicêtre, France.

## 1. Introduction

Considerable interest has been devoted in the past few years to brain microtubules : the concentration of microtubule protein is much higher in cells of the central nervous system than those of other tissues (1) ; neurotubules are components of the axone and are believed to play an important role in axone function and brain differentiation (2). However these data do not provide an answer to the following questions : is neurotubulin progressively assembled into neurotubules as soon as synthesized ? In this case the rate limiting factor would be the amount of tubulin present at each stage of development. Another possibility would be that all neurotubulin would be formed before a given period of development without undergoing massive polymerisation. In this case, at a precise stage of development, a specific signal would induce its polymerisation to neurotubules followed by rapid axone growth. An attempt was made in this work to answer these questions.

## 2. Determination of the neurotubulin content during brain development.

The colchicine binding activity of microtubule protein may be used to determine the concentration of neurotubulin (3 - 4). However, in usual buffers this activity decays in an apparent first order manner (5) ; in phosphate buffer, for instance, a number of factors (pH, ionic strength, temperature and microtubule concentration) markedly affect the decay rate. Thus, in these conditions, the values obtained must be corrected for the decay rate. Weisenberg (6) and Shelanski (7) have shown that tubulin prepared in a MES buffer and in the presence of EGTA and GTP $10^{-3}$ M is still able to polymerize *in vitro* to reconstitued microtubules. This medium probably protects the neurotubulin against denaturation. We have therefore used a MES buffer to study colchicine binding to neurotubulin. In this medium the saturating concentration of colchicine was $2.5 \times 10^{-5}$ M. Fig. 1 shows that with the 30 day old supernatant no decay of the binding activity was observed even after 2 hours of incubation. Similar results were obtained with the 9 day old preparations (not shown). Scatchard plots (not shown) of the results reported in Fig. 1 revealed that some non-specific binding was observed for colchicine concentrations higher than $2.5 \times 10^{-5}$ M. In the MES buffer at 37 C the half life (Fig. 2) of rat neurotubulin was much longer (∿ 10 hours) than that found by Bamburg et al. (8)

for the chick protein in phosphate buffer (∿ 4 hours). In addition no difference was found in the half life at different stages of development. The half life was also constant in a large range of pH (6.3 - 8) whereas in the assay used by Bamburg et al. (8) the decay rate was very sensitive to pH and varied with the stage of development.

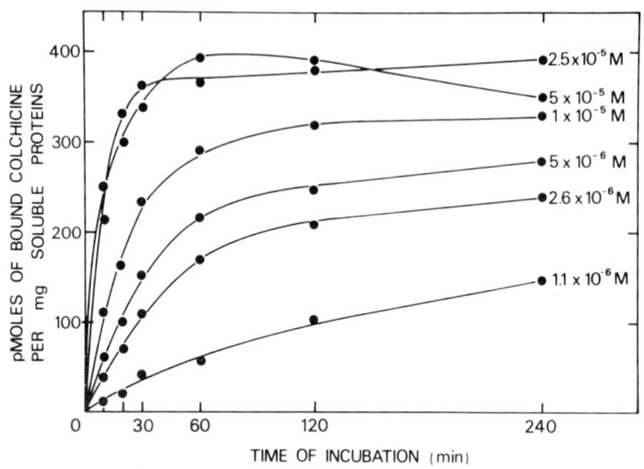


Fig. 1. Colchicine binding activity in supernatant fractions at different colchicine concentrations.
The supernatant was obtained after homogeneisation of 30 day old rat brains in the reassembly buffer of Weisenberg (6) containing 0.1 MES (pH 6.5), 1 mM EGTA, 1 mM GTP and 0.5 mM $MgCl_2$, at 4 C, and further centrifugation at 105 000 g for 1 hour at 4 C. MES-Glycerol buffer (final concentration of glycerol 4 M) was added to this supernatant ; this preparation was incubated in the presence of different concentrations of $^3$H-colchicine as indicated in the figure. Bound colchicine was measured as described by Weisenberg (12).

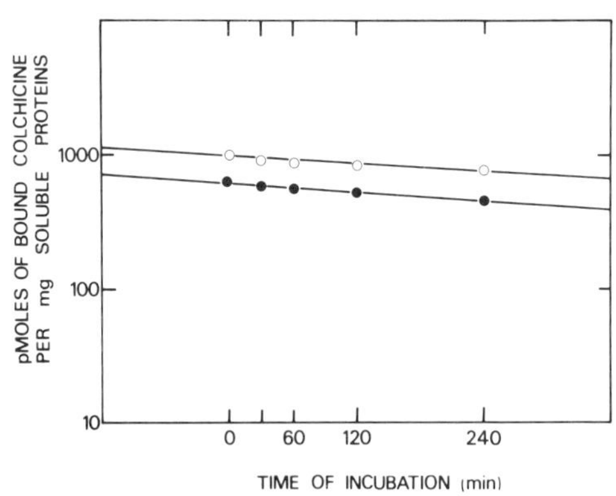


Fig. 2. First order decay of colchicine binding capacity in the 9 day and 30 day old rat brain supernatants.
Supernatant fractions prepared as described in Fig. 1 were preincubated at 37 C. At the times indicated in the figure, 2.5 x $10^{-5}$ M $^3$H-colchicine was added and incubations were continued for 2 hours. Bound colchicine was measured as described in Fig. 1. (○—○) 9 day old and (●—●) 30 day old rat brain supernatants.

The relationship between the amount of purified neurotubulin (depolymerized by cold) and colchicine binding performed in MES buffer was linear (Fig. 3) ; Similar results were found for the 30 day old and the 9 day old tubulin preparations. However, maximal binding activity at 0 C in the MES buffer was always

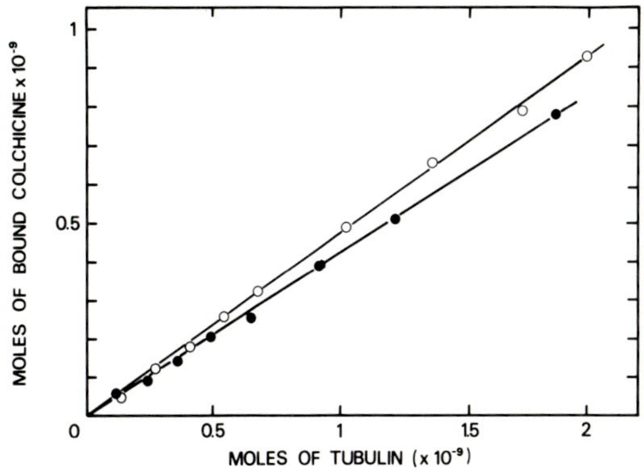


Fig. 3. Relationship between tubulin concentration and colchicine binding activity. Purified microtubules, prepared as described by Shelanski et al. (7) were depolymerized by cold ; this tubulin preparation was then incubated with 2.5 x $10^{-5}$ M $^3$H-colchicine for 2 hours at 37 C ; bound colchicine was measured as described in Fig. 1. Binding activity of tubulin prepared from 9 day (O—O) and 30 day (●—●) old rat brain.

equal to only 0.5 moles of colchicine per mole of purified neurotubulin (Fig. 3). If glycerol was added to the MES buffer this value was reduced to 0.3 moles. As shown by Shelanski (7) glycerol favours the polymerisation of neurotubulin. On the other hand it has been shown recently (9) that depolymerisation of reconstitued microtubules by cold is a slow process which is far from being complete after 1 hour. Therefore if colchicine is bound only to fully dissociated microtubules i.e. to 6 S neurotubulin (3) it is not surprising that only one fraction of the molecules is able to bind this antimitotic agent when the buffer contains MES (and glycerol). Fig. 3 shows however that the proportion of tubulin which binds colchicine is constant whatever the concentration of this protein was. Thus the assay could be used but a correction must be introduced if the dosage is expressed per mg or moles of tubulin.

Fig. 4 represents the results obtained when colchicine binding activity was determined in the 105 000 g rat brain supernatant at different stages of development. 6 days before birth (15 days of gestation) the fetal rat brain contains as much colchicine binding tubulin as the 30 day old brain. From this fetal stage till 9 - 10 days after birth the colchicine binding activity markedly increases and then decreases gradually in the adult to half of the 9 - 10 day value. Some colchicine binding activity was also present in the 105 000 g pellet of the brain homogenate which could be extracted by KCl solutions. Fig. 4 also shows that the evolution of the colchicine binding activity present in the KCl extracts is similar to that found in the supernatant. If one admits that only one half of the tubulin molecules are estimated in the MES buffer (Fig. 3) the neurotubulin content (soluble and extracted by KCl) of the brain is equal to ∿ 15 % at the fetal and adult stages and to 25 % - 30 % 9 - 10 days after birth.

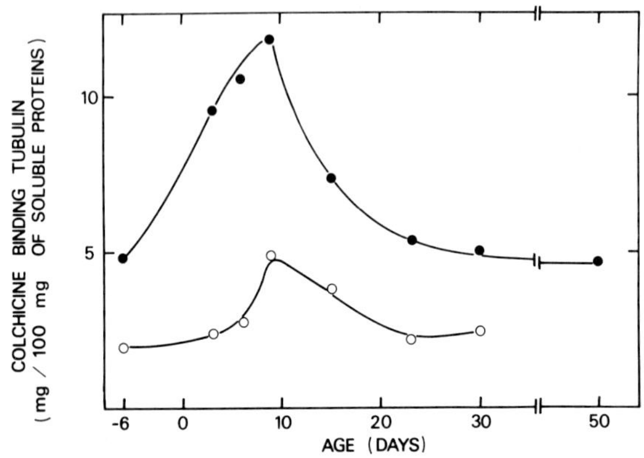


Fig. 4. Colchicine binding activity in supernatant and particulate fractions of rat brain at different stages of development.
The supernatant fraction was obtained as described in Fig. 1. The particulate fraction was obtained after extraction of the 105 000 g pellet with the Weisenberg buffer containing 0.1 M KCl. Glycerol was added to a final concentration of 4 M. Aliquots of the supernatant fraction (●—●) and of the KCl extract of the particulate (○—○) were incubated with $2.5 \times 10^{-5}$ M $^3$H-colchicine as described in Fig. 3.

3. *In vitro* polymerisation of neurotubulin prepared at different stages of brain development.

Crude soluble fractions (105 000 g supernatant of the homogenate) were prepared in the conditions described by Shelanski (7) from brains of different ages and then incubated at 37 °C. Fig. 5 depicts the time dependence of turbidimetry for assembling tubulin to neurotubules *in vitro* from the supernatants of fetal and 30 day old rat brains. Identical concentrations of colchicine binding neurotubulin were used for the two ages. Fig. 5 shows that the initial rate of polymerisation and maximal turbidimetry of the fetal preparation were much lower than those of the 30 day old preparation. In addition the lag period was much longer. Gaskin et al. (10) have shown that both the lag period and the initial rate of polymerisation probably depend on the amount of "initiator" present. Fragments of microtubules obtained by sonication may act as "initiators". Addition of small amounts of fragments of adult microtubules to adult brain supernatants resulted in a shorter lag period and in the overall increase in the rate of assembly (Fig. 6a). However the change in total turbidimetry was not very important and saturation was rapidly obtained. On the contrary addition of increasing amounts of fragments of microtubules from adult brain to the fetal preparation resulted in a dramatic change in turbidity (Fig. 6b) : the lag period was progressively decreased and the initial rate and the overall turbidity were markedly augmented. Gaskin et al. (10) have

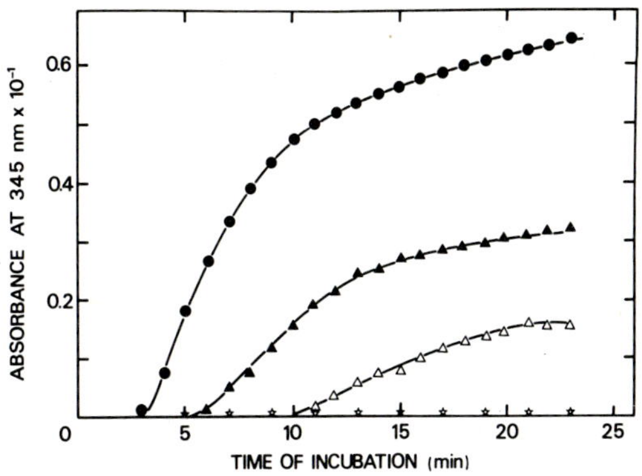


Fig. 5. Turbidimetry time course at 37 C of fetal and 30 day old rat brain supernatants.
The supernatants were prepared as described in Fig. 1. Turbidimetry measurements during the incubation at 37 C were performed at 345 nm using a Gilford 200 spectrophotometer with an automatic thermostated four sample changer. Fetal (△—△) and 30 day (●—●) old rat brain supernatants containing the same amount of colchicine binding tubulin (2100 picomoles) were compared. Turbidimetry time course experiments were also performed with a dilution of 30 day old rat brain supernatant containing only 300 picomoles of colchicine binding tubulin (☆—☆) and with a mixture (▲—▲) of fetal preparation (colchicine binding tubulin 2100 picomoles) and the same amount of 30 day old rat brain supernatant (colchicine binding activity 300 picomoles).

also studied the effect of protein concentration on equilibrium turbidimetry ; they have found that below a certain concentration (critical concentration, $C_c$) turbidimetry is the same as for the unpolymerized protein. For the fetal stage $C_c$ was 6 - 7 times higher than for the 30 day old preparation, (see Fig. 8). These results may signify that for low protein concentrations the "initiator" concentration is limiting and that the "initiator" is present in very small amounts in the fetal rat brain. Another possibility would have been that the fetal supernatant contained an inhibitor of polymerisation. Further experiments were performed to exclude this possibility : addition of small amounts of fetal brain supernatant to adult brain supernatant did not result in a decrease of the rate of polymerisation. On the contrary addition of very small volumes of adult brain supernatant to fetal brain preparations induced a marked increase in the polymerisation process (Fig. 5).

Similar experiments were performed with brain supernatants from 9 day old rats. Equilibrium turbidimetry was studied with different dilutions of soluble fractions prepared from 30 day and 9 day old rat brains. Fig. 7b shows the time dependent turbidimetry for assembling tubulin in a diluted (1177 picomoles of colchicine binding activity) supernatant from 30 day and 9 day old rat brains. In these conditions both the initial rate and maximal turbidimetry after 30 minutes of incubation, were lower for the 9 day old preparation. In addition the lag period was longer. These results might signify again that the amount of initiator is lower 9 days after birth than in the 30 old rat brain. From different kinetic

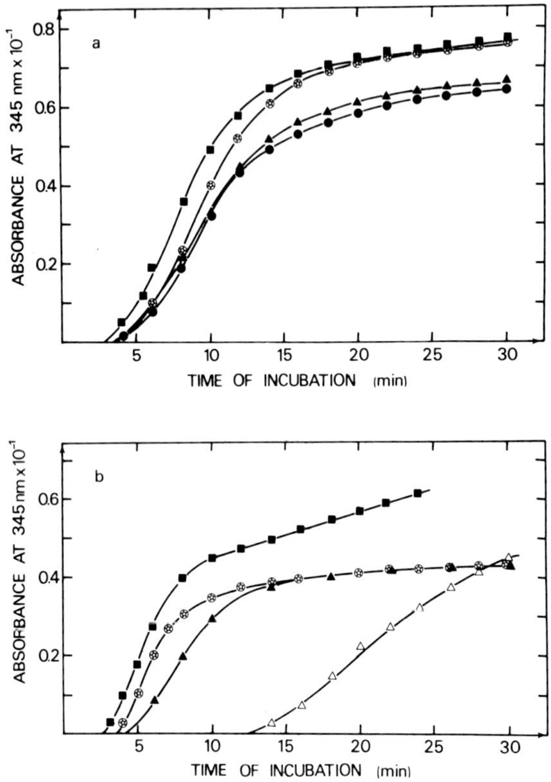


Fig. 6. Effect of 30 day old sonicated microtubules on turbidimetry time course of fetal and 30 day old rat brain supernatants.
Fetal and 30 day old rat brain supernatants were prepared as described in Fig. 1. Fragments of 30 day old microtubules were prepared according to the procedure of Gaskin et al. (10). Turbidimetry measurements were performed as described in Fig.5.
Fig. 6a. (●—●) 30 day old rat brain supernatant containing 1575 picomoles of colchicine binding tubulin to which different concentrations of sonicated microtubules (▲—▲) 32.5 µg, (☆—☆) 65 µg, (■—■) 130 µg were added.
Fig. 6b. (△—△) fetal supernatant containing 1642 picomoles of colchicine binding tubulin to which different concentrations of sonicated microtubules were added (▲—▲) 32.5 µg, (☆—☆) 65 µg, (■—■) 130 µg.

experiments performed with several dilutions of the brain soluble fractions prepared at 9 and 30 days initial rates were calculated and plotted as function of the colchicine binding activity. The critical concentration $C_c$ was 3 times higher (Fig. 9) for 9 days after birth. However, when the time dependent turbidimetry for assembling tubulin was assayed with less diluted soluble fraction (Fig. 7a) the 9 day old preparation polymerized faster than the 30 day old one. Fig. 8 also

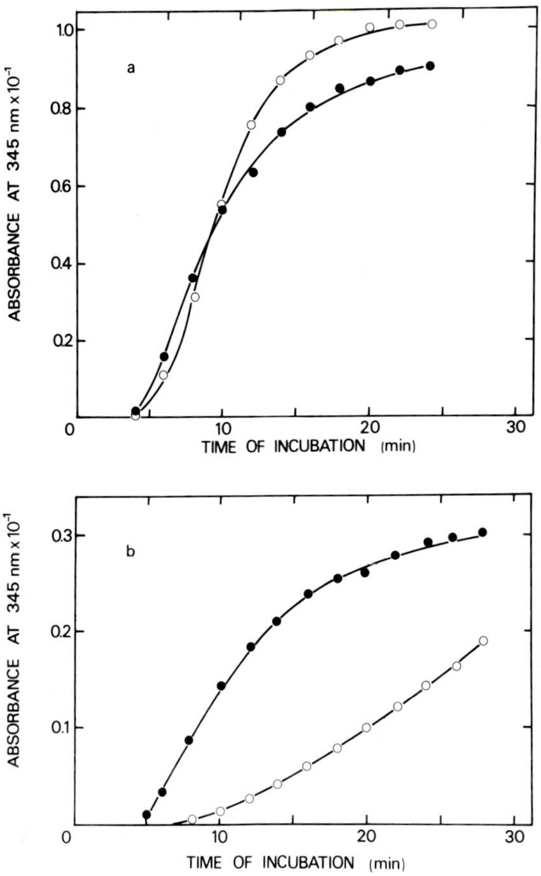


Fig. 7. Turbidimetry time course at 37 C of 9 day and 30 day old rat brain supernatants containing low and high concentrations of colchicine binding tubulin. The 9 and 30 day old rat brain supernatants were prepared as described in Fig. 1 and turbidimetry measurements were performed as described in Fig. 5.
Fig. 7a. 30 day (●—●) and 9 day (○—○) old rat brain supernatants containing an identical high amount of tubulin (3200 picomoles of colchicine binding tubulin).
Fig. 7b. 30 day (●—●) and 9 day (○—○) old rat brain supernatants containing an identical low amount of tubulin (1177 picomoles of colchicine binding tubulin).

shows that the relationship between initial rates of polymerisation and the colchicine binding activity was linear for the adult supernatant but sigmoid for the 9 day old preparation. Thus, although the concentration of "initiator" is likely to be lower in the 9 day old brain, the efficiency of polymerisation seemed better as soon as the tubulin concentration was increased. In other words the relative concentration of "initiator" seemed lower 9 days after birth than 21 days later

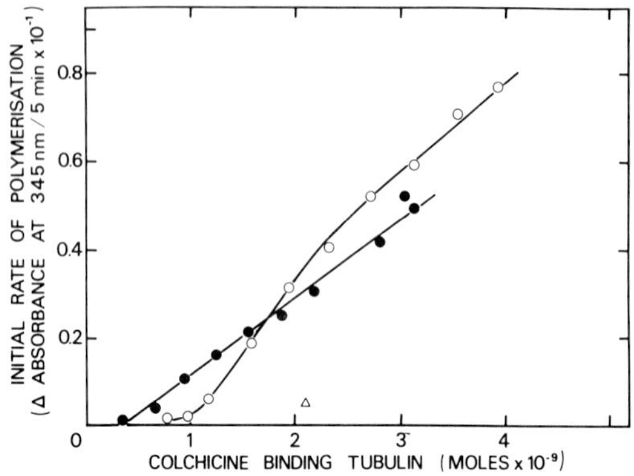


Fig. 8. Relationship between initial rate of polymerisation and colchicine binding tubulin concentrations of 30 day and 9 day old rat brain supernatants.
Initial rates (Δ absorbance at 345 nm/ 5 min.) were deduced from several kinetic experiments performed as described in Fig. 5, with various dilutions of the 30 day (●—●) and 9 day (○—○) old rat brain supernatants. Initial rate for undiluted fetal supernatant containing 2100 picomoles of colchicine binding tubulin is also represented (Δ).

but higher than in the fetal situation. In addition when the total protein was raised the concentration of initiator was no more limiting and some other unknown factor increased the rate of polymerisation. To test these assumptions microtubules assembled *in vitro* from 9 day old rat brain preparations were purified by ultracentrifugation at 20 C and dissociated by exposure at low temperature.

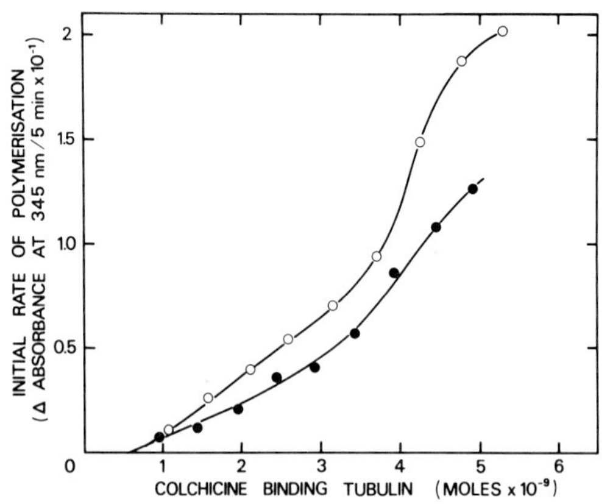


Fig. 9. Relationship between initial rates of polymerisation and tubulin concentration (colchicine binding activity).
Tubulins were prepared from 30 day and 9 day old rat brains. Purified microtubules thus obtained were depolymerized by cold and incubated again at 37 C. Turbidimetry measurements were performed as described in Fig. 5. Initial rates (Δ absorbance at 345 nm/ 5 min.) were deduced from several kinetic experiments performed with various dilutions of the 30 day (●—●) and 9 day (○—○) old rat brain preparations.

A second cycle of polymerisation was then performed at 20 C with different dilutions of this preparation. Fig. 9 shows that $C_c$ is now identical for the 9 day and 30 day preparations suggesting that the amount of "initiator" present in the polymerized microtubules is identical even if its concentration in the original supernatants was different. However the rate of polymerisation for 9 day was always higher than for 30 days. This result confirms therefore that another "factor" (activating) is present 9 days after birth. Further experiments were performed where the inhibitory effect of $Ca^{2+}$ on the rate of polymerisation of the 9 day and 30 day old preparations was studied. They suggest some difference in the microtubules obtained at these two stages. Fig. 10 shows that $Ca^{2+}$ inhibited much more

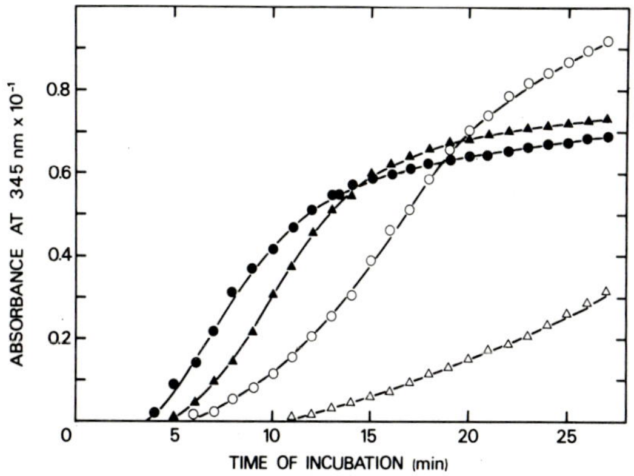


Fig. 10. Effect of $Ca^{2+}$ on turbidimetry time course in 9 day and 30 day old rat brain supernatants.
The supernatants were prepared as described in Fig. 1 ; 9 day old rat brain supernatant containing 1000 picomoles of colchicine binding tubulin with (△—△) and without (○—○) 0.4 mM $Ca^{2+}$ ; 30 day old rat brain supernatant containing 1400 picomoles of colchicine binding tubulin with (▲—▲) and without (●—●) 0.6 mM $Ca^{2+}$.

the polymerisation of tubulin from the 9 day old supernatant than that from the 30 day old one. Fig. 11 shows that the same conclusion was obtained when the effect of $Ca^{2+}$ was tested on the repolymerisation of purified neurotubulin.

## 4. Discussion.

The data reported in this work suggest that the concentration of tubulin is not a limiting factor in the process of neurotubule formation. At a late fetal stage, 6 days before birth, the amount of colchicine binding tubulin is equal to that found 30 days after birth. However this neurotubulin very poorly polymerized *in vitro* from the crude supernatant. Several data (critical concentration, lag period, initial rate of polymerisation, overall turbidimetry and effect of sonicated microtubule fragments), suggest that during the fetal stage the rate limiting factor in an "initiator" which is needed to allow the polymerisation of neurotubulin to neurotubules. The nature of this hypothetical "initiator" is unknown, but it is not excluded that it might be identical to a heat stable component (τ protein) recently described and isolated by kirshner et al. (11). Thus the results obtained in this work favour the second hypothesis made in the introduction i.e. a specific signal induces, at a precise stage of development, massive neurotubule polymerisation from a pre-existing pool of neurotubulin.

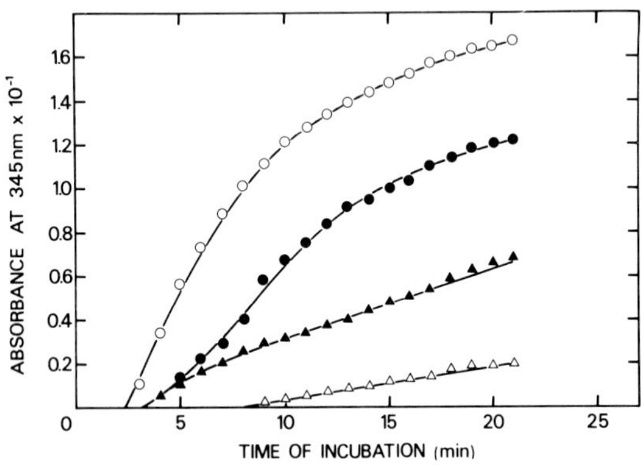


Fig. 11. Effect of $Ca^{2+}$ on turbidimetry time course in 9 day and 30 day old rat brain purified tubulins.
Purified tubulins were prepared as described in Fig. 2 ; 9 day old rat brain tubulin (1670 picomoles of colchicine binding activity) with (△—△) and without (○—○) 4 mM $Ca^{2+}$ ; 30 day old rat brain tubulin (1670 picomoles of colchicine binding activity) with (▲—▲) and without (●—●) 4 mM $Ca^{2+}$.

9 days after birth the situation seems more complicated. The data suggest that the hypothetical "initiator" is present, at a lower concentration than 21 days later. However, in the 9 days preparation the rate of polymerisation became higher than in the adult, as soon as the tubulin concentration was sufficient. The relationship between the initial rates of polymerisation and the amount of colchicine binding tubulin was sigmoïd suggesting that the process is autocatalytic. These and other data obtained by studies on the effect of $Ca^{2+}$ on the polymerisation process suggest that the microtubules obtained from 9 and 30 day old rat brains differ also in some other respect. The assumption is made that another factor is present few days after birth which enhances the rate of neurotubule growth.

References

1. Borisy, G.G. and Taylor, E.W. 1967, J. Cell Biol. 34, 525-533.

2. Wessels, N.K., Spooner, B.S., Ash, J.F., Bradley, N.O., Luduena, M.A., Taylor, E.L., Wrenn, J.T., and Yamada, K.M. 1971, Sciences, 171, 135-143.

3. Borisy, G.G. and Taylor, E.W. 1967, J. Cell Biol. 34, 535-548.

4. Taylor, E.W. 1965, J. Cell Biol. 25, 145-160.

5. Wilson, L., Bamburg, J.R., Mizel, S.B., Grisham, L.M. and Creswell, K.M. 1974, Federation Proc. 33, 158-166.

6. Weisenberg, R.C. 1972, Sciences, 177, 1104-1105.

7. Shelanski, M.L., Gaskin, F. and Cantor, R.C. 1973, Proc. Nat. Acad. Sci. 70, 765-768.

8. Bamburg, J.R., Schooter, E.M. and Wilson, L. 1973, Biochem. 12, 1476-1482.

9. Kuriyama, R. and Sakai, H. 1974, J. Biochem. (Japan) 75, 463-471.

10. Gaskin, F., Cantor, R.C. and Shelanski, M.L. 1974, J. Mol. Biol. 89, 737-755.

11. Weingarten, M.D., Lockwood, A.H., Hwo, S.Y. and Kirschner, M.W. 1975, Proc. Nat. Acad. Sci. 72, 1858-1862.

12. Weisenberg, R.C., Borisy, G.G. and Taylor, E.W. 1968, Biochem. 7, 4466-4478.

*Microtubules and Microtubule Inhibitors, eds. M. Borgers and M. de Brabander*



# QUANTITATION OF MICROTUBULES IN SECRETORY NEURONS

J.C.Sloper and Felicity Grainger

Department of Experimental Pathology,
Charing Cross Hospital Medical School, London

Summary

Microtubules are difficult to quantitate in secretory cells. Evidence of their functional role thus largely centres on the extent to which cellular activity is disturbed when microtubular structure is deranged by drugs such as colchicine. In the present study microtubular number has been analysed in unmyelinated axons of the tractus hypophyseus which convey vasopressor and oxytocic polypeptides to the posterior lobe of the pituitary. Microtubules have been counted in saline-stressed rats and in rats, newborn, 4 days old and adult, suffering from congenital diabetes insipidus (CDI). In such stressed and diabetic rats there is evidence of increased transport of polypeptides along the tractus hypophyseus. In these rats (save in newborn animals with CDI) statistically significant increases in axonal diameter and microtubular number have been found.

These increases may be essentially of structural significance, microtubular hyperplasia as it were accompanying hypertrophy of the active secretory neuron. It is noted that axons of the tractus hypophyseus are narrower than has hitherto been appreciated (331 nm in diameter), so that the neurohypopyseal polypeptides may well have to be conveyed in extravesicular as well as in intravesicular form.

## 1. Introduction

It occurred to us that, under conditions where secretory neurons were active, there might be not only hypertrophy of the perikaryon, nucleus and nucleolus, but also hyperplasia of any intra-axonal organelle concerned in transport[1]. Accordingly we studied two conditions in rats in which there is evidence of an increase in the transport of polypeptides down the tractus hypophyseus; namely in rats given hypertonic saline[2,3] which increases the secretion of vasopressin; and in rats with a form of congenital diabetes insipidus (CDI) characterised by the axonal transport of polypeptides largely devoid of antidiuretic activity (oxytocin has some slight antidiuretic activity) to the posterior pituitary[4]. The axons of this tract are unmyelinated and the great majority traversing the lower pituitary stalk appear to be concerned with conveying vasopressin and oxytocin and their related neurophysins to storage-endings in the posterior lobe of the pituitary. Axonal profiles contain microtubules (19 - 25 nm in diameter), mitochondria, microvesicles and neurosecretory vesicles with electron-dense contents (100-200 nm in diameter).

We believe that our estimations of axonal microtubular number are among the first to be made in animals subjected to physiological stress or suffering from disease. Indeed, quantitative information in normal animals seems hitherto to have been limited to estimations derived from the analysis of 80 axons in peripheral myelinated nerves of 3 rats[5]. Unmyelinated nerves of the tractus hypophyseus seem to be different in that they appear largely to lack neurofilaments. Our findings suggest that in secretory axons of the tractus hypophyseus microtubular number varies with neuronal activity.

## 2. Materials and methods

Brattleboro' rats with CDI are a mutant of the Long Evans strain. Until recently it has only been possible to breed from a homozygous male and a heterozygous female, and litters from such animals could not be tested for DI until they were three weeks old. Mrs. Grist of our animal-house has, however, succeeded in breeding from two homozygous parents, all the offspring of which have CDI.

Saline-treated rats were given 3 per cent saline instead of water for 4 days[1]. Male adult rats were anaesthetised with ether and perfused by a pulsatile pump first with saline and then with a gluteraldehyde/paraformaldehyde fixative buffered with cacodylate, these substances being injected through the left ventricle at 37°C. Newborn animals identified as male were perfused with a syringe by hand. Blocks containing the pituitary and stalk were postfixed in osmium tetroxide, dehydrated with acetone and embedded in an Araldite/Epon mixture. The stalk was sectioned transversely at a level a-a (Fig. 1) near the pituitary.

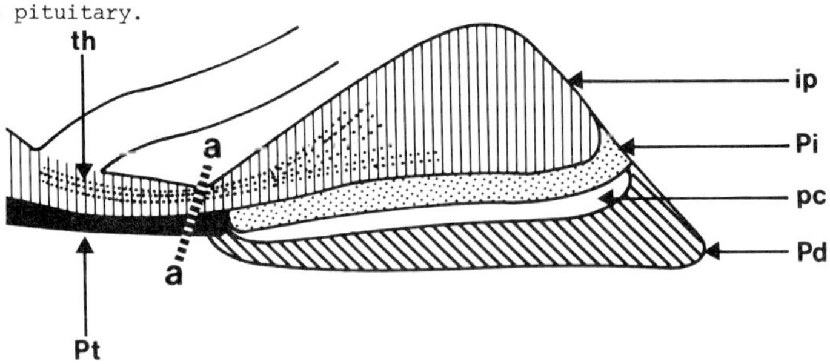


Fig.1 Diagram of sagittal section of pituitary region of the rat showing the tractus hypophyseus (th) running from the hypothalamus along the pituitary stalk to the infundibular process (ip) of the posterior lobe of the pituitary. a - a shows the level at which sections were taken for analysis. Pt-Pars tuberalis. Pi-Pars intermedia. pc-pituitary cleft. Pd-Pars distalis (anterior pituitary).
Figs.2 & 3 Transverse sections of axons in pituitary stalks of adult rats showing microtubules (mt), neurosecretory vesicles (NSV) and mitochondria(m). (Fig.2)Control rat. Note few mt. (Fig 3) Rat with CDI Note more mt in larger axons.

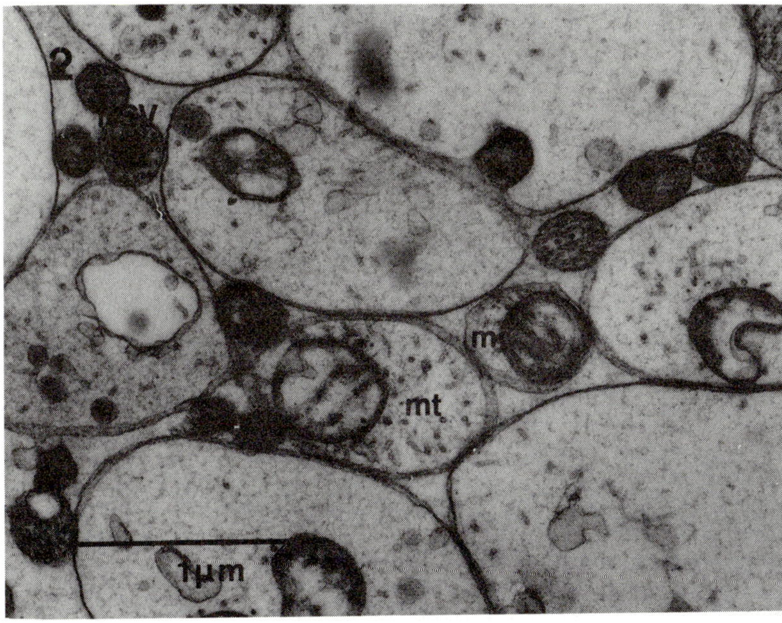
2
m
mt
1µm

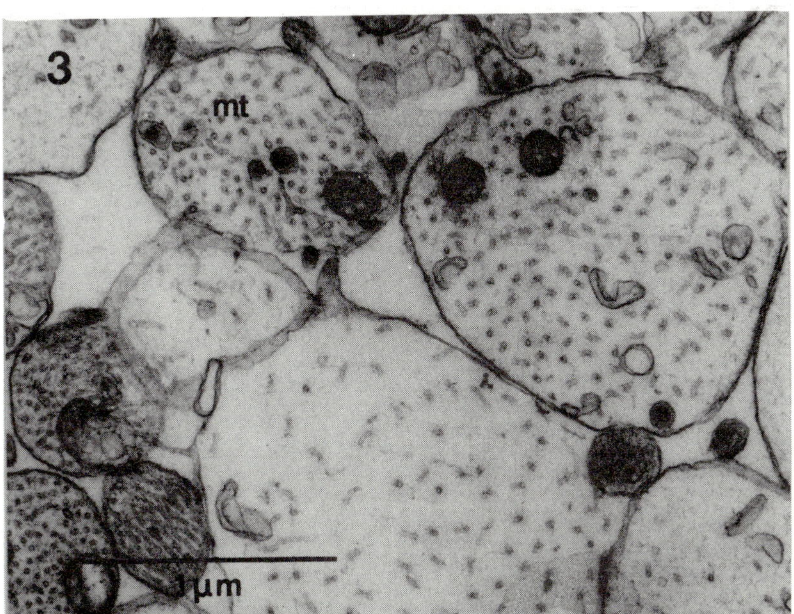
3
mt
1µm

1 um sections were stained with uranyl acetate and lead citrate. The central area of each grid window was photographed in the dorsal half of the pituitary stalk[1] and the final print was enlarged to a magnification of 35,000 times. Nerve fibres were measured across their shortest widths, and microtubules were counted.

Results have been pooled for all four animals in each experimental group. Means were calculated for axon diameter and for the numbers of microtubules/axon.

3. Results

Quantitative ultrastructural data on the tractus hypophyseus are scanty. We estimate the number of axons at some 80,000, and the mean least diameter at 331 nm $\pm$ 15.6 (2 x S.E.M.), a much smaller diameter than has hitherto been appreciated. In the normal adult animals there were 8.74 $\pm$ 1.22 microtubules per axon. In saline-stressed rats and those with CDI we found statistically significant increases in axonal diameter (372 $\pm$ 24.82 and 486 $\pm$ 20.58 respectively), and in the number of microtubules (12.48 $\pm$ 1.26 and 23.27 $\pm$ 1.76) in axons of the tractus hypophyseus, (Figs. 2 & 3). There were no obvious differences in the number of mitochondria or microvesicles in axons. Neuro-filaments (8-10 nm in diameter) were found in one of four saline-treated animals, and only in a few axonal profiles. Wispy filamentous structures were frequently present. Although it was possible to find some NSV arranged linearly in longitudinal sections of tractus hypophyseus, in transverse sections it was exceptional to see neurosecretory vesicles in contact with microtubules, save in the very small axons, e.g., those 100-150 nm in diameter

The question arose as to whether, in the case of rats with CDI, the micro-tubular changes which we observed could be interpreted as a congenital mal-formation rather than as a form of work-hyperplasia. To exclude the first possibility we had to use new-born diabetic animals bred from homozygous parents. In such animals we found no recognisable abnormality in the tractus hypophyseus (Figs. 4 & 5), but by the fourth day the number of microtubules was greater in diabetic animals than in normal animals. (During the first 5 days of life the vasopressor content of the normal rat pituitary is known to increase[6].)

Figs. 4 & 5 Transverse sections of pituitary stalks of newborn rats
(Fig.4) Control rat. Note smaller axons than in adult stalk.
(Fig.5) Rat with CDI. Similar appearance to control.

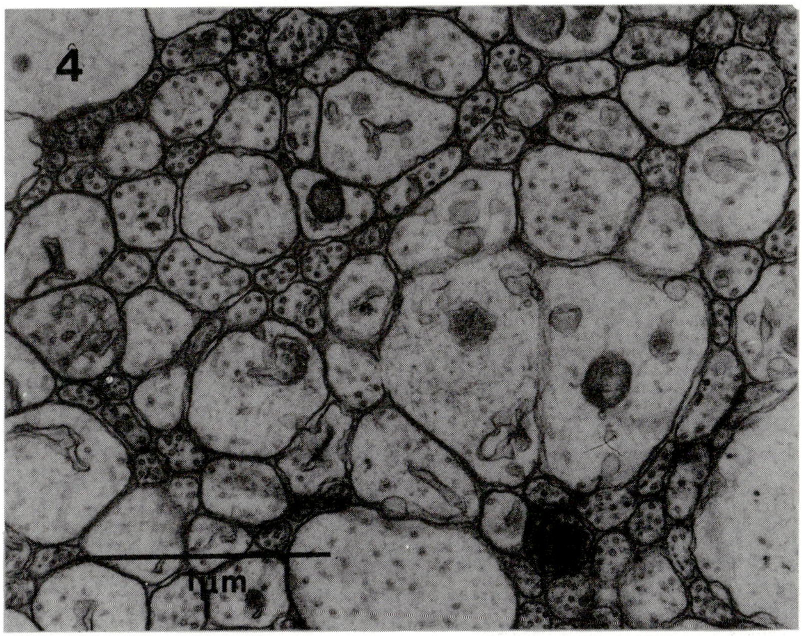
4
1 µm

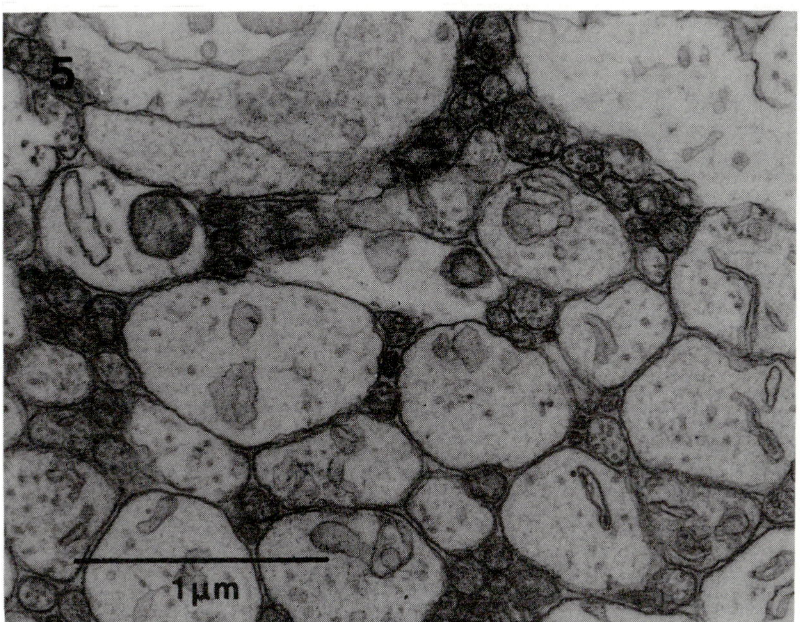
5
1 µm

4. Discussion

Our observations suggest that microtubular hyperplasia plays an important role in the axonal enlargement which accompanies secretory activity in neurons of the hypothalmo-neurohypophyseal system. There are, to our knowledge, no comparable observations on unmyelinated or myelinated nerve fibres. Indeed, quantitative data of a morphological nature on microtubules are extremely scanty.

The role of the microtubules in axons is still obscure, although, because in myelinated peripheral nerves[7] they can be shown morphologically to be destroyed after the administration of colchicine, and because in this situation there appears to be a disturbance in the transport of polypeptides along peripheral nerves, it has been supposed that microtubules play an essential part in the axonal transport or flow of polypeptides. However, the unmyelinated fibres of the tractus hypophyseus react differently, in that Norström et al.[8] and Flament-Durand and Dustin[9] have observed that transport of polypeptides in the tractus hypophyseus is interfered with in animals given colchicine, despite the absence of obvious changes in microtubules. One explanation for this apparent paradox may be that in the tractus hypophyseus there are cytoplasmic factors concerned with axoplasmic transport upon which colchicine has a more damaging effect than upon microtubules.

It should be added that it is still uncertain as to whether neuro-hormones in the tractus hypophyseus are conveyed entirely in NSV or in part in the axoplasm. For whereas in the normal subject some 90 per cent of the secretory polypeptides are intravesicular, in saline-treated subjects this figure can fall to 60 per cent[10]. The possibility that neurohormones are conveyed in part in an extravesicular manner seems the likelier in that many axons have diameters of the same order of size as neurosecretory vesicles. Gross[11] has advanced a theory which would account for the transport of vesicles and polypeptides in solution by a single mechanism.

References


1. Grainger, Felicity and Sloper,J.C., 1974 Cell Tiss.Res. 153, 101.

2. Jones, C.M. and Pickering, B.T., 1969, J. Physiol. 203, 449.

3. Norström, A. and Sjöstrand, J., 1972, J. Endocr. 52, 87.

4. Valtin, H., Sokol, H.W., Sawyer, W.H. and Harrington, A.R., 1965 Excerpta Medica Int.Congr. Ser. no. 83, vol. 2, 1287

5. Friede, R.L. and Samorajski, T., 1970, Anat. Rec. 167, 379.

6. Acher, R., Chauvet, J. and Olivry, G., 1956, Biochim. biophys. Acta (Amst.) 22, 428.

7. Wisniewski, H., Shelanski, M.L. and Terry, R.D., 1968, J. Cell Biol. 38, 224.

8. Norström, A., Hansson, H.-A. and Sjöstrand, J., 1971, Z. Zellforsch. 113, 271.

9. Flament-Durand, J. and Dustin, P., 1972, Z. Zellforsch. 130, 440.

10. Barer, R., Heller, H. and Lederis, K., 1963, Proc. roy. Soc. B 158, 388.

11. Gross, G.W., 1974, Excerpta Medica Int. Congr. Ser. no. 334, 26.

Microtubules and Microtubule Inhibitors, eds. M. Borgers and M. de Brabander



# CENTRIOLE AND CILIA FORMATION IN RAT PITUICYTES AFTER TREATMENT WITH COLCHICINE AND VINCRISTINE. STUDIES IN VIVO AND IN VITRO

P. Dustin, J.-P. Hubert and J. Flament-Durand

Department of Pathology, Université Libre de Bruxelles, Brussels, Belgium

## 1. Summary

In the pituicytes of the posterior lobe of rat's pituitary, the formation of centrioles and cilia can be induced, in the presence of colchicine, by stimulation of antidiuretic hormone secretion.

Chronic intoxication of rats by vincristine produced similar changes, without any antidiuretic stimulation. Moreover, ciliogenesis may be observed in vitro, either in the presence of colchicine and vincristine, or in Locke's solution, after an incubation of one to two hours. Macrotubules and paracrystalline structures are also produced in vitro by vincristine.

These results confirm that centriole and cilia formation may take place in the presence of microtubule poisons, and indicate that the pituicyte is particularly suitable for studying this type of microtubule assembly.

## 2. Introduction

Pituicytes are specialized glial cells located in the posterior lobe of the pituitary. Their active cytoplasm, containing some fat droplets, is closely associated with the endings of the nerve fibres transporting neurosecretory granules from the hypothalamic nuclei. The pituicytes may be involved in the release of antidiuretic hormone (ADH) carried by these granules.

In the course of experiments on the inhibition by colchicine of the axonal flow in the hypothalamic neurons[1], the effects of colchicine on the post-hypophysis were studied. Mitoses were known to increase in pituicytes after a forced secretion of ADH[2,3]. This work, which was done first on rats receiving drinking water with 2.5 % NaCl, and further after a forced stimulation of diuresis by Furosemid (Lasix®), confirmed the increased mitotic activity of the pituicytes and brought some unexpected results[4,5]. Although colchicine is known to combine with tubulin and to prevent its assembly into microtubules (MT), thus blocking the mitoses of pituicytes, several of these cells were found with many centrioles and cilia, contrary to those of control animals. The neoformation of these structures from a granular matrix was observed. It was concluded that the stimulation of the pituicytes by forced diuresis lead to a particular state in which centriolo- and ciliogenesis were stimulated, this being increased by the action of colchicine.

The results presented here, which will be reported elsewhere in more detail, confirm this observation. They indicate that the cilia formation can also take place without increased secretion of ADH, and demonstrate that similar results may be obtained in vitro by incubating the posterior lobes of rat pituitaries in Locke's solution with or without colchicine or vincristine (VCR).

## 3. Material and methods

Animals : Albino rats aged respectively 6 months and 13 days were used.

In vivo experiments : 10 µg of VCR was injected intraperitoneally daily, five days a week to adult rats. The animals were killed after a total dose of 0.85 mg/kg of body weight.

In vitro experiments : The pituitaries were incubated for one or two hours, at 37° C, in a solution of Locke to which was added $10^{-5}$ M colchicine or $10^{-5}$ M VCR (Oncovin Lilly). This procedure was chosen with the purpose of repeating the experiments of Rufener et al.[6]

Electron microscopy : Small blocks of tissue were fixed by immersion in toto in 4 % glutaraldehyde in 0.1 M phosphate buffer at pH 7.4 for 90 minutes. After washing in Millonig's buffer with 0.50 % sucrose during 24 hours the blocks were postfixed in 2 % $OsO_4$ for 30 minutes. After dehydration, the blocks were embedded in Epon. Semi-thin sections were stained with toluidine blue. Thin sections were stained with uranyl-acetate-lead. A large number of pituicytes were examined. The total number of pituitaries studied in these experiments is 32 (VCR in vivo, 4; VCR in vitro, 14; colchicine in vitro, 4; Locke's solution alone, 10). Many pituicytes of control animals of various ages have been observed : these cells may sometimes show one or two centrioles, but no formation of cilia is seen.

## 4. Results

1) Repeated injections of VCR

Previous experiments demonstrated that single doses of VCR modify the hypothalamic-pituitary system, only if injected intracisternally. In such animals, axonal flow of secretory granules is arrested, like that described with colchicine by Flament-Durand and Dustin[1]. No ciliogenesis is apparent[7] while paracrystalline MT structures are frequent. On the contrary, repeated intraperitoneal injections of VCR induced remarkable changes in the pituicytes (fig. 1). These are comparable to those described after Furosemid and colchicine[4,5]. One important difference is that no attempt was made to modify the diuresis, the rats having access to normal drinking water ad libitum. All steps of the formation of centrioles, from accumulations of a dense matrix, are visible. Some centrioles are incomplete, with less than 9 triplets. Cilia, often with an abnormal number of MT, are formed, mainly of the 9 + 0 type. They grow into ciliary vesicles. This ciliogenesis is similar to that described after Furosemid and colchicine, and resembles closely that illustrated in other ciliated cells[8,9,10].

2) VCR in vitro

Paracrystalline formations, as described by Rufener et al.[6], were observed. Moreover, some pituicytes displayed a remarkable centriolar and ciliary neoformation. This was mainly apparent in the young animals. Another change was often observed : the formation of "macrotubules", of 48 nm of diameter. Transitions between micro- and macrotubules were apparent. The macrotubules displayed an helical structure, formed by a simple helix wound regularly, at an angle of about 20° to the axis (fig. 2). This structure, contrary to the loose helix described by Behnke and Forer[11], was limited laterally by a definite line of dense material, resembling the macrotubules observed in the rat kidney after intravenous injection of VCR[12].

3) Colchicine in vitro

In adult rats, no effects of colchicine in vitro were observed; on the contrary, in the young animals, centrioles and cilia appeared to grow in spite of the linkage of colchicine with tubulins.

4) Saline solution in vitro

Surprisingly, some of the controls, which had been incubated in

Locke's solution alone, showed also centriole and cilia formation in the pituicytes. This was only apparent in young animals. As mentioned above, the selection of Locke's solution instead of a more physiological and balanced incubation fluid was the consequence of the work of Rufener et al.[6] During the incubation, the pH of our solution was observed to change from 8 to 7.4 after 1 hour and 6.8 after 2 hours. This may be significant, as slight changes in ionic composition are known to affect the assembly of tubulins into MT and other structures. This problem will be the subject of further research.

## 5. Discussion

Three points will be discussed here : the formation of centrioles and cilia in the presence of colchicine or VCR, the factors stimulating this formation and the structure of macrotubules. The results with VCR confirm that MT poisons do not necessarily prevent the assembly of tubulins into complex structures and their growth from cytoplasmic dense zones (matrices). Few other authors have observed similar facts. In the experiments of Stubblefield and Brinkley[13], the formation of centrioles and cilia took place in HeLa cells when replaced in a normal culture medium immediately after the action of colcemide. Krishan[14], and Krishan and Hsu[15], studying the action of VCR on Earle's L type fibroblasts, mentioned a multiplication of centrioles and the presence of cytoplasmic filamentous matrices. They did not give any illustrations of these changes. Wunderlich and Heumann[16] in Tetrahymena observed also the assembly of normal MT and cilia despite the presence of colchicine within the cells.

These results should be compared to the observations of Weisenberg and Rosenfeld[17] on the formation of centrioles in vitro in homogenates of eggs. Functioning microtubule organizing centres (MTOC) appeared even in eggs treated with colchicine. They conclude that this MTOC (which in some respects resembles the dense matrices observed in the pituicytes) "is not composed of tubulin, or ... is composed of tubulin in a state which is insensitive to colchicine".

In the experiments of Rufener et al.[6] no centrioles or cilia are

---

Fig. 1. Pituicyte of a rat after repeated intraperitoneal injections of VCR. L : lipids, Mt : microtubules, NSG : neurosecretory granules. Arrows : granular matrix in close relation with centrioles and cilia at various stages of formation.

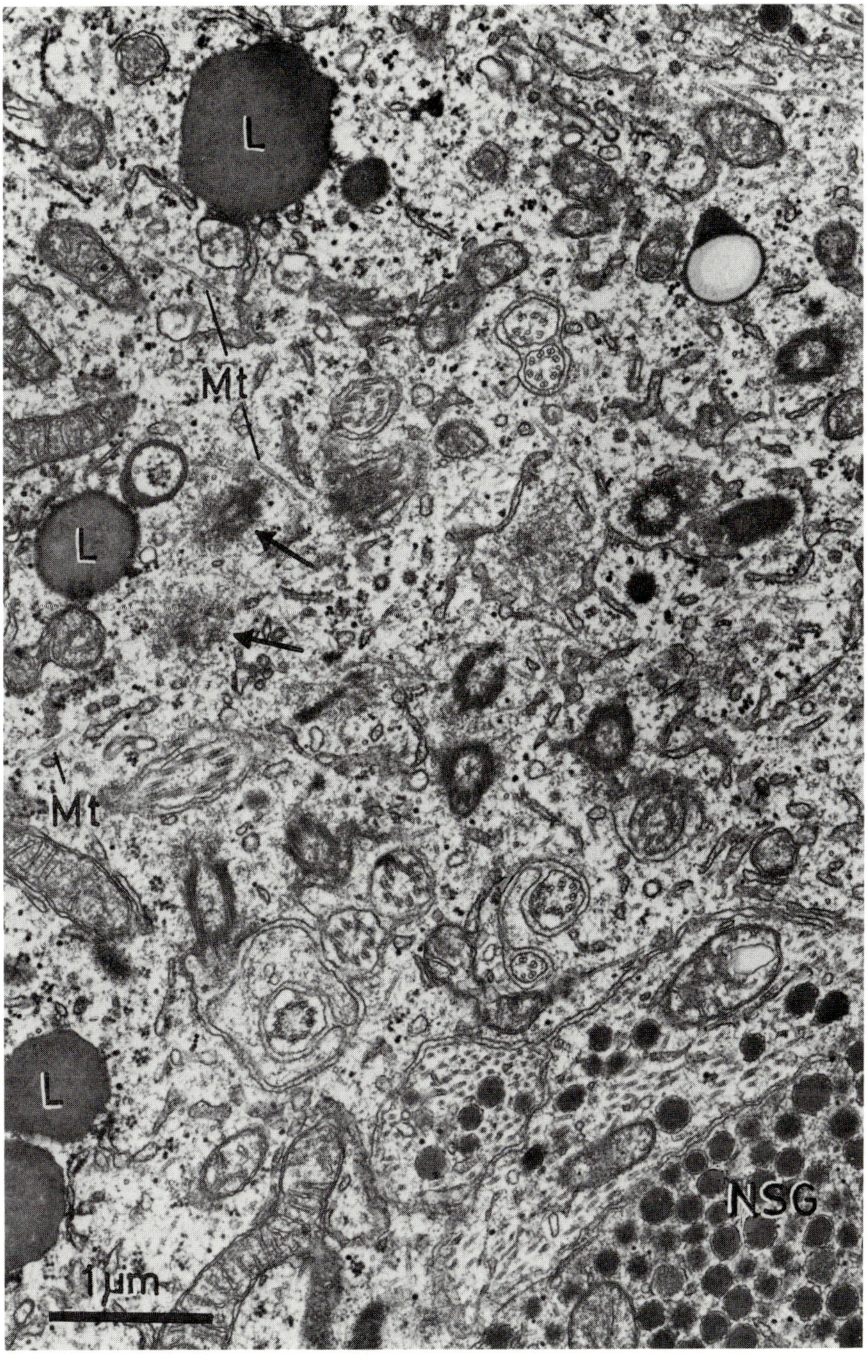
L
Mt
L
Mt
L
NSG
1 µm

mentioned. This can be explained by the doses of VCR : if all the tubulin is converted into paracrystals, none would be available for other structures. Our results with in vitro VCR incubation show that cilia are only formed when the doses are low.

The causes of the ciliogenesis remain obscure, and may be related to the physiology of the pituicytes and their relation to the secretion of ADH. In our previous results colchicine lead to the formation of many cilia only when the animals had an increased secretion of ADH, either after NaCl overloading, or Furosemid injection[4]. It has now been demonstrated that repeated injections of VCR, without any interference with urine secretion leads to identical changes. However, it should be recalled that VCR may interfere itself with ADH secretion, as indicated by the several cases of inappropriate secretion of ADH in patients treated with VCR (cf. Robertson et al.[18]). This can only be proposed as an hypothesis, as no measurements of urine excretion in our chronically VCR-intoxicated rats have been done.

More surprising was the observation that cilia formation could take place in vitro, in the presence of colchicine and VCR, and even in Locke's solution alone. This effect was limited to young animals. It is evident that incubating the posterior pituitary in Locke's solution places the pituicytes and the nerve endings in quite unphysiological conditions. At this time, we can only suggest that the changes in ionic milieu and pH during the incubation play a role in the stimulation of the pituicytes.

The formation of macrotubules after VCR raises some problems about their molecular structure. The helical structures suggest an unwinding of the MT, similar to that described by Behnke and Forer[11]. Such an unwinding into a simple helix is difficult to understand if models such as that proposed recently by Amos and Klug[19] are accepted : these models imply a triple or multiple helix, and any unwinding (by rupture of lateral bounds between successive turns of the helix) should lead to the formation of several interlocked filaments, which is not the case. It is however possible that each helical ribbon of the macrotubules may be formed by several tubulin subunits (e.g. four), although it remains to be explained how only some

---

Fig. 2. Axonal ending in the posterior lobe of a pituitary incubated for one hour in $10^{-5}$ M VCR in Locke's solution. Macrotubules with an helicoidal structure are apparent in close relation to normal neurotubules. NSG : neurosecretory granules.
Insert : schematical interpretation of macrotubule structure.

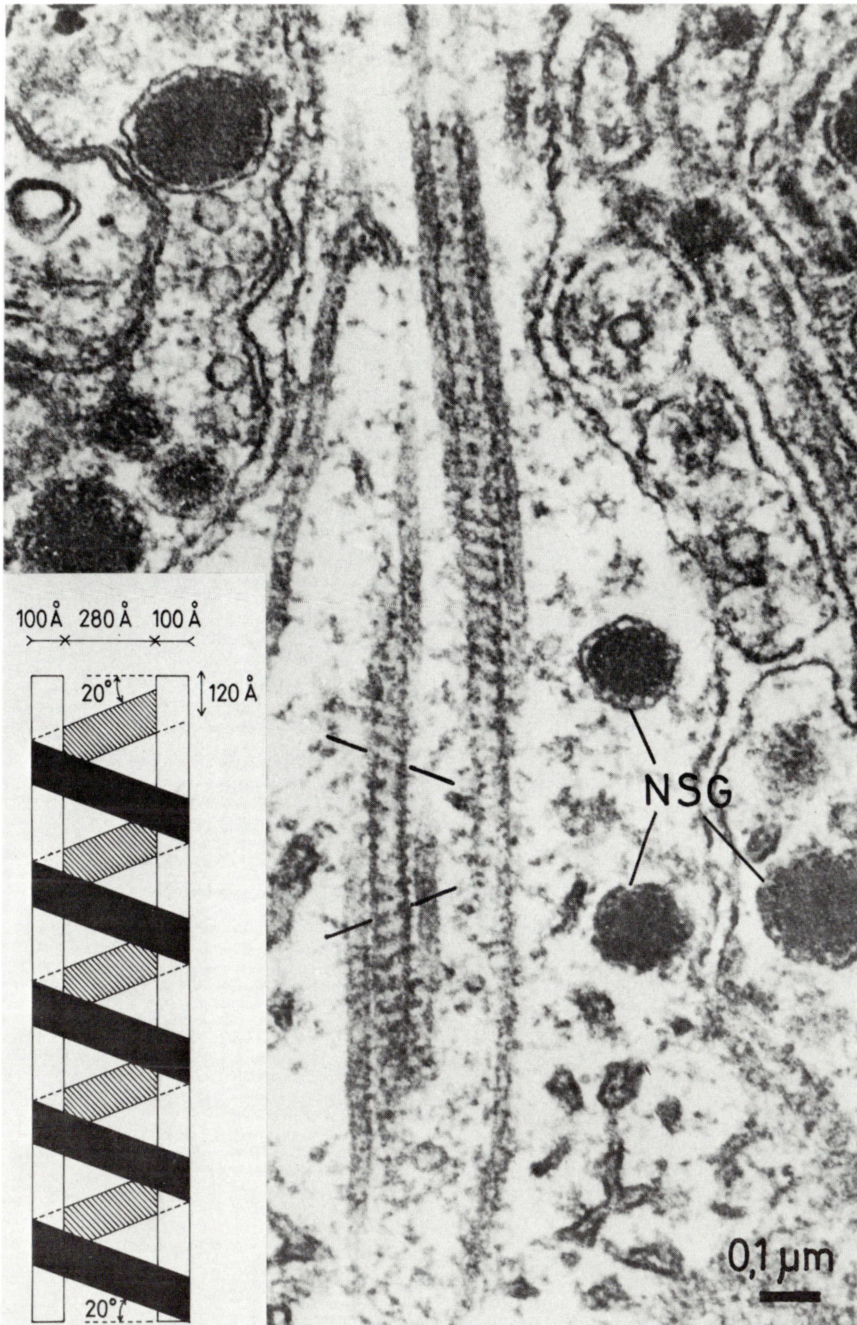
100 Å 280 Å 100 Å
20°
120 Å
20°
NSG
0,1 µm

lateral bounds between tubulin dimers could be broken. On the other hand, the explanation proposed by Warfield and Bouck[20] cannot apply to structures such as those represented in fig. 2.

Our results indicate that pituicytes are particularly interesting cells for experimental MT research. This work may also help to define better the function of these cells.

References


1. Flament-Durand, J. and Dustin, P., 1972, Z. Zellforsch. 113, 294.
2. Selye, H. and Hall, C.E., 1943, Anat. Rec. 86, 579.
3. Desclin, L., 1947, C.R. Soc. Biol. 141, 438.
4. Hubert, J.P., Flament-Durand, J. and Dustin, P., 1974, Cell Tiss. Res. 149, 349.
5. Dustin, P., Hubert, J.P. and Flament-Durand, J., 1975, Ann. N.Y. Acad. Sci., in press.
6. Rufener, C., Rouiller, C. and Orci, L., 1972, Experientia 28, 837.
7. Flament-Durand, J., Hunninck-Couck, A.M. and Dustin, P., 1975, Cell Tiss. Res., in press.
8. Sorokin, S.P., 1968, J. Cell Sci. 3, 207.
9. Dirksen, E.R., 1971, J. Cell Biol. 51, 286.
10. Staprans, I. and Dirksen, E.R., 1974, J. Cell Biol. 62, 164.
11. Behnke, O. and Forer, A., 1972, Exp. Cell Res. 73, 506.
12. Tyson, G.E. and Bulger, R.E., 1973, Z. Zellforsch. 141, 443.
13. Stubblefield, E. and Brinkley, B.R., 1966, J. Cell Biol. 30, 645.
14. Krishan, A., 1968, J. Nat. Cancer Inst. 41, 581.
15. Krishan, A. and Hsu, D., 1969, J. Cell Biol. 43, 553.
16. Wunderlich, F. and Heumann, H.G., 1974, Cytobiologie 10, 140.
17. Weisenberg, R.C. and Rosenfeld, A.C., 1975, J. Cell Biol. 64, 146.
18. Robertson, G.L., Bhoopalam, N. and Zelkowitz, L.J., 1973, Arch. Intern. Med. 132, 717.
19. Amos, L.A. and Klug, A., 1974, J. Cell Sci. 14, 523.
20. Warfield, R.K.N. and Bouck, G.B., 1974, Science 186, 1219.


Acknowledgments. The authors wish to thank Mrs. A.-M. Couck for technical assistance and Mrs. D. Libert for preparation of the manuscript. The vincristine was kindly supplied by Eli Lilly and Co.

*Microtubules and Microtubule Inhibitors, eds. M. Borgers and M. de Brabander*



# STUDIES OF MICROTUBULES IN DIVIDING AND NONDIVIDING MAMMALIAN CELLS USING ANTIBODY TO 6-S BOVINE BRAIN TUBULIN

B. R. Brinkley, G. M. Fuller and D. P. Highfield

Divisions of Cell Biology and Human Genetics

Department of Human Biological Chemistry & Genetics

Graduate School of Biomedical Sciences

The University of Texas Medical Branch

Galveston, Texas 77550

## 1. Summary

Monospecific antibody directed against bovine brain tubulin was used as an immunofluorescent probe to evaluate the distribution of microtubules in a variety of transformed and nontransformed cells in culture. Nontransformed cells were fibroblastic in appearance, contact inhibited and displayed an elaborate array of fluorescent cytoplasmic filaments termed the cytoplasmic microtubule complex (CMTC). Transformed cells were more polygonal in shape, lacked contact inhibition and contained very few cytoplasmic microtubules. The CMTC could be stimulated to appear in transformed cells grown in media containing dibutyryl-3′:5′ - cyclic AMP plus testosterone or theophylline. The appearance of a CMTC in transformed cells was accompanied by a reversion from polygonal rounded cells while piled up in culture to elongated fibroblastic cells with greater tendency to grow as monolayers. The CMTC was reversibly disrupted by microtubular inhibitors such as Colcemid (0.06 mg/ml) or cold temperatures (0°C 30 min). As cells entered mitosis, the CMTC disappeared and most of the fluorescence was confined to the mitotic spindle.

Tubulin antibody immunofluorescence permits a rapid evaluation of microtubules in cells without the need for electron microscopy. Our results provide further evidence for two separate microtubule entities in cycling nontransformed cells; a CMTC and the microtubules of the mitotic spindle. The absence of a CMTC in transformed cells is apparently due to sub-optimal levels of cAMP.

## 2. Introduction

With the development of tubulin specific antibodies in several laboratories[1-3], immunofluorescent techniques offer excellent promise as tools for the study of microtubule distribution and function in many cell types. In the present report, we describe the use of monospecific antibodies made against 6-S bovine brain tubulin as immunofluorescent probes and present evidence to support the reliability of this approach for studies of microtubules in dividing and nondividing cultured mammalian cells.

## 3. Tubulin Purification and Antibody Production

The method for purifying tubulin and the production of monospecific antibodies to the 6-S molecule has been described previously[1]. Bovine brain tubulin was prepared by a modification of the procedure described by Shelanski et al.[4] Crude tubulin was subjected to two cycles of polymerization and depolymerization precisely as described by Shelanski, et al., and Borisy et al.[31,32] The twice repolymerized tubulin was then subjected to an additional chromatographic step using Sepharose 4B equilibrated in 0.05M Tris and 0.002M $CaCl_2$pH6.8. This final step permitted the separations of 6-S tubulin from co-purifying proteins. Analysis of the 6-S tubulin by acrylamide gel electrophoresis containing sodium dodecyl sulfate demonstrated the presence of only 6-S tubulin.[5] This electrophoretically homogeneous material was used as an antigen for the production of antibodies.

Antibody Production. Approximately 1 ml of pure tubulin (10-12 mg/ml) was mixed with an equal volume of 5% glutaraldehyde and allowed to stand for 2 hours at 37°C. A thick yellow jelly-like precipitate formed. The precipitate was centrifuged, the supernatant removed and 2.0 ml of Freund's adjuvant added and the entire mixture was emulsified by forcing it through a 22 ga. hypodermic needle. The homogenization step was repeated several times. Rabbits were immunized by injecting 3.5 mg of cross-linked tubulin in the footpads followed by 2 subcutaneous booster injections of 2.0-2.5 mg each at 10 day intervals.

The antibody to 6-S tubulin was then separated from other gamma globulin by passing the rabbit serum over an immunosorbant column to which tubulin had been covalently bound by cyanogin bromide activation. Removal of the bound antisera was accomplished by treatment with 0.10 M glycine HCl buffer at pH2.6. The cross reactivity and specificity of the antibody was determined by immunodiffusion and immunoelectrophoresis. It should be pointed out that the antibody titers are low and fairly high concentration of antibodies are necessary in order to get discernable precipitin lines. This may be explained in part because of the homology of amino acid sequence in tubulin of widely divergent species.[33]

The various cell lines used in this study are shown in Table I. The cells were grown on glass coverslips. After rinsing in phosphate buffered saline (PBS), the cells were fixed for 20 min in 1.0 to 3.0 per cent formaldehyde made up in PBS. The coverslips were rinsed in PBS and subsequently fixed in absolute acetone at -10°C for 7 min. The preparations were then incubated with antibody (0.20 mg/ml in borate-saline buffer) at 37° for 30 min. The preparations were washed again in PBS and incubated in a 1:1.5 dilution of fluorescein-tagged goat antiserum against rabbit immunoglobulin G (Meloy Laboratories, Springfield, Va.

---

Fig. 1. 3T3 cell showing cytoplasmic microtubule complex (CMTC). Arrows point to microtubules parallel to cell surface. N, nucleus. Bar equals 10μ in all figures.

Fig. 2. Primary hamster embryo cell showing many cytoplasmic microtubules, N, nucleus.

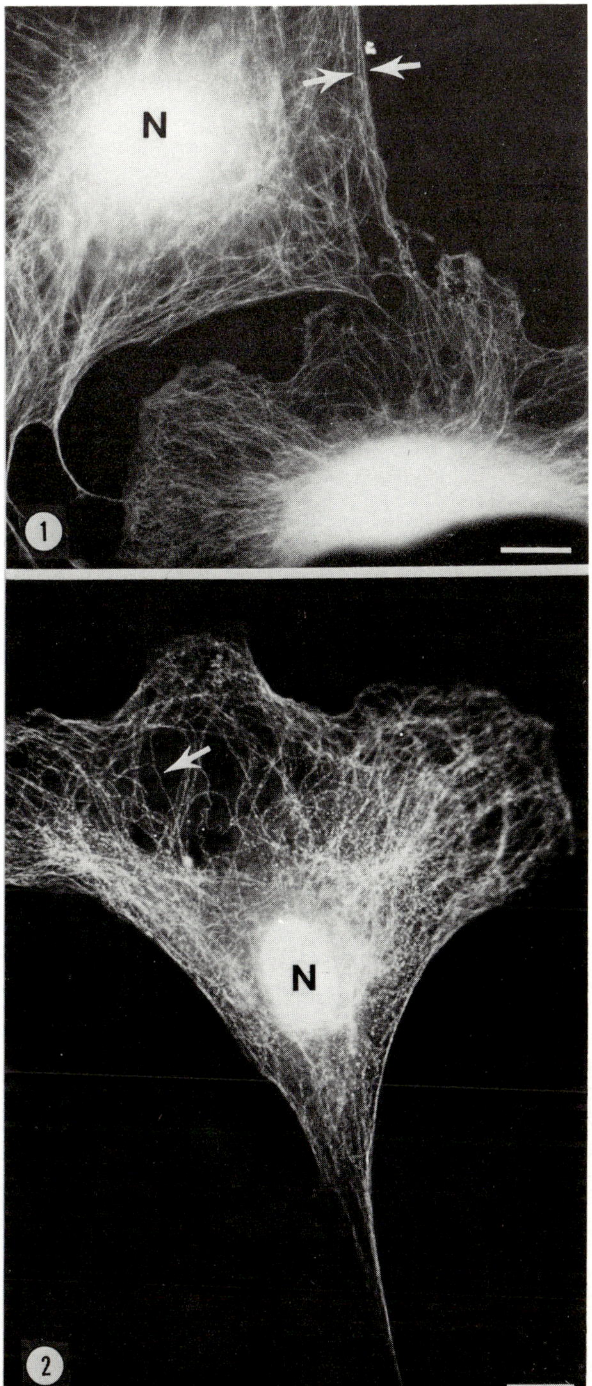
N
1
N
2

USA) for 1 hr at 37°C. The coverslips were rinsed in PBS followed by distilled $H_2O$ and mounted on glass slides in a drop of PBS:glycerol (1:1,pH8.5-10.0) for viewing in a Leitz microscope adapted for dark-field ultraviolet microscopy. Photographs were recorded on Tri-X Pan film (Kodak).

4. Cytoplasmic Microtubules in Interphase Cells

It is well established that the cytoplasm of many cells contain microtubules which are structurally similar to those of the mitotic spindle. In fact, cytoplasmic microtubules are implicated in a variety of important cellular functions including the maintenance of cell shape[6,7], axonal transport in neurons[8], cell motility[9], surface membrane motility [10], movement of pigments and other inclusions and organelles[11], mobilization and release of lysosomes[12], and the transport and release of macromolecules including hormones[13], and plasma proteins[14] from cells. In view of these findings, the detection of microtubules within the cytoplasm of cultured cells was not unexpected. We were, however, surprised to see such an elaborate array of microtubules in the cytoplasm of most cells stained for tubulin immunofluorescence. As shown in figures 1 and 2, the cytoplasm is laced with fine fluorescent filaments which, by experiments to be described later, can be shown to be microtubules. These frequently radiated out from the nuclear area of the cell and formed bundles which extended parallel with the cell axes. Often, the filaments appeared to focus on one or more specific regions of the cell presumed to be the centrosphere. The filaments frequently extended to the plasma membrane and were often bent to conform to the shape of the cell surface. Bundles of microtubules extending parallel to the cell surface (arrows figure 1) were not uncommon.

Effects of Microtubule Inhibitors. Although in vitro tests such as double immunodiffusion and immunoelectrophoresis have been used to demonstrate the specificity of our antibody for tubulin, it was important to shown that the fluorescent filaments in the cytoplasm were responsive to established microtubule inhibitors. When 3T3 cells were cultured in media containing Colcemid (0.06$\mu$g/ml) for 30 min to three hrs prior to staining, the filaments disappeared or became greatly diminished (Figs. 3 and 4). A similar response was achieved by by cooling the cells to 0°C for 30 min prior to staining. Along with the loss of cytoplasmic microtubules, the cells were converted from fibroblastic forms to more rounded amorphous

---

Fig. 3 and 4. Colcemid treated 3T3 cells showing loss of CMTC.
Fig. 5 and 6. Cytoplasmic asters (arrows) in 3T3 cells 5 min after reversal from Cocemid block.

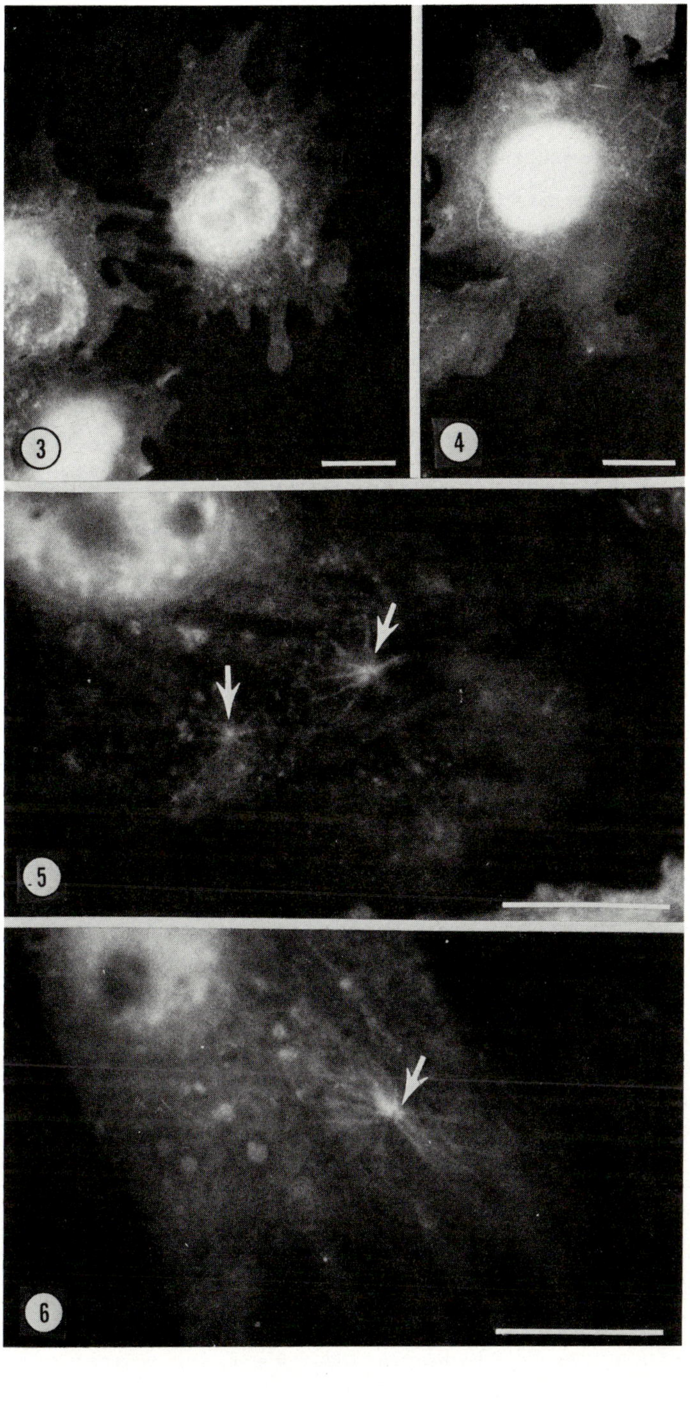
3
4
5
6

shapes. In addition, a generalized background of fluorescent material was present in the cytoplasm especially in the vicinity of the nucleus. The nuclear staining persisted and in some cells, even appeared brighter.

In both Colcemid treated cells and cells undergoing immediate reversal from Colcemid block, several small aster-like bodies were found in the cytoplasm (arrows, Figs.5 and 6). These consisted of a central bright spot with short fluorescent filaments radiating out in all directions. Not all cells displayed such asters while others contained from one to four such structures. Asters were not apparent in untreated cells and they quickly disappeared when the cells were reversed from Colcemid inhibition. Although we have not examined the asters by electron microscopy, they may represent microtubule organizing centers (MTOC)[15] and probably consist of centrioles and associated organelles. The variation in the number of asters seen in cells may reflect the number of MTOCs per cell and likely, cells with several centrioles would display several such MTOCs. Hopefully, this technique will provide more direct information on MTOC distribution and functions in mammalian cells.

Nuclear Staining. In most cells examined in the present study, fluorescent staining, either in the form of small discrete spots or generalized fluorescence throughout, was apparent in the nucleus. Although the intensity of the stain was increased in cells fixed in higher concentration of formaldehyde, staining was present even in cells fixed with 1% formaldehyde. In all instances, the nucleoli remained unstained. Further studies on the nature of the fluorescent material in nuclei are underway in our laboratory.

5. Microtubules in Transformed Cells

Transformation to malignancy of mammalian cells in tissue culture is characterized by several well known features including major alteration in cell shape, growth characteristics and cell surface properties. Typically, when cells such as mouse 3T3, with a low degree of malignancy, become transformed by SV-40 virus to highly malignant SV-3T3 cells, they are converted from fibroblastic forms which exhibit density-dependent inhibition of growth to more rounded polygonal cells which pile up in culture and lose density-dependent growth control. The involvement of microtubules in cell transformation has been suggested in several earlier studies[16,17]. Fonte and Porter[17] have shown by electron microscopy that microtubules become depleted in the cytoplasm of transformed cells.

In addition to their altered shapes, we recently discovered that transformed cells could be detected in mixed cell populations by their immunofluorescence staining patterns after

Figs. 7 and 8. Transformed SV-3T3 cells showing paucity of cytoplasmic microtubules.
Figs. 9 and 10. SV-3T3 after 6 hrs treatment with $Bt_2cAMP$. Cells become fibroblastic.

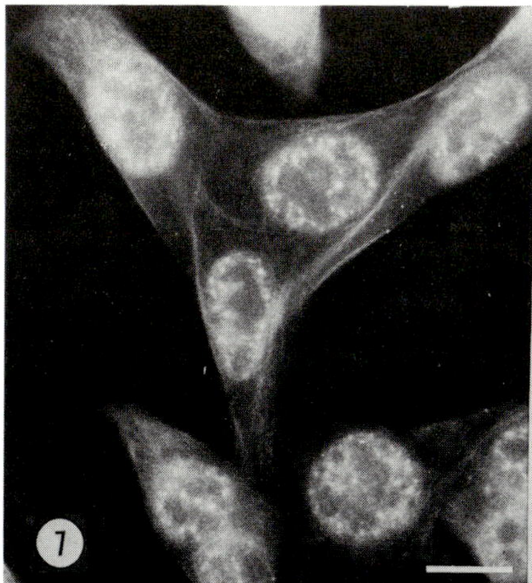
7

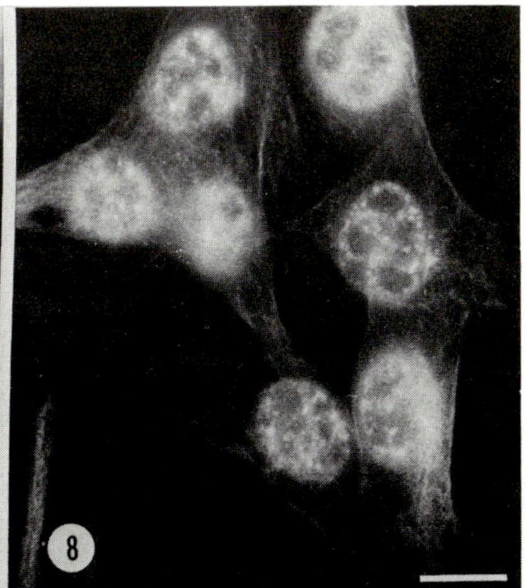
8

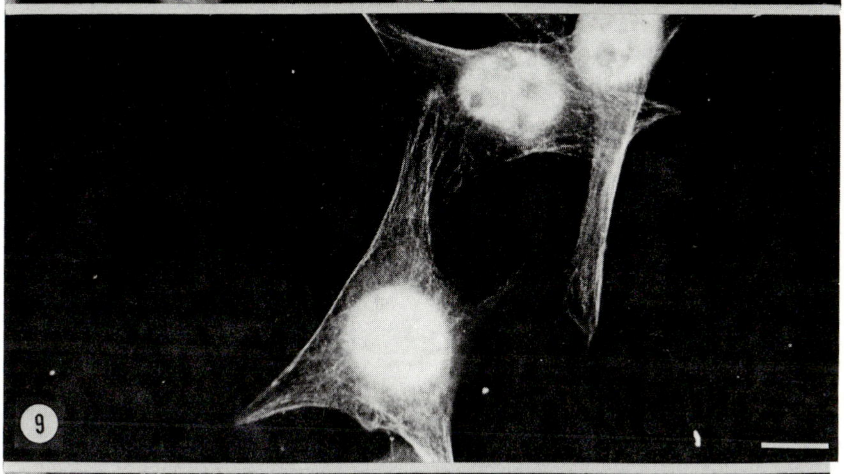
9

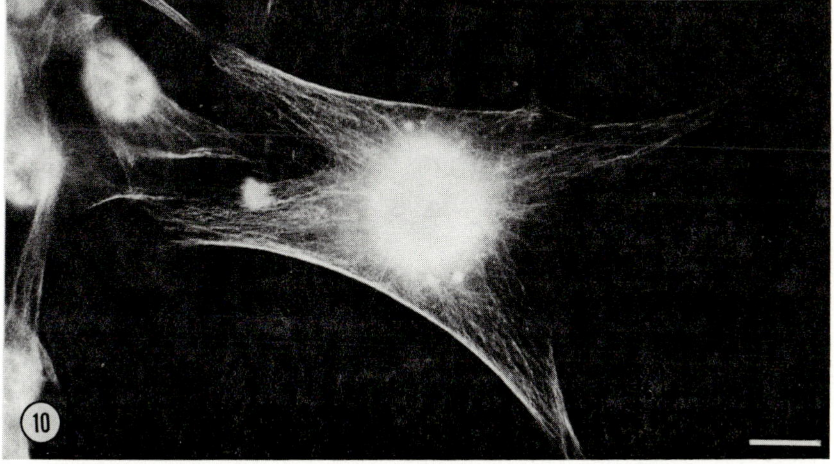
10

exposure to tubulin antibody[18]. As shown in figures 7 and 8, SV-3T3 cells are smaller, more rounded and lack a CMTC such as seen in 3T3 cells (Fig. 1). Microtubules are present in transformed cells but are less numerous and often randomly oriented. Nuclear staining of transformed cells was not unlike that of their normal counterparts. The observations on SV-3T3 cells prompted us to examine a variety of other transformed cell lines for their tubulin staining patterns. As shown in Table I, a paucity of cytoplasmic microtubules was evident in all transformed lines including those transformed by virus as well as spontaneous and chemically induced transformants. Our findings are therefore in agreement with those of Porter and co-workers[16,17] who reported fewer cytoplasmic microtubules in transformed cells.

TABLE I

Immunofluorescent Staining Properties of Various Cell Lines

| Cell Line | Cell Type | Transformed | Cytoplasmic Microtubule Complex |
|---|---|---|---|
| 3T3 | Swiss mouse embryo fibroblast | - | +* |
| PtK1 | Rat kangaroo fibroblast | - | + |
| HSF-CF | Human skin fibroblast from cystic fibrosis patient | - | + |
| HSF-Normal | Normal human skin fibroblast | - | + |
| SH1 | Primary hamster embryo | - | + |
| SV CF | Simian virus transformed human skin fibroblast from cystic fibrosis patient | + | - ǂ |
| HTC+ | Rat hepatoma | + | - |
| 6TG-11 | Rat hepatoma | + | - |
| MMT | Mouse mammary gland carinoma | + | - |
| $C_6$ | Rat glial cell | + | - |
| $A_9$ | Mouse-L cell (azaquanine resistant) | + | - |
| LM(TK$^-$) | Mouse-L cell (BUdR resistant) | + | - |
| SV-3T3 | Simian virus transformed 3T3 | + | - |
| RAG | Mouse renal adenocarcinoma | + | - |
| CHO | Chinese hamster ovary | + | - |
| HeLa | Human cervical carcinoma | + | - |

* Indicates presence of cytoplasmic microtubule complex (see Figs. 1 and 2)
ǂ Indicates absence of cytoplasmic microtubule complex (see Figs.7 and 8)

Figs. 11-16. Stages of mitosis in PtK1 cells. 11, monopolar prophase. 12, bipolar prophase. 13, monopolar prometaphase. 14, bipolar prometaphase. 15,16, bipolar prometaphase.

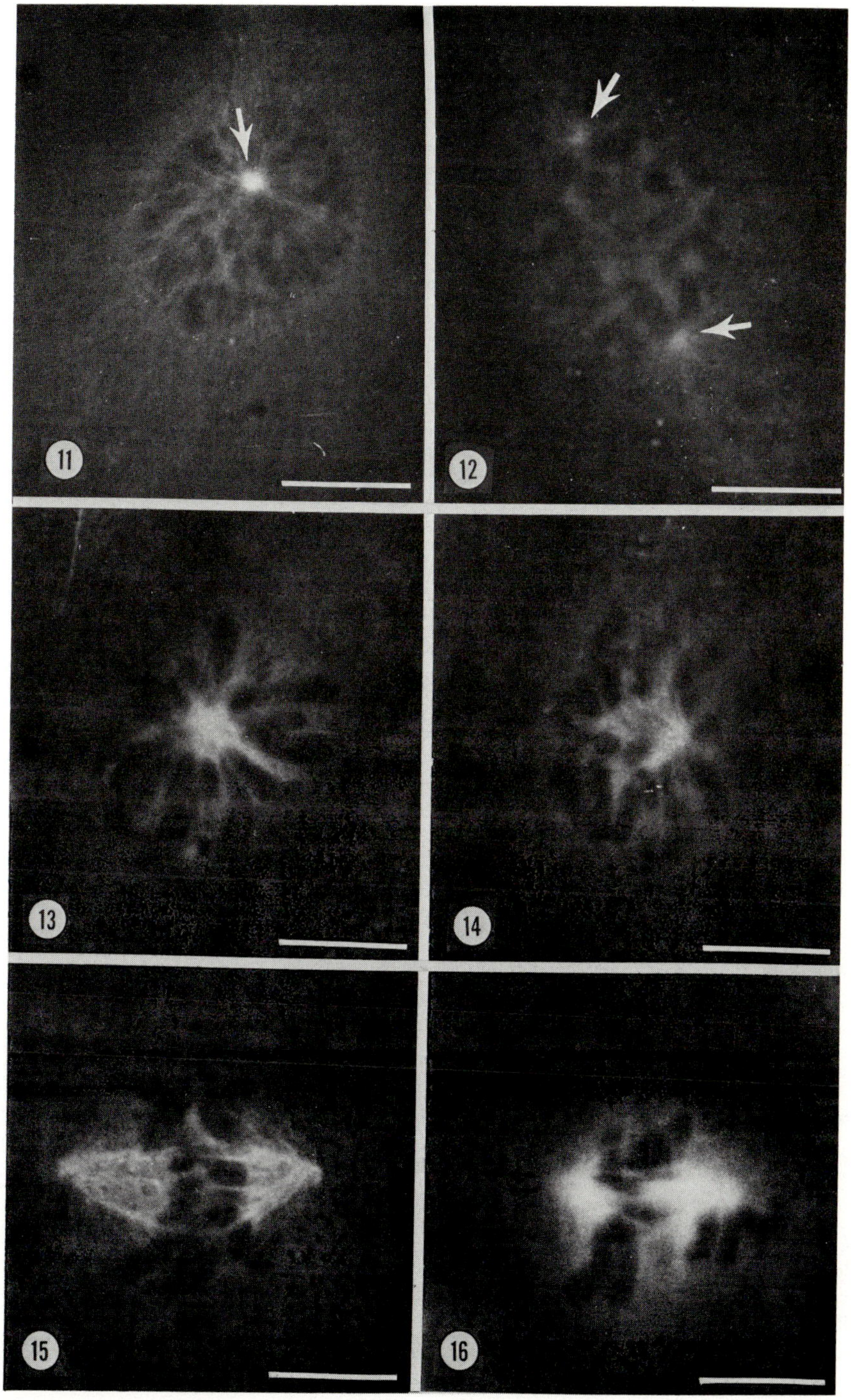
11
12
13
14
15
16

Hsie and Puck[19], Johnson et al.[20] and Sheppard[21], were able to convert transformed cells to normally appearing fibroblastic forms with derivatives of cAMP. In their experiments, treatment of cells with adenosine 3′:5′-dibutyryl cyclic AMP ($Bt_2cAMP$) plus either testosterone or theophylline led to the conversion of transformed cultures from rounded polygonal forms which grew in multilayers, to elongated fibroblast-like cells which grew as monolayers. Moreover, it was reported that treated cultures regained density-dependent growth inhibition properties[20,21]. More recently, however, Pasten et al.[22], have data which suggests that $Bt_2cAMP$ treatment may not restore inhibited growth to all transformed cells.

In view of these reports, we applied tubulin antibody immunofluorescence staining to CHO and SV-3T3 cells which had been treated with $Bt_2cAMP$ plus either testosterone or theophylline. As shown in figures 9 and 10, SV-3T3 cells assumed a more fibroblastic appearance after treatment. Along with the transition in morphology, numerous fluorescent cytoplasmic filaments appeared in the cell. After 6 hours of treatment, a complete CMTC appeared within the cytoplasm of most interphase cells in the culture. Although we did not directly evaluate the growth condition of our treated cultures, there appeared to be less tendency for cells to pile up as in the untreated SV-3T3.

Our results are therefore in agreement with the notion that microtubules are involved in cell transformation. The specific relationship of the CMTC and cyclic nucleotides to other tumor cell properties such as density-dependent inhibition of growth and altered cell surface properties requires further study. The relationship of cAMP and cGMP to cell growth characteristics, as well as their possible role in the control of microtubule assembly is under investigation in many laboratories and the literature is yet conflicting and incomplete. We can conclude, however, that the absence of a CMTC in transformed cells is due in some way to suboptimal levels of cAMP in tumor cells[23,24].

6. The Mitotic Process

The following description of mitosis is based upon observation of rat kangaroo (strain PtK1) cells which are ideally suited for such studies due to their large size, low chromosome number and flattened state during mitosis.

As cells enter prophase, specific alteration in immunofluorescence staining patterns are

---

Fig. 17. Metaphase
Fig. 18. Early anaphase, note interpolar fibers.
Fig. 19. Late anaphase.
Fig. 20. Telophase, arrow points to antibody.

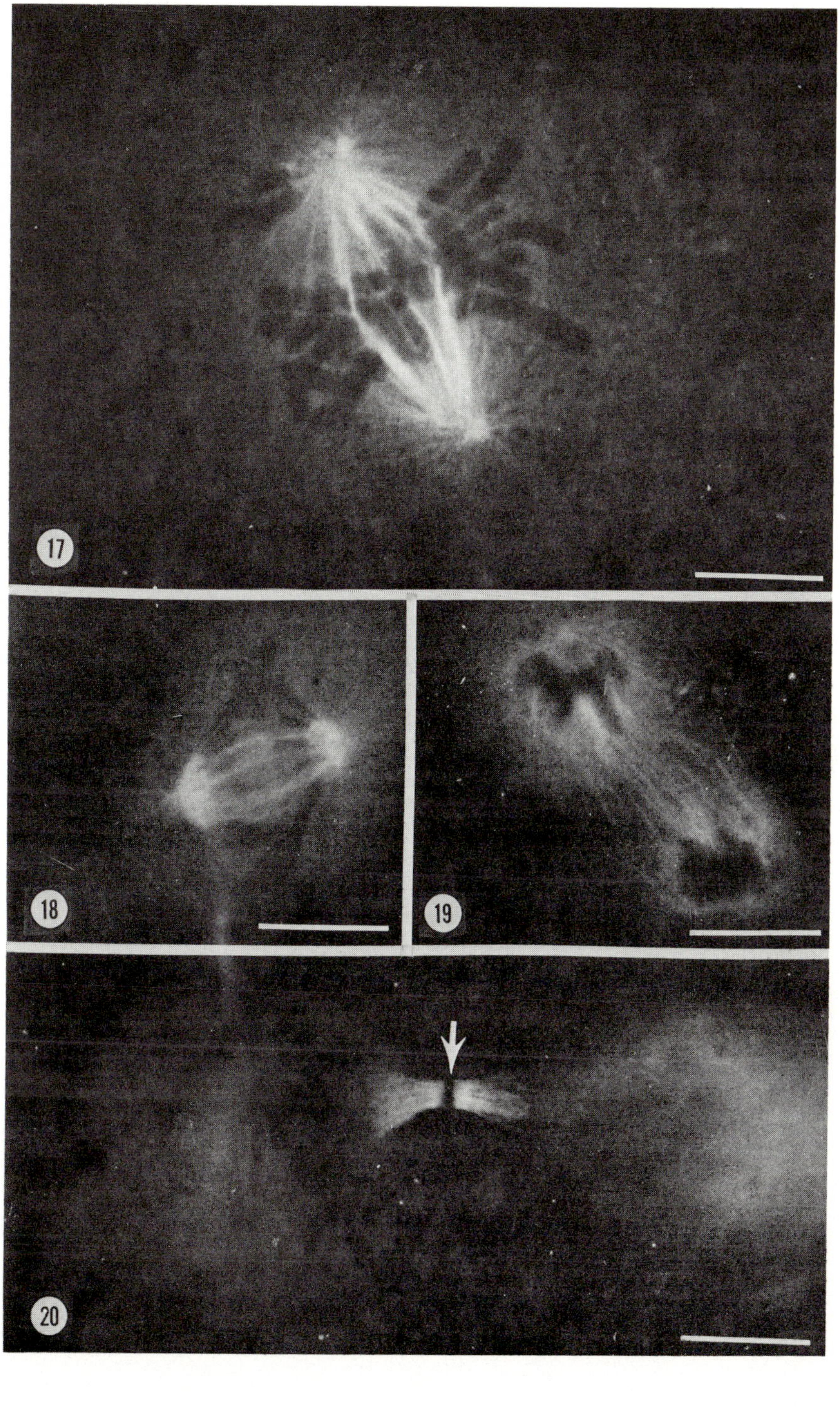
17
18
19
20

noted. The elaborate CMTC begins to disappear as cells withdraw their fibroblastic process. As mentioned earlier, the nuclear staining disappears as the chromosomes become condensed inside the nucleus. Sometimes small paired spots were apparent on each chromosome. By mid-prophase, almost all of the cytoplasmic microtubules had disappeared and most of the fluorescence was confined to either a single spot near the top of the nucleus (Fig. 11) or a pair of spots at opposite poles of the nucleus (Fig. 12). With further condensation of the chromosomes and subsequent disruption of the nuclear envelope, the cells either formed a monopolar spindle with the chromosomes grouped around one bright fluorescent pole (Fig. 13) or a bipolar spindle as shown in figure 14. Similar configurations have been noted in cells of this line by EM studies of Rattner and Berns[25] and Roos[26]. With EM, it is possible to see that in the monopolar spindle, the centriole pairs fail to separate and the chromosomes become grouped around the centrioles with microtubules radiating out like the spokes of a wheel. In the case of bipolar spindles, the centrioles migrate apart in early prophase accompanied by the appearance of interpolar microtubules between the separating centrioles. As the chromosomes become associated with the spindle at prometaphase, the latter appears as a brightly fluorescent object. At prometaphase, the spindle poles may become separated by a distance of 20$\mu$ or more and brightly fluorescent fibers are evident as chromosomes became distributed throughout the spindle (Figs. 15 and 16). Metaphase is characterized by the alignment of the chromosomes on an equatorial plate as shown in figure 17. Although the spindle shown in figure 7 is beautifully preserved and stained, it is unusually large. Typically, the metaphase spindle of diploid PtK1 cells are about 10$\mu$ in length from pole-to-pole while the spindle in figure 17 is over 20$\mu$ long. Such giant spindles are occasionally seen in this cell line and are usually found in polyploid cells. During metaphase, bundles of chromosomal filaments can be seen and the interpolar and short asteral filaments are often apparent.

As the chromosomes move to the poles in anaphase, the spindle undergoes a second elongation. The kinetochore fibers shorten as the chromosomes approach the pole and bundles of interpolar fibers can be seen spanning the interzone (Fig. 18). By early telophase (Fig. 19) the interpolar fibers become thinner and very lightly stained. Later during cytokinesis, brightly fluorescent fibers are seen in the stem korper between daughter cells. A non-fluorescent band is consistently seen in the midbody (arrow, Fig. 20). When this region is examined by EM[27] microtubules are found to interdigitate at this same area. In addition, an

Fig. 21. PtK cell treated with Colcemid. Arrows point to fluorescent spots at kinetochores.

Figs. 22-24. Normal prometaphase, metaphase and anaphase.

Figs. 25-27. Cold treated spindle.

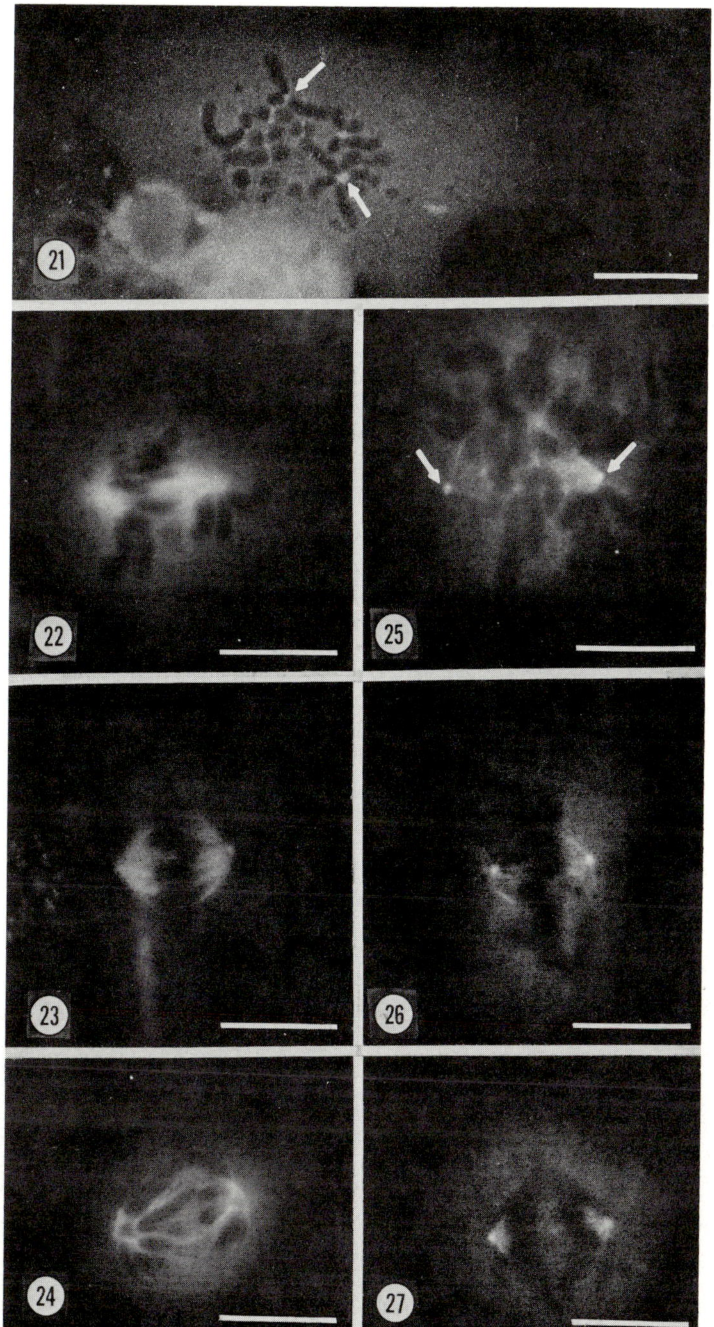
21
22
25
23
26
24
27

electron dense matrix is closely associated with the overlapping microtubules. In all of the cells studied, this was the only area known to contain microtubules which failed to show fluorescence. Apparently, the binding site for the antibody was blocked by the structural arrangement of tubules in the midbody.

In late telophase or early $G_1$ phase, the cytoplasmic microtubules reappeared in each daughter cell, first from the region of the centrosphere and then throughout the cytoplasm.

Although the previous descriptions were based on images of rat kangaroo cells, basically the same staining pattern was observed in other mammalian mitotic cells including transformed and nontransformed lines. It should be noted that no differences could be detected in the staining pattern of mitosis in normal cells and their transformed counterparts. In this regard, the first detectable difference in the staining patterns of transformed and nontransformed cells came at late telophase or early $G_1$ phase. The CMTC reappeared in normal cells at this stage by an outgrowth of microtubules from the region of the centrosphere. Transformed cells, however, formed very few cytoplasmic microtubules at this stage.

In summary, the CMTC of nontransformed cells disappears as the cell enters mitosis and most of the fluorescence is concentrated in the spindle. When the spindle disappears at the end of mitosis, the CMTC reappears. The latter is greatly diminished in transformed cells. Thus, there appears to be two functional microtubular systems in normal cells, the CMTC of interphase cells and the spindle of mitotic cells. Although the disappearance of cytoplasmic microtubules at a time when spindle microtubules are forming suggests a dynamic interchange of tubulin between the two systems, control of tubule assembly may be under separate constraints. Hence, transformed cells appear to have lost the capacity to assemble a complete CMTC but assembly of the tubules of the mitotic apparatus seems to occur normally.

7. Effects of Microtubule Inhibitors on Spindle Fluorescence.

When PtK1 cells were grown in media containing Colcemid (0.06μg/ml, 3 hrs), the spindle failed to form and the chromosomes became grouped into a cluster near the cell center. When such configurations are examined by EM, a few microtubules are associated with the kinetochores[28]. In addition, the centrioles remain near the center of the chromosome cluster. As shown in figure 21, small fluorescent spots (arrows) are seen at specific sites on the chromosomes. Since the spots correspond to the position of the kinetochores and are often paired, we can assume that the kinetochores are fluorescent under these conditions. Occasionally similar spots can be seen on late prophase chromosomes in untreated cells. The cytoplasm of Colcemid treated cells displayed an unusually bright fluorescence but no formed structures were present.

When PtK1 cells were chilled to 0°C in an ice bath according to procedures published elsewhere[29], the spindle was partially disassembled (compare Figs. 22-24 with 25-27). As previously shown by EM studies[29] cold temperatures selectively disassembles the interpolar microtubules leaving chromosomal tubules intact.

The absence of interpolar microtubules are especially apparent in figures 26 and 27. In such cold treated cells, the centriole region of the spindle are distinctly stained. This is probably due to the association of tubulin and short segments of microtubules around the centrioles.

The fluorescence pattern described for normal mitosis as well as mitotic cells treated with microtubule inhibition faithfully parallel images seen by electron microscopy. Similarly, the extensive CMTC seen in the cytoplasm of nontransformed cells has also been described in whole cells viewed by high voltage EM studies[30], thus, the tubulin antibody immunofluorescence staining method permits rapid and reliable detection of microtubules in dividing and nondividing cultured cells by conventional light microscope techniques.

Acknowledgement

We are grateful to Dr. Robert Klebe for providing many of the transformed cell lines used in this study. We thank Ms. Joan Ellison for excellent technical assistance in antibody preparation. Appreciation is extended to Mrs. Shirley Brinkley for editorial assistance. This work was supported in part by NCI Grant CA 14675 and a grant from DOW Chemical Company.

References

1. Fuller, G.M., Brinkley, B.R. and Boughter, M.J., 1975, Science, 187, 948.
2. Weber, K.,Pollack, R., and B. Bibring, T., 1975, Proc. Nat. Acad. Sci., USA, 72,459.
3. Kowit, J.D. and Fulton, C., 1974, J. Biolo. Chem. 249, 3638.
4. Shelanski, M.L.,Gaskin, F., Cantor, C., 1973, Proc. Nat. Acad. Sci., USA, 70, 765.
5. Weisenberg, R.C., 1972, Science, 177, 1104.
6. Porter, K.R., 1966. in: Principles of Biomolecular Organization, eds. G.E. Wolstenholme and M. O'Conner (Little, Brown, Boston) p. 308.
7. Tilney, L.G. & Porter, K.R., 1967, J. Cell Biol. 34, 327.
8. Smith, D.S., Järlfors, V., Cameron, B.F., 1975, Ann. N.Y. Acad. Sci., 253, 472.
9. Allison, A.C., 1973, in: Locomotion of Tissue Cells. CIBA Found. Symp. (Elsevier-Excerpta Medica. North-Holland), p. 109.
10. Berlin, R.D., 1975, Ann. N.Y. Acad. Sci., 253, 445.
11. Moellmann, G. and McGuire, J., 1975, Ann. N.Y. Acad. Sci., 253, 711.

12. Malawista, S.E., 1975, Ann. N.Y. Acad. Sci., 253, 738.

13. Wolff, J. and Bhattacharyya, B., 1975, Ann. N.Y. Acad. Sci., 253, 763.

14. Redman, C.M., Banerjee, D., Howell, K. and Palade, G.E., 1975, Ann. N.Y. Acad. Sci., 253, 780.

15. Pickett-Heaps, J., 1971, Cyto. Bios. 3, 205.

16. Porter, K.R., Puck, T.T., Hsie, A.W. & Kelly, D., 1974, Cell 2:145-162.

17. Fonte, V., & Porter, K.R., 1974, in : Eighth Int. Cong. Elect. Micro. (Aust.Acad.Sci., Canberra) p. 334.

18. Brinkley, B.R., Fuller, G.M. and Highfield, D.P., Proc. Nat. Acad. Sci., USA, (in press) 1975.

19. Hsie, A. and Puck, T.T., 1971, Proc. Nat. Acad. Sci., USA, 68,358.

20. Johnson, G.E., Friedman, R.M. & Pasten, I., 1971, Proc. Nat. Acad. Sci., USA, 68, 425.

21. Sheppard, J.R., 1971, Proc. Natl. Acad. Sci., USA, 68, 1317.

22. Pasten, I.H., Johnson, G.S. and Anderson, W.P., 1975, Ann. Rev. Biochem.,44,491.

23. Monahan, J.M., Marchand, N.W., Fritz, R.R., & Abell, C.W., 1975, Cancer Res. 35, 2540.

24. Monahan, J.M., Fritz, R.R., & Abell, C.W., 1973, Biochem.Biophysics Comm. 55, 642.

25. Rattner, J.B., and Berns, M., 1975, (personal communications)

26. Roos, V.P., 1973, Chromosoma 40, 43.

27. Byers, B., & Abramson, 1968, Protoplasma 66, 413.

28. Brinkley, B.R., Stubblefield, E. and Hsu, T.C., 1967, J. Ultrastruct. Res. 19, 1.

29. Brinkley, B.R. and Cartwright, J., 1975, Ann. N.Y. Acad. Sci., 253, 428.

30. Buckley, I.K., 1975, Tissue & Cell 7, 51.

31. Borisy, G.G., Marcum, J.B., Almsted, J.B., Murphy, D.B. and Johnson, K.A., 1975, Ann. N.Y. Acad. Sci. 253, 107.

32. Borisy, G.G., Almsted, J.R., Marcum, J.M. and Allen, C., 1974, Fed. Proc. 33,167.

33. Luduena, R.F. and Woodward, D.O., 1975, Ann. N.Y. Acad. Sci., 253, 272-283.

*Microtubules and Microtubule Inhibitors, eds. M. Borgers and M. de Brabander*




# Specific Visualization of Tubulin Containing Structures by Immunofluorescence Microscopy: Cytoplasmic Microtubules, Vinblastine Induced Paracrystals and Mitotic Figures

Klaus Weber
Max Planck Institute for Biophysical Chemistry
Goettingen, West Germany

## Introduction

Electron microscopy of animal cells grown in culture has established that at least three major fibrous systems can be detected in a wide variety of cells (1). These are the microfilaments (diameter of the individual fiber 6nm), the filaments (diameter of the individual fiber 10nm) and the microtubules (diameter most commonly 25nm). Ultrastructural studies on tissues from different organisms are in agreement with these results, and at least microfilaments and microtubules can be considered to be structures characteristic for all eucaryotic organisms.

Antibodies against the structural proteins of these fibrous systems are expected to help in the elucidation of the "proteinchemical anatomy" of these systems using immunological techniques such as immunoferitin labbeling and immufluorescence microscopy. The latter procedure was used initially to characterize the microfilament system. Thus actin, myosin and tropomyosin were shown to be localized in the microfilament bundles of non muscle cells (2,3,4).

Here we report the characterization of the microtubular system by a similar technique. An antibody was prepared against tubulin from the outer doublets of sea urchin sperm flagella (Strongylocentrotus purpuratus). This antibody crossreacts with tubulin from mammalian cells indicating the strong conservation of the tubulin amino acid sequence during evolution. Using indirect immunofluorescence we show that the antibody decorates certain well defined structures within mammalian cells: These include cytoplasmic microtubules, vinblastine induced paracrystals and the full spectrum of mitotic figures. As expected from electronmicroscopic studies the expression of cytoplasmic microtubules is abolished by exposure of the cells to colchicin or to low temperature. The antibodies against tubulin and actin decorate different fibrous systems within the cells and therefore indirect immunofluorescence can be used to distinguish the microfilament system from the microtubular organization (5,6).

## Results

### 1. Antibodies against tubulin and actin.

The preparation of the rabbit antiserum against microtubule protein using outer doublet microtubules of sperm flagella from sea urchins (Strongylocentrotus purpuratus) as an antigen has been described in detail (5). The antibody against mouse 3T3 fibroblast actin has been previously described (2). All details for the indirect immunofluorescence microscopy have been previously documented (2,5,6). Growth conditions for the cell lines used as well as the conditions for the experiments involving drugs can be found in previous reports (2,5,6).

### 2. The microtubular fibers in interphase cells.

Cells grown on coverslips were fixed in dilute formaldehyde for optimal preservation of structural organization. Exposure to cold methanol and acetone at -20$^{0}$ was used to make the plasma membrane of the cells permeable to macromolecules. Indirect rather than direct immunofluorescence was used in order to obtain the amplification effect of the first technique. Figs. 1 and 2 show the typical fluorescent pattern obtained with anti-microtubular antibody in chick embryo fibroblasts (Fig. 1). and mouse 3T3 cells (Fig. 2; see also below). The cells

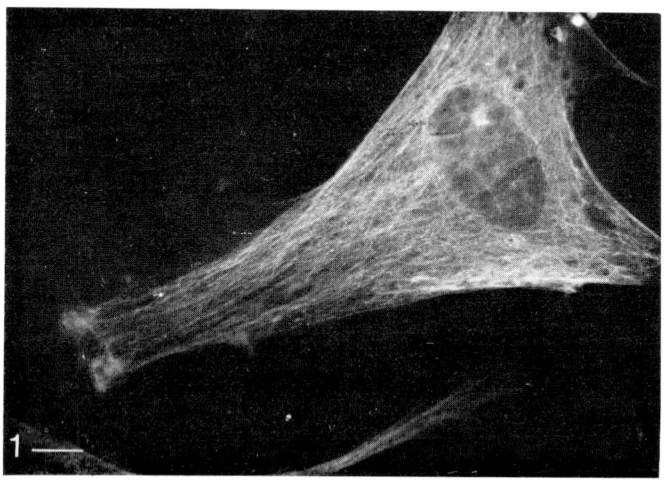

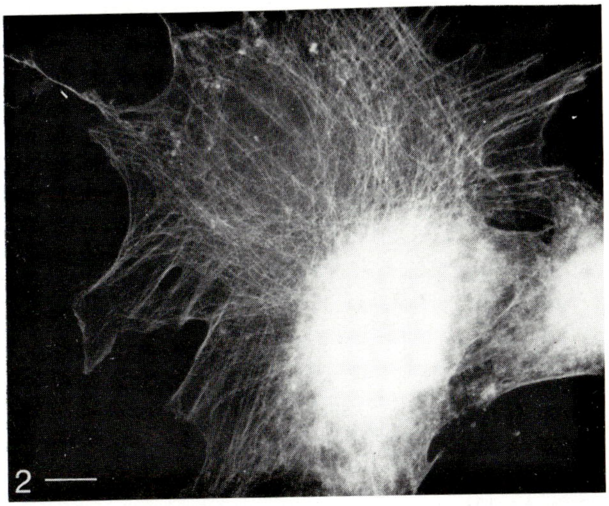


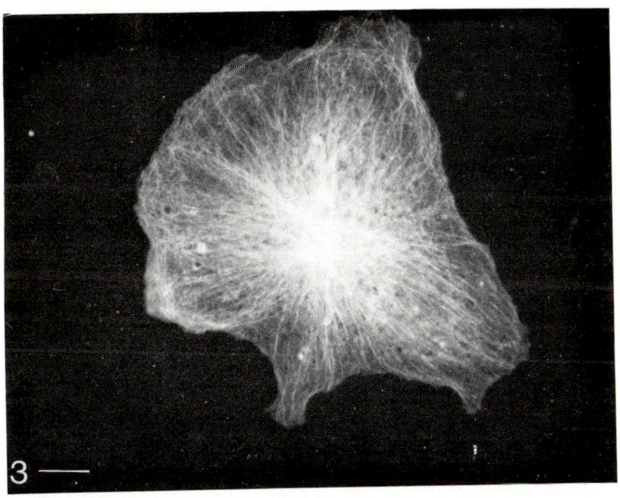


Fig. 1 - 3 Indirect immunofluorescence using tubulin antibody.
Cytoplasmic microtubules in chick embryo fibroblasts (Fig.1), mouse 3T3 cells (Fig. 2) and a monkey BSC-1 enucleate after trypsinization and replating.
The strong nuclear fluorescence is only typcial for 3T3 cells and is discussed in results. The bar indicates 10 μm.

display a multitude of very thin, separate fluorescent fibers of apparently constant diameter within their cytoplasm. These fibers traverse the whole cytoplasm and only a few fibers are arranged parallel to the margin of the cells. Differential focussing through the cells does not indicate preferential display of fibers at the bottom of the cells i.e. the fibers seem to radiate through the cytoplasm. They may end up very often at the plasma membrane, but so far we have no indication that they are arranged in close contact to or parallel to the membrane. The expression of microtubules was also studied in normal and enucleated monkey-$BSC_1$ cells. Normal $BSC_1$ cells display microtubular fibers rather similarly to the cell lines described above. After enucleation and replating enucleated $BSC_1$ cells show an extensive pattern of fragile cytoplasmic fibers decorated by the antibody. Given sufficient time to allow cell spreading of the enucleates (8-12 hrs at $37^0$) the cytoplasmic fibers traverse the entire enucleated cytoplasm (Fig. 3) Thus, the absence of the nuclear membrane in the enucleated cells does not prevent the expression of microtubular fibers (5). The intense nuclear staining seen in mouse 3T3 cells (Fig. 2) has not been found in secondary chick fibroblasts (Fig. 1), Hela cèlls, monkey $BSC_1$ cells, rat kangaroo cells or cells from rat or hamster origin. We assume that the rabbit antiserum recognizes in mouse 3T3 cells a mouse specific nuclear antigen not related to tubulin, which gives rise to the strong nuclear fluorescence.

3. Comparison of cytoplasmic microtubules and the microfilament system.

The pattern of organization of actin in a variety of cells of different origin has been described previously by us in detail (4,7,8). We have also shown a complex array of fibrous structures in living 3T3 cells with phase contrast, Normanski and polarized light optics. When these cells are stained in indirect immunofluorescence with actin antibody, the same fibers show intense fluorescence indicating that they contain actin. Electron microscopical studies revealed that these fibrous structures are submembraneous bundles of microfilaments located predominantly on the attached side of the cells inside the plasma membrane (7).

We have studied in parallel the expression of microfilament bundles and microtubular structures in interphase mouse 3T3 cells using antibodies against actin and tubulin. Fig. 4 shows a well spread 3T3 cell stained with actin antibody as originally described by Lazarides and Weber (2). Fig. 2 shows a 3T3 cell from a coverslip processed in parallel stained with antibody against tubulin. The actin and microtubule fibers in the cytoplasm are remarkably different in appearance and in arrangement. The actin fibers vary in diameter (up to 1 $\mu$) and length and are concentrated as submembraneous structures toward the adhesive side of the cell. A majority of actin fibers is arranged in parallel structures, although occasional focal points can be found. The margins of the cells often

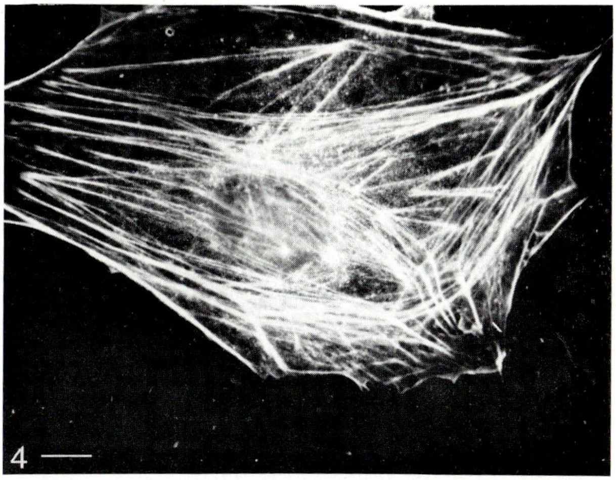


Fig. 4 Indirect immunofluorescence using actin antibody on a mouse 3T3 cell. The bar indicates 10 μm.

show strong actin fibers. In contrast to the cytoplasmic organization of the actin fibers. the microtubular fibers are very fragile structures, are separate from each other, are very thin and show no strong orientation along the boundary of the cells (5).

4. The colchicine sensitivity of cytoplasmic microtubules.

Extensive electron microscopic studies have proven that cytoplasmic microtubules disappear in the presence of the alkaloid colchicin. This *in vivo* result is in accordance with the known *in vitro* properties of microtubules (see for instance 9.) Therefore mouse 3T3 cells were plated into medium containing colchicin (15 μg /ml). Although somewhat reduced in size as a consequence of colchicin treatment the cells were well spread over the surface of the coverslip. After processing for immunofluorescence no organized fibers appeared in the cytoplasm when antibody against tubulin was used (Fig. 5). Instead there was a weak diffuse fluorescence in the cytoplasm. The fluorescence of the nucleus resembled that seen in mouse cells in the absence of colchicin and is nontubulin specific (see above and Fig. 2). When coverslips were processed in parallel with actin antibody (Fig. 6) the cells displayed the actin containing microfilament bundles (5). This result is in excellent agreement with the known colchicin-sensitivity of microtubules and the known colchicine-intensitivity of microfilament bundles (see for instance 1,9).

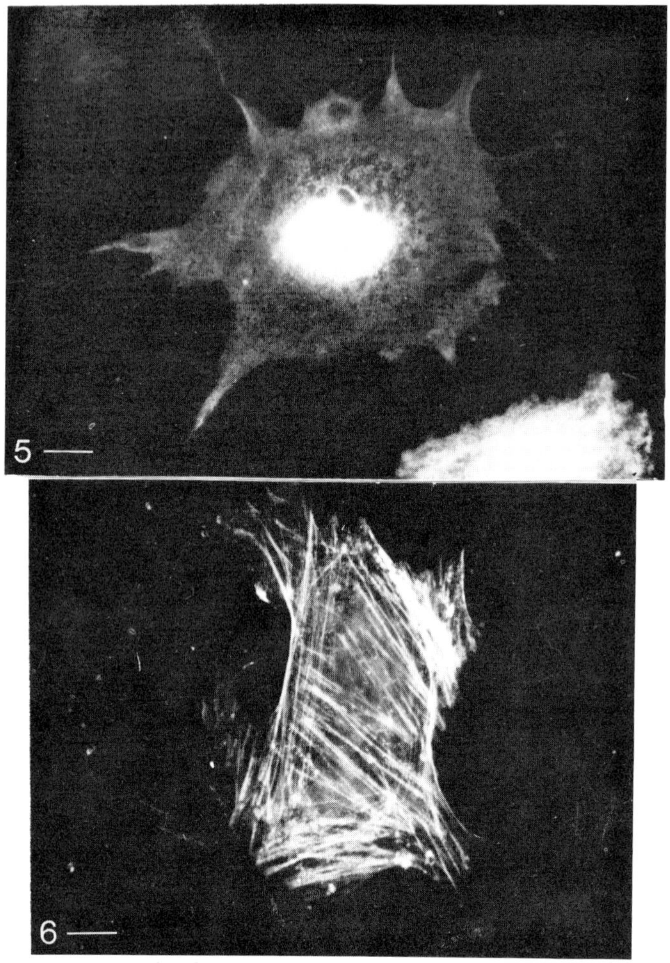


Fig. 5 - 6 **Indirect immunofluorescence using tubulin and actin antibodies.** Mouse 3T3 cells stained after colchicin treatment with tubulin antibody (Fig. 5) and actin antibody (Fig. 6). The bar indicates 10 μm.

5. Tubulin antibody decorates vinblastin induced paracrystals.

Vinblastine is known to induce the reorganization of microtubular structures in living cells. In the presence of a sufficient concentration of this alkaloid spindle microtubules as well as cytoplasmic microtubules disappear and crystalloid inclusion bodies or paracrystals appear. These paracrystals have been studied by different microscopic techniques as well as by proteinchemical studies (10-14). The experiments have indicated that the vinblastine induced paracrystals are most likely formed from homogeneous tubulin. Following the procedure of Dales and coworkers (10-12) growing mouse 3T3 and human Hela cells were treated with vinblastine (at 5 x $10^{-5}$M) for 12 to 16 hrs prior to processing for immunofluorescence. All cells were found to be devoid of cytoplasmic microtubules. Instead the cells contained several fluorescent paracrystals in their cytoplasm (Figs. 7-8). It is worthwhile to note that vinblastine treated

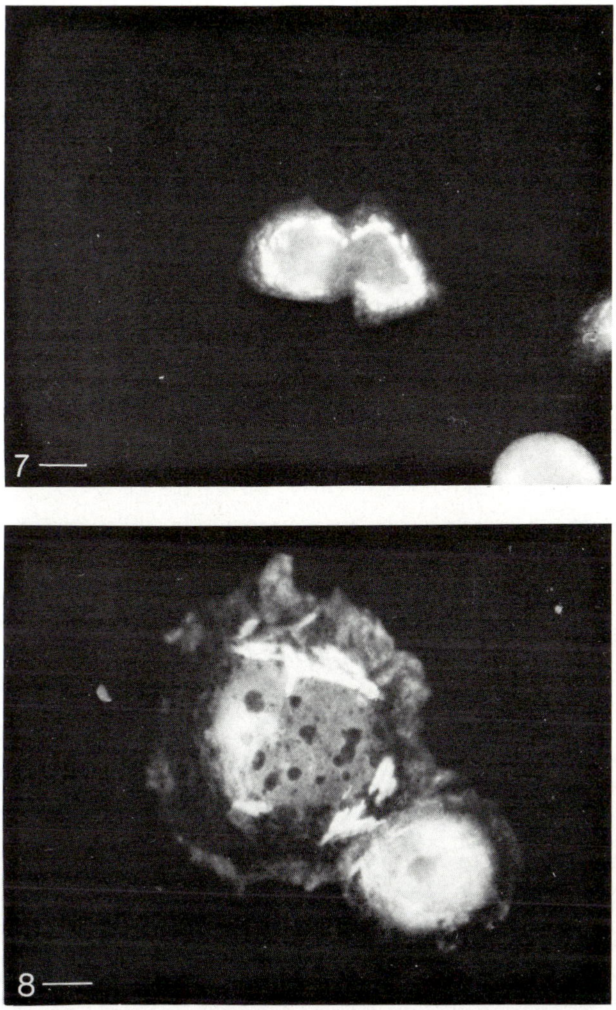


Fig. 7 - 8 Indirect immunofluorescence of vinblastine induced paracrystals using tubulin antibody. Hela cells (Fig. 7) and mouse 3T3 cells (Fig. 8) are used. The bar indicates 10 µm.

cells did not reveal crystalline inclusion bodies when stained with either antibodies to actin or antibodies to myosin (data not shown). Instead they revealed a strong fluorescence throughout the cytoplasm indicative of disorganized actin and myosin containing structures however some of the typical actin containing fibers were still present (6). Although actin as well as a variety of other $Ca^{++}$ binding proteins can be precipitated in vitro by vinblastine (15), the in vivo formation of vinblastine induced paracrystals seems to occur in a highly specific manner avoiding the incorporation of proteins other than tubulin into the crystalline matrix. This conclusion is in good agreement with the proteinchemical studies of Bryan (14) who showed that paracrystals induced in sea urchin eggs are essentially pure tubulin.

6. Mitotic Figures.

Hela cells and cells of the established rat kangaroo line $PtK_1$ were studied during mitosis using immunofluorescence microscopy (6). Fig. 9 shows a composite of mitotic Hela cells using antibody against tubulin. The different frames are arranged according to the well known phases of mitosis. The different stages of prophase can be seen and the orientation of the two centriole-connected structures is visualized outside the nuclear membrane. In pro-metaphase cells a multitude of dark chromatids and two extended fluorescent "asters" are seen. The nuclear membrane is absent in these cells. During the early part of prophase a weak background fluorescence can be seen around the condensing chromosomes.

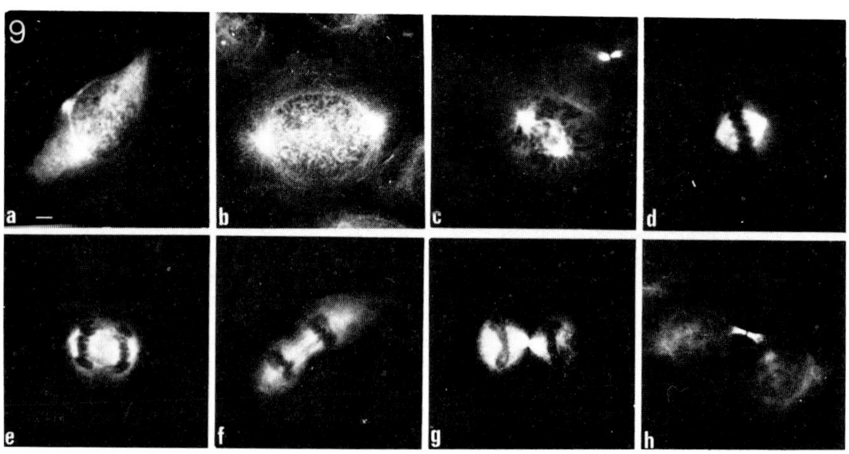


Fig. 9 A composite of Hela cells during different stages of mitosis stained in indirect immunofluorescence with tubulin antibody. The length of the bar indicates 10 μm.

Metaphase cells show the non fluorescent equatorial plate and the expression of a well developed fluorescent spindle. In a variety of metaphase and anaphase cells both the chromosomal (kinetochor to pole) and interpolar (pole to pole) fibers are apparent. With the onset of telophase there is a loss in contrast of the fluorescent structures, although the region of the interzonal fibers is still clearly visible. This effect is most likely due to an increase in cytoplasmic tubulin due to the partial disassembly of the spindle. After the clearage furrow has been well developed the two daughter cells are still connected by the highly fluorescent intercellular bridge. The position of the midbody in the bridge is generally weaker in fluorescence as two side parts of the bridge (Fig. 10).

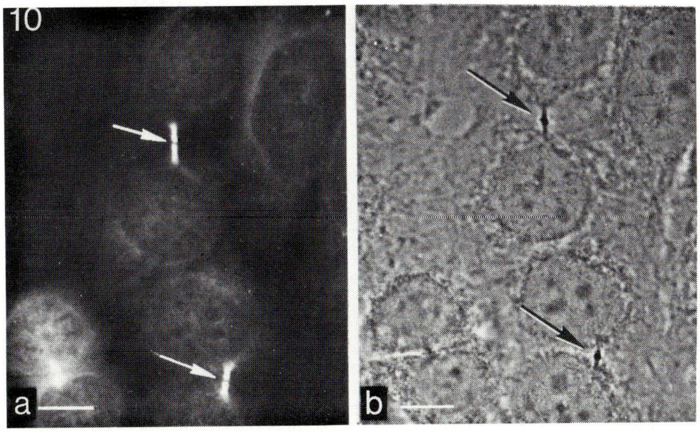


Fig. 10 The intercellular bridge between daughter Hela cells visualized by immunofluorescence with tubulin antibody (Fig. 10a) and by phase contrast (Fig. 10b). The arrows point to the midbodies. The length of the bar indicates 10 $\mu$m.

It is possible that in the midbody the antibody has less access to the microtubular structures and therefore a lower fluorescence level is obtained. Alternatively it is also possible that parts of the midbody have indeed a lower concentration of tubulin than the remainder of the intercellular bridge (6). Fig. 11 shows a typical metaphase cell of the rat kangaroo line $PtK_1$.

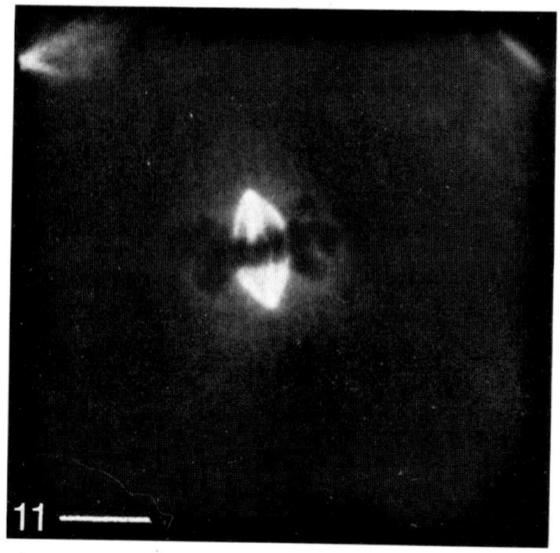

Fig. 11 A rat kangaroo $PtK_1$ cell in metaphase stained with tubulin antibody in indirect immunofluorescence. The length of the bar indicates 10 µm.

## Discussion

We have shown by indirect immunofluorescence that antibody against tubulin prepared from sea urchin outerdoublets decorates a delicate system of fibers in the cytoplasm of tissue culture cells. Mouse 3T3, monkey BSC-1 cells, normal secondary fibroblasts of human, hamster, rat and chicken origin have cytoplasmic fibers similar in structural organization and in patterns of distribution. The cytoplasmic fibers, which are decorated by the antibody preparation are sensitive to colchicin and to low temperature. These sensitivities have been found in extensive electronmicroscopic studies to be specific for microtubules but not for microfilaments (see for instance 1,9).

Several lines of evidence indicate that the microtubular fibers are distinct from the microfilament fibers, which can be visualized in immunofluorescence microscopy using the actin antibody previously described (2,7). The actin and microtubule fibers differ in diameter and length, are concentrated toward the adhesive side of the cell and are arranged predominantly parallel to each other. In contrast the microtubular fibers are separate from each other, are extremely thin structures and show no strong orientation along the boundary of the cells. The two fibrous systems are further distinguished by the sensitivity of the microtubular system against colchicin. Cells treated with colchicin display actin fibers but not cytoplasmic microtubules. A third difference is the cold sensitivity of the microtubular fibers (5). These results conform with the known property of microtubules to depolymerize in the presence of colchicin or at low temperature both *in vivo* and *in vitro* (1,9).

As expected from previous electronmicroscopical studies (see for instance 9) the antibody decorates not only cytoplasmic microtubules but also allows the visualization of the whole spectrum of *mitotic figures* (6). Thus the antigenic site of tubulin is not changed or altered in the mitotic apparatus in comparison to its arrangement in cytoplasmic microtubuli. Furthermore the antibody decorates the crystalloid inclusion bodies or *paracrystals* (6), which have been previously characterized by others using electronmicroscopy (11,13), phase microscopy (10) and immunofluorescence (10,12). Proteinchemical studies of such paracrystals obtained from sea urchin eggs have made it very likely that they are homogeneous tubulin protein (14). Our antibody preparation allows specific visualization of these paracrystals in Hela and 3T3 cells as judged by immunofluorescence. Antibody to either actin or myosin does not visualize the paracrystals. Although actin as well as a variety of other $Ca^{++}$ binding proteins (15) can be precipitated *in vitro* by vinblastine, *in vivo* formation of vinblastine induced paracrystals seems to occur in a highly specific manner avoiding the incorporation of proteins other than tubulin into the crystalline matrix.

These results strengthen the reports of Dales and coworkers on the use of vinblastine paracrystals as a useful tubulin antigen (10). They observed immunofluorescent staining of mitotic figures and vinblastine induced crystals in a variety of vertebrate cells as well as in cells from moths and mosquitos (10,12). However they seem to have been unable to visualize cytoplasmic microtubules in interphase cells. Our results show that cytoplasmic microtubules are very fragile structures and are often destroyed when standard procedures of cell fixation for immunofluorescence are used (5,6). Since special precautions were not previously taken, it seems not surprising that Dales and coworkers (10-12) revealed microtubular structures only in mitotic cells and vinblastine induced crystals but not in interphase cells. The work of Dales and coworkers (10-12) and our studies (5,6) demonstrate that microtubules contain a highly conserved antigenic structure, since antibodies against an invertebrate tubulin (sea urchin sperm outer doublets) react with human microtubules (5,6) and antibodies against vertebrate tubulin (mouse L cells) react with invertebrate tubulin (10,12). These results are in good agreement with the proteinchemical studies of Luduena and Woodward (16). They have drawn attention to the fact that both the $\alpha$ and $\beta$ polypeptide chains of tubulin have been highly conserved during evolution: the first twenty-four aminoacid residues of the $\alpha$ chains are identical for sea urchin and chicken tubulin and the same regions of the $\beta$ chains show only one amino acid replacement. The highly conservative amino acid sequence of tubulin is not unique among important proteins of eucaryotic cells. Immunological experiments have shown a similar situation for instance for RNA-polymerase II (17) and proteinchemical studies have documented the extreme preservation of the amino acid sequences of histones, for instance in the case of histone IV (18).

A monospecific antibody directed against bovine brain tubulin was recently described by Fuller et al (19). They used this antibody to visualize by indirect immunofluorescence the mitotic spindle of rat kangaroo cells ($PtK_1$) in tissue culture and demonstrated in a series of pictures rather similar to ours (Fig.9 and 11) that all stages of mitosis can be visualized. This result further emphasizes the close proteinchemical relationship of all tubulins.

The availability of antibodies against tubulin has allowed the immunofluorescent visualization of the microtubular system in tissue culture cells under a variety of different conditions. Since the experimental details for the preservation of the structure are now known it should be possible to determine whether other proteins of the cell form part of the microtubular system or are connected to it. One can expect the use of the antibody in electronmicroscopical studies using immunoferitin methods and in the elucidation of the regulation of the microtubular structures in cells and tissues.

References.


1. Goldman, R.D. and Knipe, D.M. (1972). Cold Spring Harbor Symp. Quant. Biol. 37, 523
2. Lazarides, E. and Weber, K. (1974). Proc. Nat. Acad. Sci. 71, 2268
3. Weber, K. and Groeschel-Stewart, U. (1974). Proc. Nat. Acad. Sci. 71, 4561
4. Lazarides, E. (1975). J. Cell Biology, in press
5. Weber, K., Pollack, R. and Bibring, T. (1975).Proc. Nat. Acad. Sci. 72,459
6. Weber, K., Bibring, T. and Osborn, M. (1975).Exp. Cell Research., in press
7. Goldman, R.D., Lazarides, E., Pollack, R. and Weber, K. (1975). Exp. Cell Research 90, 333
8. Weber, K., Lazarides, E., Goldman, R.D., Vogel, A. and Pollack, R. (1974). Cold Spring Harbor Symp. Quant. Biol. 39, 363
9. Olmsted, J.B. and Borisy, G.G. (1973). Annual Rev. Biochem. 42, 507
10. Nagayama, A. and Dales, S. (1970). Proc. Nat. Acad. Sci. 66, 464
11. Dales, S., Hsu, K. and Nagayama, A. (1973). J. Cell Biol. 59, 643
12. Dales, S. (1972). J. Cell Biol. 52, 748
13. Bensch, K.G. and Malawista, S.E. (1969). J. Cell Biol. 40, 95
14. Bryan, J. (1972). J. Mol. Biol. 66, 157
15. Wilson, L., Bryan, J., Ruby, A. and Mazia, D. (1970). Proc. Nat. Acad. Sci. 66. 807
16. Luduena, R.F. and Woodward, D.O. (1973). Proc. Nat. Acad. Sci. 70, 359
17. Ingles, C.J. (1973). Biochem. Biophys. Res. Commun. 55, 346
18. Delange, R.J., Fambrough, D.M., Smith, E.L. and Bonner, J. (1969). J. Biol. Chem. 244, 5669
19. Fuller, G.M., Brinkley, B.R. and Boughter, J. M. (1975). Science 187, 948

*Microtubules and Microtubule Inhibitors, eds. M. Borgers and M. de Brabander*



# MICROTUBULE MEMBRANE INTERACTIONS FLUORESCENCE TECHNIQUES

Richard D. Berlin, M.D.

Department of Physiology
University of Connecticut Health Center
Farmington, Connecticut 06032 USA

## Summary

The effects of phagocytosis on plasma membrane fluidity or microviscosity were studied by fluorescence polarization techniques. Phagocytosis of oil emulsions by rabbit polymorphonuclear leukocytes induces a marked increase in plasma membrane fluidity that parallels the extent of phagocytosis. Liposomes from membrane extracts show the same changes indicating that the alteration of fluidity results from changes in lipid composition. The decrease in microviscosity (as well as the selective movement of membrane proteins previously described) are abolished when colchicine is present during phagocytosis. Colchicine itself has no direct effect on the fluidity of isolated membranes. A theory is developed to explain MT-dependent selective partitioning of lipids and proteins during phagocytosis. The theory proposes that specific proteins that may be connected to MT induce a regional condensation of membrane lipids. The condensed phase is subsequently internalized. Recent studies using the technique of fluorescent energy transfer to explore possible MT-membrane connections are reviewed.

## Introduction

In recent years it has become recognized that the surface distribution or topography of membrane constituents is a major determinant of cell function. In addition, it is clear that surface topography is highly dynamic. Profound changes in the distribution of membrane proteins can be induced by lectin binding to surface receptors and by phagocytosis. Recent studies in our laboratory and others have indicated that microtubules (MT) play an essential role in this surface reorganization.

The phagocytic system has been particularly useful in analysis of the mechanisms that control surface topography because reorganized regions of the membrane are physically separated: thus the phagocytized particle is enveloped by membrane and internalized and the residual cytoplasmic membrane can then be examined for altered function or properties. As indicated below we believe that phagocytosis can serve as a model for surface reorganization in nonphagocytic cells wherein the regional changes are localized to smaller areas and are not separated from the plasma membrane.

The Phagocytic Model - Membrane Protein Movements

The first evidence for MT-dependent changes in surface topography during phagocytosis was obtained by studies of the redistribution of proteins. These studies have been published and so will be only briefly summarized here.

Polymorphonuclear leukocytes (PMN) and macrophages can be made to internalize a large proportion of their surfaces by phagocytosis of suitable particles. Engulfment of plastic beads or oil droplets can lead to the internalization of 50% or more of the plasma membrane.

We characterized several membrane transport systems for nonelectrolytes in rabbit PMN and alveolar macrophages and used their activity to quantify the presence of specific transport proteins.

When these cells were allowed to phagocytize extensively, surprisingly no decrease in transport was observed. We ruled out the insertion of new transport sites to replace old ones which might have been internalized (1). Thus, transport sites were preserved on the surface during phagocytosis. At low doses ($1\text{-}5 \times 10^{-6}$M) colchicine did not depress phagocytosis. However, when PMN phagocytized after colchicine treatment, transport was decreased (2). The apparent segregation of transport sites and particle-enveloping membrane was abolished. We could not distinguish in these experiments between the actual movement of transport sites out of the enveloping membrane and their prior segregation into specialized regions because there was no way of determining their exact distribution (only their quantity in residual plasma membrane).

This distinction could be made however in studies of the effect of phagocytosis on the surface topography of lectin receptors. Con A receptors are distributed uniformly over the surface. It was shown that phagocytosis leads to the selective depletion of lectin receptors from the plasma membrane, and again this selective decrease is abolished by colchicine but not by lumicolchicine (3).

Thus, lectin receptors appear to be gathered from their initially dispersed distribution and then internalized by a process that requires MT.

Direction of membrane protein movement presents a paradox.

Lectin receptors and transport proteins appear to move in opposite directions with respect to the internalized membrane (the former selectively lost, the latter preserved on the surface). This could occur if individual membrane constituents were associated with and directed by corresponding specific cytoplasmic elements, or, alternatively, if phagocytosis induced a microtubule-dependent generalized effect on membrane properties which forced a partitioning of whole classes of protein constituents. One generalized property which could affect these changes is membrane fluidity. For example, a regional decrease in membrane viscosity could favor the entrapment of proteins surrounded by phospholipids containing saturated fatty acids and the exclusion of those proteins preferentially associated with lipids that maintain their fluidity. Thus, we have studied the effect of phagocytosis on the fluidity or microviscosity of plasma membranes.

Analysis of membrane viscosity by fluorescence polarization of hydrocarbon probes

The fluorescence polarization method employed for the study of membrane viscosity has been described extensively (4) and has been previously employed with artificial liposomes (4,5) membranes (6) and whole cells (7). Essentially, the polarization of fluorescence is determined by the rotation of the molecular probe within the membrane during the lifetime of the excited state. The rotation of the probe in turn is clearly dependent on the viscosity of its environment ("microviscosity") - rapid rotation leading to depolarization. By determining polarizations of the probe in media of known viscosity a correlation between polarization values and microviscosity can be made. Unknown microviscosities of membranes can thus be inferred provided the fluorescent state lifetimes are known. The rotational depolarization of a fluorophore is described by the Perrin equation $r_0/r = 1 + C(r)\frac{T\tau}{\eta}$ where r and $r_0$ are the measured and limiting anisotropy respectively; T, the absolute temperature; $\tau$ the excited state lifetime, $\eta$ the viscosity of the medium and C(r) a parameter relating to the molecular shape and orientation of the absorption and emission dipoles and is essentially determined empirically. For our studies we used the fluorescent hydrocarbon probes perylene and DPH (1,6-diphenyl-1,3,5-hexatriene). We considered it essential to examine isolated membranes to avoid the uncertainty of localizing the probes in intact cells. Isolated membranes were isolated as before (3) and liposomes prepared from chloroform-methanol extracts of membranes by sonication.

Differences in microviscosity of control membranes reported by perylene and DPH

The results show a microviscosity of PMN membranes comparable to published data for erythrocyte ghost membranes (6) and significantly lower than apparent microviscosities of intact lymphocytes (8).

The microviscosities obtained using perylene are in general significantly lower than those obtained using DPH (Fig. 1).

Since the viscosity-polarization standard curves are obtained using the same standard oil, it is obvious that the differences in membrane microviscosities for the two probes must reflect differences in the relationship between membrane structure and the rotational motion of the probes determined by their geometry. Theoretical treatment of fluorescence depolarization by anisotropic molecular rotations (9) has shown that for planar molecules, the rotations (leading to depolarization) are dependent on the wavelength of excitation. For planar molecules such as perylene when the excitation is such that the limiting polarization approaches 0.5, then the depolarization is brought about by motion which is an average of in-plane and out-of-plane rotations. In-plane rotations dominate. DPH on the other hand, is not planar but cylindrical, and absorption and emission

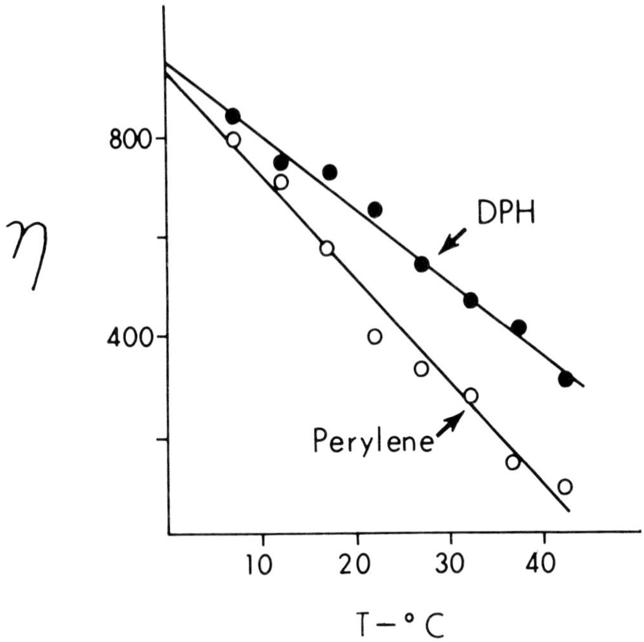


Figure 1. A plot of microviscosity η in centipoise derived from fluorescence polarization of DPH or perylene. The lines represent least squares approximations. For DPH, $\eta = 946 - 14.7T$ ($r = .993$). For perylene, $\eta = 927 - 20.8T$ ($r = .989$). T = temperature in degrees centigrade. The points represent averages of 6 experiments. (For individual experiments the plot sometimes showed a distinct curvature as suggested from the linear semilogarithmic plots shown in figure 3.)

dipoles are oriented parallel to the long, extended axis of the molecule. In this case rotation about the axis will not lead to depolarization, but tumbling motions perpendicular to the axis will cause depolarization. In a biological membrane in which the hydrophobic interior of the membrane may show some degree of order, in-plane rotations of perylene (occurring parallel to paracrystalline fatty-acid chains) may be relatively less affected than the tumbling motion required of DPH. The difference between the liposomes previously studied (8) and the natural membranes described here may result from the presence of cholesterol in the biological membrane. In any case, all of the changes reported here occurred in a qualitatively similar fashion for both perylene and DPH.

Phagocytosis is associated with an increase in fluidity of the cytoplasmic membrane.

The microviscosity of cytoplasmic membranes isolated from cells which had extensively phagocytized was markedly decreased as compared to controls (Fig. 2). This reduction was so large that values at 7° in phagocytic membranes were comparable to control membranes at 42°. This amounted to a 2-fold change at 42° and

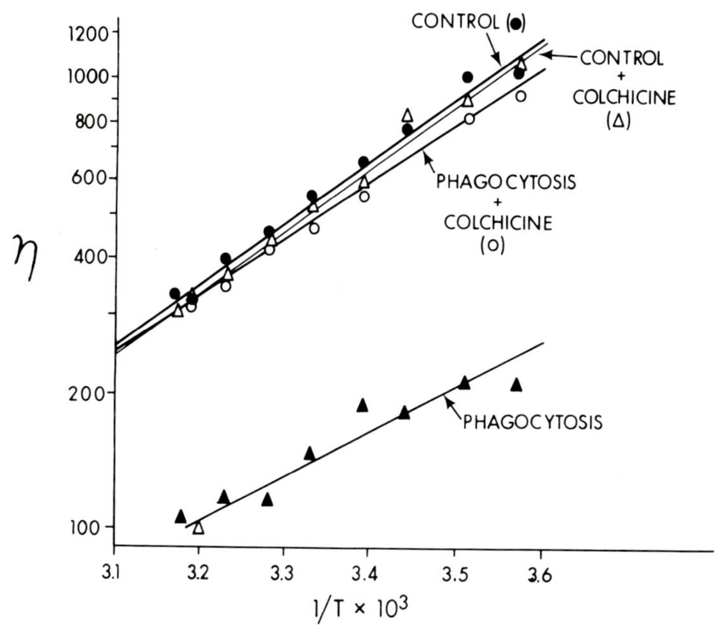


Figure 2. A semilogarithmic plot of the microviscosity, η in centipoise versus the reciprocal of the absolute temperature. A representative experiment; DPH was the fluorescent probe. Several points should be noted: 1) plasma membrane viscosity is markedly decreased after phagocytosis, 2) the presence of colchicine during phagocytosis abolishes the decrease in viscosity.

nearly 3-fold at 7°.

The decrease in viscosity paralleled the degree of phagocytosis as seen in Figure 3 in which the time course of phagocytosis previously established is shown on the same graph with viscosity determinations. In Figure 4 the data is represented in somewhat different detail. By 15' phagocytosis the microviscosity at 7° is well below that of the control at 42°.

Effect of colchicine on membrane microviscosity and on the decrease in microviscosity induced by phagocytosis.

Colchicine at 5 x $10^{-6}$M had no effect on the microviscosity of control membranes. This was true whether intact cells were incubated with colchicine and their membranes then isolated (Fig. 2.) or whether isolated membranes were subsequently incubated with colchicine.

As shown previously (3) colchicine at 5 x $10^{-6}$M had no effect on the rate or extent of phagocytosis. When membranes isolated from phagocytic cells were incubated with colchicine, there was also no effect on their lowered microviscosity (P>0.5, 9 experiments). However, when cells were allowed to phagocytize in the presence of colchicine, the characteristic decrease in membrane microviscosity did

Figure 3. Microviscosity, η, in centipoise of membranes isolated at progressive times during phagocytosis. The three upper lines show η measured at different temperatures in a single experiment. A relatively greater change in η at lower temperatures is seen. Particle uptake was quantified as before (3,10) from uptake of dye-labelled emulsion and is expressed in relative units (not shown).

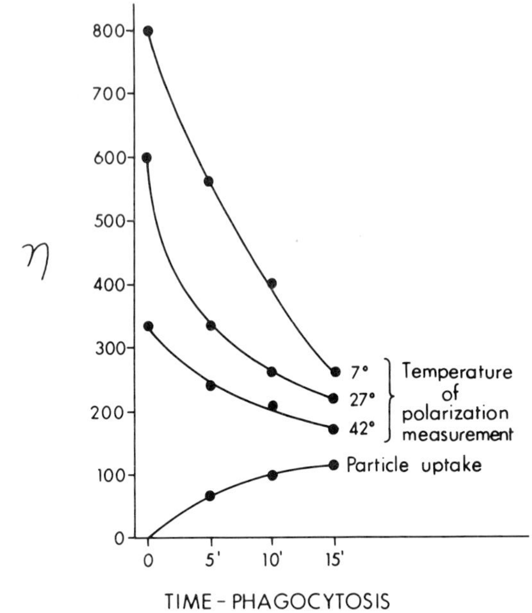


Figure 4. Microviscosity, η, in centipoise versus temperature of polarization measurement taken on plasma membranes before (0') and at successive intervals during phagocytosis. On this semilogarithmic graph the plot at 0' shows a curvature in contrast to the linearity shown in Fig. 1. Note that the absolute magnitude of the slope in the range of 20-40° falls progressively with time indicating a change in activation energy. This is discussed in detail elsewhere (11).

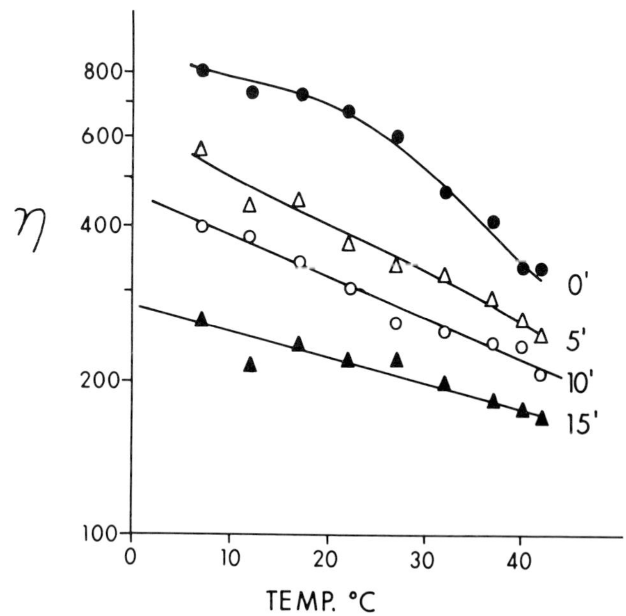

not occur. Fig. 2 shows a representative experiment. A slight decrease in microviscosity is detectable even with colchicine treatment in this experiment and in one of 5 others. The mean difference was not significant ($P<.05$).

To examine the specificity of the colchicine effect, the photo-derivative lumicolchicine was used. Lumicolchicine does not affect microtubules although it affects certain other membrane functions similarly to colchicine, the nucleoside transport system for example (12,13). Incubation of isolated membranes or phagocytizing cells prior to membrane isolation with lumicolchicine had no effect on membrane microviscosity.

Summary of evidence excluding contamination of the membrane fraction by emulsion or phagocytic vesicles.

The use of lipid emulsions for phagocytosis facilitated the high degree of membrane internalization and ready separation of membrane essential for these experiments. A potential drawback of the emulsion was its contamination of the membrane fraction since clearly the fluorescent probes would dissolve in the emulsion and report its microviscosity which is relatively low compared to membrane. Thus it was important to rule out a contribution of the emulsion to the microviscosity data under all circumstances:

a) The experiments described above showed that the oil can be internalized to the same degree in both untreated and colchicine-treated cells but that isolated membranes have a higher viscosity when colchicine is present during phagocytosis.

b) When cells were incubated with emulsion in the presence of 5 mM EDTA which blocked phagocytosis, no effect on membrane microviscosity was detectable, indicating that emulsion and membrane fraction could be effectively separated.

c) Direct determination of contamination of the membrane fraction by phagocytic vesicles was made using emulsions containing oil red O as a marker. We could have detected as little as 1% of the amount of oil red O in the vesicle fraction but could detect none in the membrane preparation.

Microviscosities determined in extracted lipid from control and phagocytic cytoplasmic membranes.

Liposomes were made from isolated lipids of control and phagocytic membranes. Qualitatively the same decrease in microviscosity of phagocytic membranes is observed as with intact membranes, Fig. 5. Moreover, the decrease is not detectable when the probe is introduced into extracted lipids from cells that have phagocytized in the presence of colchicine.

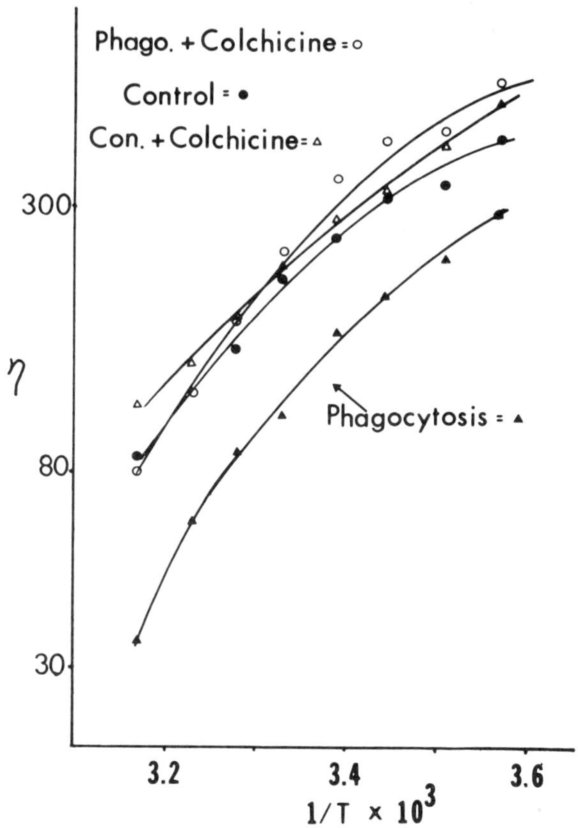


Figure 5. Semilogarithmic plot of microviscosity, η, of liposomes from extracts of plasma membranes versus the reciprocal of the absolute temperature. Data from a single experiment is shown (compare Fig. 2). To obtain scatter corrections the liposomes obtained from DPH-labelled membranes were labelled with perylene. The microviscosity from "phagocytic" liposomes was about 2-fold lower than control. A similar reduction was seen with the DPH-labelled membranes but absolute values are difficult to compare (see Fig. 1).

Interpretation of microviscosity changes.

It is clear that phagocytosis is associated with a profound increase in the fluidity of the cytoplasmic membrane. The change in fluidity parallels the degree of phagocytosis and thus does not appear to be a simple activation step resulting from particle contact, for example, but rather may result from the endocytic process itself. Since lipids isolated from cytoplasmic membranes show qualitatively the same increase in fluidity it appears that endocytosis produces a change in the lipid composition. Based on our previous work showing the selective partitioning of membrane proteins with respect to internalized membrane, it seems likely that there is a selective partitioning of lipids as well. Direct biochemical analyses of control and phagocytic membranes now in progress should clarify the nature of the partition. Alterations in either fatty acid or cholesterol content could produce the observed changes in microviscosity. Some indirect evidence not detailed here is against cholesterol.

It is important to compare the concept of partitioning with previous work suggesting a chemical remodeling of membrane fatty acids during phagocytosis. There is considerable evidence for increased though limited phospholipid turnover (14), and reacylation of lysophosphatides (15) during phagocytosis. The generation of hydrogen peroxide which can attack polyunsaturated fatty acids and possible hydrolysis of triglycerides by enzymes that could selectively mobilize saturated fatty acids (16) all suggest that saturated fatty acids might be enriched in membranes of phagocytizing cells. Smolen and Shohet (17) and Mason et al. (18) indeed found an increase in the proportion of saturated fatty acids from membranes of phagocytic vesicles, as compared to membranes of control non-phagocytizing cells. A preferential uptake of radioactive trace palmitic acid (saturated) as compared to radioactive linoleic acid (unsaturated) was also observed (17). However, the selective incorporation of saturated fatty acids into the plasma membrane of phagocytizing cells or the destruction of polyunsaturated fatty acids would lead to an increase in microviscosity and not a decrease as we observed. The presence of an increased proportion of saturated fatty acids in vesicles could be equally well explained by a membrane partitioning which removes them from the plasma membrane to form part of the vesicle structure, leaving a membrane relatively enriched for unsaturated fatty acids and therefore of lower microviscosity.

The mechanism by which colchicine-sensitive structures - presumably microtubules - can effect an increase in fluidity is unknown. Based on this report and previous observations in phagocytic cells we hypothesize that phagocytosis induces a MT-dependent regional phase change. A region of more condensed, or less fluid, lipid corresponds to the membrane region that envelops the phagocytized particle. This region is internalized and removed from the cytoplasmic membrane leaving a residual cytoplasmic membrane of low viscosity. Partitioning of membrane proteins into or out of the internalized membrane could be specified by their affinity for regional lipids. Alternatively, proteins associated with specific lipids could serve as nucleation centers for a propogated regional change. These alternatives could be viewed together as a cooperative process in which key proteins are first engaged and concentrated by MT. These proteins may condense the bilayer (as they do in model systems (see for example (19)) or they may be associated with specific lipids directly which in turn favor the association of other specific proteins and so on.

Phagocytosis appears to lead to the actual physical separation of less fluid regions from the cytoplasmic membrane. However, we suppose that it may serve as a model for other membrane effects mediated by MT in systems where membrane is not actually separated. In non-phagocytic cells condensation of a lipid-protein complex would not lead to a change in the overall composition of the membrane but could still determine the mobility of membrane constituents. For example, binding of concanavalin A to receptors on lymphocytes could induce a small regional condensation of membrane lipids which could be locally propagated, leading to the partitioning of various membrane proteins. Classes of proteins could be thus immobilized in the condensed lipid phase surrounding the receptor, by a microtubule dependent process as described by Yahara and Edelman (21). There is evidence employing spin-labels that mitogenic lectins induce a rapid increase in membrane fluidity (21). Further, Romeo et al. (22) have shown that the contact of intact PMN with polystyrene beads brings about an immediate increase in fluorescence intensity of 1-anilino-8-naphthalene sulfonate (ANS). These rapid changes could correspond to initial steps in the development of regional phase condensations.

The theory expounded here could also account for some of the changes in lymphocyte surface properties that occur at long intervals after mitogen binding. For example, Resch has described changes in membrane lipid composition of lymphocytes correlated with an increase in fluidity of liposomes made from lipid extracts of membranes several hours after exposure to mitogens (23). Possible biosynthetic mechanisms responsible for bringing about the appropriate changes in fatty acid composition were identified. As an alternative to the biosynthetic explanation, it seems possible that pinocytic internalization of selected membrane lipids may be induced by the bound mitogen. This could account in part for the observed remodeling of the membrane lipid composition.

Molecular interactions between MT and membranes.

A major problem in understanding the role of MT in surface phenomena is how a submembranous system of MT (and microfilaments) could interact with membrane components to specify their movement. Our hypothesis implies that only limited connections are required. These connections could be to part of the lipid matrix itself or alternatively and more likely to specific proteins in the membrane whose orientation would determine specific lipid associations that nucleate a regional phase change and thereby partition whole classes of membrane proteins. Numerous morphological studies indicate that a MT-microfilament network does indeed abut the membrane. Suggestions that membrane may even nucleate MT polymerization have also been made. But direct demonstration of associations between membranes and MT has been difficult.

We have recently published some evidence for relatively specific associations between membranes and MT (24). The technique employing involving resonance energy transfer is sufficiently promising that it bears some description. Reference to more detailed theory will be found in (24). Resonance energy transfer occurs when a fluorescent chromophore which is excited by the absorption of energy transmits a part of the energy to another (acceptor) chromophore some distance away. It does not involve the trivial reabsorption of light emitted by the donor. Rather, the excited electron of the donor imposes a field of a frequency proportional to the energy gap between its ground and excited states. If this frequency corresponds to a frequency required for excitation of an acceptor, energy may be transferred by resonance. This is analogous to the transfer of a radio signal to an appropriately tuned receiver. In a molecule there is a whole set of frequencies (due to vibrational states) resulting in excitation; hence one obtains continuous absorption and emission spectra. Resonance transfer thus requires the overlap of the donor emission spectrum with the acceptor absorption spectrum. The extent of energy transfer depends on the degree of this overlap and on several other factors the most critical of which is the distance separating the donor and acceptor chromophores. The efficiency of resonance energy transfer falls off rapidly with this distance and for most practical purposes does not occur beyond 100 Å. Quantitative theory allows estimation of the distance separating chromophores from the efficiency of energy transfer, and the method has been exploited extensively in recent years to study conformation changes and other properties of proteins. At this point the utility of the approach for the examination of polymerization and the interaction of tubulin with membranes becomes apparent. Essentially polymerization or membrane association is a physical approximation of elements. If the elements can be separately labelled with different chromophores whose spectra overlap, then resonance energy transfer will occur only when the elements are brought into proximity (or contact). The extent of resonance transfer in this case will be proportional to the number of units brought together. For studies of polymerization for example, aliquots of tubulin are separately labelled with fluorescein isothiocyanate (FITC) or rhodamine isothiocyanate (RITC) unbound dye removed, the differently labelled tubulins mixed and allowed to polymerize. Energy transfer is progressive with polymerization. When polymerization is complete a mean distance between chromophores can be calculated and some information on their geometry inferred. Similarly, membranes can be labelled with one chromophore and tubulin with another. Again energy transfer occurs only when tubulin is bound. With monophoton counting and other techniques it is possible to analyze for the proportion of total tubulin bound or, to show if there is more than one population of binding site.

Resonance energy transfer between MT and membranes.

Cytoplasmic membranes were isolated from rabbit polymorphonuclear leukocytes as for the microviscosity studies and labelled covalently with fluorescein isothiocyanate (FITC). The labelled membranes were then washed free of unreacted

FITC by recentrifugation through a discontinuous sucrose gradient. Rhodamine labelled tubulin was prepared by incubation of isolated tubulin with rhodamine isothiocyanate (RITC). The excess RITC was removed by passage of the dye-tubulin mixture over Sephadex. Membranes (FITC-labelled) and tubulin (RITC-labelled) were then mixed under various conditions. At 4°C virtually no energy transfer took place between FITC-labelled membranes and RITC-labelled tubulin. At 37°C polymerization of the rhodamine-tubulin occurred, and resonance energy transfer between fluorescein-membranes and rhodamine-tubulin developed in parallel with the increase in light scatter. This transfer was not diminished by the addition of unlabelled tubulin subsequent to polymerization. Similar experiments using rhodamine-labelled bovine serum albumin and fluorescein-labelled membranes did not reveal energy transfer at any temperature. These experiments suggest a protein-specific association of polymerized MT with membrane. Energy transfer between tubulin and membranes also increases progressively with time in the presence of colchicine. However, addition of unlabelled tubulin to the membrane-tubulin complex formed in the presence of colchicine ($10^{-4}$M) rapidly reduced the resonance transfer. Thus, in the presence of colchicine the association of tubulin with membranes, seems readily reversible and is distinct from the association formed in the absence of the alkaloid. The high efficiency of energy transfer observed (>50%) indicates considerable interaction of the tubulin and membrane, and suggests that some labelled constituent of the membrane preparation could serve as a nucleation center for MT polymerization.

## Concluding remarks.

It is clear that during phagocytosis there is an extensive reorganization of protein and lipid components of membranes. We hypothesize that this process involves the partitioning of lipids and proteins into a condensed region which becomes internalized. Only limited connections between MT and specific protein elements may be required for this to occur. We suppose that similar processes leading to regional condensations but without internalization result from binding of surface receptors by lectins such as Con A. In preliminary experiments we have also demonstrated a relatively specific association between isolated membranes and MT. Our studies emphasize the capacity of fluorescence measurements to probe highly dynamic systems.

## Acknowledgement

Supported by National Institutes of Health grant# CA 15544.

References


1. Tsan, M.F. and Berlin, R.D, 1971, J. Exp. Med. 134, 1016-1035.
2. Ukena, T.E. and Berlin, R.D., 1972, J. Exp. Med. 136, 1-7.
3. Oliver, J.M., Ukena, T.E. and Berlin, R.D., 1974, Proc. Nat. Acad. Sci. USA 71, 394-398.
4. Shinitzky, M., Dianoux, A.-C., Gitler, C. and Weber, G., 1971, Biochemistry 10, 2106-2113.
5. Cogan, U., Shinitzky, M., Weber, G. and Nishida, T., 1973, Biochemistry 12, 521-527.
6. Rudy, B. and Gitler, C., 1972, Biochim. Biophys. Acta 288, 231-236.
7. Shinitzky, M. and Inbar, M., 1974, J. Mol. Biol. 85, 603-615.
8. Shinitzky, M. and Barenholz, Y., 1974, J. Biol. Chem. 249, 2652-2657.
9. Weber, G., 1971, J. Chem. Phys. 55, 2399-2407.
10. Stossel, T.P., Pollard, T.D., Mason, R.J. and Vaughan, M., 1971, J. Clin. Invest. 50, 1745-1757.
11. Berlin, R.D. and Fera, J.P. To be published.
12. Mizel, S.B. and Wilson, L., 1972, Biochemistry 11, 2573.
13. Berlin, R.D., 1973, J. Biol. Chem. 248, 4724-4730.
14. Elsbach, P. and Levy, S., 1968, J. Clin. Invest. 47, 2217.
15. Paul, B. and Sbarra, A.J., 1968, Biochim. Biophys. Acta 156, 168.
16. Elsbach, P. and Farrow, F., 1969, Biochim. Biophys. Acta 176, 438.
17. Smolen, J.E. and Shohet, S.B., 1974, J. Clin. Invest 53, 726-734.
18. Mason, R.J., Stossel, T.P. and Vaughan, M., J. Clin. Invest. 51, 2399.
19. Phillips, M.C., Graham, D.E. and Hauser, H., 1975, Nature 254, 154-156.
20. Yahara, I. and Edelman, G.M., 1975, Proc. Nat. Acad. Sci. USA 72, 1579-1583.
21. Barnett, R.E., Scott, R.E., Furcht, L.T. and Kersey, J.H., 1974, Nature 249, 465-466.
22. Romeo, D., Cramer, R. and Rossi, F., 1970, Biochem. Biophys. Res. Commun. 41, 582-586.
23. Resch, K. and Ferber, E., 1975, in Immune Recognition pp.281-312, Acad. Press, New York.
24. Becker, J.S., Oliver, J.M. and Berlin, R.D., 1975, Nature 254, 152-154.

*Microtubules and Microtubule Inhibitors, eds. M. Borgers and M. de Brabander*
© 1975, North-Holland Publishing Company - Amsterdam


# DEFECTS IN CYCLIC GMP GENERATION AND MICROTUBULE ASSEMBLY IN CHEDIAK-HIGASHI AND MALIGNANT CELLS

Janet M. Oliver, Ph.D.

Departments of Physiology and Pathology
University of Connecticut Health Center
Farmington, Connecticut 06032 USA

## Summary

It has proven difficult to select for cells with genetic defects in microtubule (MT) structure or function. However, cells from the mutant beige or Chediak-Higashi (CH) mouse have properties that indicate abnormal MT function. Previous investigators have described impairment of chemotaxis and lysosomal degranulation in polymorphonuclear leukocytes (PMN) from CH mice. Our studies have shown that the mobility of membrane proteins on CH PMN is similar to that on colchicine-treated normal mouse PMN. Thus, normal cells do not form surface caps with Concanavalin A (Con A) except after treatment with colchicine, which inhibits MT assembly, whereas CH PMN cap spontaneously. CH mouse primary embryonic fibroblasts also form caps with Con A and are readily rounded up and detached by colchicine. By contrast, normal embryonic or 3T3 fibroblasts are relatively insensitive to this microtubule agent. Interestingly, SV3T3 fibroblasts are similar to CH fibroblasts: they show shape changes and Con A cap formation after colchicine treatment, suggesting that MT stability may be reduced by viral transformation.

The apparent MT defect in CH cells appears to be related to inadequate generation of cyclic GMP. Enhanced Con A capping on both CH PMN and colchicine-treated normal PMN is prevented by cyclic GMP and by the cholinergic drugs carbamylcholine and carbamyl β-methylcholine, that promote cyclic GMP generation. These agents also prevent capping and colchicine-induced shape changes in CH fibroblasts. Several peptide hormones which increase cyclic GMP in their target tissues mimic the cholinergic drugs. In addition, the giant lysosomal granules characteristic of CH cells are replaced by morphologically normal granules in CH fibroblasts cultured in the presence of cholinergic drugs. Thus, the giant granules that are pathognomonic of CH syndrome may reflect the proposed cyclic GMP-MT defect.

## Introduction

Analysis of microtubule (MT) function *in vivo* has been limited largely to study of the effects of antimitotic drugs. Those studies have provided evidence for a role of cytoplasmic MT in chemotaxis (1), lysosomal degranulation (2), axoplasmic transport (3), secretion of proteins and hormones (4), etc. At the subcellular level, MT have been implicated in the selective movement of membrane proteins and lipids during phagocytosis in polymorphonuclear leukocytes (PMN) (5,6,7) as well as in regulation of the mobility of receptors for surface-bound ligands (lectins, antibodies) in PMN (8) lymphocytes (9) and fibroblasts (10).

However, attempts to consolidate and extend these data by genetic selection of MT-defective cell lines in which MT function could be studied without drugs or questions as to drug specificity, have been unsuccessful. It has proved possible, for example, to isolate colchicine-resistant mammalian cells but these have consistently shown impaired membrane permeability to colchicine rather than altered MT protein (11). Similarly, virus-transformed fibroblasts show evidence of impaired MT function (high susceptibility to colchicine-induced shape changes and surface cap formation with lectins) (10) but many other cellular properties are modified by transformation, so that these cells are poor models of MT-defective cells.

The experiments described here provide preliminary evidence for a MT defect in an inherited disorder, the Chediak-Higashi (CH) syndrome. The evidence obtained so far suggests that the primary defect is not in the MT protein *per se* but in the subcellular mechanisms that regulate the assembly of cytoplasmic MT. CH cells may provide a useful system in which to further characterize MT function and to study regulation of MT assembly *in vivo* in mammalian cells.

## Chediak-Higashi Syndrome

CH syndrome is an inherited disease of man (12) and certain mutant strains of animals (mice (13), mink (14), cattle (15), cats (16)). It is characterized by partial albinism, high susceptibility to infection and the presence of aberrant giant granules in all granule-containing cells. CH PMN show impaired bactericidal activity, that is correlated with reduced lysosomal degranulation during phagocytosis. Chemotactic activity of CH PMN is also impaired both *in vivo* and *in vitro*. Since colchicine has been reported to supress both lysosomal degranulation (2) and chemotaxis (1) in normal PMN, we proposed that MT function may be defective in these cells.

## Evidence for impaired MT function in CH cells.

MT function was compared between peripheral blood PMN from normal C57/6J black mice and from the C57/6J beige mouse, a spontaneous mutant of the black mouse and a homologue of the CH syndrome in man. The test system employed measures the ability of cells to form surface caps with the plant lectin, Concanavalin A (Con A). It has been established in a number of cell types that the mobility of Con A on cell surfaces is a function of MT assembly and MT membrane interactions. Thus PMN (8), SV40 transformed fibroblasts (10) and lymphocytes exposed to Con A above a critical concentration (5 μg/ml) (9) bind the lectin to surface glycoprotein receptors and the lectin-receptor complexes remain fairly randomly distributed in the membrane. However, when the cells are pretreated with colchicine a large proportion of cells now form surface caps with Con A. From this, it is considered that colchicine-sensitive structures, presumably MT, normally restrain the mobility of lectin-receptor complexes in cell membranes. Inhibition of MT assembly with colchicine removes these restraints and permits movement of lectin-receptor

---

Fig. 1. The distribution of F-Con A on mouse PMN. (a) Normal PMN, diffuse labelling predominates; (b) CH PMN, most cells are capped; (c) Colchicine-treated normal PMN, the cells are capped; (d) cyclic GMP-treated CH PMN, labelling is diffuse. See ref. 8 for details

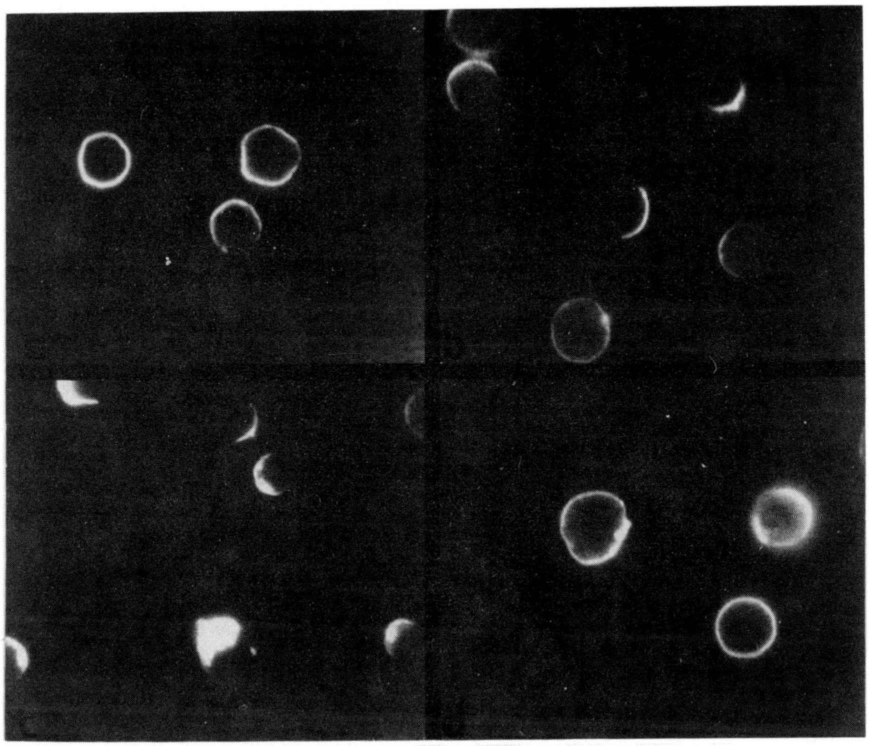

complexes to form patches and caps on the membrane. In another test system that has been studied in our laboratory, the differentiating chick erythrocyte, Con A forms surface caps on proerythroblasts, which contain MT randomly distributed in the cytoplasm but not on erythroblasts, where MT are organized into a marginal band in close approximation to the membrane (17). This is also suggestive of a role of MT-membrane interactions in controlling the mobility of lectin receptors on cell surfaces.

The relative ability of PMN from normal and CH mice to form caps with Con A is illustrated in Fig. 1 and quantified in Table I (normal PMN) and Table II (CH PMN). When peripheral blood PMN from normal mice are incubated with fluorescein isothiocyanate-conjugated Con A (F-Con A) for 10 minutes at 37°C, then rinsed, fixed and examined by fluorescence microscopy, it is seen that the majority of cells show a random distribution of lectins and very few form surface caps (Fig. la; Table 1, line 1). However, when normal cells are pretreated with $10^{-6}$M colchicine for 30 minutes prior to labelling, there is a significant change in the distribution of Con A, with the majority of cells now appearing in the capped configuration (Fig. lc; Table 1, line 2). Lumicolchicine, a photochemical derivative that shares certain properties with colchicine (e.g., inhibition of membrane transport of nucleosides (18)) but does not bind with tubulin, has no effect on Con A cap formation (Table 1, line 3). Thus, colchicine-induced capping is most likely a consequence of inhibition of MT assembly.

By contrast with the normal cells, CH mouse PMN form Con A caps spontaneously on a high proportion of cells (Fig. Ib; Table 2, line 1) and colchicine does not further increase the proportion of capped cells (Table 2, line 2).

## The Biochemical Basis of the Proposed MT Defect

The similarity between the distribution of Con A on untreated CH PMN and colchicine-treated normal PMN provides evidence for an alteration in the MT-dependent series of events that control the distribution and mobility of membrane proteins in CH cells.

The simplest explanation of these data would be a structural alteration in the cytoplasmic tubulin that prevents or impairs its polymerization to form MT. However, this explanation seems unlikely to be correct. We have described a fluorescence technique which allows study of the copolymerization of small amounts of tubulin from one source with large amounts of tubulin obtained from another source (19). The methodology is described in this volume (7). When rabbit brain tubulin was used as a 'carrier' to look at polymerization of MT out of a fluorescein isothiocyanate (FITC)-prelabelled mouse brain homogenate *in vitro*, we found no difference in copolymerization between normal black and CH mouse brain with rabbit tubulin.

Therefore we considered an alternative hypothesis, that a factor or factors involved in controlling MT polymerization may be absent or defective in CH cells.

Distribution of F-Con A on Mouse PMN

| Table I Line | Preincubation (30 min.) | Normal PMN Random | Capped | Patched |
|---|---|---|---|---|
| 1 | Buffer | 62 | 11 | 27 |
| 2 | Colchicine ($10^{-6}$M) | 23 | 55 | 22 |
| 3 | Lumicolchicine ($10^{-6}$M) | 73 | 11 | 16 |
| 4 | cAMP ($10^{-3}$M) + theophylline ($10^{-3}$M) | 67 | 8 | 25 |
| 5 | cGMP ($0.5 \times 10^{-4}$M) | 70 | 8 | 22 |
| 6 | cGMP + colchicine | 72 | 12 | 16 |
| 7 | Carbachol ($10^{-5}$M) | 87 | 2 | 11 |
| 8 | Carbachol + colchicine | 76 | 10 | 14 |
| 9 | PMA (10 ng/ml) | 77 | 6 | 17 |
| 10 | PMA + colchicine | 75 | 14 | 11 |

| Table II | | CH PMN | | |
|---|---|---|---|---|
| 1 | Buffer | 33 | 44 | 23 |
| 2 | Colchicine ($10^{-6}$M) | 32 | 50 | 18 |
| 3 | cAMP ($10^{-3}$M) + theophylline ($10^{-3}$M) | 41 | 38 | 21 |
| 4 | $PGE_1$ ($2.8 \times 10^{-6}$M) + theophylline | 48 | 32 | 20 |
| 5 | cGMP ($0.5 \times 10^{-4}$M) | 76 | 10 | 14 |
| 6 | Carbachol ($10^{-5}$M) | 72 | 8 | 20 |
| 7 | Bethanechol ($10^{-5}$M) | 69 | 17 | 14 |
| 8 | PMA (10 ng/ml) | 69 | 6 | 25 |
| 9 | CCK octapeptide (1 μg/ml) | 65 | 15 | 20 |
| 10 | Gastrin (I + II) (10 μg/ml) | 69 | 25 | 6 |
| 11 | C5a (10 μg/ml) | 64 | 18 | 18 |
| 12 | Guanine ($0.5 \times 10^{-4}$M) | 25 | 46 | 29 |
| 13 | Guanosine ($0.5 \times 10^{-4}$M) | 30 | 46 | 24 |
| 14 | GMP ($0.5 \times 10^{-4}$M) | 35 | 47 | 18 |

Monolayers of peripheral blood PMN were labelled with F-Con A (5 μg/ml, 10-15 min) and the distribution of bound lectin examined by fluorescence microscopy as previously described (8). Random, a uniform distribution of fluorescence; capped, fluorescence concentrated in a polar shell or knob; patched, fluorescence inside the cells in a patchy distribution due to endocytosis of lectin-receptor complexes.

Recent evidence indicates that the degree of MT assembly and MT-membrane interaction varies in a dynamic manner in intact cells. Thus, Reaven and Axline (20) have shown MT and MF in close association with areas of a macrophage membrane in contact with a glass surface or a phagocytic particle and absent from resting areas of the membrane. Similarly Weissmann and coworkers (21) have demonstrated increased numbers of MT profiles in PMN after exposure to zymosan particles or the C5a component of complement. Edelman et al. (9) have reported that a critical amount of Con A must bind to lymphocyte surface receptors before Con A-receptor complexes are immobilized; it seems likely that this amount of Con A is the minimum required to assemble MT.

It has also been shown that certain substances, particularly cyclic nucleotides appear to either enhance or decrease MT assembly in intact cells. Thus, the increase in number of MT profiles and degree of lysosomal degranulation caused by serum-treated zymosan or C5a in PMN can be reduced by colchicine or cyclic AMP, and enhanced by $D_2O$, which improves MT stability, and cyclic GMP (2,21,22).

Therefore, we tested the effects of cyclic nucleotides on the distribution of Con A on normal and CH PMN. The results obtained are summarized in Tables I and II.

Cyclic AMP and prostaglandin $E_1$ ($PGE_1$) which raises cyclic AMP levels in PMN (2), had no significant effect on capping on either normal (Table 1, line 4) or CH (Table 2, lines 3 and 4) PMN.

However, cyclic GMP caused a dramatic reduction in Con A capping in CH PMN to the levels seen in normal cells (Table 2, line 5). This effect is illustrated in Fig. 1. Cyclic GMP treated CH PMN (Fig. 1d) appear identical to untreated normal PMN (Fig. I,a).

A number of agents that have been shown to increase cyclic GMP generation in PMN also reduced Con A capping in CH cells to normal levels. These include two analogues of acetylcholine, carbamylcholine (carbachol) and carbamyl β-methylcholine (bethanechol) and the cocarcinogen phorbol myristate acetate (PMA) (Table 2, lines 6,7,8) (23). Other compounds that reduced capping in CH PMN are the C5a component of complement, which raises cyclic GMP (24) and stimulates MT assembly (21) in normal PMN (Table 2, line 11) and two gastrointestinal hormones, the C-terminal octapeptide of cholecystokinin and gastrin (Table 2, lines 9 and 10). These hormones promote secretory events in their target cells, accompanied by increased cyclic GMP generation (25).

---

Fig. 2. Colchicine-sensitivity of primary mouse embryonic fibroblasts. (a) CH cell, showing large granules; (b) CH cells exposed to colchicine ($10^{-6}$M, 30 minutes), the cells are rounded up and beginning to detach from the substratum; (c) normal mouse cell; (d) normal cell exposed to colchicine; (e) CH cell grown to near-confluency in carbachol, showing normal granules; (f) CH cells grown in carbachol and exposed to colchicine. Photographed using Nomarski illumination. Initial magnification x 500.

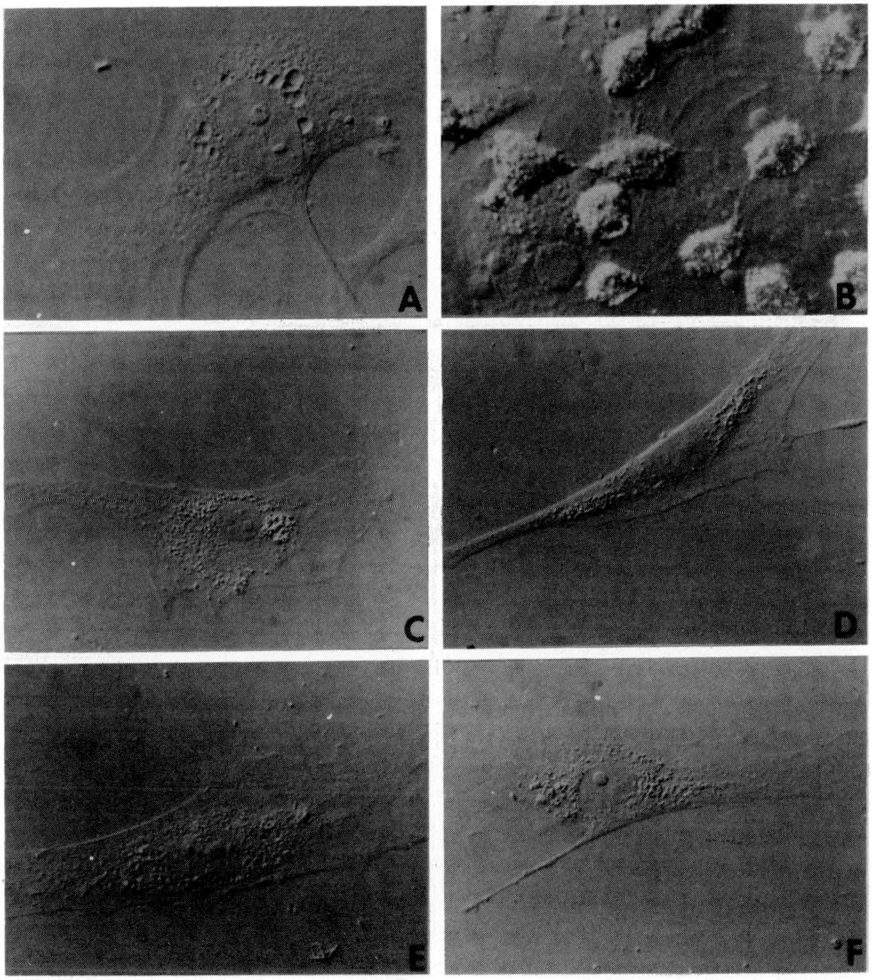
A
B
C
D
E
F

On the other hand, the potential degradation products of cyclic GMP, guanine, guanosine and GMP did not affect Con A capping on CH cells (Table 2, lines 12,13, 14). Thus, the effect appears to be specific for the cyclic nucleotide.

Further experiments with normal mouse PMN support the hypothesis that cyclic GMP inhibits capping via an effect on MT assembly. From Table 1, lines 5,7 & 9, it can be seen that cyclic GMP and agents that promote its generation do not have any direct effect on the distribution of Con A on normal cells. However, they antagonize the increase in capping that follows colchicine treatment in normal mouse PMN (Table I, lines 6,8 & 10). This strongly suggests that MT stability is enhanced by cyclic GMP.

From consideration of all these data, we propose that MT function is indeed impaired in CH cells. However, the basic defect appears to lie in the inability to generate suitable amounts of a compound that regulates MT assembly, cyclic GMP, after lectin binding rather than in a defect in the tubulin itself.

Long Term Effects of Cyclic GMP on CH Cells

The presence of giant lysosomes in all lysosome-containing cells is pathognomonic of CH syndrome. Since secondary lysosomes are derived in part from plasma membrane that is endocytized, and since we have previously established that the membrane composition of endocytic vesicles in colchicine-treated cells differs from that of untreated normal cells (5,6,7), it seemed possible that morphologically and functionally abnormal lysosomes could result from a chronic impairment of MT assembly. By extension, agents that appear to promote MT assembly might be predicted to normalize lysosome morphology in CH cells.

To test this, we isolated primary embryonic fibroblasts from CH and normal mice and grew the cells in tissue culture until they reached confluency. The cells were stored frozen and were thawed at intervals to give a supply of rapidly growing cells capable of steady division for about 3-4 weeks. The two cell lines grew at similar rates but differed in at least three respects.

First, the CH fibroblasts were highly susceptible to shape changes induced by colchicine; exposure to $10^{-6}$M colchicine for 30 minutes caused these cells to become rounded up and to detach from the culture dishes (compare fig. 2a, CH cells without colchicine, to fig. 2b, the same cells exposed to colchicine for 30 minutes). By contrast, cells from the black mouse showed no morphological change during acute exposure to colchicine, indicating a much greater degree of MT stability (compare figs. 2c, without colchicine, with 2d, exposed to colchicine). In

Fig. 3. Lysosome morphology in primary mouse embryonic fibroblasts. (a) Normal cells, showing small granules dispersed through the cytoplasm; (b) CH cell, with huge granules clustered in the perinuclear area of the cell; (c) CH cell grown to confluency in the presence of carbachol, showing morphologically normal lysosomal granules. The cells were stained with acridine orange (1in $10^6$ in phosphate-buffered saline 3 minutes exposure at 37°C) and examined by fluorescence microscopy. Initial magnification x 500.

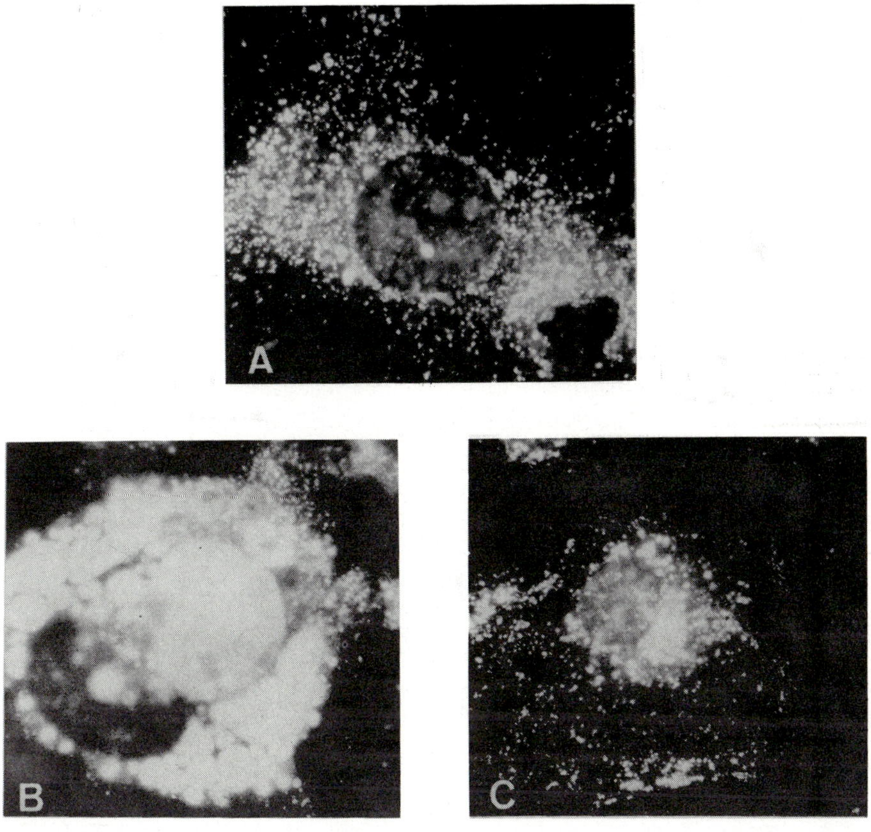
A
B
C

this property, CH cells are similar to SV40 transformed 3T3 cells which also show shape changes and loss of adhesiveness during acute exposure to colchicine (10).

Second, the CH fibroblasts showed spontaneous surface cap formation with F-Con A (25 μg/ml for 5 minutes at 37°C) on a high proportion of cells (48%). By contrast, normal fibroblasts showed some surface patching and pinocytic internalization of Con A but large aggregates or caps were seen on less than 10% of cells (26). This property of CH fibroblasts is an exaggeration of an established property of SV3T3 fibroblasts: transformed cells, unlike their parental 3T3 line, form surface caps with Con A after brief exposure to colchicine (26,10).

Third, the CH cells developed giant lysosomal granules, as shown by Nomarski illumination in Fig. 2a and after vital staining for 3 minutes with the fluorescent lysosomotropic dye, acridine orange, in Fig. 3b. These abnormal structures appear to arise by successive fusion of smaller granules to produce large granules since they are considerably larger and more prominent in confluent cultures than in growing cultures (compare the CH cell in fig. 2a from a growing culture and containing a mixture of large and small granules with the cell in fig. 3b, from a confluent culture). At confluence, at least 60% of cells contain these giant granules and the remainder of cells have granules that are larger than normal. A similar evolution was previously documented for the giant granules in PMN of mink _in vivo_ (15).

To test the effect of cyclic GMP on these properties, cells were plated sparsely and allowed to grow in Dulbecco's modified Eagle's Medium (DMEM) containing 10% fetal calf serum (FCS) plus carbachol ($10^{-5}$M) or PMA (10 ng/ml). The media were changed daily. These drugs were chosen for use rather than cyclic GMP itself because they were highly effective in preventing Con A capping in PMN and because cyclic GMP would be predicted to be destroyed rapidly by enzymatic action.

Neither carbachol nor PMA affected the growth characteristics of the cells. The growth rate, determined by counting cells on petri dishes at daily intervals between plating and confluency, was essentially identical in treated and untreated dishes and cell density at confluency was unchanged. However, dramatic changes in colchicine-sensitivity and lysosomal morphology were observed.

Figure 2e and 2f show CH fibroblasts cultured for 6 days in carbachol and then incubated for 30 minutes in phosphate-buffer saline with and without $10^{-6}$M colchicine, followed by fixation and examination. The cells are no longer rounded or detached by colchicine and cannot be distinguished from normal cells. Con A cap formation was also reduced in the drug-treated CH fibroblasts. After a 5 minute labelling period with FITC-Con A (25 μg/ml) only 11% of PMA-treated cells and 19% of carbachol-treated cells formed caps, compared with 48% of untreated cells.

When the same cells were stained with acridine orange for 3 minutes and then examined with the fluorescence microscope further modifications were seen. Figure 3a shows the appearance of lysosomal granules in normal fibroblasts. The cells contain large numbers of small, randomly-distributed granules that fluoresce a bright orange against a background of green cytoplasm and nucleus. Figure 3c, as well as 2e and 2f show the typical appearance of carbachol or PMA treated CH fibroblasts. Approximately 90% of cells have only small granules at confluence and the remaining 10% of cells have only one or two large granules. These may reflect dilution of granules in the cells that were originally plated. Thus, CH cells grown with these agents develop granules that are morphologically similar to those in normal fibroblasts.

## Discussion

Cells from animals or man with CH syndrome may provide a valuable model system for the study of MT function and the control of MT assembly and MT-membrane interactions *in vivo*.

CH mouse PMN show an exaggerated surface capping response after Con A binding. A similar response is induced in normal PMN with colchicine, suggesting a MT defect in the CH cells. Copolymerization studies indicate that enhanced capping is not due to the absence of tubulin or to the presence of tubulin that cannot polymerize under *in vitro* polymerization conditions. In any case the cells can divide, indicating that at least the spindle MT are normal. Therefore, we favor the hypothesis that the abnormal surface properties of CH PMN reflect a defect in the ability to assemble cytoplasmic MT *in vivo* in response to certain stimuli.

This defect appears to lie in the sequence:

```
                 cyclic GMP generation
                           ↓
ligand binding ──────────────→ MT assembly ───→ receptor immobilization
                            and MT-membrane
                              interaction
```

In normal mouse PMN, Con A binding *per se* may increase cyclic GMP generation and so initiate this series of events. In CH cells Con A is an inadequate stimulus, but addition of other ligands that are probably more potent stimulators of cyclic GMP generation (as judged from their ability to antagonize the effects of colchicine in normal cells) can initiate the sequence that leads to immobilization of Con A-receptor complexes in CH PMN.

There is precedent for this proposed cyclic GMP-MT association in intact cells. Thus, other leukocyte functions thought to depend on MT, such as chemotaxis (27) and lysosomal degranulation (2), have been shown to be enhanced by cyclic GMP. It has also been shown that Con A binding with lymphocytes and interaction of serum-treated zymosan with PMN promotes cyclic GMP generation (24,28). In PMN, addition of serum-treated zymosan plus acetylcholine causes a 9-fold increase in cyclic GMP compared with a 4-fold increase with zymosan alone (24). Furthermore, addition of C5a in the presence of PMA gives an enhancement of numbers of visible MT over the numbers seen with C5a alone (22).

At present it is not possible to define the role of cyclic GMP in MT assembly. Using purified rabbit brain tubulin, we do not find any increase in polymerization or in MT-membrane association when cyclic GMP is added to a polymerization mixture. Nor can cyclic GMP substitute for GTP as a substrate for polymerization. However, the in vitro system could very well lack a cyclic GMP-dependent protein kinase that, for example, plays a role in determining the rate or extent of MT assembly in vivo. Alternatively, it may contain a guanylcyclase that generates sufficient cyclic GMP from the GTP present in the polymerization mixture.

It is not established that the abnormal lysosomal morphology in CH cells is a consequence of defective MT assembly. However, an attractive case for this possibility can be made. Our previous studies have shown that when cells phagocytize there is selective removal of certain proteins and lipids from the cell surface into phagocytic vesicles and selective retention of other components in the plasma membrane (5,6,7,26). Colchicine does not inhibit phagocytosis but abolishes this selective rearrangement of membrane components that normally accompanies phagocytosis. Thus, the membrane composition of phagocytic vesicles from colchicine-treated cells is different from that of normal cells. Since the membrane of secondary lysosomes must be derived at least in part from endocytic vesicle membrane (29), it is reasonable to suppose that lysosomal membrane composition may be different between CH and normal cells. Perhaps a recognition signal normally present in internalized membrane is lost so that the lysosomes fuse with each other to form giant granules. In the continued presence of carbachol or PMA, MT stability may be increased so that the internalized membrane has the proper composition and the lysosomal granules develop a normal morphology.

The most important properties of the CH fibroblasts may be those that resemble characteristic properties of malignant cells. Both CH and SV40 transformed 3T3 fibroblasts (SV3T3) are highly sensitive to shape changes and detachment from the substratum by colchicine. Both form surface caps with Con A, spontaneously in the case of CH fibroblasts and after brief exposure to colchicine in the case of SV3T3 (26). Both can be protected against colchicine by carbachol. The corresponding normal cells (primary embryonic fibroblasts from black mice and 3T3 fibroblasts) are not susceptible to shape change during acute (30 minutes) exposure to $10^{-6}$M colchicine, and do not cap with Con A even after incubation with colchicine (10, 26). Thus, CH and SV3T3 fibroblasts behave like MT-defective cells, with the CH cells showing an exaggeration of the defect seen in SV3T3 cells.

The biochemical basis of this similarity may be defective generation of cyclic GMP in these two cell types. Rudland and his coworkers have compared the levels of cyclic GMP and activities of membrane-associated guanylcyclase between 3T3 and SV3T3 cells (30,31). They report that cyclic GMP levels are highly regulated in 3T3 whereas they are poorly modulated, if at all, in SV3T3 cells under a range of incubation conditions. Similar studies with CH cells are in progress in our laboratory. We propose that abnormal regulation of cyclic GMP generation in SV3T3 cells, as in CH cells, may cause impaired MT assembly and so be responsible in part for the altered membrane properties of transformed cells. Thus, the CH fibroblast may be a valuable tool for study of specific aspects of the transformed state.

References


1. Bandmann, U., Rydgren, L. and Norberg, B., 1974, Exp. Cell Res. 88, 63.
2. Zurier, R.B., Weissmann, G., Hoffstein, S., Kammerman, S. and Tai, H.H., 1974, J. Clin. Invest. 53, 297.
3. Wooten, G.F., Kopin, I.J. and Axelrod, J., 1975, Annals. N.Y. Acad. Sci. 253, 528.
4. Poisner, A.M. and Cooke, P., 1975, Annals. N.Y. Acad. Sci. 253, 653.
5. Ukena, T.E. and Berlin, R.D., 1972, J. Exp. Med. 136, 1.
6. Oliver, J.M., Ukena, T.E. and Berlin, R.D., 1974, Proc. Nat. Acad. Sci. USA 71, 394.
7. Berlin, R.D. and Fera, J.P., 1975, In this publication.
8. Oliver, J.M., Zurier, R.B. and Berlin, R.D., 1975, Nature 253, 47.
9. Edelman, G.M., Yahara, I. and Wang, J.L., 1973, Proc. Nat. Acad. Sci. USA 70, 1442.
10. Rosenblith, J.Z., Ukena, T.E., Yin, H.H., Berlin, R.D. and Karnovsky, J.J., 1973, Proc. Nat. Acad. Sci. USA 70, 1625.
11. Till, J.E., Baker, R.M., Brunette, D.M., Ling, V., Thompson, L.H. and Wright, J.A., 1975, Fed Proc. 32, 9.
12. Wolff, S.M., Dale, D.C., Clark, R.A., Root, R.K. and Kimball, H.R., 1972, Ann. Intern. Med., 76, 293.
13. Gallin, J.I., Bujak, J.S., Patten, E. and Wolff, S.M., 1974, Blood 403, 201.
14. Leader, R.W., Padgett, G.A. and Gorham, J.R., 1963, Blood 22, 477.
15. Padgett, G.A., 1968, Adv. Vet. Sci. 12, 239.
16. Kramer, J.W., Davis, W.C. and Prieur, D.J., 1975, Fed. Proc. 34, 861.
17. Chan, L.-N. and Oliver, J.M., 1975, In preparation.
18. Berlin, R.D., 1973, J. Biol. Chem. 248, 4724.
19. Becker, J.S., Oliver, J.M. and Berlin, R.D., 1975, Nature 254, 152.
20. Reaven, E.P. and Axline, S.G., 1973, J. Cell Biol. 59, 12.
21. Weissmann, G., Goldstein, I., Hoffstein, S. and Tsung, P.-K., 1975, Annals. N.Y. Acad. Sci. 253, 750.

22. Goldstein, I., Hoffstein, S., Gallin, J. and Weissmann, G., 1973, Proc. Nat. Acad. Sci. USA 70, 2916.

23. Goldberg, N.D., Hadden, M.K., Hartle, D.K. and Hadden, J.M., 1973, Proc. 5th Int. Cong. Pharm. p. 146.

24. Ignarro, L.J., Lint, T.F. and George, W.J., 1974, J. Exp. Med. 139, 1395.

25. Robberecht, R., Deschodt-Lanckman, M., DeNeef, P., Borgeat, P., and Christophe, J., 1974, Febs. Letts. 43, 139.

26. Oliver, J.M. In Immune Recognition, Ed. A.S. Rosenthal, Academic Press. p.445 (1975).

27. Estensen, R.D., Hill, H.R., Quie, P.G., Hogan, N. and Goldberg, N.D., 1973, Nature 245, 458.

28. Hadden, J.M., Hadden, E.M., Haddox, M.K. and Goldberg, N.D., 1972, Proc. Nat. Acad. Sci. USA 69, 3024.

29. Cohn, Z.A. and Benson, B., 1965, J. Exp. Med. 121, 835.

30. Rudland, P.S., Gospodarowicz, D. and Seifert, W., 1974, Nature 250, 741.

31. Rudland, P.S., Seeley, M. and Seifert, W., 1974, Nature 251, 417.

Acknowledgement

This work was supported by grants BC-179 from the American Cancer Society and 1 R01 ES 01106-01 from the National Institutes of Health.

*Microtubules and Microtubule Inhibitors, eds. M. Borgers and M. de Brabander*



# THE EFFECT OF MICROTUBULAR INHIBITORS AND CYTOCHALASIN B ON AGGREGATION OF CONCANAVALIN A RECEPTORS IN MOUSE PERITONEAL MACROPHAGES

A.N.Bhisey and N.G.Mehta,
Cancer Research Institute, Tata Memorial Centre,
Bombay 400 012, India.

## Summary

The Concanavalin A-induced redistribution of the membrane receptors leads to the formation of a central cap in cultured mouse macrophages. However, when the microtubules are depolymerised, the cells fail to form a cap but show patchy distribution. The degree of agglutination, determined by an indirect haemagglutination assay, is slightly enhanced. When the microtubules are stabilized with treatment of the cells with heavy water, the aggregation of receptors is like that in controls, but is much weaker. These cells show practically no agglutination. Removal of microfilaments from normal or microtubule-deprived cells brings about complete inhibition of receptor aggregation, and agglutination is totally abolished. The microtubules seem to be required for cap formation in these cells, and the microfilaments appear to be necessary for aggregation. Since aggregation does not occur in the absence of microfilaments, their role in cap formation cannot be determined in this system.

## Introduction

Cultured mouse peritoneal macrophages exhibit membrane ruffling which requires the presence of cytoplasmic microtubules and microfilaments. Depolymerization of the tubules by colchicine or vinblastin induces ameboid movement in these cells. If normal cells or cells treated with microtubule-inhibitors are deprived of their microfilaments, they cease membrane activity except zeiotic blebbing. The relationship between these movements and the presence of cytoplasmic microtubules and microfilaments have been elucidated by us[1, 2].

In recent years, malignant cells have exhibited a generally uniform pattern of behaviour which is due to alterations in their membrane properties. This behaviour, viz. agglutinability by plant lectins, is not shown by normal cells[3, 4]. An observation made by several investigators is that aggregation of lectin receptors in the membrane is a prerequisite for agglutination[5-8]. Thus, the ability of lectin receptors to aggregate seems to be a fundamental change occurring in cancer.

Some workers have noted that microtubules and the network of microfilaments underlying the cell membrane play a role in the aggregation of membrane receptors[9-20]. Since the effect of the microtubule- and microfilament-depolymerising agents in the macrophage system is well characterised, we have used it to study the role of these structural elements in the redistribution of Concanavalin A (Con A) receptors.

Material and Methods

Chemicals and their sources are as follows : colchicine, Calbiochem; vinblastin, Eli Lily; cytochalasin B, Imperial Chemical Industries; heavy water of 99.9% purity, Bhabha Atomic Research Centre, Bombay; fluorescein isothiocyanate, Sigma Chemicals; and highly purified soluble Con A, Biochemicals Unit, V.P.Chest Institute, New Delhi prepared by Surolia et al.[21]. Fluorescein isothiocyanate-conjugated Con A (Fl-Con A) was prepared according to Shoham and Sachs[22].

Peritoneal macrophages were collected from 10-12 week old random-bred Swiss albino mice. The methods of isolation and culture have been described earlier[1]. The cells were cultured in plastic petri-dishes on 10 x 20 mm glass coverslips for 24 to 72 hr.

Con A induced agglutination of cells was studied by microhemadsorption method as described by Yin et al.[19], at 10, 25, and 50 μg/ml of the lectin. Con A binding and further incubation were done at 37°. Fresh, thrice washed 'O' group RBCs were used at 0.5% haematocrit. Colchicine (CLC, $10^{-4}$M) or vinblastine (VLB, $10^{-5}$M) treatment was given for 2.5 hr to disorganise microtubules. Microfilaments of untreated or CLC-treated cells were disrupted by treatment with cytochalasin B (CB), 2 μg/ml, for 1 hr. The drugs at the given concentrations were incorporated in PBS used for washing the cells.

For heavy water treatment, cells were incubated in medium made up of 80% heavy water ($D_2O$) for 3-4 hr. Subsequent washings were done with the $D_2O$ medium without serum. Cells after fixation in gluteraldehyde were examined under phase contrast microscope, and scored as: free cells, cells with 1-2 RBCs attached, cells with 3-5 RBCs and cells with more than 5 RBCs.

For fluorescence studies, the cells were washed twice with PBS and incubated with 25 μg/ml Fl-Con A for 1 min at room temperature. After washing thrice with PBS, the cells were incubated in a medium without glucose, supplemented with dialysed serum at 15% level. Cultures were fixed at 0, 10, 25 and 45 min and observed in Nikon fluorescence microscope with an HBO-200 lamp.

The cell suspension obtained from the peritoneal cavity was centrifuged and the pellet was washed twice with PBS. The cells were suspended in 25 μg/ml Fl-Con A in PBS and incubated at 37° for 10 min. The cells were washed with PBS and observed under the fluorescence microscope.

For reversal studies the Con A-treated cells were incubated with $10^{-1}$ M glucose in PBS for 20 min at 37°. The cells were washed three times with PBS and used for either agglutination assay or fluorescence observation.

Fixation of cells was done in 1% glutaraldehyde for at least 15 min.

Results

Freshly isolated mouse peritoneal macrophages are not agglutinated by Con A concentrations upto 200 µg/ml. We have, therefore, used the indirect agglutination system of Yin et al.[19]. In this assay system, the number of cells binding RBCs increases linearly upto Con A concentration of 25 µg/ml. Saturation occurs at 50 µg/ml of the lectin (Fig. 1). Pretreatment of cells with CLC or VLB did not alter this relationship. With increasing concentrations of Con A, there was an increase in the number of RBCs adhering to macrophages. We have, therefore, expressed the results as per cent cells without RBCs, with 1-2 RBCs, with 3-5 RBCs and with more than 5 RBCs.

When cells were treated with $10^{-1}$ M glucose following their incubation with 50 µg/ml Con A prior to RBC binding, agglutination obtained was exactly comparable to that obtained in absence of Con A treatment, indicating the specificity of the assay.

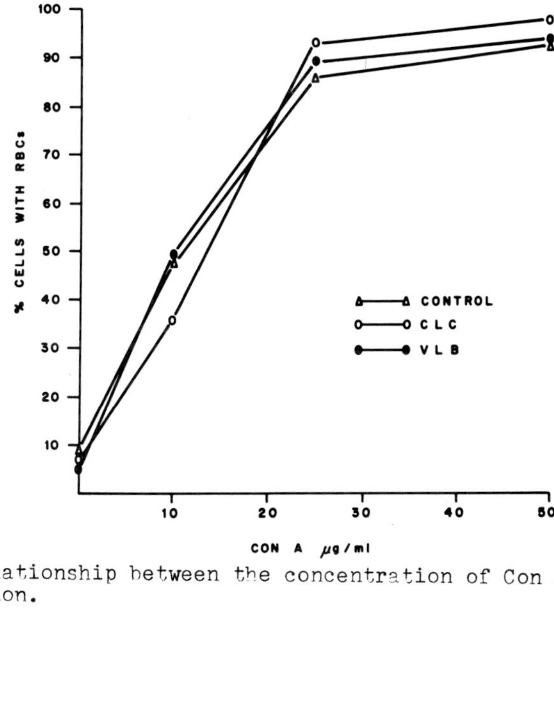


Fig. 1. Relationship between the concentration of Con A and per cent agglutination.

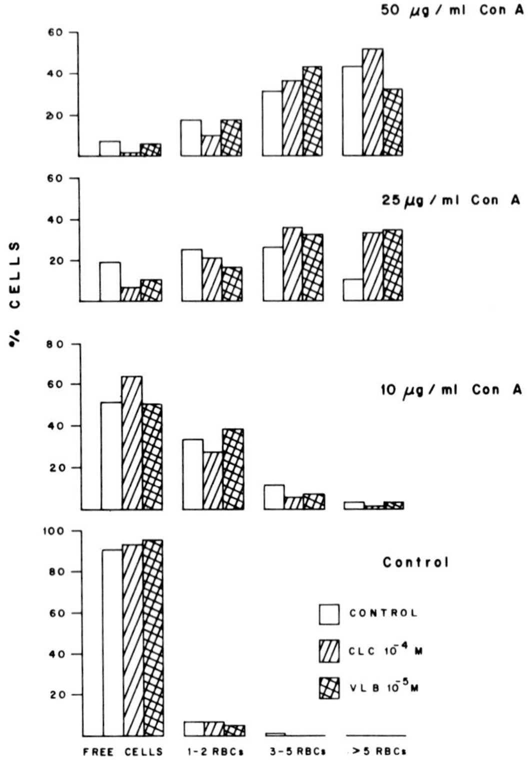


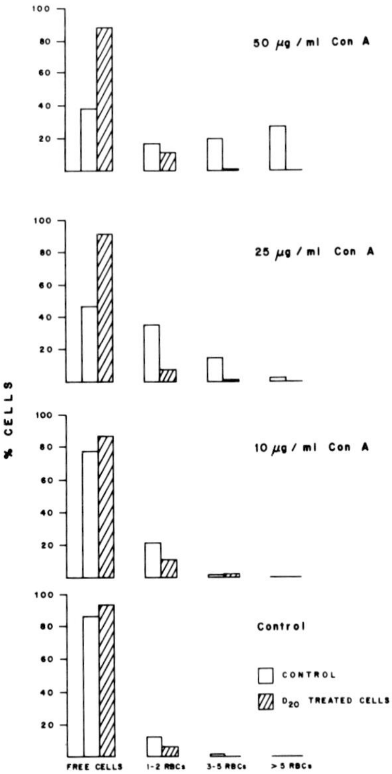


Fig. 2. Per cent cells binding different number of RBCs in control, CLC- and VLB-treated cells at different concentrations of Con A.

Fig. 3. The effect of $D_2O$ treatment on agglutination of macrophages at various concentrations of Con A.

Fig. 2 shows the effect of pre-treatment of cells with CLC or VLB on Con A-induced agglutination. There was no marked effect of micro-tubule deprivation on agglutination at 10 µg/ml of Con A. At 25 and 50 µg/ml concentrations, the number of free cells and cells with 1-2 RBCs was smaller than in controls, while the number of cells with 3-5 and more than 5 RBCs was higher. Thus, the effect of microtubule-inhibitors seems to slightly enhance agglutination.

When cells were treated with the $D_2O$ medium there was marked effect on agglutination (Fig. 3). Even at Con A concentration as high as 50 µg/ml, the cells practically did not bind any RBCs.

Treatment of cells with CB also decreased agglutination in the presence of Con A. At 25 µg/ml of the lectin, only 16% of the cells had 1-2, and practically none had more RBCs adhering (Fig. 4). The

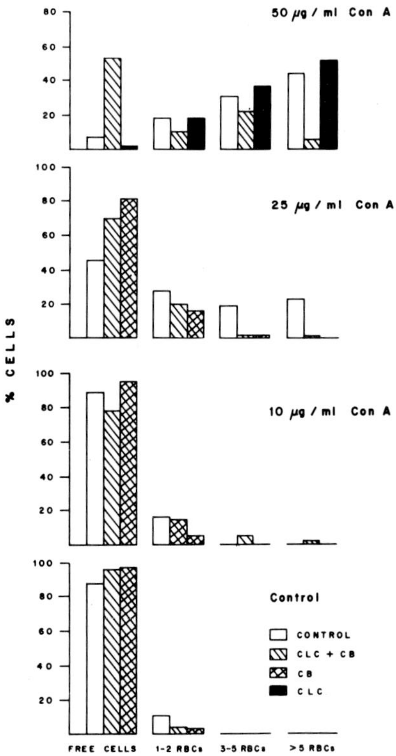


Fig. 4. The per cent distribution of cells binding different number of RBCs in control, CB-treated and CLC+CB-treated cells at 0. 10 and 25 µg/ml Con A. Also comparison of effect of CLC- and CLC+CB treatment at 50 µg/ml Con A.

enhancement in agglutination caused by the removal of microtubules was also abolished by further treatment of the CLC-treated cells with CB (Fig. 4). This is evident also at 50 µg/ml of Con A.

Incubation period of 1 min was chosen since initial experiments suggested redistribution of Con A-bound receptors if longer incubation periods were used.

In the control macrophages, after 1 min treatment with Fl-Con A, the entire cell, including the ruffled membranes, showed uniform fluorescence; the processes, however, were only weakly fluorescent

(Fig. 5). Upon further incubation, the intensity of peripheral fluorescence decreased, while the central area became brighter, suggesting the tendency towards formation of a central cap (Fig. 6). The caps were usually formed in 30-35 min and beyond 45 min, no further redistribution was observed. It was of interest to note that the typical patches seen, e.g. in lymphocytes which precede cap formation[23] were not observed. The cap occupies a fairly large area over the cytoplasm. In different experiments, 40-60% of the cells showed the central cap. However, this has to be an underestimate as, a large proportion of the remaining cells were rounded and hence were uniformly fluorescent. Rounded up cells on the coverslips, like the freshly isolated macrophages, never showed any redistribution.

In agglutination experiments, the RBCs predominantly occupied areas over the cell body (Fig. 11). Only occasionally, a long process was seen to bind RBCs. This again indicated a greater concentration of Con A receptors away from the periphery.

In the cells treated with CLC or VLB, the distribution of Fl-Con A was initially uniform. In contrast to the controls where the fluorescence was always seen to redistribute centripetally without the formation of patches, the CLC- or VLB-treated cells showed distinct patches in about 10 min (Fig. 7 and 8). Further incubation

---

Fig. 5. Fl-Con A-treated control macrophages at 0 min, showing uniform fluorescence.

Fig. 6. Redistribution of fluorescence into a cap after 45 min in the control cell.

Fig. 7. Uniform binding of Fl-Con A to CLC-treated cells at 0 min.

Fig. 8. Aggregation of fluorescence into patches in 10 min after FL-Con A binding to microtubule-negative cells.

Fig. 9. Uniform fluorescence distribution on CB- or CLC+CB-treated cells after 45 min.

Fig.10. Fluorescence redistribution over $D_2O$-treated macrophages at 45 min.

Fig.11. Agglutination of control macrophages by RBCs. Note the predominant location of erythrocytes over the cell body.

Fig.12. RBCs adhering to CLC-treated cells after treatment with Con A. The RBCs mainly bind to peripheral regions.

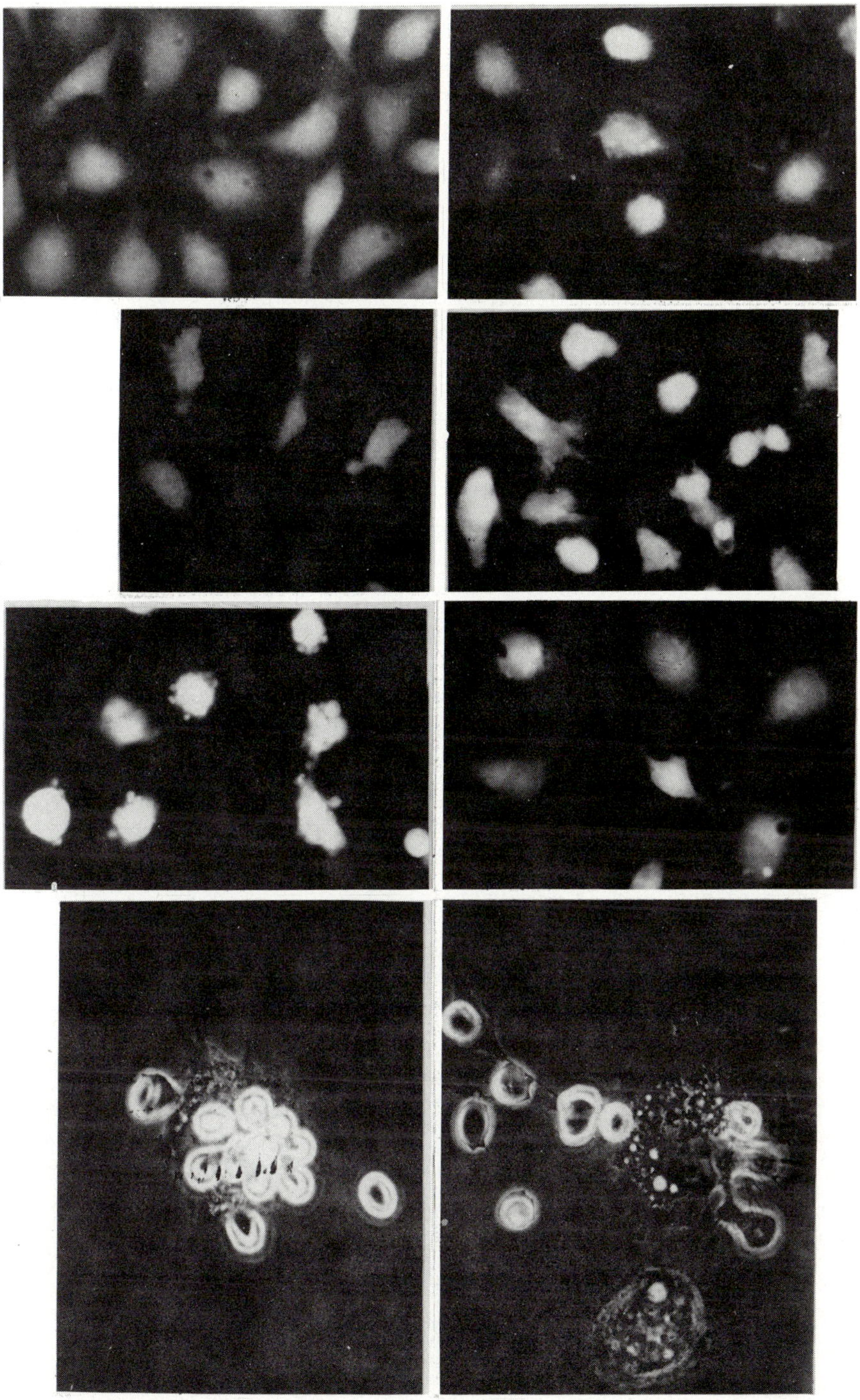

did not alter this pattern. About 50% of the cells showed patchy distribution of fluorescence, the remaining uniformly fluorescent ones were mainly the rounded cells. The entire cell body was fluorescent in the cells showing patches, but certain areas were more intense than others. In some cells, a patch was seen in the trailing end, while in others, patches were located in the pseudopods. Many times, the patches were localised at the sides of either the trailing end or the pseudopod. Unlike the controls, the CLC- or VLB-treated cells showed intense fluorescence in their processes.

The distribution of bound RBCs on the ameboid cells was also different than in controls. In these cells, RBCs were regularly seen to adhere to processes (Fig. 12), especially at the trailing end of the cell. They were also found to occupy an area on the cell body, but not in as great a number as in controls.

Cells treated with CLC+CB or CB alone did not show any redistribution of fluorescence whatsoever. The blebs, present in these cells were as intensely fluorescent as the remainder of the cell (Fig. 9).

The microfilament-disrupted cells did not bind more RBCs in presence of Con A. Whatever few RBCs were seen to bind, did not show any preferential location.

The fluorescence distribution pattern of $D_2O$-treated cells was somewhat like that obtained in controls. Following incubation, redistribution towards the centre was seen (Fig. 10). However, very few caps (4-7%) were observed. Despite this centripetal movement, the redistribution was much slower than in controls. Longer incubation upto 75 min did not alter this pattern. The $D_2O$-treated cells did not bind any RBCs in the presence of Con A.

## Discussion

Redistribution of membrane receptors into clusters seems to be required prior to agglutination of the cells by lectins. If the membrane proteins are crosslinked with bifunctional reagents, inhibiting their movement , the agglutination of cells is also inhibited[5-8]. A variant of plasmacytoma cell strain, non-agglutinable by high concentration of Con A, was shown to have lectin receptors randomly arranged in the membrane, whereas the agglutinable parent cell had clustered receptor distribution[24]. This is another evidence indicating the relationship between aggregation of receptors in the membrane and agglutination of cells.

Redistribution of surface receptors induced by polyvalent ligands, e.g. antibodies or lectins, initially occurs into discrete patches, which upon further incubation aggregate into a single localised cap[8, 15]. The patch formation is due to cross-linking of lectin-receptor complexes in localised regions of the membrane. A number of studies have indicated that patch formation is an energy-independent phenomenon, whereas capping is prevented when cells are treated with metabolic inhibitors[11, 12, 14-17].

Apart from the lateral movement of the receptors, and requirement of energy, subsurface elements like microfilaments and microtubules also seem to play a role in the phenomena of patching and capping. Microtubules may be involved in patch formation[9, 10], as well as in the formation of caps[10, 12], although some workers have failed to observe any effect of microtubule-depolymerization on the phenomenon of cap formation[14, 15, 17, 18]. Except for the reports of Unanue et al.[13] and Ryan et al.[14], many studies have shown that removal of microfilaments abolishes, at least partially, the formation of caps[15, 15-18, 25].

Involvement of microtubules[19] and microfilaments[20] has been shown also in lectin-induced agglutination of cells. In the present study the untreated, well spread macrophages invariably showed central caps, which occupied quite a large area on the cell body. This is in contrast to rather compact and small caps observed in other cell types[14, 23]. The macrophages also take much longer time to form the caps, about 30-35 min in contrast to 5 min in lymphocytes[17] and PMN leucocytes[14]. The RBC distribution on these cells was always in the central region, again indicating the central location of the bound Con A.

The removal of microtubules shows that (i) the receptors aggregate into patches, which never cap; (ii) the patches are found in different regions of the cells; and (iii) there is a slight increase in agglutination. On the other hand, as a consequence of $D_2O$ treatment, there is (i) little and apparently slow aggregation, and (ii) no agglutination. Thus, microtubule elimination seems to increase aggregation, whereas an increase in their number (as a consequence of stabilization and hence shift in equilibrium), retards this process. It thus seems, that the movement of receptors is in some way restricted by microtubules. It may be suggested that a portion of the receptor molecule is in contact with the microtubules lying beneath the membrane. Ji and Nicolson[26] have shown that the portion of glycophorin of the erythrocyte membrane exposed to the cytoplasm is associated with spectrin located on the inner face of the membrane Normally, the glycophorin molecules are unable to move in the membrane, but when spectrin is removed, the glycoprotein becomes free to move and aggregate when appropriate stimulus is applied[27]. If some of the Con A receptors are in contact with microtubules, their movement may be restricted, while the rest of the receptors are free to aggregate. Upon the removal of the tubules, all receptors become free to move, increasing the number of aggregates leading to higher agglutination. When greater number of microtubules is produced, more receptors are now fixed by these, leading to decrease in agglutination. Thus, despite any direct evidence of association between membrane proteins and microtubules, one may propose that aggregation of receptors, necessary for agglutination is determined by receptors not associated with microtubules. Recently, Baker et al.[28] have provided evidence of interaction between microtubular subunits and the membrane.

Localised clustering (patching, not distinctly seen in control macrophages) and capping are believed to be two distinct phenomena, although the latter has to follow the former. It is often suggested that the cap results due to movement of the cell in a direction[14,17,18]. The cross linked aggregates are held back when the membrane flows in a particular direction, e.g. the tail cap in PMN leucocytes [14], and the cap in the uropod region[17]. The peritoneal macrophages, which normally exhibit ruffling, start ameboid locomotion when their microtubules are depolymerised[1]. Despite this directed movement, we have never seen a cap, although there may be greater concentration of receptors in some regions. Locomotion may have some role in concentrating some of receptors in such regions, it does not produce caps. Since the cells lack microtubules, we feel that these may play a role in cap formation also. The PMN leucocytes studied by Karnovsky's group[14] exhibit ameboid locomotion, and this is in the presence of microtubules. They invariably find all receptors concentrated in the tail end into a cap; we did not observe this, and the only difference is the presence of microtubules in their cells and their absence in ours. This leads to the conclusion that microtubules play a positive role in the localisation of all receptors into a cap. How this occurs is unclear to us.

Microfilaments were absent in cells treated with CB, and ruffling and ameboid movements were totally inhibited[2]. In the present study, CB abolished the redistribution of receptors and agglutination. Microfilaments have been shown to interact with membranes[29], which may influence mobility of membrane receptors. Another possibility is that a distinct CB-sensitive system may be present in the membrane. Membrane associated actin-like substances have been demonstrated in amebae[30], erythrocytes[31] and in 3T3 and HeLa cells[32]. Myosin-like proteins have also been found to be present in the membranes[33, 34]. It is possible that the CB effect observed in the present studies may be mediated via such a system.

Since the treatment of cells with CB prevents mobility of Con A receptors, it seems the CB-sensitive system is involved in the regulation of movement of the receptors. The receptors can move freely even when incorporated in artificial liposomes[35]. However, our present results indicate that for the movement of Con A receptors in mouse macrophage membranes, the presence of a CB-sensitive system is essential. This is difficult to visualise. One plausible explanation is that the microfilament or microfilament-like material keeps the receptors in the fluid region of the membrane. The membrane lipids are known to be heterogeneous in the distribution with regards to fluidity[36]. Alternatively, the interaction of the CB-sensitive system and the receptor may bring about localised increase in fluidity as found in the case of lectin-receptor binding[37].

References

1. Bhisey, A.N. and Freed, J.J., 1971, Exp. Cell Res. 64, 419.
2. Bhisey, A.N. and Freed, J.J., 1975, in: Regulation of Growth and Differentiated Function in Eukaryotic Cells, ed. G.P.Talwar, (Raven Press, N.Y.) p. 155.
3. Inbar, M. and Sachs, L., 1969, Proc. Nat. Acad. Sci. 63, 1418.
4. Burger, M.M. and Goldberg, A.R., 1967, Proc. Nat. Acad. Sci. 57, 359.
5. Inbar, M., Huet, C., Oseroff, A.R., Ben-Bassat, H. and Sachs, L., 1973, Biochim. Biophys. Acta. 311, 594.
6. Noonan, K.D. and Burger, M.M., 1973, J. Cell Biol. 59, 134.
7. Nicolson, G.L., 1973, Nature 243, 218.
8. Rosenblith, J.Z., Ukena, T.E., Yin, H.H., Berlin, R.D. and Karnovsky, M.J., 1973, J. Cell Biol. 70, 1625.
9. Edelman, G.M., Yahara, I. and Wang, J.L., 1973, Proc. Nat. Acad. Sci. 70, 1442.
10. Yahara, I. and Edelman, G.M., 1973, Nature 236, 152.
11. Edidin, M. and Weiss, A., 1972, Proc. Nat. Acad. Sci. 69, 2456.
12. de Petris, S., 1974, Nature 250, 54.
13. Unanue, E.R., Karnovsky, M.J. and Engers, H.D., 1973, J. Exp. Med. 137, 675.
14. Ryan, G.B., Borysenko, J.Z. and Karnovsky, M.J., 1974, J. Cell

Biol. 62, 351.

15. Taylor, R.P., Duffus, P., Raff, M.C. and de Petris, S., 1971 Nature New Biol. 233, 225.

16. Loor, F., Forni, L. and Pernis, B., 1972, Eur. J. Immunol. 2,203.

17. Raff, M.C. and de Petris, S., 1973, Fed. Proc. 32, 48.

18. Rutishauser, U., Yahara, I. and Edelman, G.M., 1974, Proc. Nat. Acad. Sci. 71, 1149.

19. Yin, H.H., Ukena, J.S. and Berlin, R.D., 1972, Science 178, 867.

20. Kaneko, I., Satoh, H. and Ukita, T., 1973, Biochim. Biophys. Res. Commun. 50, 1087.

21. Surolia, A., Prakash, N., Bishayee, S. and Bachhawat, B.K., 1973, Ind. J. Biochem. Biophys. 10, 145.

22. Shoham, J. and Sachs, L., 1972, Proc. Nat. Acad. Sci. 69, 2479.

23. Inbar, M. and Sach, L., 1973, FEBS Letters 32, 124.

24. Guerin, E., Zachowski, A., Prigent, B., Paraf, A., Dunia, I., Diawara, M-A.and Beneditti, E.L., 1974, Proc. Nat. Acad. Sci. 71, 114.

25. Loor, F., 1973, Exp. Cell Res. 82, 415.

26. Ji, T.H. and Nicolson, G.L., 1974, Proc. Nat. Acad. Sci. 71, 2212.

27. Elgsaeter, G. and Branton, D., 1974, J. Cell Biol. 63, 1018.

28. Baker, J.S., Oliver, J.M. and Berlin, R.D., 1975, Nature,254, 152.

29. Perdue, J.F., 1973, J. Cell Biol. 58, 265.

30. Clarke, M., Schatten, G., Mazia, D. and Spudich, J.A., 1975, Proc. Nat. Acad. Sci. 72,

31. Avissar, N., de Vries, A., Ben Shaul, Y. and Cohen, I., 1975, Biochim. Biophys. Acta 375, 35.

32. Gruenstein, E., Rich, A. and Weihing, R.R., 1975, J. Cell Biol. 64, 223.

33. Willingham, M.C., Ostlund, R.G. and Pastan, I., 1974, Proc. Nat. Acad. Sci. 71, 444.

34. Painter, R.G., Sheetz, M. and Singer, S.J., 1975, Proc. Nat. Acad. Sci. 72, 1359.

35. Grant, G.W.M. and Mc Connel, H.M., 1974, Proc. Nat. Acad. Sci. 71, 4653.

36. Horwitz, A.F., Hatten, M.E. and Burger, M.M., 1974, Proc. Nat. Acad. Sci. 71, 3115.

37. Barnett, R.E., Scott, R.E., Furchet, L.T. and Kersey, J.H., 1974, Nature 249, 465.

*Microtubules and Microtubule Inhibitors, eds. M. Borgers and M. de Brabander*



# CULTURED HAPLOID CELLS RESISTANT TO ANTITUBULINS

Jerome J. Freed

The Institute for Cancer Research, Fox Chase Cancer Center,
Philadelphia, Pennsylvania, USA

and

Betsy M. Ohlsson-Wilhelm

Department of Microbiology, University of Rochester
School of Medicine and Dentistry
Rochester, New York, USA

## SUMMARY

Variants from cultured animal cell lines that have been selected for resistance to antitubulins have been shown in several instances to owe their resistance to reduced uptake or accelerated efflux of the drug. This form of resistance is termed drug-exclusion; it is non-specific, in that the variants also acquire resistance to drugs unrelated in both molecular structure and cellular target system. In this article, we describe several antitubulin-resistant strains that we have selected from cell line ICR B20, a haploid isolate derived from the stable haploid cell line ICR 2A. In these, resistance is specifically directed toward the alkaloid used as the selecting agent while the cells remain as sensitive as the parental stock to other antitubulins. In one strain, selected for resistance to podophyllin, cross-resistance to colchicine has been detected, although the cells remain sensitive to vinblastine. This phenotype is of particular interest, since podophyllin and colchicine appear to bind to a common site on the tubulin molecule while vinblastine binds independently. The specificity of resistance in the haploid cell variants suggests that they may arise independently of the more common, non-specific, drug-exclusion mechanism.

## 1. Introduction

The concern of this meeting is with the organization and function of cellular ultrastructures, the microtubules, and with the way in which their constituent proteins, the tubulins, interact with drugs. For such investigations, genetically mutant cells could provide useful tools. Using the procedures of cell culture genetics, we and others have sought to isolate cell variants with alterations that involve microtubules, using two general strategies: 1) to isolate temperature-sensitive variants that can be conditionally blocked in some aspect of microtubule function, e.g., in mitosis, and 2) to isolate variants with altered tubulins <u>via</u> selection for resistance to the stathmokinetic effects of microtubule-depolymerizing drugs, commonly called antitubulins.

Temperature-sensitive variants that are conditionally blocked in their progress through the cell cycle have been extensively studied in yeast[1], and similar cell-cycle mutants have been sought in animal cell cultures. Strains that block at cycle-specific points have been described: i.e., at cytokinesis[2], mitosis[3,4] or at specific points during interphase[5,6]. Most efforts have been devoted to obtaining heat-sensitive variants selected by the scheme described by Thompson et al.[7], but there have been some attempts to obtain cold-sensitive, assembly defective strains[8,9]. While cold-sensitive variants were obtained, it is not clear that any of these express a cold-sensitive defect in microtubule assembly.

In this presentation our concern is with the second approach: the direct isolation of variants resistant to antitubulins. By this method we seek to isolate cells whose microtubule proteins are altered in such a way that they no longer interact with the drug. Such genetically altered tubulins would be useful for the analysis of the mode of action of antitubulins in interfering with the microtubule assembly process. We will review some of the properties of antitubulin-resistant variants that have been obtained from animal cell cultures and we will give a preliminary account of variants that we have isolated from the unique haploid cell line ICR 2A (haploid Rana pipiens embryo)[10]. These haploid variants are of special interest since they appear to express a more specific resistance to the selective agent than do comparable isolates from ordinary cultured cell lines.

## 2. Properties of antitubulin resistant cultured animal cells

Mammalian cell cultures that are exposed to selection with inhibitory concentrations of colchicine give rise to strains that show increased tolerance of the drug, i.e., they can multiply in mass cultures or give rise to colonies in dilute cultures at higher concentrations of drug than can the parental cells. In the cases that have so far been described, the mechanism of resistance appears to be a reduced uptake of the drug, the intrinsic colchicine-binding capacity of tubulins from these cells remaining unaltered[11,13]. The increase in resistance at a single selective step is relatively small, usually involving a less than 10-fold increase in the tolerated concentration of drug; more resistant strains can be developed by successive selections at progressively higher concentrations of the drug. Further, the reduced capacity for uptake does not appear to result from an alteration that is specific for the drug used as the selective agent. Rather, the resistant strains also become tolerant to a variety of unrelated drugs; these may be antitubulins with different binding specificity (e.g., vinblastine) or antibiotics of unrelated structure and dissimilar target molecule (e.g., actinomycin D). An exception to this generalization was found in cells derived from the Syrian hamster, which express a colchicine-specific resistance that is characteristic of the species[12].

It is doubtful that the origin of this type of variant involves a simple mutational event. Although Harris[13] found that the incidence of vinblastine resistance in Chinese hamster cell cultures was increased by the mutagen nitrosoguanidine, the response was similar in diploid and tetraploid cell populations. Since vinblastine resistance in these cells had been shown to behave as a dominant phenotype in cell fusion hybrids[14], it would have been expected that a larger yield of mutants would have been obtained from the cell strain of higher ploidy if a simple dominant mutation were involved. Colchicine resistance was found by Minor and Roscoe[12] to occur at a high frequency that was unaffected by treatment with the mutagen ethyl methane sulfonate. They concluded from this and other evidence that their cells were unlikely to be mutants in the bacterial sense. While Ling and Thompson[11] considered that the stability of resistance in their colchicine-resistant Chinese hamster cells suggested a mutational origin, the large number of levels of resistance that could be obtained in repeated selections would make it necessary to invoke a large number of genetic elements.

In general, it seems likely that the antitubulin resistance of mammalian cells described to date is an expression of a non-specific alteration in cellular uptake of drugs which may be termed the "drug-exclusion mechanism" (see below). This alteration may not be mutational in origin, but since the variant phenotype is propagated from cell to cell through mitosis as a moderately stable inherited character it is not easy to rule out either genetic or physiological mechanisms by techniques that are presently available. However, since it seems likely that both mutational and epigenetic phenomena can give rise to variant cell strains *in vitro*, it appeared worthwhile to screen for antitubulin-resistant variants in a cell strain with a single chromosome set. In such a haploid strain, it might be possible to recover a class of recessive changes leading to resistance that would remain undetectable in ordinary diploid or heteroploid cultures.

## 3. Material and Methods

*Cell cultures* were propagated at 25°C in growth medium composed of diluted Leibovitz L-15 medium supplemented with 10 per cent fetal bovine serum as described elsewhere[15]. The parental cell strain, ICR B20, is a haploid thymidine transport-deficient variant isolated as a colony resistant to 15 µg/ml 5-bromodeoxyuridine (BUdR)[16] after treatment of the stable haploid cell line ICR 2A[10] with the acridine half-mustard mutagen ICR 340[17]. A stock of these cells was recovered from exposure to the acridine half-mustard ICR 191 and found to contain an increased incidence of cells resistant to 300 μg/ml BUdR, most of which were thymidine-kinase deficient[18]. Although the mutational nature of this loss of enzyme expression is not established[19], we chose a frozen sample of this stock of mutagen-treated cells to screen for the induction of other variant phenotypes, i.e., antitubulin resistance in the work presented here.

Selection experiments were carried out by exposing portions of the mutagen-treated cell population to vinblastine sulfate (Velban, E. Lilly and Co.) and portions to podophyllin U.S.P. (Nutritional Biochemicals Corp.). The drugs were added to growth medium and maintained at the desired concentration by twice-weekly feeding with drug-supplemented medium. During the selections, the cultures were protected from light to avoid photolytic decomposition of the selecting agent. The isolates whose properties are discussed below were derived as follows: pod B1 was isolated as a colony surviving podophyllin at 0.04 μg/ml, and was subsequently recloned; pod C1 arose in a separate flask, but was derived from the same pool of cells that gave rise to pod B1. Pod C1 was also recloned. Vin 1 (0.2) and vin 2 arose in separate flasks after vinblastine selection at 0.2 μg/ml. Vin 1 (0.4) was derived from vin 1 (0.2) by a subsequent vinblastine selection at 0.4 μg/ml. Vin 2h was similarly derived by a second round of selection from vin 2.

Resistance was assayed by determining the efficiency of plating in growth medium supplemented with graded concentrations of drug. To 25 $cm^2$ plastic cell culture flasks or 2-ounce prescription bottles that contained 5 ml of complete growth medium plus drug, a cell suspension was added that had been adjusted to contain 200 to 1,000 cells, depending on the expected efficiency of plating. Each drug concentration was assayed in triplicate, in some cases at two sizes of inoculum. After incubation for 18 to 21 days, the flasks were stained with Giemsa and scanned with a stereo dissecting microscope to determine the number of colonies developed. The surviving fraction was calculated and the efficiency of plating relative to drug-free medium was determined (relative plating efficiency, RPE). The approximate degree of resistance is expressed as the ratio of $ID_{50}$ (the drug dose producing one-half the control RPE) for the variant relative to the $ID_{50}$ for the parental ICR B20 cells determined with the same drug solutions.

## 4. Antitubulin resistance of haploid frog cell variants

Haploid frog cells selected for resistance to antitubulins appear to acquire a resistance that is more specific than that which has been reported in similar experiments with mammalian cell lines. That is, the acquired resistance is directed to the selecting drug or to an agent having a similar mode of interaction with the biochemical target within the cell. As shown in figure 1, pod B1 appears to express this latter form of resistance. It is resistant to about 3 times the concentration of podophyllin that inhibits the ICR B20 parental cells, and is about 2.5 times as resistant to colchicine as the unselected cells. However, pod B1 has acquired no resistance to vinblastine: colonies were formed at 0.01 g/ml, but not at higher concentrations: this is essentially the same behavior shown by the parental strain. The extent of resistance of cells selected with vinblastine may be seen in Figure 3; pod B1 is clearly not cross-resistant.

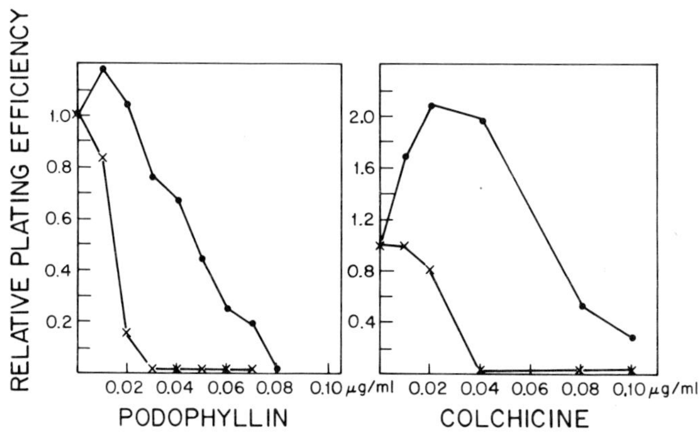


Fig. 1. Resistance of pod B1 is directed both to podophyllin, the selecting drug, and to colchicine, a functionally related antitubulin. Plating efficiency was determined as described in Materials and Methods and is expressed relative to the plating efficiency in drug-free medium. X——X, parental strain ICR B20; ●——●, selected strain pod B1.

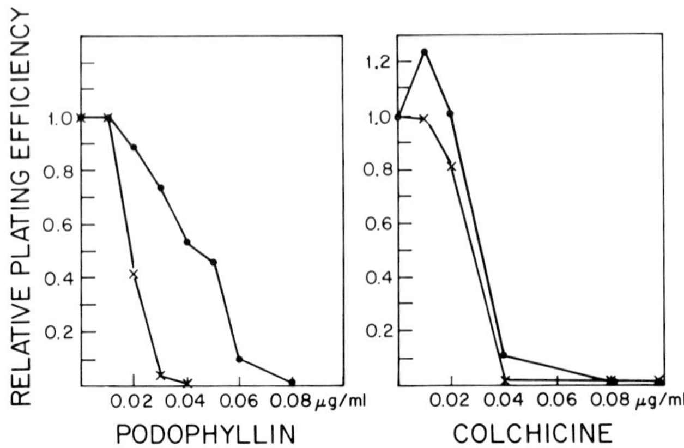


Fig. 2. Resistance to pod C1 is specific for podophyllin, the selecting drug. Procedure as in Fig. 1. X——X, parental strain ICR B20; 0——0, selected strain pod C1.

Pod C1 appears to express greater specificity of resistance than pod B1. As shown in figure 2, it is about 2.4 times as resistant to podophyllin as is ICR B20, but does not appear to differ from the parental strain in its resistance to colchicine. As in the case of pod B1, no resistance to vinblastine could be demonstrated.

The strains selected from exposure to vinblastine appear to be specifically resistant to this agent. As expected from the work of others, it was possible to increase the extent of resistance by cycling, i.e., by exposure to successively higher concentrations of vinblastine. As shown in figure 3, vin 1(0.2), which was obtained from a single round of selection, is about 3.3 times as resistant as the parental cells; its derivative vin 1(0.4), obtained from a second round of selection, became about 4.4 times as resistant as the ICR B20 cells. Vin 2 showed a greater resistance after the first selection, becoming about 16 times as resistant as the parental cells, while vin 2h obtained from a second round of selection had become about 31 times as resistant as the original parental culture. None of the vinblastine-selected strains appears to have acquired detectable resistance to podophyllin: when assayed for cross-resistance, they formed colonies at 0.1 μg/ml but not at 0.4 μg/ml podophyllin. From the curves for ICR B20 shown in figures 1 and 2, it will be seen that they are no more resistant than unselected cells. It is noteworthy that even vin 2h, the most resistant strain, shows no cross-resistance to podophyllin.

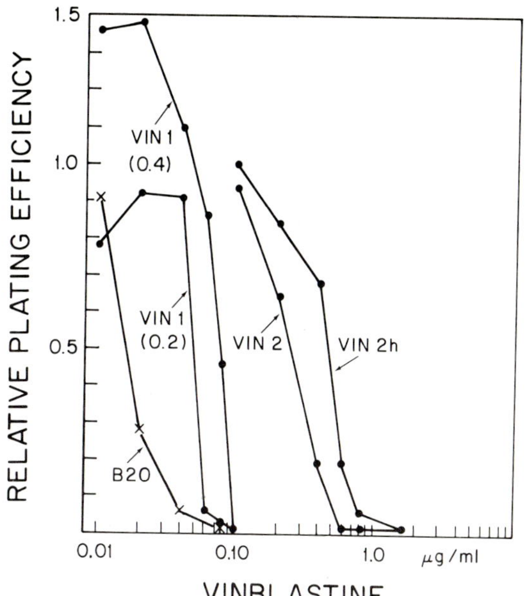


Fig. 3. Various levels of resistance to vinblastine are demonstrated in strains selected in this drug. X——X, parental strain ICR B20, ●——●, strains selected for resistance to differing concentrations of vinblastine as discussed in the text.

The variants derived from selection with antitubulins all appear to have retained the BUdR-resistant character of the original ICR B20 cells. The RPE of the several strains was determined at 15 μg/ml BUdR and compared to the RPE of ICR 2A1, a subclone of the wild-type haploid cell line ICR 2A; the wild-type failed to give detectable colony formation in BUdR. ICR B20 was resistant (RPE 1.0) to this concentration of the drug, which appears to be a toxic level only if a functional thymidine-transport reaction is present[18]. The podophyllin and vinblastine-resistant isolates all retained the ability to grow in BUdR (RPE from 0.65 to 1.5) and are presumed to have retained the transport-deficient character.

The pattern of resistance to other antitubulins and antitubulin synergists displayed by these frog cell lines is to be explored in future work.

5. Antitubulin resistance in frog cells in relation to the drug-exclusion phenomenon

Reports in the literature suggest that the resistance acquired by mammalian cells after selection with antitubulins is a manifestation of an acquired capacity to exclude a variety of structurally dissimilar, physiologically active compounds. This response may be termed the drug-exclusion phenomenon and it is likely that an essentially similar phenotype is elicited when cells are selected with a variety of cytotoxic agents, notably inhibitors of transcription and translation, but including antitubulins. This generalization is supported by the observations that 1) selection with one of these agents results in some degree of acquired resistance to others[11,12,20,21,22,23,others] and 2) that the mechanism of resistance in each case can be shown to involve reduced uptake of the selecting agents[11,12,20,23,24,25]. Thus, the cells need not be exposed to antitubulins to become antitubulin-resistant, as was observed by Burchenal _et al_. in their studies of leukemia cells resistant to phthalanilide antibiotics[26].

Puromycin-resistance appears to be another manifestation of this phenomenon: originally proposed as a useful genetic marker[27], the expression of puromycin resistance proved to be in large measure non-genetic[28]. Resistance in frog cells was shown to be derived from a restricted uptake of drug[29]. Cass[30] has shown that uptake is decreased by crowding of unselected pig cells on the growth surface of monolayer cultures.

The drug-exclusion phenomenon can also be expressed in frog cell cultures, since Mezger-Freed[29] has described stably puromycin-resistant frog cell strains that appear to behave in this way. These were shown to incorporate little tritiated puromycin, the extent of uptake was most strikingly reduced in cells that had been previously propagated in a tolerated concentration of the drug.

The mechanism that produces this exclusion is not well understood, but may involve the cell surface coat (the glycocalyx). Bosmann[31] observed that in actinomycin D-resistant cell lines glycoprotein synthesis was increased and greater amounts of glycoprotein and glycolipid were found in the plasma membranes. He proposed that such changes imply altered membranes and altered membrane synthetic apparatus, and that these changes may be responsible for decreased drug permeation and thus for resistance.

Further evidence that resistance of this type stems from exclusion of the drug comes from the observation that in some instances the presence of surface-active material (e.g., Tween-80) has been shown to overcome resistance or enhance uptake[11,32].

It must be recognized that the apparent reduction in rate of accumulation of labelled drugs in resistant cells may stem from an increased rate of _efflux_. The cells that have acquired resistance may transport drug out of the cell by an active process (a "pump") at a rate sufficient to limit toxicity. Such increased rates of efflux were observed in daunomycin-resistant mouse leukemia cells; in these, both sensitive and resistant cell lines showed comparable rates of uptake, but loss of drug was more rapid in the resistant variants[23]. These cells were cross-resistant to vinblastine. It has also been shown that extrusion of drug is responsible for the recovery of the ciliate Tetrahymena from prolonged incubation with drugs, e.g., cycloheximide[33] or colchicine[34].

The drug-exclusion phenomenon may be a specific kind of physiological activity rather than a consequence of simple impermeability of the cell surface. A barrier to the accumulation of daunomycin[23] and other phthalanilide antibiotics[35] could be demonstrated in growing, but not in stationary-phase, cells of resistant lines. Carlsen et al.[36] have recently reported that the apparent uptake of colchicine into the normal and colchicine-resistant CHO cells of Ling and Thompson[11] is stimulated by cyanide and azide as well as by vinblastine and Tween-80. Thus, the drug-exclusion phenomenon might require an energy-dependent physiological removal process that is induced in the selected cells.

It is important to distinguish the drug-exclusion phenomenon from the specific changes that produce resistance to purine and pyrimidine analogues in cultured cells. Strains resistant to high concentrations of drugs like BUdR, azaguanine or cytosine arabinoside are widely employed in cell culture genetics; these characteristically lack activity of a phosphorylating enzyme that forms part of a salvage pathway for base reutilization. However, in certain variants, resistance is associated with loss of thymidine uptake, the capacity to phosphorylate thymidine being retained[18,37]. ICR B20 has been characterized in some detail: it has been shown that there is a loss of a saturable permease function specific for thymidine and its analog BUdR and that the defect may result from loss of a thermolabile form of cytosol thymidine kinase (Freed and Hames, in preparation). Thus, as noted by Subak-Sharpe[38], such analogue-resistant variants behave quite differently from those selected with antibiotics. In ICR B20, BUdR enters by physical diffusion, so that the internal and external concentrations of drug approach equality; as a result these cells remain sensitive to high concentrations of BUdR (i.e., 1 mM) unless further selection at this drug concentration results in loss of thymidine kinase[18]. Puromycin-resistant frog cells, on the other hand, maintain an internal concentration of puromycin and its derivatives that remains less than the external concentration during 150 minutes of continuous exposure to the labelled drug[29].

The haploid antitubulin-resistant cell strains discussed in this article may stem from a mechanism of resistance more specific than the drug-exclusion phenomenon. While we cannot yet specify whether resistance is associated with reduced uptake, the mechanism appears to be directed against the selecting agent only. Even the most resistant strain in this group, vin 2h, showed no measurable resistance to podophyllin. It may be significant that in the instance in which we observed cross-resistance (pod B1) the two agents involved have been shown to compete for binding to a common site on the tubulin protein[39]. Vinblastine, which does not have a competitive effect on the binding of colchicine to tubulin, in no instance induced measurable resistance to podophyllin. Thus, whatever the mechanism of resistance, it may, as in the case of tubulin, have the ability to discriminate between the two types of microtubule-depolymerizing agents. We would like to suggest that tubulin or a tubulin-like molecule might form part of a transport system that brings anti-tubulins into the cell or sequesters them for "pumping out".

The specificity of antitubulin resistance in our haploid cells, in contrast to the less specific resistance of mammalian cells, may stem from the ability of the former to express phenotypically recessive alterations. Drug exclusion would appear, from the results of Harris[14] with vinblastine, to behave as a phenotypic dominant. The more specific form of resistance we have recovered may be related to the haploid nature of our cells, rather than to physiological differences between amphibians and mammals. Colchicine-resistant mutants of the haploid alga Chlamydomonas were shown by Adams and Warr[40] to segregate as simple mendelian factors; Flavin and Slaughter[41] found that a highly colchicine-resistant Chlamydomonas mutant was resistant to epipodophyllotoxin, but only equivocally resistant to vinblastine. Other mutants, less resistant to colchicine, remained resistant both to epipodophyllotoxin and to vinblastine. Thus the variants elicited from the algal haploid system seem to be similar to ours.

This work was begun while Dr. Ohlsson-Wilhelm was a research associate at The Institute for Cancer Research, and was supported by grants RR-05539, CA-05959 and CA-06927 from the U.S. National Cancer Institute, contract E(11-1)3110 with the U.S. Energy Research and Development Administration (Report COO-3110-20) and by an appropriation from the Commonwealth of Pennsylvania. Additional support, for work at the University of Rochester, came from American Cancer Society grant VC-126 to Dr. Ohlsson-Wilhelm.

References


1. Hartwell, L.H., 1974, Bact. Rev. 38, 164.
2. Smith, B.J. and Wigglesworth, N.M., 1972, J. Cell. Physiol. 80, 253.
3. Wang, R.J., 1974, Nature 248, 76.
4. Wang, R.J., Yin, L., DuMontier, A. and Sheridan, W., 1974, J. Cell Biol. 63, 365a.
5. Liskay, R.M., 1974, J. Cell. Physiol. 84, 49.
6. Burstin, S.J., Meiss, H.K. and Basilico, C., 1974, J. Cell. Physiol. 84, 397.
7. Thompson, L.H., Mankovitz, R., Baker, R.M., Till, J.E., Siminovitch, L. and Whitmore, G.F., 1970, Proc. Nat. Acad. Sci. (U.S.A.) 66, 377.
8. Ohlsson-Wilhelm, B.M., Freed, J.J. and Perry, R.P., 1974, J. Cell Biol. 63, 248a.
9. Farber, R.A. and Unrau, P., 1975, Molec. gen. Genet. 138, 233.
10. Freed, J.J. and Mezger-Freed, L., 1970, Proc. Nat. Acad. Sci. (U.S.A.) 65, 337.
11. Ling, V. and Thompson, L.H., 1973, J. Cell. Physiol. 83, 103.
12. Minor, P.D. and Roscoe, D.H., 1975, J. Cell Sci. 17, 381.
13. Harris, M., 1974, J. Natl. Cancer Inst. 52, 1811.
14. Harris, M., 1973, J. Natl. Cancer Inst. 50, 423.
15. Freed, J.J. and Mezger-Freed, L., 1970, in: Methods in cell physiology, vol. 4, ed. D. M. Prescott, Academic Press, New York, p. 19.
16. Mezger-Freed, L., 1972, Nature new biol. 235, 245.
17. Creech, H.J., Preston, R.K., Peck, R.M., O'Connell, A.P. and Ames, B.N., 1972, J. Med. Chem. 15, 739.
18. Freed, J.J. and Mezger-Freed, L., 1973, J. Cell. Physiol. 82, 199.
19. Mezger-Freed, L., 1974, in: Chemical carcinogenesis, eds. P.O.T'so and J.A. Di Paolo, Marcel Dekker, New York, p. 591.
20. Biedler, J.L. and Riehm, H., 1970, Cancer Res. 30, 1174.
21. Riehm, H. and Biedler, J.L., 1971, Cancer Res. 31, 409.
22. Simard, R. and Cassingena, R., 1969, Cancer Res. 29, 1590.
23. Kessel, D., Botterill, U. and Wodinsky, I., 1968, Cancer Res. 28, 938.
24. Kessel, D. and Bosmann, H.B., 1970, Cancer Res. 30, 2695.
25. Kessel, D., 1970, in: Proc. 10th Int. Cancer Congress, Medical Arts, Houston, p. 422.
26. Burchenal, S.H., Coley, U., Purple, J.R., Bucholz, E., Lyman, M.S. and Kreis, W., 1963, Cancer Res. 23, 1364.
27. Lieberman, I. and Ove, P., 1959, Proc. Nat. Acad. Sci. (U.S.A.) 45, 867 & 872.
28. Harris, M., 1967, J. Natl. Cancer Inst. 38, 185.

29. Mezger-Freed, L., 1971, J. Cell. Biol. 51, 742.
30. Cass, C.E., 1972, J. Cell. Physiol. 79, 139.
31. Bosmann, H.B., 1971, Nature 233, 566.
32. Riehm, H. and Biedler, J.L., 1972, Cancer Res. 32, 1195.
33. Frankel, J., 1970, J. Cell. Physiol. 76, 55.
34. Rosenbaum, J.L. and Carlson, K., 1969, J. Cell Biol. 40, 415.
35. Yesair, D.W., Kohner, F.A., Rogers, W.I., Baronowsky, P.E. and Kensler, C.J., 1966, Cancer Res. 26, 202.
36. Carlsen, S.A., See, Y.P., Ling, V. and Till, J.E., 1974, Can. J. Genetics Cytol. 17, 704 (abstract).
37. Breslow, R.E. and Goldsby, R.A., 1969, Exp. Cell Res. 55, 339.
38. Subak-Sharpe, H., 1965, Exp. Cell Res. 38, 106.
39. Wilson, L., Bamburg, J.R., Mizel, S.B., Grisham, L.M. and Creswell, K.M., 1974, Federation Proc. 33, 158.
40. Adams, M. and Warr, J.R., 1972, Exp. Cell Res. 71, 473.
41. Flavin, M. and Slaughter, C., 1974, J. Bacteriol. 118, 59.

*Microtubules and Microtubule Inhibitors, eds. M. Borgers and M. de Brabander*


# STUDIES ON THE INHIBITORY ROLE OF CALCIUM IN THE REGULATION OF MICROTUBULE ASSEMBLY IN VITRO AND IN VIVO

G. M. Fuller, J. J. Ellison, M. McGill, L. A. Sordahl and B. R. Brinkley

Department of Human Biological Chemistry and Genetics

Graduate School of Biomedical Sciences

The University of Texas Medical Branch

Galveston, TX 77550


## 1. Summary

Bovine brain tubulin, rat liver mitochondria, and calcium were mixed in an in vitro assembly system and the polymerization of tubulin into microtubules measured by viscometry. Mitochondria were capable of removing sufficient amounts of calcium from the mixture to allow tubulin polymerization to proceed. Mitochondria exposed to oligomycin and KCN, however, could not remove enough calcium to allow polymerization to occur. These studies suggest that mitochondria are capable of sequestering sufficient quantities of calcium from the medium to play a role in the formation of microtubules in vitro.

Electron microscopy and immunofluorescence studies using antitubulin have shown striking differences in microtubule patterns and distributions between 3T3 and their transformed counterpart SV3T3 cells.[8,15] Normal 3T3 cells display an elaborate array of cytoplasmic microtubules whereas SV3T3 cells have a greatly diminished number of microtubules. Isolation of tubulin from the two cell types, however, showed that the tubulin was present in sufficient quantity and competency to form microtubules in an in vitro assembly system. Atomic absorption and spectrophotometry shows significantly more calcium in the SV3T3 cells than in 3T3 cells. Taken together, the in vitro and in vivo studies seem to indicate that calcium plays an important role in regulating the formation of cytoplasmic microtubules.


## 2. Introduction

One of the tantalizing properties of the molecule tubulin is its capacity to undergo an ordered assembly to form cytoplasmic and spindle microtubules. Within the last few years, a number of investigations have shown that under certain conditions[1-3] assembly of tubulin into microtubules can occur in vitro. Interestingly, the ubiquitous and important divalent cation calcium has been found to be an inhibitor of the in vitro assembly process.[1,3] In fact, a number of papers have suggested that intracellular calcium may play an important role in the in vivo regulation of the formation of microtubules.[4,5,6] In this system then, calcium would be acting as an inhibitor and upon removal of the cation the assembly process would occur. If this hypothesis is operant, it is important to (a) demonstrate that when intracellular levels of calcium are high, there are indeed reduced numbers of formed microtubules, and

(b) demonstrate that those organelles known to be involved in calcium uptake and release can regulate amounts of calcium sufficient to influence tubulin assembly.

3. Tubulin Preparation

Fresh calf brains were obtained from a local slaughterhouse within 10–15 min following death. The entire brain was immediately immersed in ice cold buffer (0.25 M glycerol, 0.1 M Mes, 2 (N-morpholino)-ethanesulfonic acid, 1.0 mM EGTA, and .5 mM $MgCl_2$, pH 7.0) and remained immersed for ca. 45 min to insure complete chilling of the brain. The superficial blood vessels and meninges were then removed and the tissue diced, drained of excess buffer, weighed, and then homogenized in a Waring blender using the same buffer at a weight to volume ratio of 1 gm : 1 ml. The homogenate was then centrifuged at 14,000 rpm for 15 min and the supernatant saved. A high speed centrifugation was then carried out at 100,000 xg and the supernatant designated as crude tubulin was divided into smaller aliquots and stored at -70°C.

Mitochondrial Preparation. Rat liver mitochondria were isolated by the methods described by Sordahl et al.[7] Rats were anesthetized and the livers perfused via the portal vein with 0.25 M sucrose, 1 mM tris, 1 mM EDTA, pH 7.2. Following perfusion of 50 ml's of buffer, the liver was removed, blotted dry and weighed. The liver was then diced and homogenized in a glass homogenizer with a teflon pestile using a volume to weight ratio of 10 ml : 1 gm. A low speed centrifugation 2,500 rpm was carried out for 10 min. The supernatant was saved and subjected to a second centrifugation of 8,800 rpm for 12 min. The pellet was resuspended in half the original volume and further homogenized followed by recentrifugation at 10,000 rpm for 10 min. The pellet was recovered and the last step repeated two more times. The final pellet (mitochondria) was suspended in 2.0 ml of the following buffer: 0.25 M glycerol, 0.1 M Mes, 1.0 mM EGTA, 0.5 mM $MgCl_2$, 1.0 mM GTP. Under certain conditions 5.0 mM ATP was also added to the buffer. Mitochondrial protein concentrations were determined using a Biuret Assay with bovine serum albumin as a standard.

Viscometry Assays. Quantitative measurements of tubulin polymerization were carried out using Ostwald viscometers immersed in waterbath calibrated to $30 + 0.01^{o}C$. Usually 0.6 ml of tubulin (ca. 18 mg/ml) plus buffers were placed in a viscometer at $0^{o}C$ and then transferred to the waterbath. Outflow times were measured at 5.0 min intervals for 60 min. Specific viscosities were calculated based on the following relationship:

$$\eta \text{ specific} = \frac{\text{Outflow time extract - outflow time buffer}}{\text{Outflow time buffer}}$$

All measurements were done in duplicate using three different calibrated viscometers. When mitochondria were added to the system they were added at $0^{o}C$ usually to a final volume of 1-3 ml. Six-tenths of a milliliter was then pipetted into the viscometer, and the initial reading was taken immediately. All experiments were repeated with more than one tubulin preparation to insure against variations in the tubulin's capacity to polymerize.

Calcium Determinations:

Mitochondria. Total amounts of calcium present in mitochondria were determined using a Varian Techtron Model 1200 Atomic Absorption Spectrophotometer. Complete incubation mixtures containing tubulin ($\sim$18.0 mg/ml) and mitochondria 0.7 mg/ml plus 3 mM sodium succinate, 1 mM $K_2H\ PO_4$ in a final volume of 2.6 ml were incubated for varying time periods. Following incubation, the mixtures were centrifuged at 10,000 rpm at $0^{o}C$ for 15 min and the mitochondrial pellet recovered. The mitochondria were washed twice in calcium free saline; following the final wash the mitochondria were suspended in 1.0 ml of 0.1 M lanthinum HCl, boiled for 5 min, and then centrifuged for 15 min at 15,000 rpm. The supernatant was taken for calcium determinations.

Cells. SV3T3 and 3T3 cells were grown to confluency using McCoy's modified media. The cells were removed by trypsinization, washed in fresh media, and counted using a Coulter Automatic Cell counter. Samples containing $6 \times 10^6$ and $15 \times 10^6$ of each cell type were washed two times in $Ca^{++}$-free buffered saline. Following the last wash, the cells were taken up in 1.0 ml of 2x glass distilled water and sonnicated to insure complate disruption of all cells. Fifty microliters from each sample were removed for protein determinations. The cellular homogenates were then used to determine the total calcium content in an analogous manner to that described for isolated mitochondria. Results are reported as nM of $Ca^{++}$/mg protein.

Tubulin Reassembly in SV3T3 and 3T3 Cells. Confluent cultures of cells were removed from flasks with a rubber policeman. The cells were washed twice with reassembly buffer at 0°C. Following the last wash the cells were resuspended in 1.0 ml of reassembly buffer and homogenized by sonication for 30 sec at 0°C. Small drops of cell homogenate were pipetted on parafilm sheets which had previously been placed in 30 mm Petri dishes containing moist filter paper. The drops were incubated at 37°C for 30-45 min. Following incubation, collodian-coated 300 mesh copper grids were gently placed on the drops, then removed, blotted, rinsed in distilled water, and stained with 1% uranyl acetate for 15 min. The grids were examined on a Phillips 300 Electron microscope.

4. Effect of Calcium on Tubulin Polymerization

The capacity of calcium to inhibit polymerization of bovine brain tubulin is shown in Fig. 1. In these experiments the concentration of magnesium was held constant and the

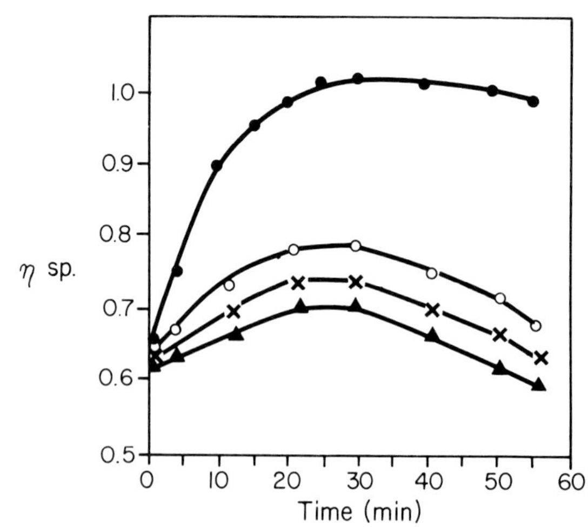


Fig. 1. Tubulin polymerization in the presence of calcium chloride. ● tubulin alone; o tubulin + 125 $\mu$M $CaCl_2$; ✕ -tubulin + 250 $\mu$M $CaCl_2$; ▲ tubulin + 500 $\mu$M $CaCl_2$. Protein concentration 17.8 mg/ml.

calcium concentrations increased. It is apparent that 500 μM calcium is sufficient to inhibit polymerization by 60-70% As the concentration of calcium is decreased, the degree of polymerization increased.

Effects of Calcium on Tubulin Polymerization in the Presence of Mitochondria. In another series of experiments freshly isolated mitochondria were added to the polymerizing system. The viability of the mitochondria in the modified polymerization buffer was determined by monitoring the ability to utilize oxygen from the media using a Gilson Oxygraph. Figure 2 shows the polymerization curve of the mixture of tubulin plus mitochondria. The polymerization curve, although somewhat higher in the initial viscosity, has the same shape as that of tubulin alone. Upon addition of 500 μM of calcium the initial rate and extent of polymerization is almost identical to that of tubulin plus mitochondria for the first 25-30 min. The viscosity begins to decline at the end of 30 min and continues for the next 15 min. Monitoring the viscosity beyond 60 min indicates that by the end of 2 hours the viscosity is only slightly greater than that of the initial value. This suggests that the mitochondria can rapidly remove calcium from the medium, but after 40-50 min the calcium begins to leak out of the mitochondria.

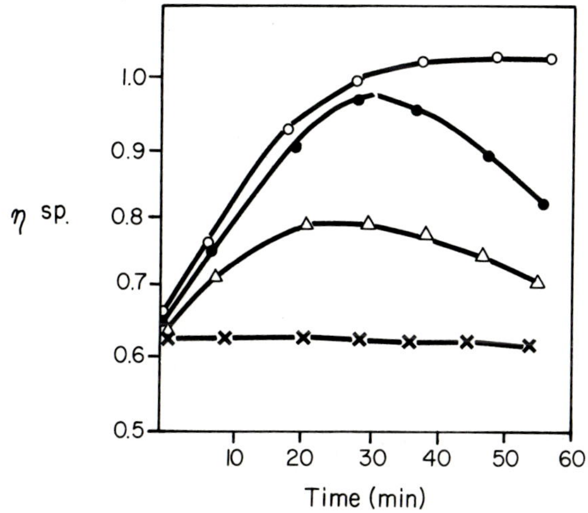


Fig. 2. Tubulin polymerization in the presence of mitochondria and calcium. o tubulin + mitochondria; ● tubulin + mitochondria + 500 μM $CaCl_2$; △ tubulin + 500 μM $CaCl_2$; ✖ tubulin + colchicine.

Since uptake of calcium into the mitochondria is an active process and the total amount of calcium in the media is quite large, the organelle simply cannot hold the ion beyond 40-50 min. Figure 3 shows the results of the experiments in which two mitochondrial inhibitors oligomyocin and KCN are added to the polymerization system. If the uptake of mitochondrial uptake system are poisoned by these two agents, calcium removal from the medium is greatly retarded and, consequently, the polymerization profile is almost identical to that of tubulin + $Ca^{++}$ in the absence of mitochondria.

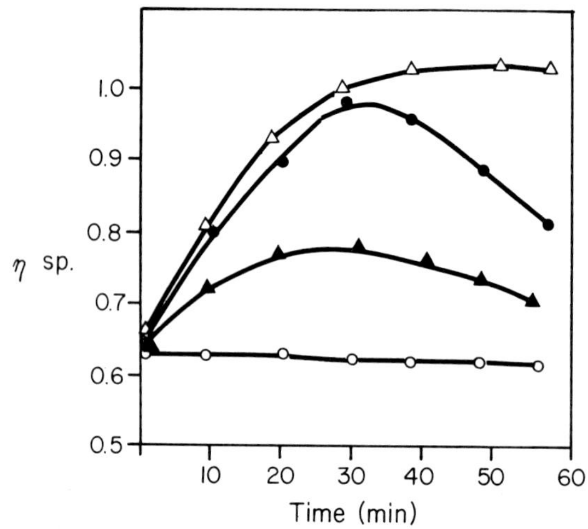


Fig. 3. Effects of mitochondrial inhibitors on tubulin polymerization. △ tubulin + mitochondria + oligomycin + KCN; ● tubulin + mitochondria + 500 μM $CaCl_2$; ▲ tubulin + mitochondria + 500 μM $CaCl_2$ + oligomycin + KCN; o tubulin + colchicine.

Figure 4 shows that calcium is actually being taken into the mitochondria. In these experiments mitochondria were isolated from the polymerization media at various times and the total amount of calcium determined. There is a rapid uptake, a plateau and then decrease in the amount of calcium in the mitochondria. The polymerization curve follows the calcium curve almost precisely and lends support to the suggestion that mitochondria can remove calcium but cannot continue to sequester it beyond 30-40 min in this system.

Does calcium play a role in microtubule formation in vivo? In the foregoing experiments, it was demonstrated that calcium could inhibit the assembly of tubulin into microtubules.

Further, it was also shown that a cellular organelle has the capcity to remove calcium from a polymerizing media and thereby allowing for the formation of microtubules.

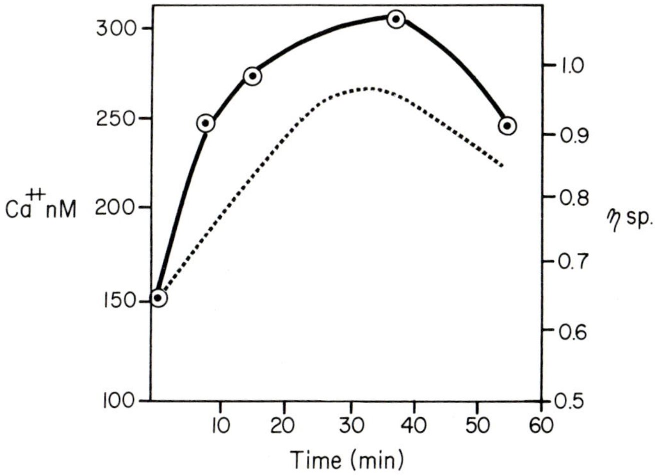


Fig. 4. Uptake of calcium by mitochondria. Mitochondria were isolated from a polymerization mixture and the amount of calcium determined by atomic absorption spectrophotometry (solid line). Dashed line is a measure of tubulin polymerization in a duplicate experiment.

It has been shown[8,9] that virally transformed mouse fibroblasts (SV3T3) have a significant dimunition of cytoplasmic microtubules as compared to their normal nontransformed counterparts (3T3 cells). Furthermore, it has been shown that when SV3T3 cells are exposed to the analogue of the cyclic nucleotide adenosine monophosphate (dibutryl cyclic adenosine monophosphate), the morphology of the cells changes significantly and the number of cytoplasmic microtubules are greatly increased.

In order to demonstrate that the tubulin in SV3T3 and 3T3 were competent, that is, capable of undergoing *in vitro* polymerization to form microtubules, approximately $10^8$ SV3T3 and $10^7$ 3T3 cells were harvested and disrupted into a polymerizing medium. Figure 5 shows whole mount preparation of the *in vitro* assembled microtubules. Reassembled bovine brain tubulin is also shown for comparative purposes. Figure 5 (d) is a thin section of SV3T3 microtubules. The results shown in these experiments demonstrate that tubulin from SV3T3 cells is capable of undergoing assembly and, by morphological comparison, these microtubules are indistinguishable from both 3T3 and bovine brain microtubules.

Total amounts of calcium in SV3T3 and 3T3 cells were determined in three different experiments by atomic absorption spectrophotometry. Table 1 shows the average of two different experiments. These results indicate that SV3T3 cells have significantly more calcium than do their normal counterparts.

TABLE 1

Total Amounts of Calcium in 3T3 and SV3T3 Fibroblasts

| | n Moles $Ca^{++}$/mg Protein | |
|---|---|---|
| Experiment | 3T3 Cells | SV3T3 Cells |
| 1 | 37.8 | 52.3 |
| 2 | 44.5 | 81.1 |
| Average | 41.15 | 66.7 |

5. Discussion

The results of experiments presented here demonstrate that the divalent cation calcium may play a fundamental role in controlling the ability of cytoplasmic pools of tubulin to undergo assembly into microtubules. The molecular mechanism of this inhibition, however, remains to be elucidated. The finding that mitochondria can, in fact, sequester sufficient quantities of calcium to permit assembly does suggest that another cytoplasmic organelle can indirectly affect the formation of microtubules.

It is well known that under in vivo conditions most microtubules are extremely labile structures. For example, during the mitotic process microtubules are assembled into a mitotic apparatus which serves to bring about chromosome movement. At the end of telophase, most of the spindle microtubules break down. Inoue and co-workers[10,14] have shown that a number of factors influence the assembly and disassembly of spindle microtubules. It seems from the data presented here that calcium could also play an important role. For example, if at the onset of mitosis calcium were sequestered by mitochondria and smooth endoplasmic

---

Fig. 5. Whole mount preparation of reassembled tubulin from (A) bovine brain, 110,000 x, (B) SV3T3 cells 99,000 x, (C) 3T3 cells 140,000 x. D is a thin section of SV3T3 microtubules, 78,600 x.

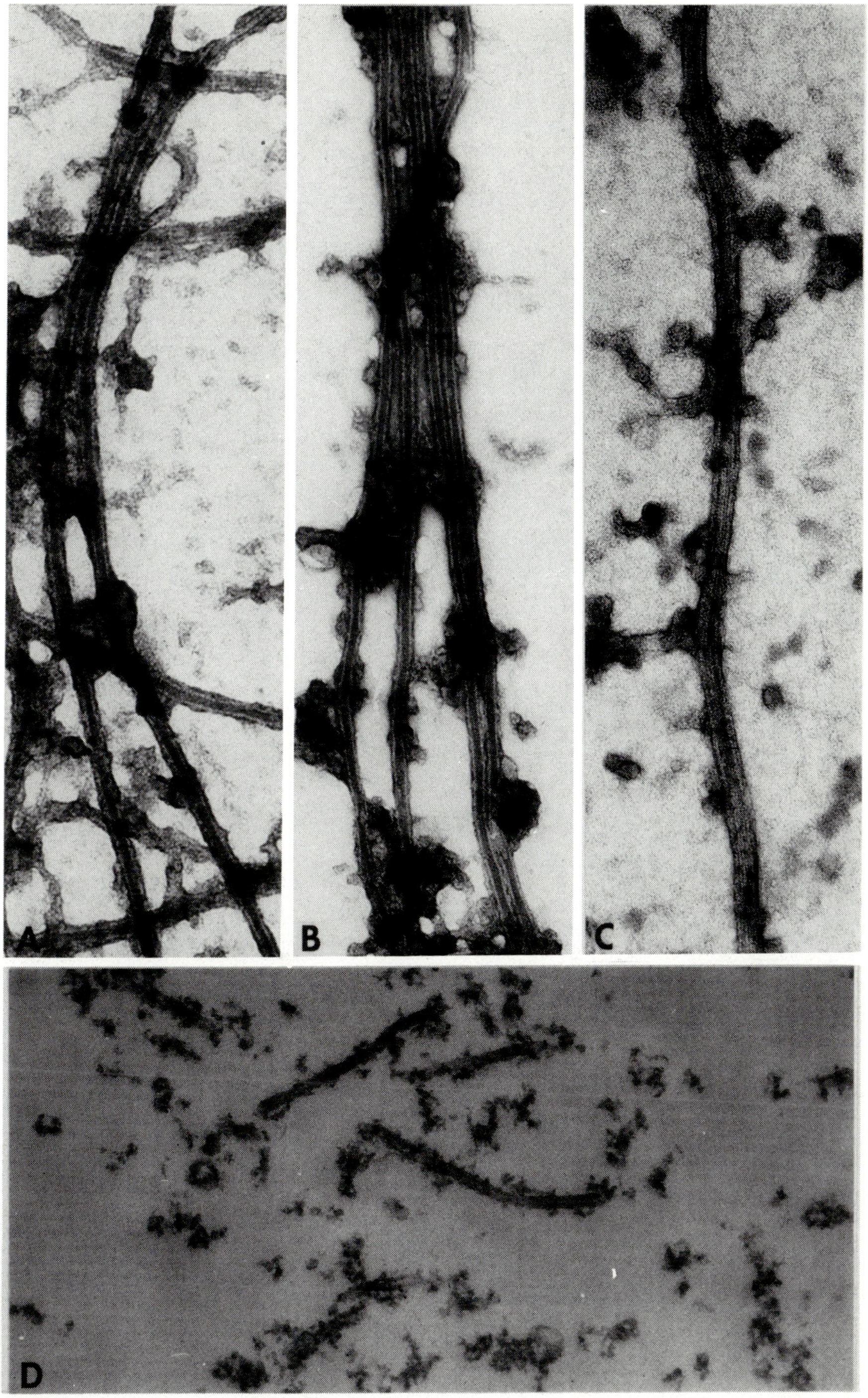
A
B
C
D

reticulum in the vicinity of microtubule organizing centers (MTOC) such as the centrioles and kinetochores, microtubules could be assembled to form the spindle. At the end of mitosis, the release of calcium could lead to the disruption of the spindle. As demonstrated by Brinkley et al. the disassembly of the mitotic apparatus is followed by the reassembly of the cytoplasmic microtubules complex (CMTC), hence calcium would again be sequestered at the MTOC's to initiate the assembly process. Thus, the fluctuation of polymerized tubules to tubulin and vice versa would be in part regulated by cyclic fluctuation of ionic calcium at the organizing centers. This idea is illustrated diagrammatically in Fig. 6.

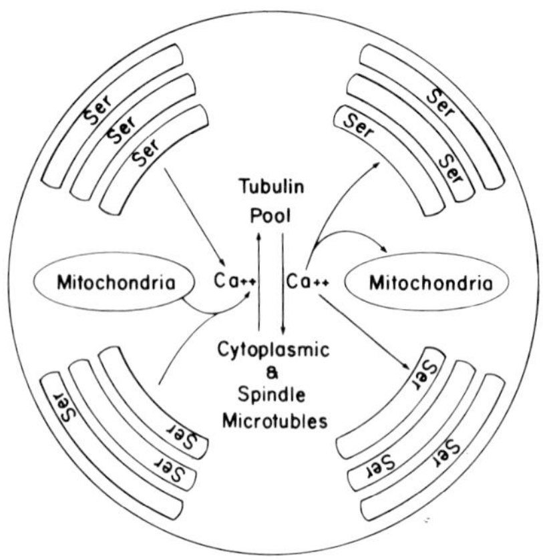


Fig. 6.

Transformed cells appear defective in their capacity to assemble a complete cytoplasmic microtubules complex. The finding of increased calcium may account for the reduced number of cytoplasmic microtubules in SV3T3 cells. Interestingly, mitotic spindle microtubules are assembled in SV3T3 cells and mitosis proceeds normally. Thus, it appears that tubule assembly may be under separate controls in the two systems. Microtubules of the spindle could have $Ca^{++}$ sensitivities different from those of the CMTC's or $Ca^{++}$ levels in SV3T3 may be elevated only during interphase and not during mitosis. There is growing evidence for multiple microtubule population within cells.[10,11,12,13] Thus, it would not be unexpected to see variable sensitivities to different calcium levels.

As yet, no direct evidence exists for the specific role of calcium in regulating microtubule assembly. Indeed the recent work of Olmsted and Borisy have shown that a $Mg^{++}$-GTP complex is necessary for assembly under *in vitro* conditions. Further, they caution against employing calcium as an *in vivo* regulator.

The evidence presented in this study does show calcium to have a profound influence on microtubule formation. There is a need for additional data before the precise molecular role of this cation on the assembly-disassembly process can be firmly established.

Acknowledgment

We thank Mr. Don Highfield, Mr. Michael Stewart, and Ms. Janis Moore for their help in growing the cells and in the calcium determinations. This work was supported in part by the National Institutes of Health grant, CA 14675.

References

1. Weisenberg, R.C., 1972, Science, 177, 1104.

2. Shelanski, M.L., Gaskin, F., Cantor, C., 1973, Proc. Natl. Acad. Sci., U.S.A. 70, 765.

3. Olmsted, J.B., Borisy, G.G., 1973, Biochemistry 12, 4282.

4. Shelanski, M.L., 1973, J. Histochem. and Cytochem. 21, 529.

5. Shelanski, M.L., 1973, in Proteins of the Nervous System, Raven Press, Ney York, p. 227.

6. Rasmussen, H. and Goodman, D.B.D., 1975, N.Y. Acad. Sci. 253, 789.

7. Sordahl, Louis A., Johnson, Carl, Blailock, Zack R., and Schwartz, Arnold, 1971, in Methods in Pharmacology, Vol. I, ed. Arnold Schwartz, Appleton Century Crofts, New York, p. 247.

8. Brinkley, B.R., Fuller, G.M., and Highfield, D., Proc. Natl. Acad. Sci., U.S.A (In Press).

9. Brinkley, B.R., Fuller, G.M., and Highfield, D. (These proceedings).

10. Inoue, S., 1964, in Princ. Motile Systems in Cell Biology, R.D. Allen and N. Kamiya, eds., Acad. Press, Inc.

11. Handel, M E. and L.E. Roth, 1971, Develop. Biol. 25, 78-95.

12. Mole-Bajer, J., 1967, Chromosoma 22, 465.

13. Roth, L.E. and Shigenaka, 1970, J. Ultrastruct. Res. 31, 356.

14. Inoue, S. and H. Sato, 1967, J. Gen. Physiol. 50, 259.

15. Fonte, V., and Porter, K.R., 1974, in Eighth Int. Cong. Elect. Micro. (Aust. Acad. Sci. Canberra), p. 334.

# NEW ASPECTS OF MICROTUBULE FUNCTION IN MITOTIC CELLS

*Microtubules and Microtubule Inhibitors, eds. M. Borgers and M. de Brabander*



# LATERAL INTERACTION OF MICROTUBULES AND CHROMOSOME MOVEMENTS

A. S. Bajer*, Molè-Bajer*, and A. M. Lambert**

*Department of Biology, University of Oregon, Eugene, OR 97403 USA

**Institut de Botanique, Univ. Louis Pasteur, 67083 Strasbourg-Cedex, FRANCE

CONTENTS

I. INTRODUCTION:

The history of science shows that reinvestigation of basic problems and checking of generally accepted "truths" very often supplies new insights. If the problem is reinvestigated from a different angle, i.e., utilising different theoretical assumptions, new aspects become apparent.

In this paper, preliminary results are reported on the effects of low temperature, colcemid, and chloral hydrate. The first two are believed to disorganize microtubules. This undoubtedly is the case under conditions of prolonged exposure. However, it is not known how the spindle MTs are disorganized on the fine structure level during the initial stages of their action and how this effects chromosome movements. It was asked, therefore, whether any conclusions concerning the mechanism of chromosome movements can be derived from the precise correlation of fine structure with studies on living cells, in the early stages of action.

Reinvestigation of these problems was undertaken to check some of the theoretical predictions of the zipper hypothesis (Bajer, 1973). The zipper hypothesis assumes that MTs both assemble-disassemble and interact with each other. It was speculated, therefore, that if the rate of disassembly is increased, especially in stages when assembly normally predominates, this might be reflected both in the fine structure and in chromosome movement. The results obtained surpassed expectations - in certain conditions kinetochore fibres of totally different structure have been produced experimentally. At the same time the movement of chromosomes has undergone pronounced changes. More detailed theoretical remarks (Bajer, in preparation) and more extensive data on the effect of colcemid, chloral hydrate (Molè-Bajer) and low temperature (Lambert and Bajer) will be published separately.

Terminology and abbreviations

Due to the unclear relations in respect to the origin and turnover of spindle microtubules (MTs), different terms are used by various authors. The following terms are proposed here:

Kinetochore fibre (k-fibre) composed of kinetochore MTs (k-MTs)-embedded in kinetochore "ball" and "cup" (Bajer and Molè-Bajer, 1972).

Penetrating MTs (Luykx, 1970) which go through the chromosome body in different regions.

Non-chromosomal fibres - composed of MTs not connected with chromosomes. They comprise continuous and mantle MTs.

Interzonal fibres - composed of interzonal MTs during anaphase. These are, at least in the initial stages, non-chromosomal MTs. New MTs are formed in this region. They contribute to the formation of the phragmoplast in plants, and possibly the stem body in animal cells.

Coupled MTs - two MTs, parallel to each other, spaced precisely at a distance equal to double the electron transparent halo. This term is proposed to avoid the term "doublets" used in relation to cilia and flagella.

II. ARRANGEMENTS OF MICTROTUBULES DURING MITOSIS

GENERAL REMARKS

Detailed analysis of the rearrangement of the MTs of kinetochore fibres and of the interzonal region during anaphase in plant and animal cells led to the development of the zipper hypothesis (Bajer, 1973; Lambert and Bajer, 1972). The data are consistent with the assumption that a specific type of interaction across the spindle (transversed to the long axis of the spindle), namely zipping between MTs, supplies the motive for chromosome movements.

What are the fine structural features which lead to assumptions concerning lateral interaction? The major points are as follows:

Detailed analysis of the fine structure combined with time lapse studies reveal that the arrangement of MTs parallel to the long axis of the spindle is never precise during mitosis. The stages when some MTs tend to be more parallel are transitory and occur generally only during or after movement. A regular parallel arrangement seems, therefore, to be the consequence and not the cause of the movement. The kinetochores themselves move obliquely across the spindle (prometaphase) and it is hard to visualize how a regular parallel arrangement would allow for such movements. Kinetochores seem to move only when MTs are rapidly rearranged from irregular to regular or opposite. One would expect, therefore, the existence of a force which acts across the spindle and is responsible, at least partly, for MT rearrangement. Before discussing how such a force can move the kinetochores, some features of spindle fibres will be mentioned briefly.

A. FINE STRUCTURE OF THE SPINDLE AND KINETOCHORE FIBRE IN ANASTRAL SPINDLE

The technique of directional tilt (Couderc, 1974) shows several features of the organization of the spindle and kinetochore fibre as expected on the basis of zipping. In this technique, the micrograph is viewed at 30-40$^{o}$ and rotated in the plane of observation. Such viewing changes the perspective, eliminates the background noise and enables the curvatures of the MTs to be followed easily. It demonstrates the following features: (Pl. III, Fig. a)

a) MTs of the kinetochore fibre are not straight, but diverge and tend to be arranged in arcs.

b) The majority of MTs which compose the kinetochore fibre end in the kinetochore region (ball and cup-Bajer and Molè-Bajer, 1972). Long MTs of the kinetochore fibre are more parallel and more densely arranged close to the kinetochore, and in the center of the fibre, than at the edges.

c) Coupled MTs are found abundantly on the periphery of the kinetochore fibre. This is especially clear on sections which are not precisely parallel to the long axis of the kinetochore fibre. MTs in the polar regions of each half-spindle often cluster together and form larger or smaller regions where many MTs or their segments are parallel. Thus the diffused polar region of the anastral spindle is, in fact, composed of numerous "mini-poles" or mini-centers", when MTs are packed together closely.

d) Non-kinetochore MTs in the half-spindle fan out and intermingle with neighboring kinetochore fibres.

Such organization of the spindle suggests answers to several questions concerning the functional aspects of the spindle and permits several conclusions to be drawn:

1. The MTs which form the kinetochore fibres are not the only MTs which are in contact with the chromosomes. Numerous MTs penetrate the chromosomal body (see review of Fuge, 1974). In Haemanthus such MTs very often form bundles very close to the kinetochores and intermingle with the kinetochore fibres close to the pole (Jensen and Bajer, 1973; Lambert and Bajer, 1975), or run through the chromatin of the sister chromosome. Although the precise length of penetrating MTs is not known, most of them certainly pass the equator, but probably very few, if any, run from pole to pole. Their function may be to disrupt the parallel arrangement of MTs in the center of the kinetochore fibre. Consequently during metaphase the k-fibres are bullet shaped (Plate II, Fig. b) with the curvature of MTs arched toward the equator. Such a shape suggests the existence of a transverse force acting in the spindle which tends to widen and disrupt the fibre perpendicular to its long axis. Such a shape is incompatible with any assumption that the pulling force is located at the distal ends of MTs (close to the pole). If this were the case, MTs would be straight and not arched.

Non-kinetochore and non-chromosomal (continuous) MTs must either break or pull out (slide out) from each other during anaphase. This necessarily will lead to a more parallel arrangement of MTs in the interzonal region between the kinetochores. This may explain why, during anaphase, there are so many MTs just below the kinetochore and gradually fewer towards the equator. It can be concluded therefore that no regular and predictable number of MTs would be found in cross sections of the spindle in anaphase as compared to metaphase. The number may be a little more predictable in small rapidly dividing cells (McIntosh and Landis, 1971; Brinkley and Cartwright, 1971), although even in such cases precise correlation does not exist.

2. The regions of the kinetochore fibre where MTs are clustered together (Pl. III, Fig. a) are interpreted as being regions of intense interaction and the sites of force production. There may be several such small centers ("minipoles") within each kinetochore fibre, or only one, with a considerable range. Thus either a large portion of the kinetochore fibre is involved in force production (divergent kinetochore fibre, as in Haemanthus), or only the end, or polar region of the fibre (straight kinetochore fibre).

3) The kinetochore fibres and the whole spindle are composed of interdigitating MTs. Some MTs of the kinetochore fibre may end at a certain distance from the kinetochore. They originate from the breakage of other MTs and are incorpor-

ated into the kinetochore fibre. Such MTs must be anchored within the kinetochore fibre to MTs which are anchored themselves to the kinetochore.

B. REARRANGEMENT OF MTs DURING COURSE OF MITOSIS

Analysis of fine structure during normal cell division shows that the basic changes of MT arrangement are similar in all types of mitosis of higher, and many lower, organisms. Modifications of mitosis may result from various factors, some of them mentioned here, including the possibility of the existence of different mechanisms.

If we assume that two basic processes occur within the spindle in the turnover of MTs (assembly-disassembly) and their rearrangement in relation to the long axis of the spindle, then the spindle cycle may be characterized by definite stages when the intensity of these processes varies.

1. Prophase-prometaphase. MTs become arranged more parallel during the program of prophase (Pl. II, Fig. a). Prometaphase starts with the breaking of the nuclear envelope. The tendency towards a more parallel arrangement is retained throughout prometaphase and leads to a gradual slowing down of movement in the majority of cells. The movements are slowest when MTs become most regular at metaphase. MT assembly exceeds disassembly throughout this stage *i.e.* the amount of MTs increases.

2. Metaphase. This is a comparatively stable stage where the spindle is in equilibrium. Rearrangement of MTs may lead to a gradual decrease of spindle length and increase in its diameter (*Haemanthus*, crane fly).

3. Anaphase. The stable, more parallel arrangement of MTs in metaphase is disorganized during anaphase. Rearrangement of MTs, from regular to irregular, takes place. The chromosome velocity decreases gradually when the spindle is disassembled. The number of MTs decreases rapidly due to the movement of kinetochores to the poles, as MTs behind the kinetochores (*i.e.* MTs in the interzonal region) are automatically arranged more regularly. The regular arrangement in this region is increased during telophase.

4. Telophase. Remnants of MTs are left behind the kinetochores. Clusters and groups of parallel MTs are very often found. Rapid assembly of new MTs takes place. New structures such as the phragmoplast and stem body are formed. They have little morphological continuity with the metaphase--anaphase spindle, and are characterized by an approximately parallel arrangement of MTs. Assembly prevails at this stage. There are no bodies (chromosomes) tending to distort the parallel arrangement of MTs which, therefore, arrange themselves regularly more easily than in metaphase. Such arrangement is more stable; the structure persists for a long time and disassembles slowly.

III. THE ZIPPER HYPOTHESIS

A. LATERAL INTERACTION, ZIPPING, BREAKAGE AND DISASSEMBLY OF MTs.

To simplify further discussion, the terms 'lateral interaction' and 'zipping' will be defined.

Lateral interaction is defined as sequential and progressive interaction between beighboring MTs, in which no longitudinal shift (sliding) of the interacting MTs occurs. There are several possible types of lateral interaction (Bajer, 1973).

'Parallel zipping' is the coming together of two MTs which bond laterally and progressively along their length in such a way that the two ends come closer together. This type of zipping is therefore a specific, more restricted, case of lateral interaction. It is comparatively simple geometrically but certainly not on the molecular level (Pl. I).

'Touch and go' zipping can be defined as parallel zipping followed by breakage of the bonds between the zipped MTs rather than of the MTs themselves. This type of zipping will result in jerks if the bonds are short lasting. If the bonds last for a longer time the result will be tension within the spindle (Section III, B). During 'touch and go' zipping, one MT may zip in several regions simultaneously, without breakage and disassembly. The final result will be the same as in parallel zipping.

It is assumed that both types of zipping, parallel-break and parallel touch-and-go, may occur at the same time. It is impossible at present to discriminate between these two types on the basis of the morphology of MT arrangements. Therefore 'zipping' in further discussion will refer to these two types.

In the zipper hypothesis it is postulated that parallel zipped MTs are not involved in the production of the motive force.

Photographs of models (Pl. I) show how zipping may move chromosomes, and simultaneously account for elongation of the spindle, increase in divergence of kinetochore MTs, and their gradual decrease in number. Each zipping interaction can move a kinetochore only a minute distance. MTs of each kinetochore fibre must be involved in hundreds, if not thousands of interactions (parallel, touch-and-go or both) to account for the overall movement from the equator to the pole or vice versa.

---

PLATE I

Geometrical relations between zipping, kinetochore movements and spindle elongation during anaphase. The simplest relations (a, c) are when light MTs zip laterally to dark ones - kinetochores (circles) are pulled to the poles and the central core MTs (2 whole MTs) are pushed toward the poles and contribute to the elongation of the spindle.

The more complex relationships which are closer to the structure of the kinetochore fibre are shown on b, d. Some k-MTs are not anchored at kinetochores - they are zipped to anchored MTs. MTs passing through the chromosomal body (P-MTs-penetrating MTs) either break or slide out passively during anaphase. They may play a role in "unzipping" of zipped k-fibre (d). The spindle forms a framework where zipped MTs (f) do not permit collapsing of the structure during the movement.

PLATE I

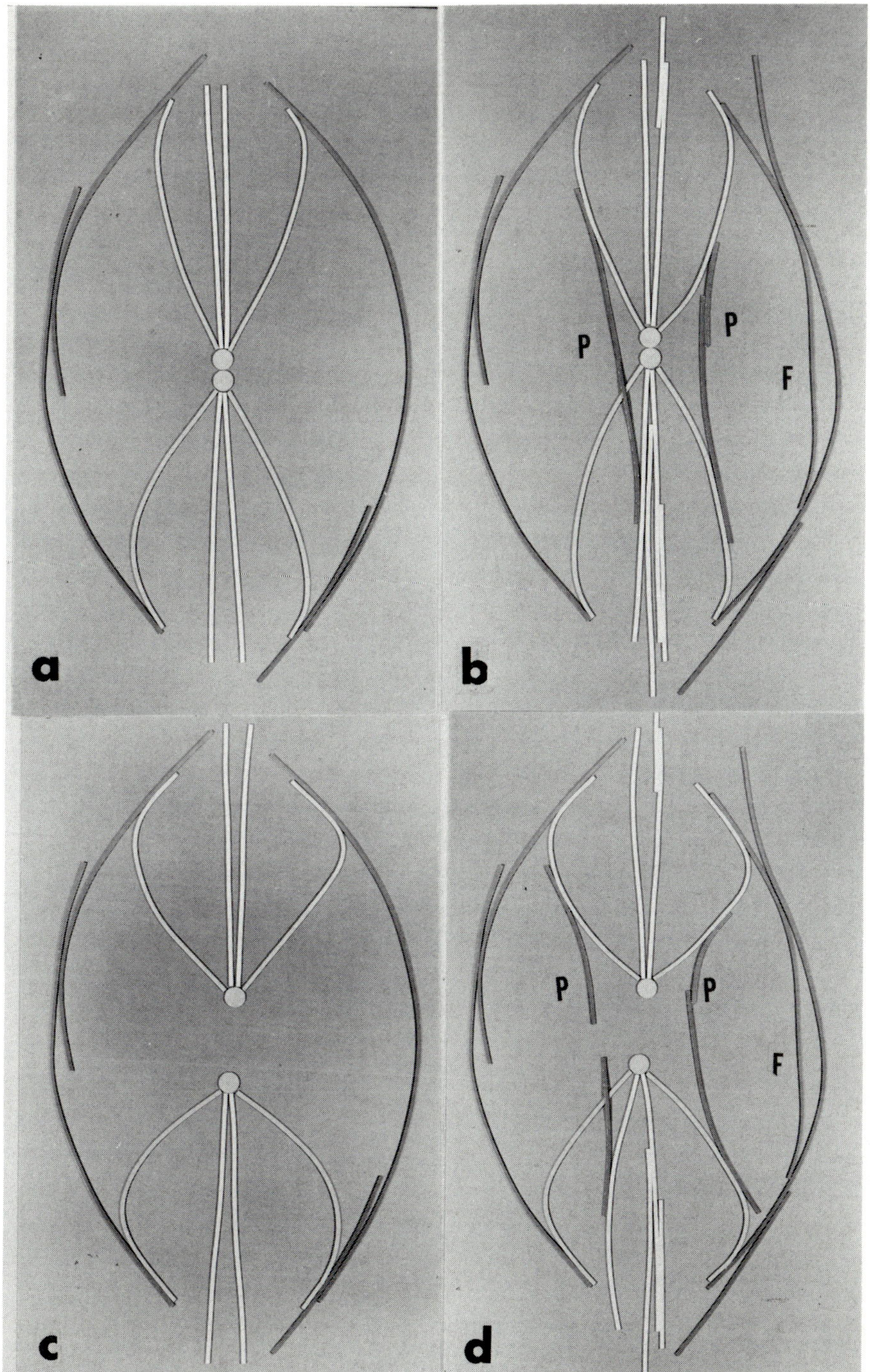

MTs can zip only a certain distance before further progressive zipping between longer segments is prevented by such conditions as interference of other MTs. For the movement to continue (for other MTs to zip) the MTs which zipped first must either disassemble, or bonds must be broken. Movement depends, therefore, on the ability of MTs or bonds (in zipped segments) to disassemble both locally (breakage) or over longer distances. It is postulated therefore, that assembly-disassembly of MTs is a factor regulating the rate of zipping and thus the rate of movements. Some theoretical aspects of this problem will be discussed elsewhere (Bajer, in preparation). This assumption would also explain why the data on birefringence fit well with the rates of chromosome movements (summarized by Inoué and Sato, 1967 and Inoué and Ritter, 1975). It should be realized, however, that in certain cases, *e.g.* during disruption of the spindle, chromosomes may change their position as a result of disassembly combined with other factors such as elasticity of chromosomes themselves, etc.

If the MTs break after zipping, the resulting fragments either must be removed from the spindle, or disassemble. If this process did not occur, the fragments, if abundant, would make further zipping difficult, if not impossible. Data from *Haemanthus* thin sections (Jensen and Bajer, 1973) indicate that MTs are irregularly arranged at the periphery of the kinetochores during anaphase. On the other hand, high voltage electron micrographs (Ris, Molè-Bajer and Bajer, unpublished. See also Pl. III, Fig. a) indicate that a surprising number of MTs in a single section of a *Haemanthus* cell, are over 20 μm long. This may indicate either that zipped segments persist for a considerable time, or that "touch-and-go" zipping prevails. Thus immediate breakage may not be necessary for chromosome movements, especially in long, thin kinetochore fibres where zipping occurs in a comparatively small region of the spindle.

B. ANCHORAGE OF MTs AND SPINDLE TENSION

One assumption of the zipper hypothesis is that all MTs are anchored, but to different extents, by zipping to each other and/or frictional resistance of the spindle matrix. Linkage of MTs by less well defined material (Allen, 1974) or by arms (cross-bridges) may be sufficient to anchor MTs. The information concerning bridges in the spindle is scarce (see literature, McIntosh, 1974). However, MTs

---

PLATE II

*Haemanthus katherinae* - Normal untreated division. (a) - early prophase - MTs arranged in parallel bundles in some regions of the clear zone. Such arrangement is found only in rare mechanical conditions. (b) Metaphase - k-fibres are "bullet shaped" (upper left). There are fewer MTs between k-fibres. They often form parallel bundles composed of 2 or more MTs. MTs within k-fibres are more parallel close to the kinetochore. Intermingling and irregular arrangement further from the kinetochore is clearly seen.

PLATE II

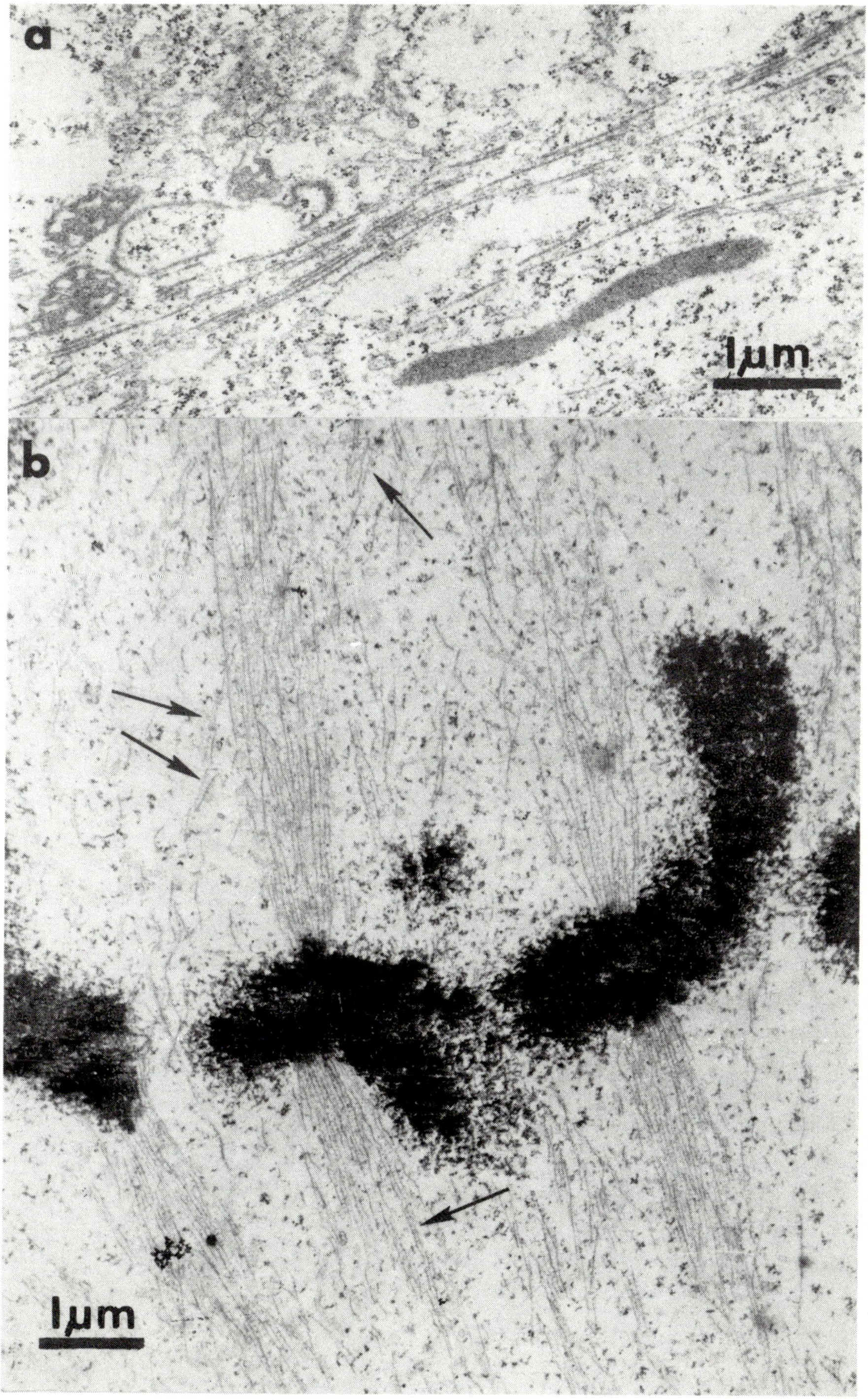

in several systems are linked by periodic, or non-periodic bridges. Intertubule links may resist considerable tension, as in tentacles of the suctorian Tokophyra (Tucker, 1974), where they stretch and still hold MTs together. It is quite conceivable that the bridges in the spindle form thousands of links between MTs.

It is therefore postulated that zipping (by means of bridges or other undefined elements (Pl. III, Fig. b, c, d) is the major factor in the production of tension and thrust within the spindle. Tension is understood as a pulling force on segments of MTs between the zipped segment and the point of attachment, or between two zipped segments of the same MT. By this definition not all segments of MTs are under tension and zipping will therefore tend to bend MTs. Thrust occurs due to the tendency of MTs to straighten.

The internal tension shortens and widens the whole spindle (La Fountain, 1972) due to zipping during metaphase because the two half-spindles are "joined" at the equatorial plane when the kinetochore regions are not divided. The only way to release the tension is to decrease the diameter of the metaphase plate. This can occur only when the kinetochores start their movement during anaphase. Anaphase can therefore be considered as a general result of the release of tension between MTs within the spindle.

Assuming that all MTs are anchored by zipping to other MTs or to membranes (Lambert, 1970; Franke, 1971) leads to the following conclusion:

Each half-spindle is structurally and functionally an independent unit with the kinetochores. Anaphase can therefore proceed without any interaction between MTs in the interzonal region, and the presence of MTs between separating chromosome groups interzonal MTs is not needed for the movements. It is an open question whether such interzonal MTs have any other function e.g. structural support.

C. GRADIENT OF MT DISTRIBUTION, DIRECTION OF ZIPPING

On cross sections of the Haemanthus spindle in metaphase there are fewer MTs in the polar region than at the equatorial plane. Thus a density gradient of MTs exists in each half-spindle. It is generated during prometaphase and plays an important role in establishing the metaphase plate and determining the direction of the chromosome movement. If there is no density gradient, or MTs do not break (disassemble), equilibrium in the spindle is stable and the metaphase plate may

PLATE III

Haemanthus katherinae. (a) - High voltage EM. 1 μm section of mid-anaphase. Intermingling long MTs are closer together in several regions marked by the arrows. This is very clear if the micrograph is viewed at an angle and rotated slowly. It is interpreted that MTs in such regions interact laterally and represent several "mini-poles" of the anastral spindles. Parallel arrangement of MTs in cells treated with colcemid (b) chloral hydrate (c) and after UV microbeam (d-e) - b segment of k-fibre in metaphase. (c and d-e) anaphase - d-e: MTs in half-spindle are probably in the process of rapid disassembly (fixed about 4 min. after UV microbeam irradiation).

PLATE III

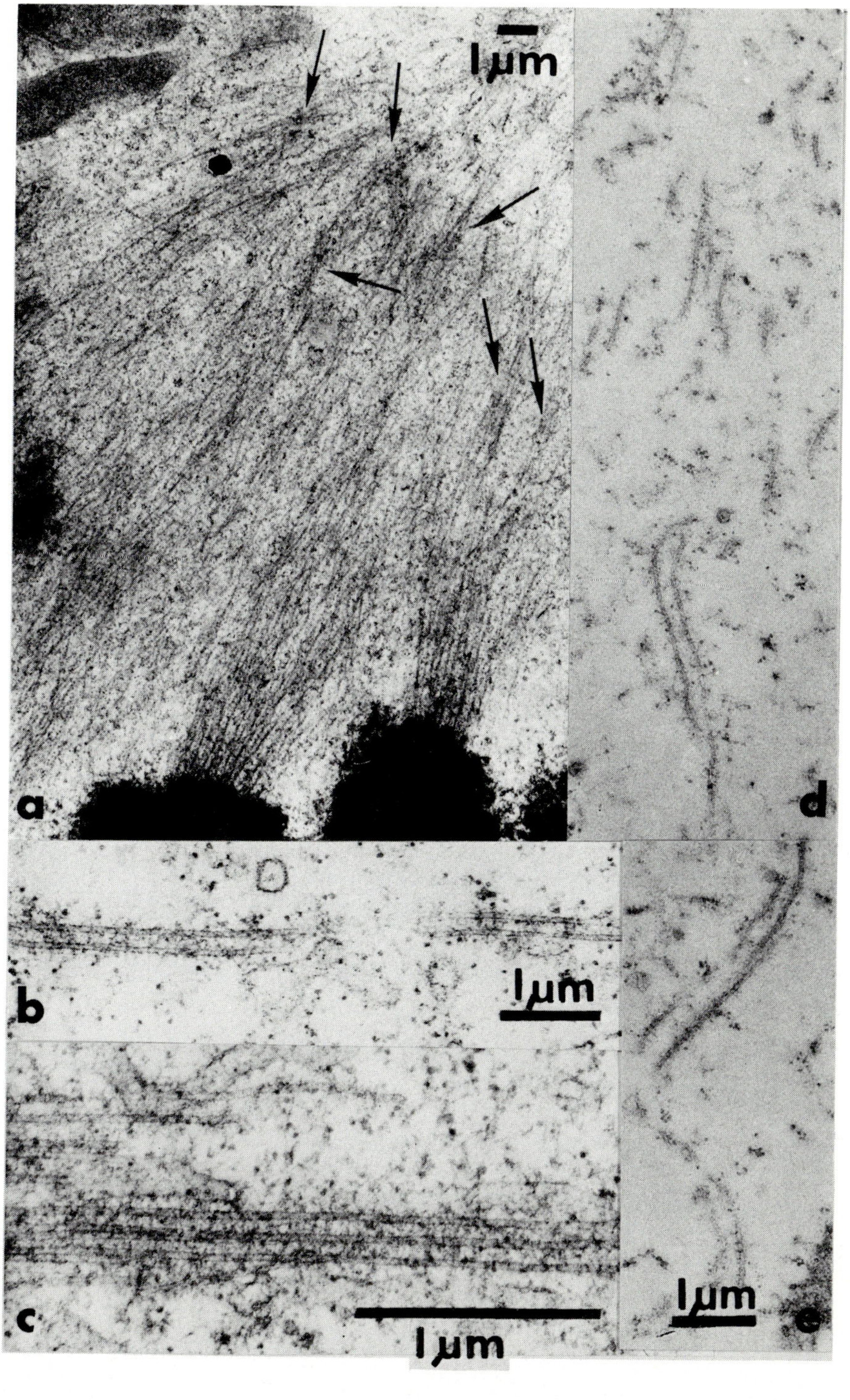

persist for a very long time. This may actually occur in the eggs of some marine organisms where the density of MTs is uniform throughout the spindle (Sato, 1969).

Three main factors lead to the establishment of such a gradient in prometaphase:

1. The presence of chromosomes with sister kinetochore fibres pointing in opposite directions. The growth of kinetochore fibres must result in the breakage of certain bonds, such as those between kinetochore MTs, and an increase in divergence of the MTs. The latter is due to the presence of MTs which cross the equator and zip to the kinetochore MTs.
2. The shortening of some MTs by breakage or disassembly, a process undoubtedly more active in the polar regions, which in turn may result in shortening of the spindle during prometaphase. It is expected, therefore, that if the kinetochore fibre did not grow during prometaphase-metaphase, the spindle would disassemble completely.
3. The occurrence of zipping between all types of MTs with the prevalent direction being determined by the presence or absence of mechanical obstacles.

If MTs grow from the kinetochore poleward, then the presence of kinetochore fibres on opposite sides of the equatorial plane is a factor which is sufficient to produce a distribution gradient of MTs. More parallel MTs will be found close to the equator while toward the poles the arrangement becomes progressively less and less regular. Such a gradient may determine the degree of zipping towards or away from the equator in various regions of the spindle and, consequently, the general transport properties of the spindle, including non-kinetochore transport (Bajer, 1973). The zipper hypothesis makes no postulate regarding the polarity of MTs. It must be assumed therefore that zipping can occur towards both the poles and the equator. In the functional spindle the intensity of zipping towards the pole must be lower than that towards the equator to account for the direction of chromosome movement.

## IV. LATERAL INTERACTION OF MTs IN EXPERIMENTAL CONDITIONS

### GENERAL REMARKS

The rearrangement of MTs during normal division in Haemanthus endosperm is comparatively well known (see review Bajer and Molè-Bajer, 1972). Several features of kinetochore fibres are also mentioned in the theoretical part of the this paper. Plates II-IIIshow some characteristic features of spindle fibre structure during normal mitosis.

All factors studied (low temperature, chloral hydrate and colcemid) disassemble spindle MTs under certain conditions. The arrangement of MTs within k-fibres may undergo pronounced changes in comparison with untreated cells. As expected, differences are found in the effects of the various agents.

A. LOW TEMPERATURE:

Methods:

a. Endosperm preparations were placed in Petri dishes and placed above a tray of ice ($0^{o}C$) in an open container. Preparations were kept at temperatures ranging from $0^{o}C$ to $+7^{o}C$. Most experiments were done at temperatures between $+3^{o}$ and $4^{o}C$. Fixation with glutaraldehyde (methods of Molè-Bajer and Bajer, 1967) was performed either with the fixation at the experimental temperature or at $22^{o}C$ room temperature. Some preparations were kept at low temperatures for as long as 24 hours. Temperatures were monitored with calibrated thermocouples (details below).

b. Preparations were mounted in a special perfusion chamber (Lambert and Bajer, in preparation). This permitted observation in the LM, both of effects of low temperature and recovery, and enabled cells with precisely known chromosome behavior to be fixed. The temperature in the chamber was lowered from $22^{o}C$ to the experimental temperature (i.e. $5^{o}C$) within 3 to 5 minutes. Thus, the change of temperature was gradual, the rate decreasing with time, possibly exponentially. The temperature in the perfusion chamber was monitored by two thermocouples (Yellow Spring Instruments): YSI 423 mounted in the inflow of the perfusing water-ethanol mixture (5% ethanol); and the other (YSI 421) on the top of the cover slip with the cells.

Selected cells were sectioned for EM studies. The results obtained were the same, independent of the technique used.

Results: Observation in vitro.

At the time of writing the experiments are in progress and precise data on chromosome behavior in the living cells, exposed to low temperatures are not yet available. Preliminary observations indicate that low temperature slows down considerably the normal rate of all chromosome movements.

It is not yet known whether and in what conditions the start of anaphase is blocked. Comparison with the newt cells (Rieder, unpublished) indicates that one can expect a variety of complex events. It is not surprising, therefore, that in *Haemanthus*, irregular anaphases were observed in preparations kept as long as 24 hrs. at $4^{o}C$ to $+5^{o}C$.

It is not yet known to what extent effects differ due to variation of exposure time and the rate at which the temperature is lowered. Adaptation of cells to prolonged low temperature may, for instance, liberate suppressed mechanisms not fully operational in cells at room temperature.

Fine Structure:

Short exposure to low temperature ($+4^{o}C$) has pronounced effects on MTs. Typical changes do not differ considerably from those found in cells exposed to such temperatures for 30-60 mins. The effect on MTs is obviously more advanced in cells exposed to low temperature for a longer time. Al observations are consistent and no pronounced differences were found between cells.

Prometaphase-Metaphase: After about 10 mins. and probably less, nearly all non-chromosomal MTs disappear. Chromosomal MTs persist but are considerably reduced in number and length. Most of the chromosomal MTs are k-MTs. In normal untreated cells the number of k-MTs may exceed 100 per kinetochore (Jensen and Bajer, 1974). Although precise counts in low temperatures have not been made, the number of MTs seems to be reduced by more than half. It is not known precisely how long the k-MTs are, but they probably exceed 25 μm in untreated cells of Haemanthus. In low temperatures they are often shorter than 15 μm. The most characteristic feature is that MTs are arranged precisely parallel (Pl. IV, Fig. a). The kinetochore bundles, comprising k-MTs and penetrating MTs, have a larger diameter close to the kinetochore than near the pole (spear shaped). The arrangement of MTs suggests that neighboring MTs are progressively bound to those in the center of the fibre. Hence, frayed ends of MTs are seen often at the edge of the fibre (Pl. V, Fig. b). The distance between parallel MTs corresponds to twice the width of a MT halo (closest possible packing) and indistinct side arms seem to connect MTs. Efforts to find out more precisely the spacing between side arms are inconclusive. However, existence of some regularity is highly probable.

Preliminary studies show that if the cells are exposed to low temperature at metaphase for sufficiently long periods the kinetochore regions split, but regular movements do not take place. It is not yet known, under what conditions the kinetochores move or cease their movement.

During recovery from low temperature treatment of 1 hr. or less the number of MTs increases rapidly and seems to surpass the number in untreated cells. Recovery from low temperature shocks of longer duration have not yet been studied.

During recovery, MTs fan out and the normal course of anaphase follows. It has not been established at what stage of MT rearrangement kinetochores resume their normal migration.

Anaphase: Rapid lowering of the temperature during mid and early anaphase produces changes similar to those during metaphase (Pl. V, Figs. a-b). Preliminary data indicate that chromosomes continue to move at +5 to +7$^{\circ}$C, but at much lower speed. During recovery at 22-23$^{\circ}$C after low temperature treatment (10-60 mins.)

---

PLATE IV

Haemanthus albiflos. Effects of low temperature. (a) - metaphase. Temperature lowered from room temperature (22$^{\circ}$C) to +4$^{\circ}$C within less than 1 min. Cell exposed to +4$^{\circ}$C for 20 min. and fixed immediately - only MTs of k-fibres are present. They are composed of parallel bundles. Inset (b): cell in plastic. (d-e): recovery. Cell in vitro after 30 minutes at 5$^{\circ}$C (d). Cell in plastic-in late prometaphase 12 min. after raising of temperature to 22$^{\circ}$C. (e) - Central part of the spindle in EM. Numerous kinetochore and non-kinetochore MTs.

PLATE IV

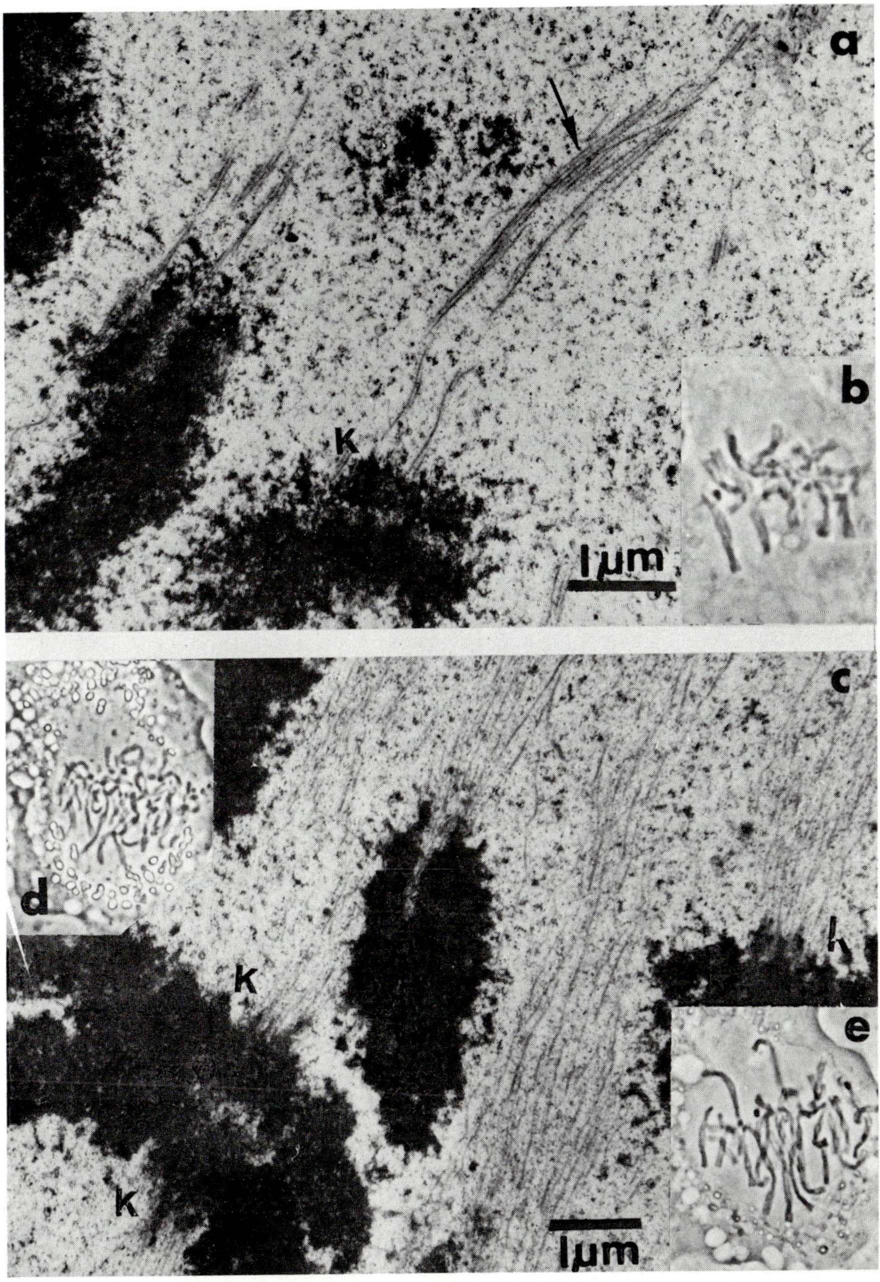

the course of anaphase is resumed with normal speed (Pl. V, Figs. c-e). MTs fan out and the number of k-MTs increases rapidly. It should be stressed that non-k-MTs, which were not detected at low temperatures, now appear abundantly. This indicates that such MTs can form independently of k-MTs.

Conclusions. The data on the effect of low temperature on fibre structure in endosperm closely resembles that of the rat kangaroo (Brinkley and Cartwright, 1975). In both cases characteristic clustering of MTs into bundles, with similar spacing, is observed. However, since the results concerning low temperature effects on the rat kangaroo cells are derived from cross sections, and our results are mostly from longitudinal sections, comparison of the effects is difficult.

In endosperm the effects of low temperature can be produced rapidly (certainly less than 4 minutes). Progressive clustering of MTs forming k-fibres, which are always to some extent divergent, and disappearance of nearly all non-chromosomal MTs, is characteristic. Preliminary data on endosperm show that chromosomes slow down or even stop when MTs become parallel, and move normally when MTs fan out.

For comparison, data on cells in tissue culture (*Taricha granulosa* obtained by C. Rieder, unpublished) indicate that: (1) at +4°C cells are blocked at metaphase, anaphase does not start and a restitution nucleus is formed. (2) The rate of movement during anaphase in cells exposed to such a rapid change of temperature is slowed down about 4 times (from about 2 µm/min. at +25°C to about 0.5 µm/min. at +2.5° to +4.5°C).

B. CHLORAL HYDRATE

Methods: The same methods were used as previously (Molè-Bajer, 1967, 1969). In this technique chloral hydrate, at a concentration of $1.2 \times 10^{-5}$M, is not removed from the medium during the experiments. Chloral hydrate is a very stable compound and therefore the concentration does not change during the experiments. This property of chloral hydrate must be taken into account in any interpretation concerning its action.

---

PLATE V

*Haemanthus albiflos* - effects of low temperature. a - beginning and b - very early anaphase. Temperature lowered from 22°C (room temperature) to +4°C (within less than 1 minute). Cells kept 5 mins. at 4°C and 20 mins. at 7°C, and fixed immediately. Only k-fibres and penetrating MTs are present. MTs often arranged parallel to long ER cisternae (arrow in a). Arrangement of some MTs (arrow in b) makes impression of progressive lateral interaction (parallel zipping). c-e - Recovery from the effects of low temperature (the same cell). The cell - (e) 12 min. after recovery at temperature 22°C after 5°C for 30 min (d).

C-cell in plastic EM micrograph of the region marked by arrow on d-e. Numerous kinetochore and non-kinetochore MTs. Some of them show pronounced divergence (double arrows). The number of MTs seem to be higher than in normal untreated cells

PLATE V

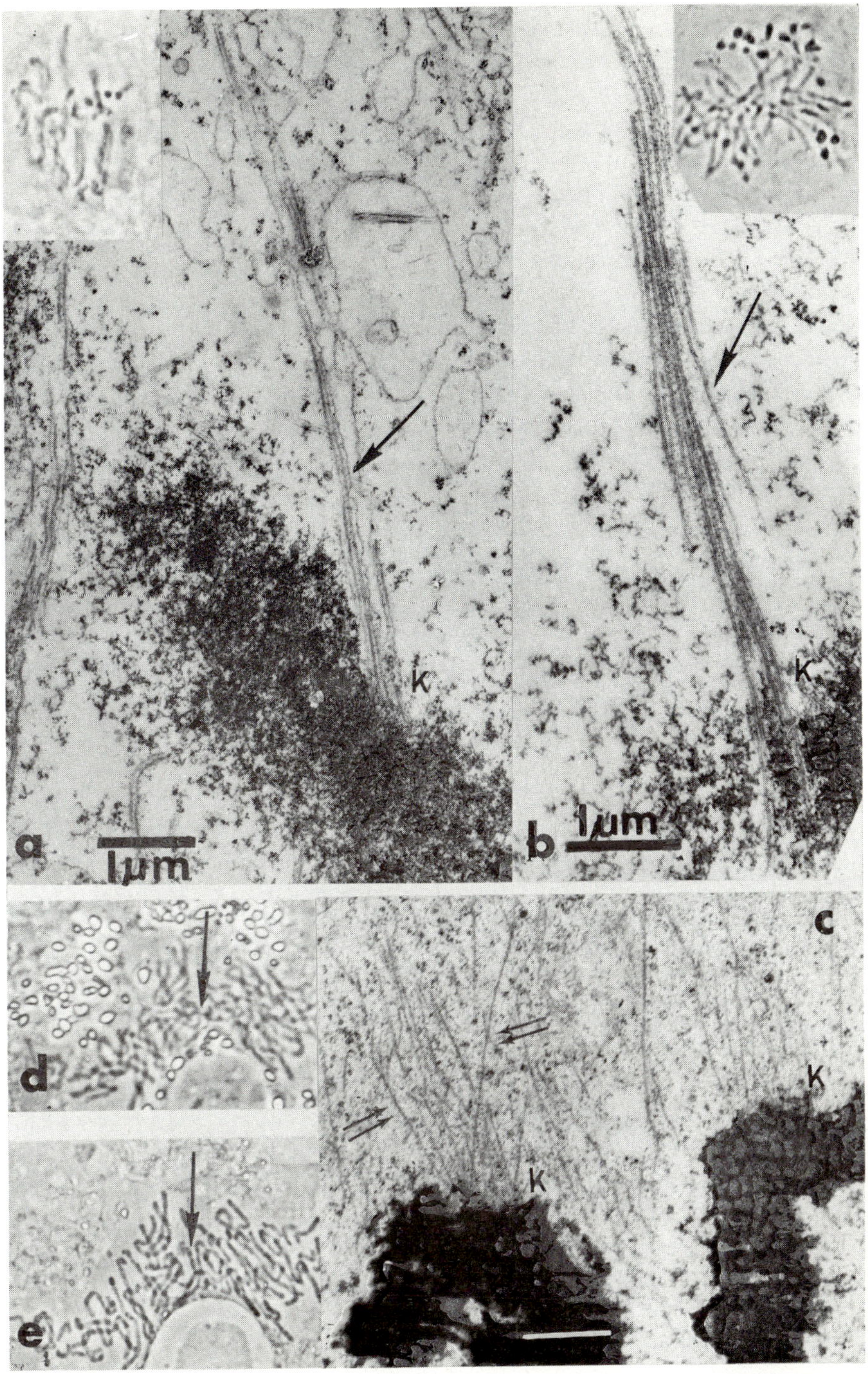

Results: Studies of the effect of chloral hydrate *in vitro* and on the fine structure have been presented (Molè-Bajer, 1967, 1969). The present investigation has supplied data concerning the gradual disorganization of the spindle in the initial stages of action. Precise conditions necessary for the s art of movement were established and analysis of the factors regulating the distance of the movement of particular kinetochores was performed.

Observations in vitro: After prolonged action, which varies for individual cells from 1½ to 4 hrs., the effect is very similar to that of colcemid, and no kinetochore movements take place. In such cells typical c-metaphase is followed by c-anaphase.

With shorter exposure, anaphase is usually disturbed but completed. Bridges are often observed, the maximum separation is much shorter than in normal cells, and bipolar anaphases may be followed by the formation of multipolar phragmoplasts. Effects on cells which were in premetaphase or early metaphase at the start of chloral hydrate action, are of special interest. In such cells, typical bipolar movements are gradually lost and the metaphase plate is either not formed, or is disorganized. Finally the chromosome configuration resembles that of c-metaphases. The kinetochore region splits but no movement occurs for several minutes, occasionally up to 1 hr. During this period, sister chromosomes were parallel to each other and uniformly separated by aoµbt 4 µm. Often they are shifted slowly longitudinally, *i.e.*, sister kinetochores are not precisely opposite each other. Finally movement of the kinetochores starts (chromosome bends at the kinetochore. Text figs. 1 and 2--see also Pl. VI, Fig. e).

Text Fig. 1. Anaphase movements in endosperm cell of *Haemanthus katherine* under the action of chloral hydrate. Time in minutes counted from the moment the kinetochores split. Independent movement of sister chromosomes clearly visible. Graphs of kinetochores (1-2) movements given in Text Fig. 2.

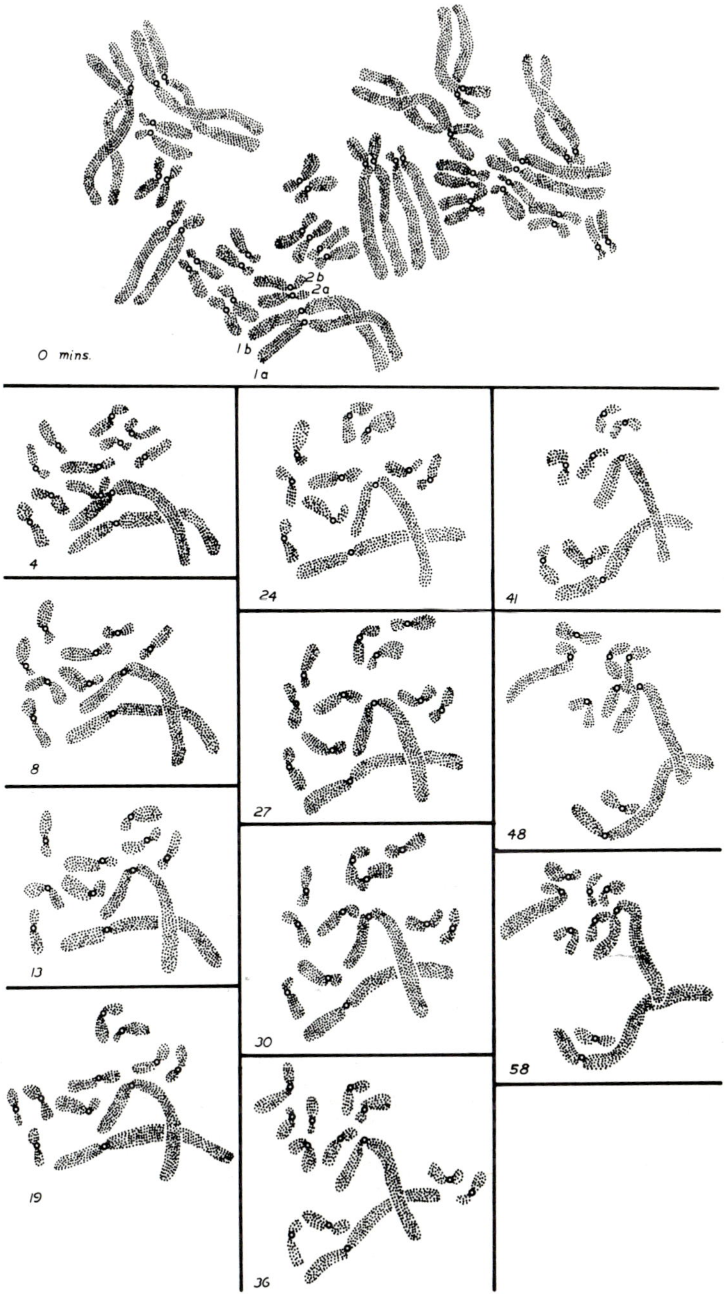
2b
2a
1b
1a
0 mins.
4
8
13
19
24
27
30
36
41
48
58

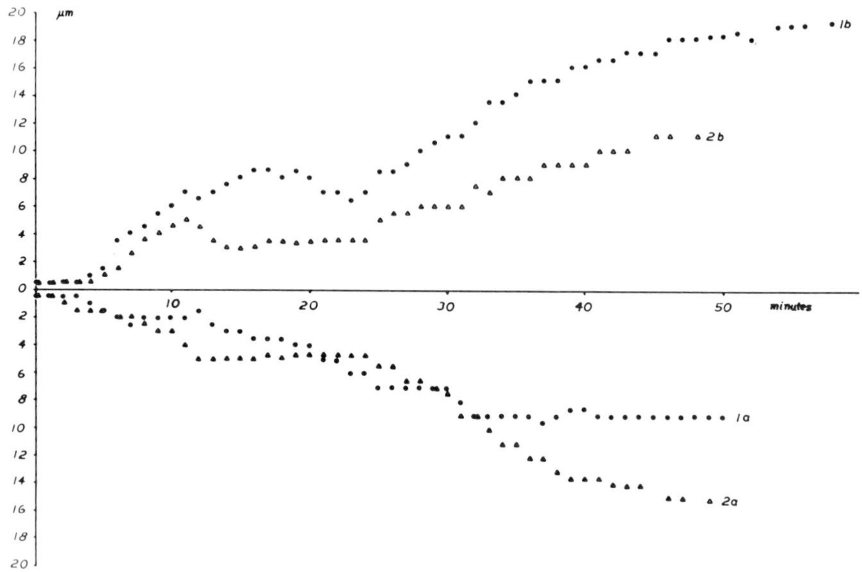


Text Fig. 2. Movement of two kinetochore pairs in anaphase. Distance m measured along the path of the movement plotted against the time.
Curves 1a and 1b represent the movements of two sister chromosomes. Curves 2a and 2b show the movement of the second pair of chromosomes. (Text Fig. 1.)

The movements begin at different times for different kinetochores. Often one sister kinetochore moves over a considerable distance, while the other is nearly motionless. The distance travelled is always shorter than during normal anaphase. Also the speed of different kinetochores varies considerably. The movements give the impression of being either multipolar, or without defined poles. However, there are a few regions (centers) to which some, but not all, kinetochores converge. The more clear the center, the more efficient the movements. If the

PLATE VI

Haemanthus katherinae - effects of chloral hydrate on metaphase (a,c) and anaphase (d,e). Micrographs of cells in plastic before sectioning: insets b (metaphase) and e (anaphase).

a, c. MTs are in the process of assembly. K-MTs are clustered together (zippered) and show closest possible packing. Elongated ER cisternae are often parallel to MTs.

Anaphase. Kinetochores are oriented toward several "centers" or "minipoles" Only K-MTs are present. They no longer zipped like in metaphase, but diverge and intermingle with MTs of other kinetochores.

PLATE VI

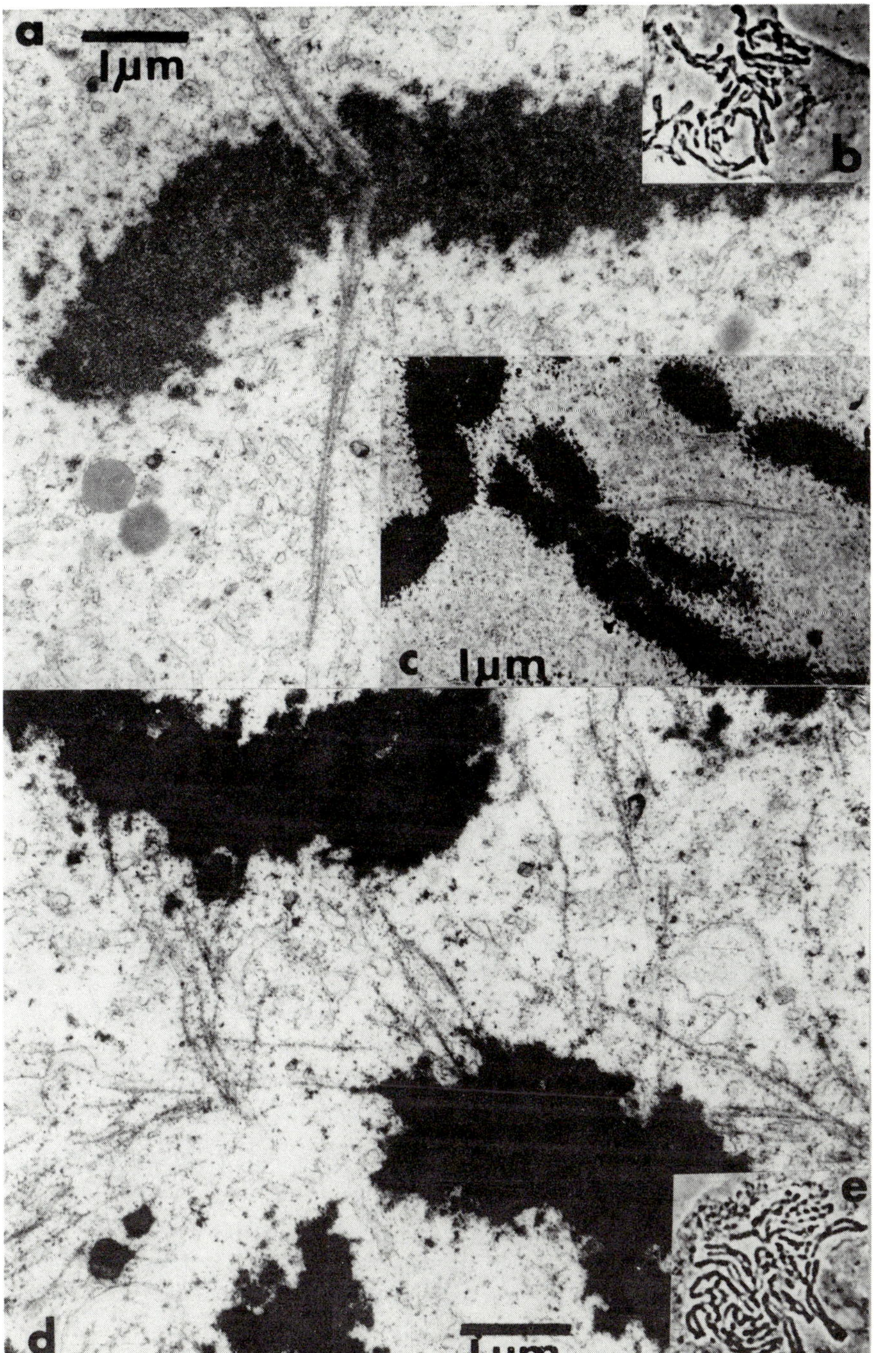

chromosome is well separated from the others, usually either no movement, or movement over a very short distance, takes place. Analysis of some aspects of such movements is given in Text Figs. 1-3.

Fine structure: The effects of short exposure (20-30 mins.) to chloral hydrate are reflected in the fine structure. The overall number of MTs in the spindle at any stage decreases. It is evident from prolonged action that non-chromosomal MTs disassemble first while k-MTs persist for a long time. No MTs are detectable after 4-8 hrs. and division resembles c-mitosis. Cells after 2-4 hrs. of treatment are of special interest. In all such cells MTs disassemble, but k-MTs reappear (Pl. VI, Fig. a-b) during metaphase configuration. The first sign is the appearance of oblong cisternae of ER which radiate from the kinetochore. Such cisternae often persist until telophase. MTs appear parallel to such cisternae. They are short at the beginning and grow longer in later stages. The number of MTs connected with each kinetochore is reduced to approximately 0.5 to 0.25 as compared with untreated cells. MTs are arranged parallel or are slightly convergent, (not divergent as in normal division) and the characteristic spear shape of the bundle of MTs forming k-fibres is even more pronounced than under cold treatment. As a rule the closest possible packing between neighboring MTs is observed. The overall arrangement of chromosomes resembles that of typical c-mitosis. The same MT configuration is observed after sister chromosomes split in the kinetochore regions but no movement takes place. When the kinetochores change their position, k-MTs fan out (Pl. VI, Fig. d and Text Fig. 3) and the movements start. Several short fragments of MTs appear within the k-fibres. However no evidence has been found that MTs other than k-MTs are present. K-fibres are not visible in phase contrast time lapse film. However, indistinct waving zones often seem to be present in the region where one would expect k-fibres to be present.

Conclusion: The studies on fine structure combined with time lapse observations give a strong impression that MTs behave like tentacles moving sideways, trying to find an attachment or anchorage. The k-fibre undergoes a pronounced change after the kinetochore regions split. MTs of the spear-shaped k-fibre (initially parallel) diverge. This indicates that a fibre with parallel MTs is not capable of pulling, while the fibre with diverging MTs is potentially active. It should be stressed, however, that it is not yet known whether k-fibres with diverging MTs can be inactive. Movement occurs when active k-fibres are (1) either in closer contact with other MTs (2) if ER is abundant around MTs. This suggests that either other MTs or membrane systems can be utilized as a site for the attachment of MTs during pulling. K-MTs of one k-fibre can serve as an anchorage for the MTs of other k-fibres. Thus only one type of MT is needed for movements to occur.

C. COLCEMID

Methods: The technique used was the same as in the studies on chloral hydrate.

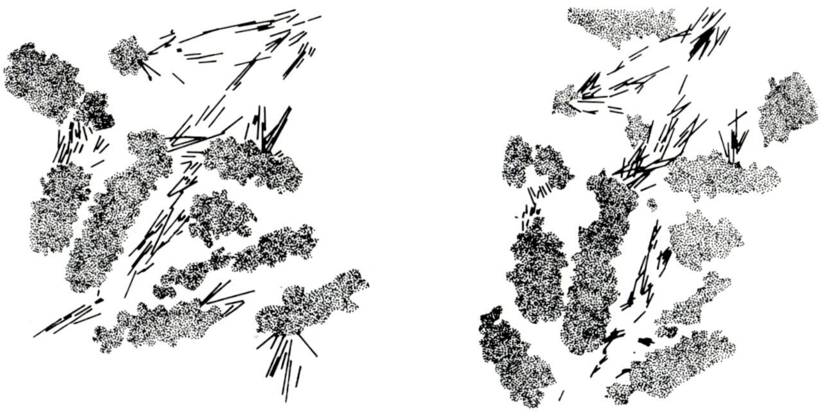

Text Fig. 3. Two serial sections of anaphase in a cell similar to that shown in Text Fig. 1 of chloral hydrate. Chromosomal fibers with divergent MTs, as well as intermingling of MTs from different kinetochore fibers, are clearly visible.

Concentration used was 1.25 x $10^{-6}$M. Cells were fixed 5 mins. to 6 hrs. after application of colcemid. Colcemid is not a stable substance and disintegrates under the influence of UV. It is unlikely, however, that in these short lasting experiments the initial concentration has changed.

Results: Observation in vitro - Under the action of colcemid regular movements of the kinetochore cease rapidly with considerable variation in time between different cells. Cells treated at early anaphase usually complete anaphase, phragmoplasts do not form, and the sister nuclei often fuse. Cells treated at late metaphase often start anaphase, although, as a rule, is not completed. Kinetochore movements are arrested in the second part of anaphase and eventually a restitution nucleus is formed. Earlier stages (prometaphase, early metaphase) form typical c-metaphases. Occasionally however, movements can be observed as long as 45 mins. after the application of colcemid.

Fine Structure: Effects on the fine structure during beginning of action were studied. MTs disintegrate rapidly, those not connected with chromosomes being affected first. The results are basically similar to those on chloral hydrate and low temperature: MTs tend to form parallel complexes. However, the process of disintegration advances continuously (Pl. VII-VIII) and no sign of adaptation to colcemid has been seen. Due to progressive disassembly of MTs parallel bundles of MTs either cannot be formed or do not persist. The MTs shorten and laterally interacting eomplexes are composed mostly from 2 MTs only (coupled MTs). As under the action of chloral hydrate, MTs are very close and parallel to the ER. Some pictures suggest that they may be transformed into membrane complexes (Pls. VII-VIII). The metaphase configuration is preserved for a considerable period although there are no more non-chromosomal MTs.

During anaphase, which is very seldom completed, MTs fan out perpendicular to the poles. Kinetochore fibres such as those in Plates VII, VIII (with fewer MTs than in control spindles) are most probably still capable of pulling kinetochores, but are rapidly losing this ability. Persisting MTs of k-fibres are often straight and form long parallel complexes (Pl. VIII). It should be stressed that MTs arranged parallel over such long distances have never been observed during normal division.

V. DISCUSSION AND CONCLUSIONS:

The theoretical part of this paper draws attention to the complexity of the factors involved in chromosome movement. Only a few selected problems and conclusions will be mentioned below.

The experiments reported here are in full agreement with the generally accepted view that MTs are not only involved in, but are instrumental to chromosome motion. MTs can assemble very rapidly (cold recovery) and disassemble in the living cells under experimental conditions. The comparison of our data with those of C. Rieder (unpublished) on the behavior of chromosomes of Taricha in vitro and EM studies of Brinkley and Cartwright (1975) on the rat kangaroo suggest that the general effects of low temperature are not restricted to endosperm. They also show, as expected that the effects are complex and that some variation exists between different materials. This concerns, for instance, the resumption and cessation of movement at the normal rate. If disassembly is not instantaneous, as in experiments with UV, longer segments of MTs can interact laterally and form very regualr parallel complexes (low temperature and chloral hydrate). Such complexes either persist for a long time (low temperature) or disassemble further (colcemid). All experiments combined leave little doubt that MTs within the living cells can interact laterally. Brinkley and Cartwright (1975) report similar clustering of MTs under the influence of much lower temperature ($0^{o}$-$4^{o}$C) in rat kangaroo fibroblasts. The ability of MTs to form parallel complexes to most likely one of their general characteristics. The data from experiments on assembly of MTs

---

PLATE VII

Haemanthus katherinae - effects of colcemid (35 mins.) on metaphase (c). MTs in this cell are in a process of disassembly (a-b). Some of them seem to be progressively bonded together (parallel zipping, arrow in b). Often MTs are also arranged parallel to elongated ER cisternae. Numberous MTs fragments seen in all regions of the spindle.

PLATE VII

from tubulin subunits in vitro (Weisenberg and Rosenfeld, 1974) show also that MTs tend to form parallel complexes. However, the mechanism of lateral interaction is far from being clear. It is difficult to visualize that such characteristic rearrangement a) can be the result of passive interaction and b) would be absent in the normal untreated cell.

Analysis of fine structure in normal untreated cells shows that the same process of lateral interaction, or more precisely, parallel zipping, operates in living cells. Arguments to the contrary, that experimental conditions stimulate or liberate a mechanism non-existent in normal cells are, according to the authors, only remotely possible. Observations of mitotic stages where MTs assemble and rearrange rapidly, while not being attached to chromosomes (early prophase, formation of the phragmoplast) also suggest lateral interaction (Pl. II, see also Lambert and Bajer, 1972). Therefore, the data presented are considered as experimental evidence for the existence of active lateral interaction of MTs in the living cells. The data are consistent with the zipper hypothesis (Bajer, 1973). Parallel and touch-and-go zipping, regulated by controlled assembly-disassembly, can well explain complex chromosome movements.

Observation on chloral hydrate show that the movements begin when the MTs fan out. The kinetochores do not move when their MTs are precisely parallel (are zipped). In such conditions the movements occur, most likely with only k-MTs present. Fine structure studies suggest strongly that interaction occurs between MTs and membranes i.e., pulling MTs are anchored to the membranes.

If parallel zipped MTs do not contribute to normal chromosome movements, as indicated by presented experiments, they still may, and probably do, play a very important role in the spindle structure. They serve as mechanical links which hold the spindle elements together. Self-supporting spindle models can be constructed easily on such assumptions. Such models incorporate several properties of the spindle. They are elastic, permitting considerable deformation, and will not collapse during movement. Furthermore, such models support the assumptions that the mechanism of pull is located at the front of the kinetochore, and that each half spindle is a functionally independent unit.

It should be stressed, however, that the data on MT interaction in the living cells should be interpreted cautiously and further experiments are necessary. The effects of short and long term action and the way the agents are applied may be different. The cell and cell components often adapt to unfavorable conditions,

PLATE VIII

Haemanthus katherinae - effects of colcemid on early anaphase. a-c the same cell (inset-cell in plastic). a-b 2 serial sections. MTs diverge but are reduced in number. Some MTs are considerably longer and are arranged parallel to elongated ER cisternae. Arrangement of k-MTs (parallel coupled MTs) for initial action of colcemid is very characteristic.

PLATE VIII

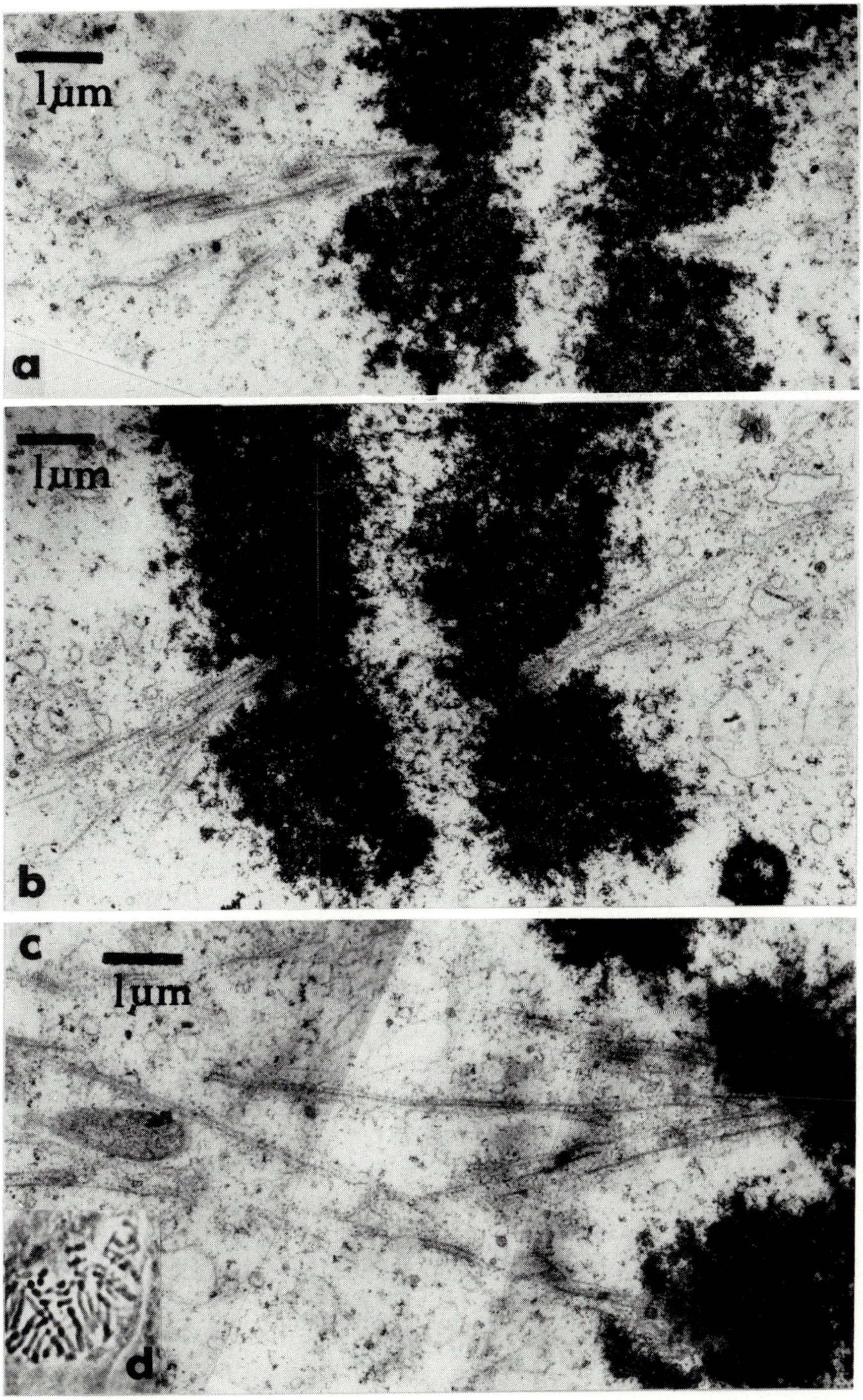

and may perform, at least partly, their function. This has been reported in experiments involving high hydrostatic pressure for spindle MTs (Salmon, 1975; a, b) and axonemal MTs of Echinosphaerium (Tilney, Hiramoto and Marsland, 1966). Experiments on the action of such agents as low temperature on the axopodia (Tilney and Porter, 1967), colcemid on the spindle asters (Brinkley, Stubblefield and Hsu, 1967) and chloral hydrate reported here, also show clear adaptation. The most possible explanation is that, in such experimental conditions, although the equilibrium is shifted considerably towards disassembly, the process may be, in fact, much more complex. Several factors within the cell may be involved in such regulation. It is possible that the mechanism of kinetochore movements observed in some experimental conditions is not the same as that observed during normal division. This may occur in both cases even if MTs are involved and might be reflected, for instance, in chromosome motion of a different order of magnitude.

The most important question is to what extent, in such experimental conditions, the mechanism of movement is comparable to that occuring in the normal course of mitosis. It is conceivable that movement or its lack, can occur due to the liberation of new or hidden mechanisms or by blockage of certain side processes or reactions. One may even speculate that MTs at normal and low temperatures are physiologically different (cold resistant MTs as suggested by Brinkley and Cartwright, 1975). The data presented here is only preliminary. Only extensive experiments is precisely difined conditions, especially in relation to low temperature, can supply the answer to the basic question: Which arrangement of MTs is necessary for movement and which prevents movement?

## VI. SUMMARY

Overwhelming evidence based on a variety of studies indicates that microtubules (MTs) play an essential, although still largely undefined, role during chromosome movement. The spatial arrangement of MTs undergoes pronounced changes during mitosis. MTs are arranged at random during early prophase and are most regularly arranged just before the start of anaphase. During anaphase, the spindle is almost completely disrupted and the regular arrangement of MTs is lost.

Studies of MTs arrangement, precisely correlated with various types of chromosome movement under normal and experimental conditions, lead to the following conclusions:

A) Kinetochore motion occurs only during a change in the arrangement of MTs from a parallel to an irregular state, or opposite.

B) The extreme cases of parallel or completely irregular arrangements are characteristically found when movement has ceased, or slowed down considerably.

The date on the rearrangement of MTs is consistent with the assumption that the motive force for chromosome movement is generated by specific lateral interactions between MTs (parallel and/or touch-and-go zipping). Experimentally induced rearrengement of MTs, under physical and chemical factors, agrees with theoretical predictions from the zipper hypothesis. Conclusions must be drawn very carefully because it is not known whether movement in experimental conditions operates on the same principle as during normal mitosis. The factors studied (low temperature, chloral hydrate, colcemid) disrupt the spindle in specific but different ways, particularly in the initial stages of their action. During disruption, continuous (non-chromosomal MTs) disassemble first. MTs of single kinetochore fibres interact laterally, cluster and tend to form parallel bundles with the closest possible packing. Such bundles may either persist or disassemble after prolonged treatment. Although numerous neighboring MTs may interact with each other, the basic unit of interaction is 2 MTs (coupled MTs). MTs can zip to other structures such as membranes. Movements in certain conditions can occur without continuous MTs. The experiments on the initial stages of MT disassembly within the living cell open a new approach to the study of MT interaction during mitosis.

## VII. REFERENCES


Allen, R. D., 1975, Evidence for firm linkages between microtubules and membrane-bound vesicles, J. Cell Biol., 64, 497-503.

Bajer, A. S., 1973, Interaction of microtubules and the mechanism of chromosome movement (zipper hypothesis), 1. General principle, Cytobios, 8, 139-160.

Bajer, A. S., and J. Molè-Bajer, 1972, Spindle dynamics and chromosome Movements, Cytology, Suppl., 3, 1-271, (Academic Press New York).

Brinkley, B. R. and Cartwright, J. Jr., 1971, Ultrastructural analysis of mitotic spindle elongation in mammalian cells in vitro: Direct microtubule counts, J. Cell Biol, 50, 416-431.

Brinkley, B. R. and J. Cartwright, Jr., 1975, Cold-labile and cold-stable microtubules in the mitotic spindle of mammalian cells, Ann. N.Y. Acad. Sci, 253, 428-439.

Brinkley, B. R., E. Stubblefield, and T. C. Hsu, 1967, The effects of colcemid inhibition and reversal on the fine structure of the mitotic apparatus of Chinese hamster cells, J. Ustra. Res., 19, 1-18.

Couderc, H., 1974, Anamorphoses des images D'objets biologiques tridimensionnels observes au M.E.B., Jeol News, 12, 11-15.

Forer, A., 1966, Characterization of the mitotic traction system and evidence that birefringent spindle fibres neither produce nor transmit force for chromosome movement, Chromosoma, 19, 44-98.

Franke, W.W., 1971, Cytoplasmic microtubules linked to endoplasmic reticulum with cross bridges, Exp. Cell Res., 66, 486-489.

Fuge, H., 1974, Ultrastructure and function of the spindle apparatus. Microtubules and chromosomes during nuclear division, Protoplasma, 82, 289-320.

Inouè, S. and H. Ritter, 1975, Dynamics of mitotic spindle organization and function, Society of general physiologists symposium on molecules and cell movement, In press.

Inouè, S. and H. Sato, 1967, Cell motility by labile association of molecules, J. Gen. Physiology, 50, 259-292.

Jensen, C. and A. Bajer, 1973, Spindle dynamics and arrangement of microtubules, Chromosoma (Berl.), 44, 73-89.

La Fountain, J. R., 1972, Spindle shape changes as an indicator of force production in crane fly spermatochyes, J. Cell Sci., 10, 79-83.

Lambert, A., 1970, Etude de structures cinétiques en rapport avec la rupture de la membrane nucléaire en début de méiose chez Mnium hornum, L. Organisation des centromères, C. R. Acad. Sc. Paris, t.270 sér.D, p.481-484.

Lambert, A. M. and A. S. Bajer, 1972, Dynamics of spindle fibers and microtubules during anaphase and phragmoplast formation, Chromosoma, 39, 101-144.

Lambert, A. M. and A. S. Bajer, 1975, Dynamics and fine structure of prometaphase spindle, J. de Micr., In the press.

Luykx, P., 1970, Cellular michanisms of chromosome distribution, Int. Rev. Cytol. Suppl., 2, 173, (Academic Press, New York).

McIntosh, J. R., 1974, Bridges between MTs, J. Cell Biol., 61, 166-187.

McIntosh, J. R. and S. C. Landis, 1971, The distribution of spindle microtubules during mitosis in cultured human cells, J. Cell Biol., 49, 468-497.

Molè-Bajer, J., 1967, Chromosome movements in chloral hydrate treated endosperm cells _in vitro_, Chromosoma 22, 465-480.

Molè-Bajer, J., 1969, Fine structural studies of apolar mitosis, Chromosoma, 26, 427-448.

Rieder, C. L., 1975, Experimental cold shock: _In vitro_ chromosome movement at temperatures below 4$^{o}$C, In preparation.

Salmon, E. D., 1975, a. Pressure-induced depolymerization of spindle microtubules. II. Thermodynamics of _in vivo_ spindle assembly, J. Cell Biol., 66, 114-127.

Salmon, E. D., 1975, b. Spindle microtubules: thermodynamics of _in vivo_ assembly and role in chromosome movement, Ann. N. Y. Acad. Sci., 253, 383-406.

Sato, H., 1969, Analysis of form birefringence in the mitotic spindle, Amer. Zool., 9, 592.

Tilney, L., Y. Hiramoto, and D. Marsland, 1966, Studies on the microtubules in _Heliozoa_. III. A pressure analysis of the role of these structures in the formation and maintenance of the axopodia of _Actinosphaerium nucleofilum_ (Barrett), J. Cell Biol., 29, 77-95.

Tilney, L. G. and K. R. Porter., 1967, Studies on the microtubules in _Heliozoa_. II. The effect of low temperature on these structures on the formation and maintenance of the axopodia, J. Cell Biol., 34, 327-343.

Tucker, J. B., 1974, Microtubule arms and Cytoplasmic streaming and microtubule bending and stretching of intertubule links in the feeding tentacle of the suctorian ciliate _Tokophrya_, J. Cell Biol., 62, 424-437.

Weisenberg, R. C. and A. Rosenfeld, 1975, In vitro polymerization of microtubules into asters and spindles of surf clam eggs, J. Cell Biol., 64, 146-158.

ACKNOWLEDGEMENTS

The authors would like to thank Dr. P. McNeill of the University of Oregon for constructive criticism, and to C. Rieder sharing with us his unpublished results.

The work was partly supported by NIH Research Grant, GM21741 (A. Bajer). The John Simon Guggenheim Fellowship (1975) and the CRNS appointment (1974/75), (A. Bajer) were instrumental in carrying out this joint work.

EM micrograph Pl. IIa was made with cooperation of Dr. H. T. Bonnett of the University of Oregon. Plate IIIa was made with cooperation of Dr. H. Ris of the University of Wisconsin (Madison).

NOTE:

Experiments on the effects of low temperature (with cooperation of Dr. H. Sato of the University of Pennsylvania at Philadelphia) now in progress indicate that: 1. Birefringence is rapidly reduced if temperatures are lowered to $+4^{o}C$ to $+6^{o}C$, but k-fibres do not disappear completely, 2. Chromosomes continue to move at a temperature of $+4^{o}C$, but at a slower rate, 3. Birefringence increases during recovery at room temperature ($22^{o}C$) in 20 to 30 sec., 4. It is concluded, therefore, that precise monitoring of the temperature is necessary during fixation procedure. Rapid increase of the birefringence suggests that even several seconds increase of the temperature before the cell is killed, may result in pronounced differences in the number and distribution of the MTs.

# APPLIED ASPECTS OF MICROTUBULE FUNCTION

*Microtubules and Microtubule Inhibitors, eds. M. Borgers and M. de Brabander*
© 1975, North-Holland Publishing Company - Amsterdam 

# MICROTUBULES AND ANTIMITOTIC SUBSTANCES.

Guy Deysson

Laboratoire de Biologie cellulaire,

Faculté des Sciences pharmaceutiques et biologiques

4, Avenue de l'Observatoire - 75270 Paris Cedex 06 - France

Summary.

This lecture is divided into 3 parts.

The first one is a general survey of mitoclasic agents which have been intensively studied since the rediscovery of the mitoclasic action of colchicine by A. Dustin and his team (1937). The characteristics of stathmokinesis or c-mitosis and of its variations are described and the main categories of mitoclasic agents are reviewed. Finally, conclusions obtained from a comparative study of the antimitotic properties of various compounds on animal cells and plant cells are summarized.

The second part describes the results of personal ultrastructural studies carried out on plant cells. They include researches on the normal behaviour of microtubules in these cells, on the development of c-mitosis, on macrotubules resulting from a prolonged colchicine treatment, on some abnormal anaphases (double star anaphase and " floating poles " anaphase) and on the influence on microtubules of cytokinesis inhibitors and of cycloheximide.

The third and last part deals with the action mechanism of mitoclasic compounds, especially colchicine. Experiments carried out allow to define the active part of the molecule and to put forward the hypothesis of an allosteric effect on the tubulin subunit.

## 1. Mitoclasic agents

From the end of the $19^{th}$ century, the attention of certain researchers was drawn by the abnormal appearance of mitotic figures brought about by chemical substances. Thus, as early as 1889, Pernice [1], observed the unusual state of the mitoses in the intestine of a dog with colchicine poisoning. Studies of the effect of narcotics (O. Hertwig and R. Hertwig [2]) and of chloral hydrate (Von Wasielewski [3]) on mitosis date from the same period. However, little attention was paid to these results and it was only after the rediscovery of the antimitotic action of colchicine by Albert Dustin and his team [4] in 1937 that the period of intensive research on antimitotic agents began.

Amongst these agents, those which first drew the attention of cytologists were those which disturb the spindle, causing the apparition of spectacular mitotic patterns, stathmokinesis or c-mitosis. To summarize, the characteristics of these mitoses are as follows : a normal prophase is followed by stathmometaphase (c-metaphase) with chromosomes split into chromatids. These chromatids shorten progressively but remain attached by the region adjacent to the kinetochore " in characteristic cross-shaped pairs, often referred to as c-pairs " (Levan [5]).

Chromosomes are scattered over the cell, generally at the surface of a spherical or ovoid zone corresponding to the spindle but showing no signs of orientation : this is the pseudo-spindle. Stathmometaphase is longer than normal metaphase. The next stage is stathmoanaphase (c-anaphase) which is characterized by the complete separation of the two adjacent chromatids of each chromosome, thus forming the characteristic " ski-pair " pattern. Finally in stathmotelophase (c-telophase) the 4 n chromosomes form a single nucleus of irregular shape.

Different variations of this stathmokinesis have been described, in which the chromosomes may be scattered (exploded type) or, conversely, closely grouped in the centre of the cell (clumped type or ball metaphase). Finally, the chromosomes may form a star, their kinetochores meeting at the same point (star metaphase).

Besides colchicine, whose action mechanism will be discussed in greater detail later, mitoclasic properties were observed in a great number of substances, so varied that it is useless to try to give a general classification. Mitoclasic action is always coupled with a diminution (to a varying degree) in the number of cells entering mitosis and not all mitoclasic compounds can render the spindle completely inactive; this led Simonet and Guinochet [6] to distinguish between the colchicine effect and the p-dichlorobenzene effect, although this distinction was soon shown to be too absolute (Gavaudan [7]).

Without wishing to develop on the theme of these results, already out-dated by the discovery of microtubules, here are the main categories of mitoclasic agents : (more details being available from the following papers : Mangenot [8], Ostergren [9], Biesele [10], Deysson [11], etc.).

a. Mineral and Organomineral derivatives. In 1936, Ludford [12] demonstrated the blocking of metaphase mitosis by the absence of spindle due to the presence of arsenic in the cultures of cancer cells in mice. Dustin [13] subsequently showed the similarities between abnormal mitoses caused by sodium cacodylate and those induced by colchicine. Later works supplemented these early data and mitoclasic properties were also observed by numerous authors in mercury, copper, nickel and lead salts etc. These properties can usually be explained by the sulphydryl reaction of these metallic ions.

b. Organoaliphatic compounds. As already stated [3], chloral hydrate was one of the first antimitotic agents to be studied. Many authors resumed the study of the effects of chloral hydrate. All noted that the threshold of mitoclasic effect of this substance was very close to the threshold of toxicity. It is interesting

---

Fig. 1. Aspects of c-mitosis in plant and animal cells. (A) dispersed c-metaphase (Allium sativum root cell). (B) star metaphase (Allium cell). (C) highly polyploid c-metaphase (Allium cell). (D) dispersed c-metaphase (He La cell). (E) ball c-metaphase (He La cell).

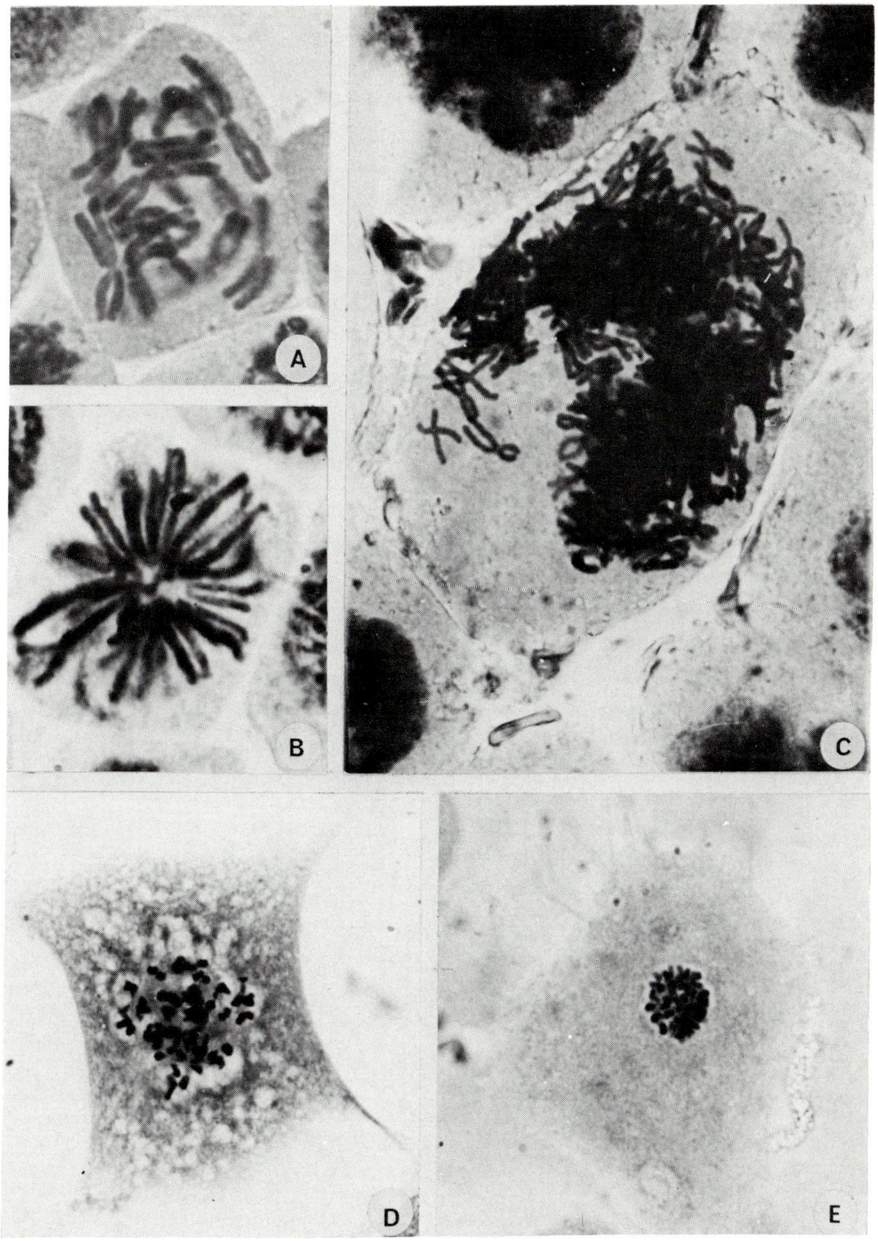
A
B
C
D
E

to note that chloralose, which is clearly mitodepressive as well as being rather toxic, has practically no mitoclasic effect (M. Deysson [14]).

Ether and chloroform were also the subject of many early studies (Gavaudan [15], Deysson [16]). It may be concluded that the mitoclasic effect of these substances is always accompanied by an intense mitodepressive effect and, in the case of chloroform, the complete mitoclasic effect is only obtained in conditions which are almost lethal.

c. Aryl organic compounds. The list of these compounds is extremely long, ranging from hydrocarbures like benzene or acenaphthene to the most varied substituted compounds, chlorinated, nitro and phenol compounds. One of the most active compounds worth mentioning is hexanitrodiphenylamine, the ammonium salt of which is a stain called aurantia (Faworsky [17], Gavaudan [15], Deysson and Khessaïri [18]) and trifluraline or trifluoro 2-6-dinitro-N-N-dipropyl p-toluidine (Lignowski and Scott [19], Delcourt and Deysson [20]).

d. Alkaloids. Many alkaloids have been demonstrated as having mitoclasic properties but the most noteworthy are vinblastine and vincristine, the Vinca rosea alkaloids, which have proved most promising in the treatment of cancerous diseases (Beer [21], Cutts, Beer and Noble [22]). The cytological observations of different authors (Cardinali et al.[23], Siebs [24], Sentein [25]) led to the comparison of their effects with that of colchicine. In the Allium test (Deysson and Truhaut [26]) the required concentrations are about the same but there is a distinct difference : the mitodepressive effect, which accompanies the mitoclasic

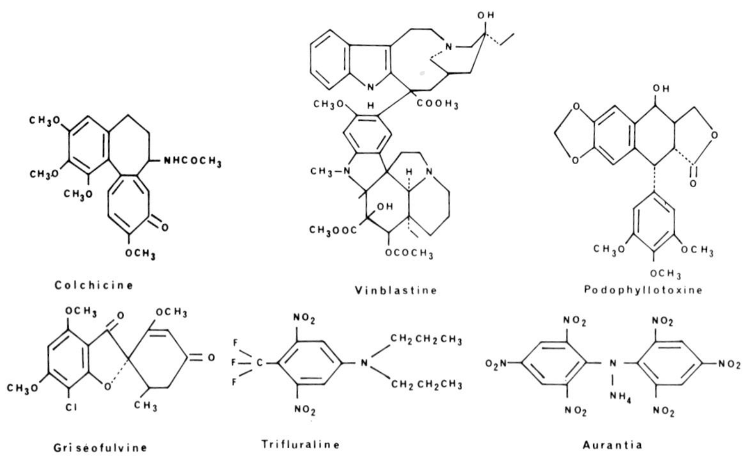


Fig. 2. Chemical structure of several c-mitotic substances.

effect is much more marked with vinblastine. This compound never brings about substantial increases in the mitotic index as colchicine does, for whenever the mitoclasic effect is considerable, it is closely followed by a marked decrease in the number of cells entering prophase. Furthermore, when treatment is stopped after a mitoclasic effect, there follows a mitodepressive period, or even a complete disappearance of mitosis, which is never observed after a colchicine treatment.

e. Podophyllin. The active constituents of Podophyllum peltatum resin (podophyllin, α and β -peltatins) used as a remedy for benign papillomas of the skin also possess mitoclasic properties similar to those af colchicine (cf. Kelly and Hartwell [27], Biesele [10], Maturova et al.[28], etc.).

f. Griseofulvin. The antimitotic activity of this antibiotic was observed by Paget and Walpole [29] simultaneously in animal and plant cells. In a comparative study [30], I noticed that it was much more active in the He La test than in the Allium test but that in both cases its action was of the colchicine type.

g. Heparin and anticoagulants. Heparin possesses mitoclasic properties which Heilbrunn [31] ascribes to its anticoagulant power. However, an accurate study (Deysson and Longevialle [32])has shown that, if many anticoagulants do have antimitotic properties, there is no relation between their antimitotic and anticoagulant action. Moreover, in the case of the heparinoid anticoagulants, the antimitotic properties seem to be related, not to the anticoagulant action, but to the clearing activity of these substances (di Marco et al.[33], Deysson [34]).

As the evaluation of mitoclasic effects has been made partly on plant and partly on animal cells, the question remains : are the results obtained in these two cases really comparable? Since 1962, simultaneous studies have often been carried out in my laboratory of the antimitotic properties of various compounds on animal cells (foetal diploid fibroblasts of human skin or lung, heteroploid cells of a cancerous He La line) and plant cells (root meristem cells of Allium sativum). From these studies, the following general conclusions may be drawn.

In the Allium test, the mitoclasic effect resulted mainly in stathmokinesis with scattered chromosomes, lasting longer than normal mitosis but resulting finally in a doubling of chromosome stock; star mitoses were also observed, although less frequently, with the chromosomes kinetochores meeting at the same point in the centre of the cell. A moderate mitoclasic effect is shown by the appearance of multipolar mitoses.

In human cells cultured in vitro, star or ball metaphases are the most frequently observed. Usually these mitoses are not completed and degenerate by pyknosis. This difference in sensitivity seems to be linked with the blocking of RNA synthesis during mitosis : the disappearance of one or more essential messengers RNA is felt more rapidly by the animal cell than by the plant. De Harven and Dustin [35] have shown that, in these star metaphases, the point of attraction of the kinetochores corresponds to the centrioles which are not separated and are placed in the centre of the cell; much less frequently, metaphases of the scattered type may be observed, as for plant cells. Finally, a moderate mitoclasic effect results in the appearance of multipolar mitoses, again like the plant cells.

From a quantitative standpoint, the minimum concentration inducing a partial and a full mitoclasic effect in the Allium test, on the He La cells and even on the diploid human fibroblasts was determined. These results are given in Table 1.

Table 1

Comparison of the effect of mitoclasic compounds on normal plant cells. (Concentrations expressed in moles per milliliter; treatment lasting 3 days except in the case of antipyrine where it lasts 24 hours).

| | Mitoclasic threshold | Full mitoclasic effect |
|---|---|---|
| Colchicine | | |
| Allium cells | $1.25\times10^{-7}$ | $2.5\times10^{-7}$ |
| He La cells | $2.5\times10^{-12}$ | $5\times10^{-12}$ |
| Diploid human fibroblasts | $5\times10^{-12}$ | $10^{-11}$ |
| Vincristine | | |
| Allium cells | $2.25\times10^{-8}$ | $4.4\times10^{-7}$ |
| He La cells | $2.5\times10^{-12}$ | $10^{-10}$ |
| Griseofulvin | | |
| Allium cells | $2.5\times10^{-8}$ | impossible |
| He La cells | $2.5\times10^{-10}$ | $2.5\times10^{-9}$ |
| Hexanitrodiphenylamine (ammonium salt) | | |
| Allium cells | $10^{-9}$ | $2.5\times10^{-9}$ |
| He La cells | $10^{-8}$ | impossible |
| Diploid human fibroblasts | $10^{-8}$ | impossible |
| Antipyrine | | |
| Allium cells | $3.9\times10^{-5}$ | |
| He La cells | $1.06\times10^{-5}$ | |

Generally, animal cells prove more sensitive than plant cells : about 100,000 times more in the case of colchicine, 10,000 times for vincristine and 100 times for griseofulvin. One exception was however, noted : hexanitrodiphenylamine or

Fig.3. Microtubules in prophase (N = nucleus, cw = cell wall). (A) Allium cell entering prophase with an equatorial ring of microtubules. (B) Part of A showing microtubules (arrows). (C) More advanced prophase : equatorial microtubules have disappeared and a perinuclear " clear zone " made of microtubules has been formed. (D) Part of C showing perinuclear microtubules (arrows).

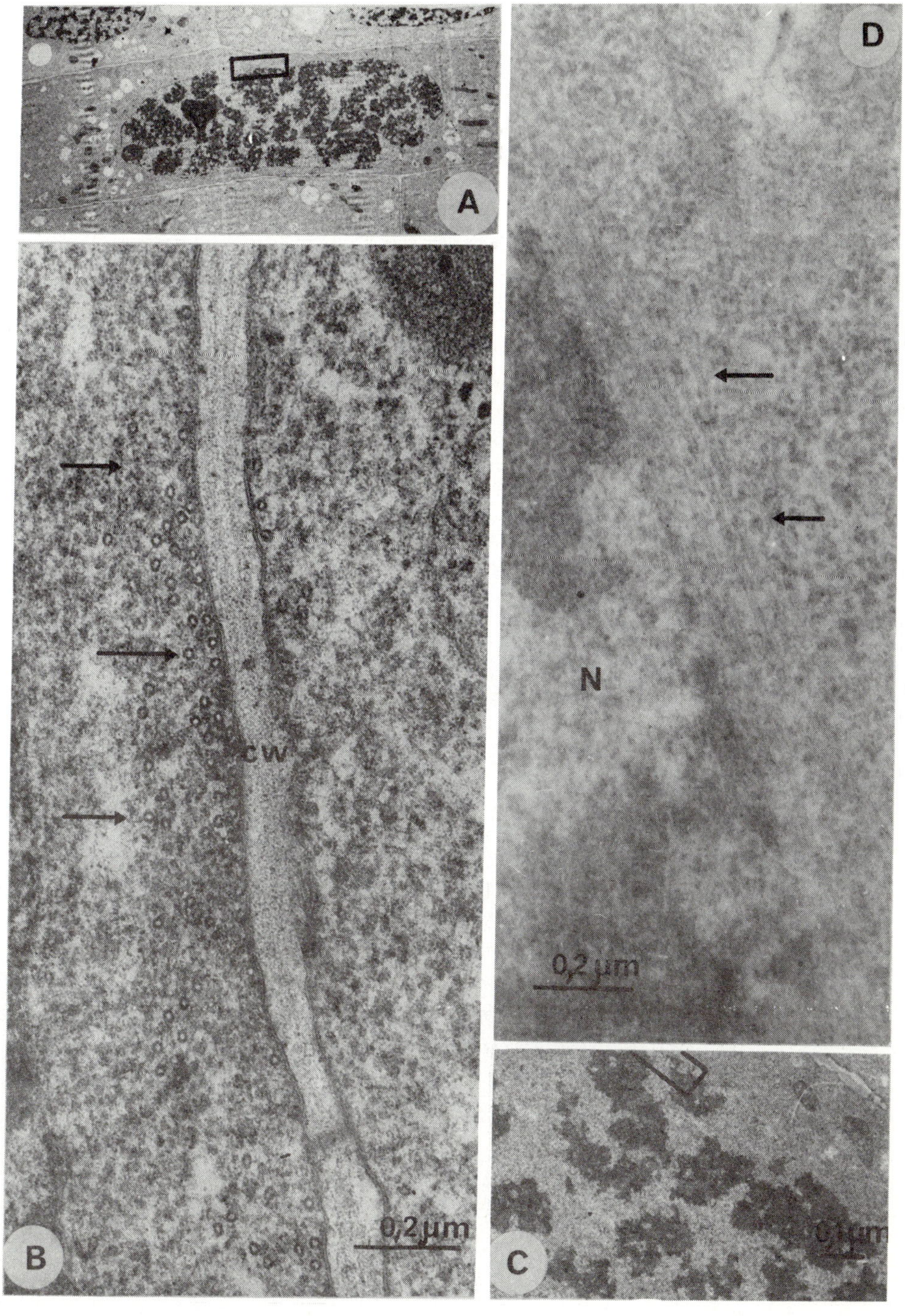
A
B
C
D
CW
N
0,2 µm
0,2 µm
0,1 µm

aurantia which affects plant cells in weaker solutions than colchicine, but human cells are about 10 times less sensitive to it than plant cells. Another exceptional case is that of antipyrine which, whilst mainly inhibiting the onset of mitosis, has, in stronger solutions, mitoclasic properties, with almost identical threshold concentrations for Allium and He La cells.

2. Ultrastructural study : microtubules

Since 1963, when Ledbetter and Porter [36] first called attention to these new structures, microtubules have been observed in different locations in interphasic cells. When the cells divide, the spindle fibres seem to be made up of microtubules resembling the others. It has been observed, both in plant and animal cells, that all the microtubules are now in the spindle.

Since all our ultrastructural studies have been carried out on plant cells, I shall first describe the characteristics of these microtubules during the plant cell cycle before going on to the observations of antimitotic effects.

a. Microtubules of root meristem cells. In the interphasic cells, they form a frame just inside the ectoplasmic membrane but, whereas they are in line along the side walls (which are to become elongated), they have no semblance of order at the top and bottom. Pickett-Heaps and Northcote [37], then Burgess and Northcote [38], studied the behaviour of microtubules during mitosis and described the temporary formation, during preprophase, of an equatorial ring of microtubules which disappeared during prophase, whilst others appeared deeper in the cytoplasm and perpendicular to the first set.

In 1968, with M-C. Benbadis [39], we studied the setting up of the spindle apparatus in four new species. At the onset of prophase whereas the microtubules in the region of the ectoplasmic membrane disappear almost completely, a group of similarly orientated microtubules appears in the equatorial region of the cell. The number of microtubules observed in these groups, corresponding to the ring described by Pickett-Heaps and Northcote, varies from 46 to 114. It is difficult to ascertain at what precise moment this ring is formed. The grouping of microtubules at the equator of cells was sometimes seen to begin whilst the nucleus was still clearly in interphase. However, whenever there were a lot of microtubules grouped in more than two lines, the nuclei already showed signs of having entered prophase from the appearance of the chromatine and the nucleoli.

---

Fig. 4. Microtubules in late telophase. (A) Allium cell in late telophase. (B)(C)(D) Part of A showing the setting up of microtubules of the interphase cell, which reappear deep in the cytoplasm (arrows).

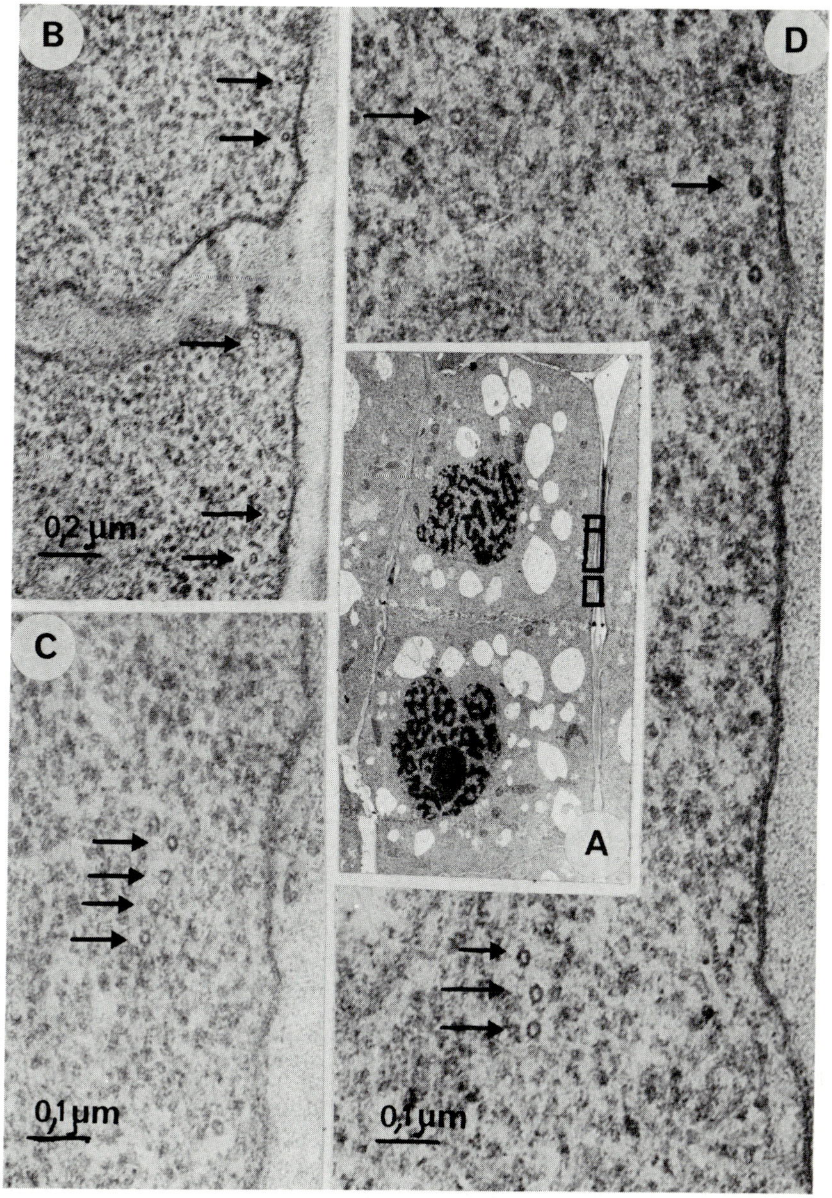
B
D
0,2 µm
C
A
0,1 µm
0,1 µm

At the onset of prophase, the cell contained only the microtubules making up the ring, whose position seems closely linked to the ulterior position of the spindle. When prophase is under way, there appears, in the region of the nuclear membrane, a ring of microtubules, which becomes more and more numerous.

These microtubules are perpendicular to the others, which disappear, corresponding to the Bajer's clear zone (Deysson and Benbadis [39]) and insuring the formation of the continuous spindle fibres when the nuclear membrane disappears. It should be noted that, while the nuclear membrane has not completely disappeared, there is no trace of microtubules in the nucleus. In our paper, published in 1968, we showed, with the aid of a photograph, that it was likely that the same microtubules gave rise to the two successive groups by migration. The possibility of such a migration was not then accepted but Pickett-Heaps [40] came to the same conclusion in 1969.

At the end of mitosis, the microtubules of the interphase cells reappear deep in the cytoplasm then move to the cell wall. This formation appears when the microtubules of the cytokinesis spindle have completely disappeared and the chromatin has resumed its interphase appearance (Deysson and Benbadis [41]).

b. Ultrastructural study of colchicine stathmokinesis. In 1964, Robbins and Gonatas [42] discovered the absence of microtubules in He La cells treated with colchicine, and Pickett-Heaps [43] made a detailed ultrastructural study of the effect of colchicine on wheat cells. With concentrations of up to 0.3%, that is, far higher than the threshold of mitoclasic effect, he observed an almost complete lack of microtubules in cells which were at rest or in mitosis, whilst other cellular organelles showed no change. The changes in endoplasmic reticulum and in mitochondria described by Mesquita [44] and by Rodkiewicz and Mikulska [45] seem to be independent of the mitoclasic action, having been observed only after the use of high concentrations.

The development of stathmokinesis induced by hexanitrodiphenylamine in the root meristem cells of Allium sativum L. has been described (Deysson, Benbadis and Khessaīri [46]) as follows.

Interphase cells do not usually have ectoplasmic microtubules, which are found in normal cells. However, occasionally, some may remain, although no explanation can be found for this resistance.

---

Fig. 5. C-mitosis induced by hexanitrodiphenylamine ($2.5\times10^{-9}$ mol/ml. 24 hrs) in Allium cells. (A)(B)(D) c-metaphases : mitochondria, proplastids and vacuoles are pushed away at the periphery and do not penetrate into the pseudo-spindle. (C) Part of B showing some short microtubules attached to a kinetochore and orientated in anyone direction. (E) Part of D. id.

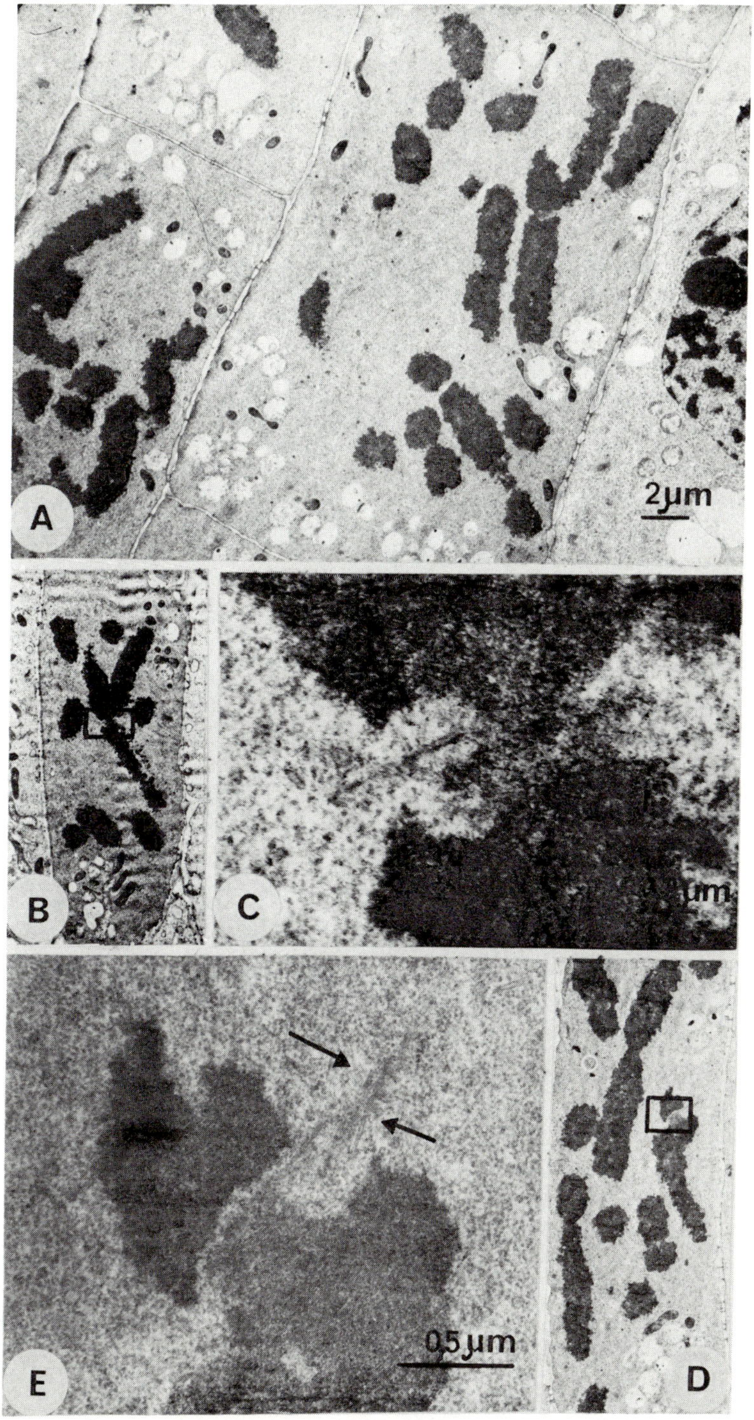
A
2µm
B
C
µm
E
0.5µm
D

At the onset of prophase, the equatorial ring of microtubules, which usually appears, and the perinuclear microtubules are absent. Moreover, prophases continue normally, which confirms that the role of the equatorial ring is linked solely with the formation of the spindle.

In the stathmometaphase and stathmoanaphase patterns, the pseudo-spindle is easily identified, with the chromosomes seemingly scattered irregularly over it. This region, which is devoid of microtubules, contains many ribosomes and sometimes some dictyosomes but the bulkier organelles such as mitochondria and proplasts are absent, as are vacuoles. In most stathmokineses, there are no microtubules. However, a few short ones have sometimes been observed attached to the kinetochores. All are orientated in any one direction whereas, in normal mitosis, they are always directed towards the poles. The occasional presence of these microtubules shows that normally the kinetochores are indeed where microtubules are formed and not only a meeting point for existing microtubules.

Finally, during stathmotelophase, the space corresponding to the pseudospindle is gradually invaded by cytoplasmic organelles. At no time, however, have the structure, quantity and disposition of cytoplasmic organelles been found to be abnormal.

c. Prolonged effect of colchicine : macrotubules. As has already been stated, a prolonged colchicine treatment does not prevent the onset of prophase in plant cells and the dividing cells continue with mitosis, free of microtubules. These c-mitoses are characterised by a complete mitotic depolarisation and the blocking of cytokinesis, and culminate in the formation of a highly polyploid nucleus. An electron microscope study of such cells (Deysson and Benbadis [47]), showed, after a three-day treatment with a 0.02% solution of colchicine, that abnormal tubular structures appeared in the polyploid cells of the subterminal meristematic zone, whereas these cells remained free of any spindle or cytoplasmic tubules. In the cytoplasm, large flexible tubules, of indeterminate length and approximately 600 Å in diameter, were observed. These macrotubules are usually in bundles or tangled clumps without any special orientation but are rarely found singly. They are made up of a thick membrane in which no subunits comparable to these found in microtubules are discernible. As opposed to normal microtubules which occur singly and are often surrounded by a clear halo, macrotubules may be joined. When a cytoplasmic zone is invaded by macrotubules, there are no organelles except ribosomes. Moreover, it has been observed that macrotubules are often found either near the ectoplasmic membrane or near the nuclear membrane, with which they may be closely linked.

Fig.6. Allium cell after prolonged colchicine treatment (0.02% - 3 days).
(A) General view of a cell in which 2 clumps of macrotubules are seen. (B)(C) Macrotubules arranged in bundles without any special orientation. (D) Normal microtubules at the same magnification.

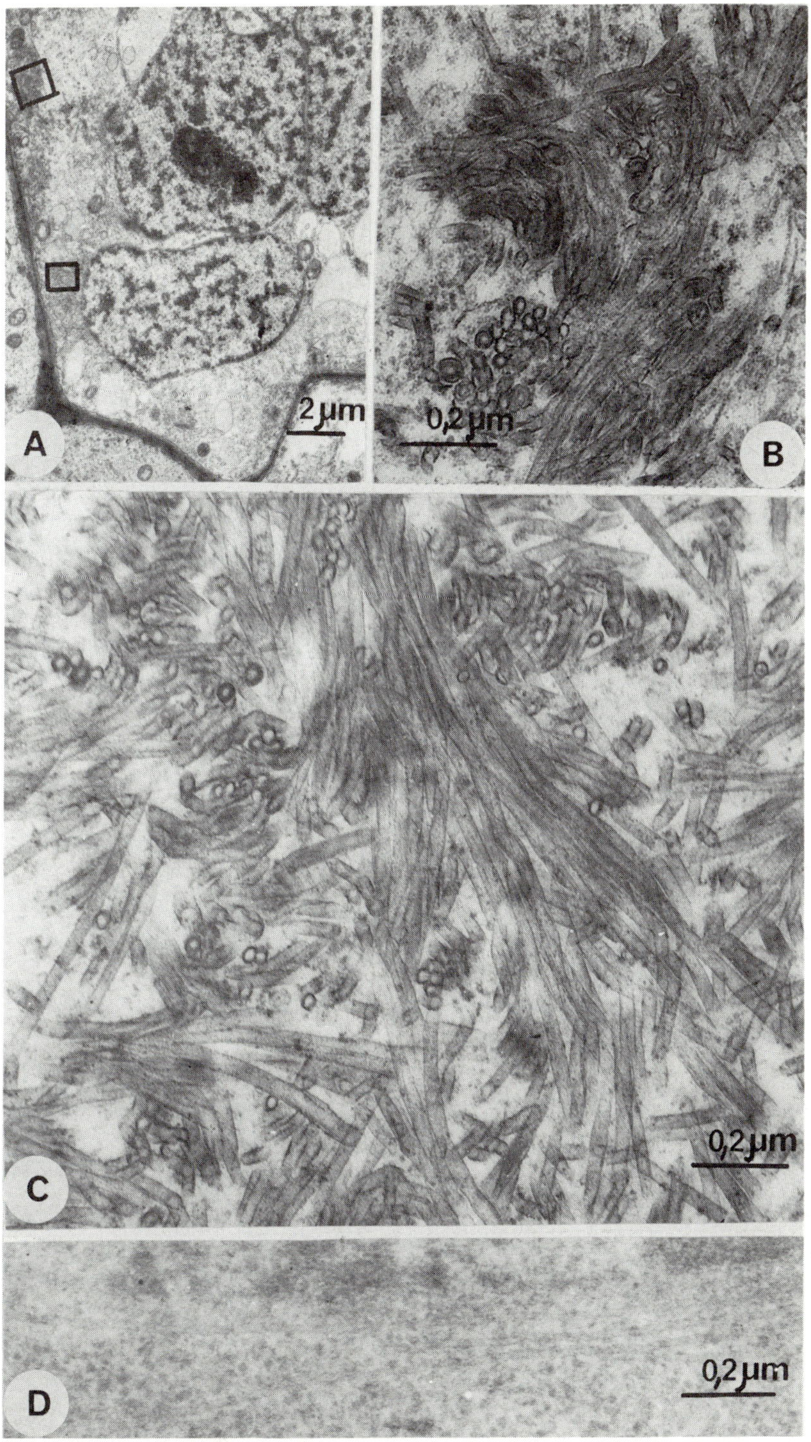

A
2 µm
B
0,2 µm
C
0,2 µm
D
0,2 µm

Macrotubules can only be observed after aldehyde fixation as osmic fixation alone does not conserve their tubular structure and this suggests that they may be made of protein. Moreover, a partial extraction of this abnormal structure has been obtained with 0.5% pronase for one hour.

Experiments have been carried out on cells returned to their normal medium after a three-day treatment of colchicine. This showed that normal microtubules resume their shape 12 to 18 hours after the end of the treatment whereas macrotubules disappear.

These macrotubules resemble these described by Hanzely and Olah [48] in c-mitosis induced by digitonine treatment of plant meristems. They may thus be thought to be the result of the abnormal polymerisation of protein monomeres which normally become microtubules.

d. Aster appearance in mitosis of higher plants : double star anaphases. The anastral structure of the mitotic spindle in higher forms of plant life has often been contrasted with the amphiastral structure of the mitotic spindle in animals. This difference is equated with the presence of centrioles in animal cells and their absence in higher plants.

However, in 1953, Ostergren et al. [49] described in Allium cepa L., during amidopyrine treatment, the formation of double star anaphases and found that the examination of the slides, with phase contrast microscope, shows that chromosomes are localized in asters existing at the poles, at that stage. According to these authors, less obvious asters are formed temporarily in the course of normal mitosis, during the passage from anaphase to telophase but they admit the objection that this may be the result of a fixation artefact. In our laboratory, we have succeeded several times, in the course of experiments, in inducing the formation of double star anaphases using different compounds like antipyrine or tubulosine which are mitoclasic in large quantities. However, even under favourable conditions, these figures remain quite rare, which explains why they have never been studied under the electron microscope. This study was undertaken on meristematic cells which had previously been synchronised using hydroxyurea which permits an accumulation of these patterns. We were thus able to prove the existence of microtubules radiating from all the poles of the star anaphases we examined [50]. This led us to try to find the outline of asters at the onset of telophase, as described by Ostergren et al.. Although microtubules are less frequent then, than in the poles of double star anaphase, their existence is beyond doubt.

---

Fig. 7. Double star anaphase (Allium cell - Tubulosine 0.047% 2 hrs 30 mn). (A) General view. (B) Part of A : some microtubules are outlined. (C) View of the central part of an aster, showing microtubules radiating from the pole.

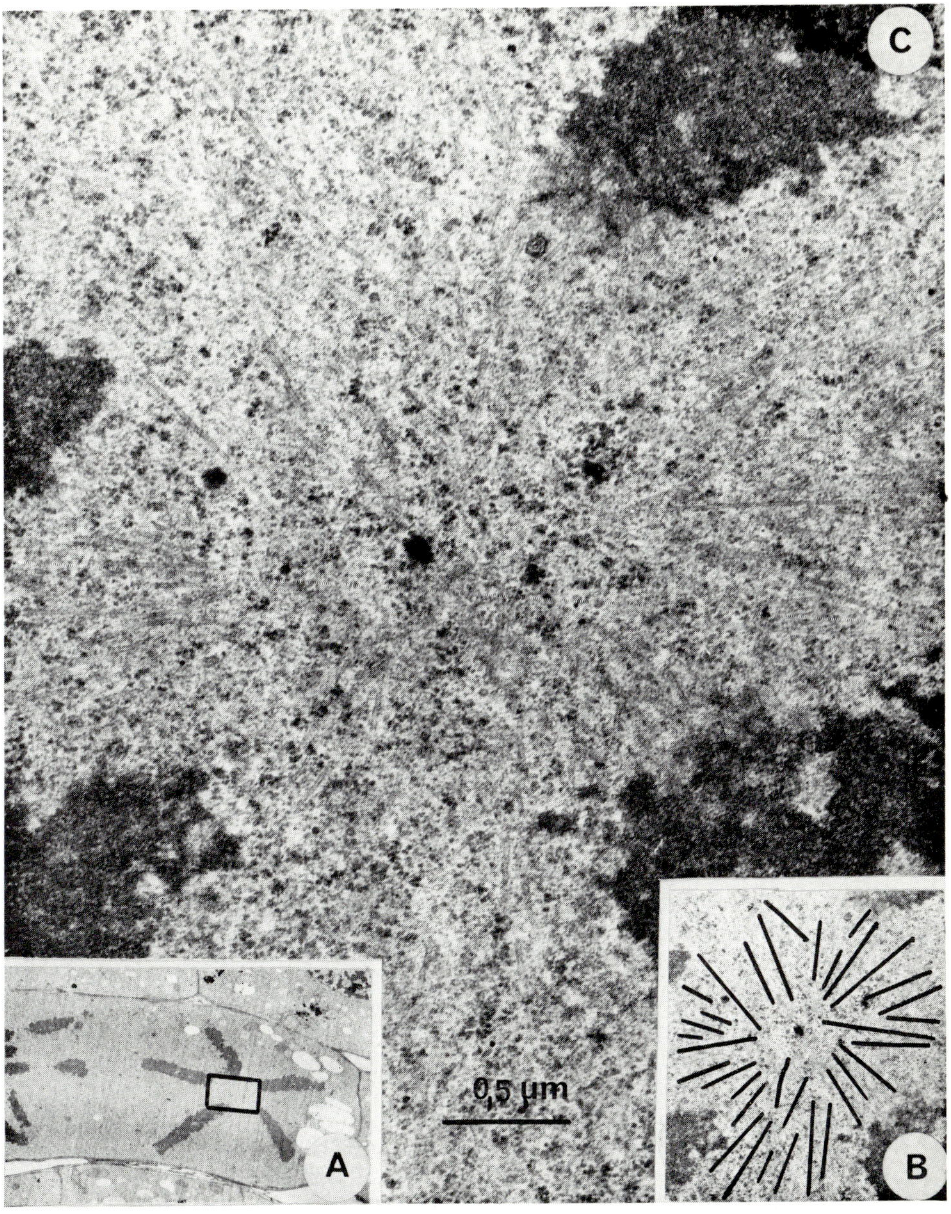
C
A
B
0,5 µm

Thus, the use of mitoclasic compounds having only a partial effect allowed us, by slowing down the process of anaphase, to prove that the double star anaphases are the result of the formation of asters which under normal conditions only appear at the beginning of telophase, and then only temporarily.

e. Destruction of continuous spindle fibres : " floating poles " anaphases. Under the conditions necessary for the formation of double star anaphases, that is with a mild mitoclasic treatment, anaphases with " floating poles " also appear, that is, instead of being face to face, the two poles of the anaphase are both turned in any direction, some anaphases even having both poles turned the same way [51]. These patterns only seem comprehensible if it is accepted that spindle fibres were normal, throughout the ascension of the chromosomes to the poles and that continuous fibres were destroyed, thus freeing the poles which " float " in the cytoplasm. The ultrastructural study carried out, as before, on meristems whose cells had previously been synchronised by the use of hydroxyurea showed that, in fact, there are no fibres left between both groups of chromosomes whereas at the poles, there are microtubules attached to the kinetochores, representing the remains of chromosomic fibres. These " floating poles " anaphases give rise to binuclear cells as the spindle fibres of cytokinesis are not formed.

These " floating poles " anaphases may be the result of selective destruction of the continuous fibres by the mitoclasic compound but there is also another possible explanation : the fact that the anaphase lasts longer does not necessarily mean that the continuous fibres do, and they would thus disappear before the telophasic nuclei are formed.

f. Influence of the cytokinesis inhibitors on microtubules. A certain number of compounds, like caffeine and quinazolone derivatives which only inhibit cytokinesis, are already known. The ultrastructural research we have carried out [41] shows that this blockage is not due to the destruction of interzonal fibres (of the cytokinesis spindle). These fibres can even be found in post-telophase, later than in normal cells, but this does not prevent peripheral microtubules reappearing. In the cases studied until now, cytokinesis inhibition had no effect on microtubules.

g. Influence of prophasic inhibitors : case of cycloheximide. Some agents, like cycloheximide [52], prolong prophases which are already under way at the beginning of the treatment then bring about the reversal of these prophases to $G_2$. It has

---

Fig. 8. " Floating poles " anaphases (Allium cell - Tubulosine 0.047% 3 hrs). Arrows show the orientations of the poles.

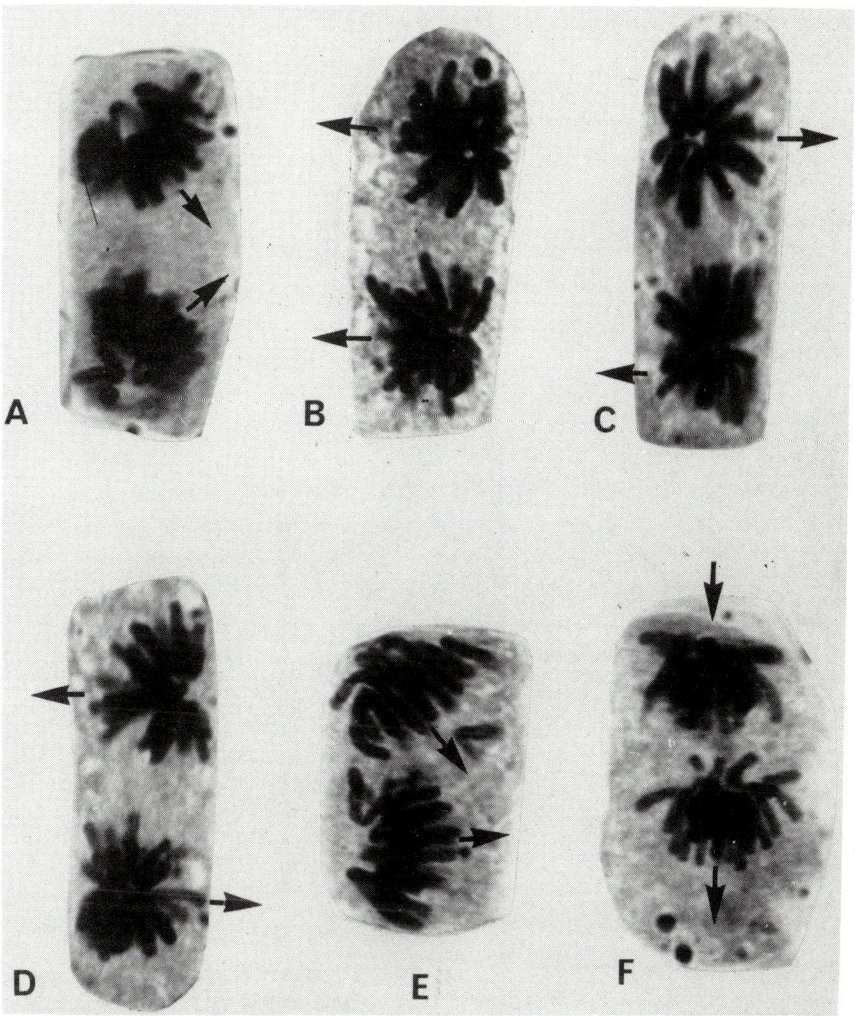
A
B
C
D
E
F

been observed that, in this case, the ring of preprophasic microtubules is formed in the normal way by peripheral microtubules. During normal mitosis, this preprophasic ring would be temporary and would disappear, whilst the microtubules would gather round the prophasic nucleus to form the " clear zone ". This second stage is hardly even begun when cycloheximide is present. Although the outlines of several small microtubules are observed, sometimes attached to the external membrane of the nuclear enveloppe, they are few in number and the microtubules of the preprophasic ring always remain simultaneously : moreover, the prophasic cells are often completely lacking in perinuclear microtubules. Later, the nuclear membrane takes on an undulating appearance but never disappears thus causing the regression of chromatin to the interphasic state.

3. Action mechanism of mitoclasic compounds

Having described the ultrastructural characteristics of the effect of different compounds on cellular microtubules, I shall now turn to the biochemical mechanism of their action. Out of all the mitoclasic compounds, very few have been studied from the point of view of the mechanism of their effect. Here, I am concerned mainly with colchicine.

An accurate comparison of the mitoclasic effect of 15 colchicine derivatives led me to conclude, in 1960, that substitutions on carbons 14 and 15 are determining for the mitoclasic properties and to suggest that it is probably in this region that the colchicine molecule becomes attached to a cell componant [53]. The only compound which is totally lacking in mitoclasic effect out of all those studied is aminocolchimidazole in which a molecule of guanidine is attached to carbons 14 and 15. My supposition also took into account the previous findings of Steinegger and Levan [54,55] who showed that isocolchicine (inversion of carbons 14 and 15 substitutions) is about 100 times less effective than colchicine and that colchiceine (demethylation of C14) is almost ineffective. In 1965, Schindler [56], who did not seem to know of my results, was led to conclude that the presence of a carbonyl group at C15 is of great importance, for several derivatives, in which the carbonyl group was eliminated or shifted to C13 or C14, were found to be ineffective. Later, Wilson et al.[57] found that lumicolchicine derivatives, in which the change is a conversion of ring C of colchicine to two

---

Fig. 9. Incomplete blockage of cytokinesis in Allium cell (Caffeine $2.1\times10^{-6}$ mol/ml 24 hrs). (A) General view. (B)(C)(D) Part of A. One can see that microtubules of the cytokinesis spindle are still somewhat numerous (D) and that this does not prevent peripheral microtubules reappearing (B and C).

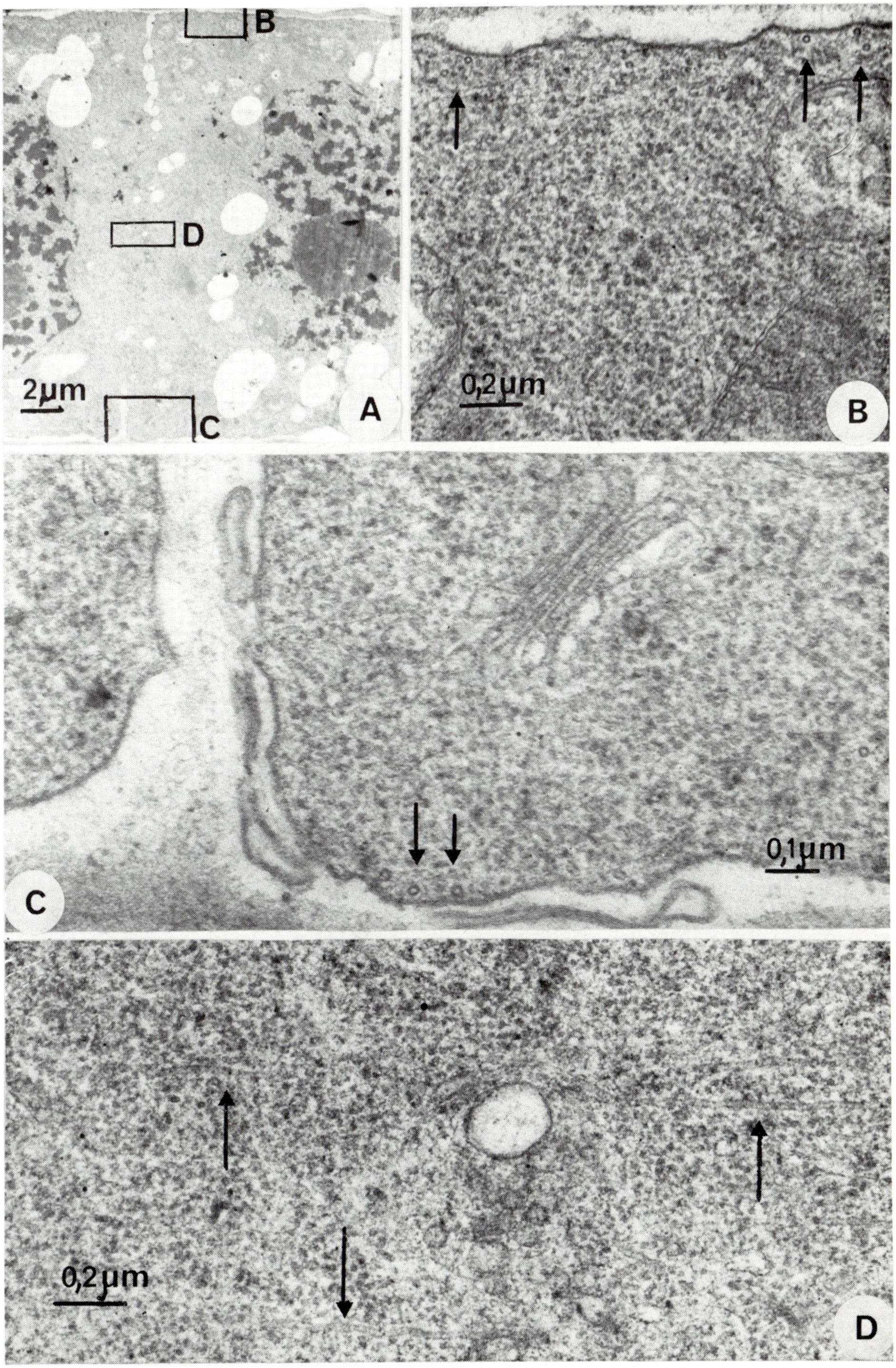
B
D
2µm
C
A
0,2µm
B
0,1µm
C
0,2µm
D

smaller rings, are inactive as antimitotic agents. Hence, the integrity of the 7-membered ring C of colchicine appears to be of critical importance for the ability of colchicine to disrupt microtubules. Schindler and Wilson's results confirm the conclusions I drew in 1960 as to the role of C14 and C15. I also observed in 1952 that when pea seedlings have absorbed colchicine, this substance cannot be identified by its specific nitric reaction, which means that colchicine does not remain in its purest form in the cells [58]. From 1965, Taylor, Borisy, Shelanski, Wilson, etc... showed, throught detailled research, that colchicine becomes attached to tubulin thus preventing the formation of dimeres which lead to the formation of microtubules. Since these results have just been explained by Wilson, I do not need to discuss them any further.

Whilst carrying out this research, I also studied one aspect of the mechanism of the colchicine effect, i.e. the influence of different compounds on this effect. Here are some of the results I obtained, as they supplement the findings of the biochemists.

The works of Rapkine [59], Stern [60] and Sakai [61] having shown the importance of SH-groups in the formation of spindle fibres, and also of the compounds blocking these groups such as lead nitrate, phenylmercury hydroxide or N-ethylmaleimide, which have mitoclasic properties, I studied the influence of these compounds on the mitoclasic effect of colchicine (Deysson and Delcourt [62]). In each case, the addition of a compound affecting SH-groups causes, at the most, a slight strengthening of the colchicine effect, that is to say this is simply a phenomenon of addition and not of potentiation. This means that either the mechanisms of the effect of these two compounds are identical, or, on the contrary, there is no relation between them. Since colchicine is not a reactive of SH-groups, this leads us to the conclusion that the mitoclasic action of these two types of compounds must come from two separate independent metabolic sources.

Griseofulvin, on the other hand, which has mitoclasic properties, potentiates the effect of colchicine (Deysson [63]) which shows that these two compounds have different mechanisms but these may be complementary. Moreover, Wilson has recently shown that griseofulvin does not prevent the formation of microtubules as colchicine does, but the way in which it deters the movement of chromosomes still remains obscure.

---

Fig. 10. Allium cell. Cycloheximide 0.001% 6 hrs. (A) Cell entering prophase, with an abnormal nucleolus (nu) and an equatorial ring of microtubules (mt). (B) Part of A, showing microtubules. (C) Prophase : a few microtubules are seen near the nuclear envelope. (D) Prophase regressing to the interphasic state. The nuclear envelope is undulating but does not break. Abnormal nucleolus (nu) with segregation of granules (g) and fibrilles (f).

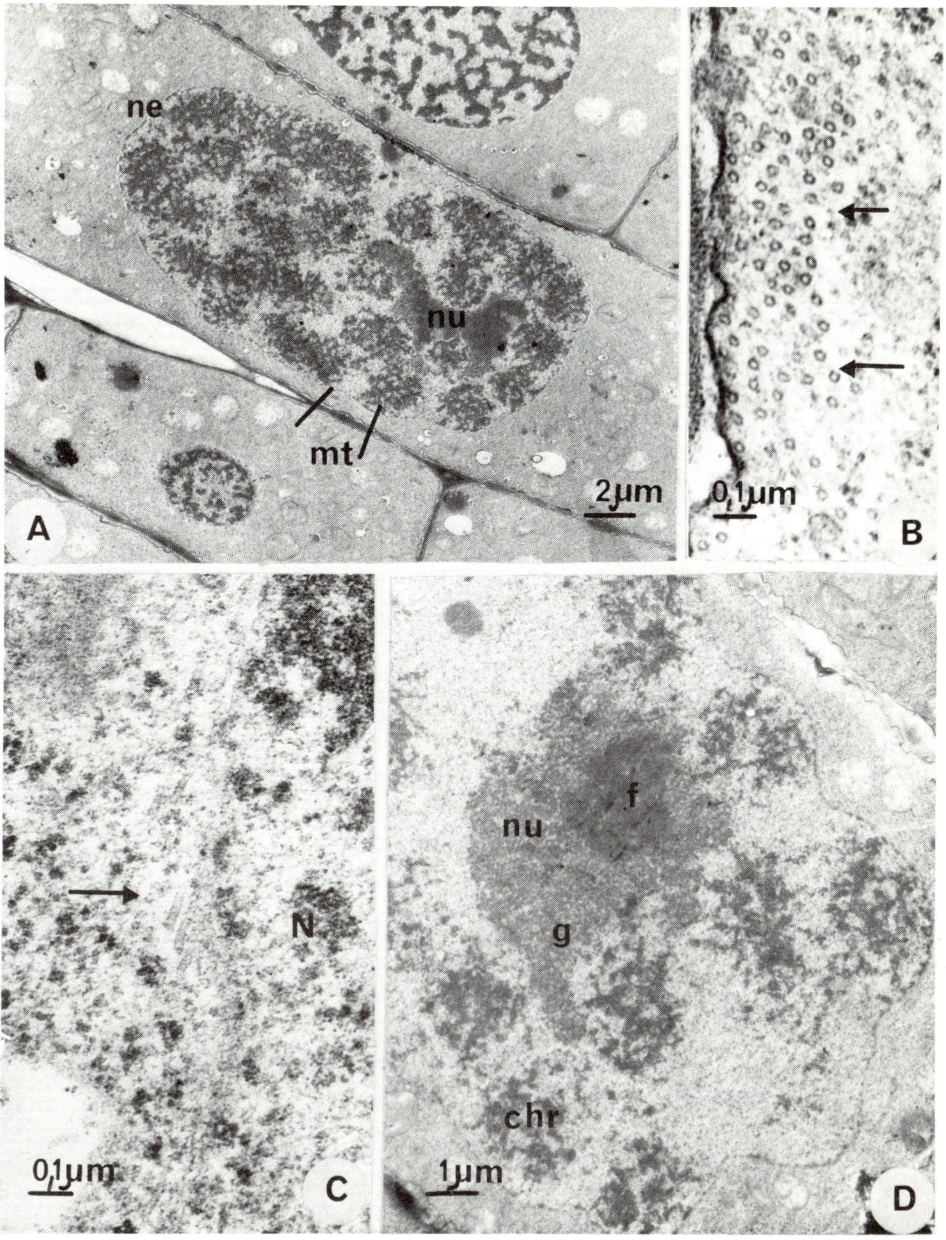
ne
nu
mt
2µm
A
0,1µm
B
N
0,1µm
C
f
nu
g
chr
1µm
D

In the course of research carried out in 1965 (Deysson and Delage [64]), we tried to modify the effect of colchicine with natural extracts and we found that an aqueous extract of yeast totally prevents colchicine from exercising its mitoclasic properties on plant cells. The yeast extract must, however, be administered before colchicine for its effect is only preventive, not curative.

We then showed that the protective properties of the yeast were due to the presence of both nicotinamide and riboflavin. Nicotinamide alone already had protective properties but mainly in higher concentrations. This substance is known to form part of the dehydrogenase coenzymes NAD and NADP, which also have distinct protective properties, although they are no stronger than those of nicotinamide, if used in oxydized form. Conversely, 6-aminonicotinamide which is structurally similar to nicotinamide and is moreover mitoclasic in strong doses, considerably strengthens the mitoclasic effect of colchicine (Deysson and Delcourt [65]). This synergism, of a potentiative nature, seems to indicate that the mechanisms of the mitoclasic effect of colchicine and 6-aminonicotinamide are not identical but take place in two successive stages.

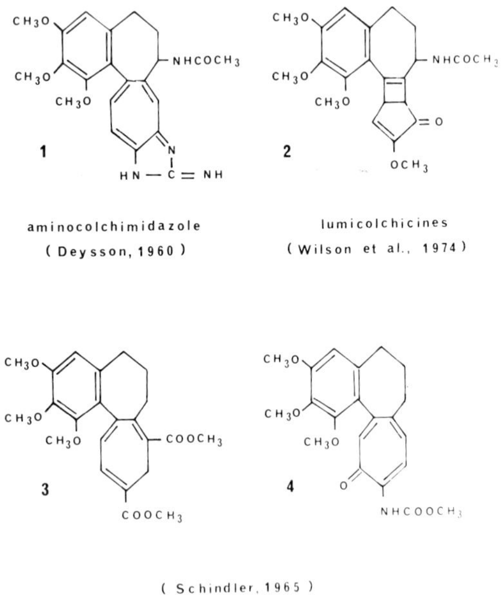


Fig. 11. Colchicine derivatives devoid of mitoclasic effectiveness.

The nature of the link between colchicine and tubulin remains unknown. Taylor [66] suggests two hypotheses : the fixing of colchicine on " a protein monomer unit at one such site that it could prevent or reduce greatly the interaction with a second subunit " or the fixing of colchicine in another site which would lead, by an allosteric reaction, to a change in the configuration of the protein in a way that reduces the protein-protein interaction. To explain the facts I have just discussed, I put forward the following hypothesis : the fixing of colchicine to tubulin would modify, by an allosteric reaction, the site where the hydrogen carrier should be attached in order to allow the oxydation reaction of the SH-group indispensable to the formation of dimeres to take place. This fixing of the hydrogen carrier would become difficult but not totally impossible. It would become possible in the presence of an excess of carrier (the case of the addition of nicotinamide or yeast extract. It would however be completely impossible with a physiologically modified carrier as in the case of the addition of 6-aminonicotinamide.

4. Conclusions

If so much research has been done on mitoclasic agents, it is pricipally because of their possible contribution to tumour therapy, and in this respect, vinblastine is one of the compounds which have proved most useful. However, as Dustin [67] says " mitotic inhibitors are more than possible drugs for cancer chemotherapy; they are tools which enable the cytologist to dissect the interrelated phases of the mitotic cycle ". From this point of view, the study of the effect of mitoclasic compounds has proved most fruitful, as, after having helped us to understand more clearly the role of the spindle in mitosis, it has helped us to progress in our study of the structure of microtubules.

Finally, it seems important to point out that the practical value of mitoclasic compounds is not restricted to their effect on mitosis. It is well known that microtubules have many different roles in cellular physiology, especially that they are numerous in nerve cells. The discovery of the mitoclasic properties of any given compound should therefore encourage us to study the effect of that compound on other cellular actions involving microtubules. I shall simply remind you of the hypothesis put forward by Allison and Nunn [68], that the effect of general anaesthetics on microtubules would be a possible means of anaesthetizing or the most recent observations of Haschter et al. [69] who found that the local anaesthetics lidocaine, procaine and etidocaine have a direct effect on tubulin monomers and prevent the formation of microtubules.

Références.

1. Pernice, B., 1889, Sicilia Med. 1, 265.
2. Hertwig, O. and Hertwig, R., 1887, Iena Z. Naturwiss. 20, 120.
3. Wasielewski, W. Von, 1902, Jahrb. Wiss. Botan. 38, 377.
4. Dustin, A.P., Havas, L. and Lits, F., 1937, C.R. Assoc. Anat. 32, 170.
5. Levan, A., 1938, Hereditas 24, 471.
6. Simonet, M. and Guinochet, M., 1939, C.R. Soc. Biol. 131, 222.
7. Gavaudan, P., 1943, C.R. Soc. Biol., 137, 571.
8. Mangenot, G., 1941, Rev. Cytol. Cytophysiol. veget. 5, 169.
9. Ostergren, G., 1944, Hereditas 30, 429.
10. Biesele, J.J., 1958, Mitotic Poisons and the Cancer Problem (Elsevier, Amsterdam).
11. Deysson, G., 1968, Internat. Rev. Cytol. 24, 99.
12. Ludford, R.J., 1936, Arch. Exper. Zellforsch. 18, 411.
13. Dustin, A.P., 1938, C.R. Assoc. Anat. 33, 205.
14. Deysson, M., 1954, C.R. Acad. Sci. 238, 1064.
15. Gavaudan, P., 1945, Thesis, Fac. de Pharm., Univ. de Paris.
16. Deysson, G., 1948, Thesis, Fac. de Pharm., Univ. de Paris.
17. Faworsky, M.V., 1939, C.R. Acad. Sci. URSS 25, 71.
18. Deysson, G. and Khessaïri, H., 1968, C.R. Soc. Biol. 162, 843.
19. Lignowski, E.M. and Scott, E.G., 1971, Plant and Cell Physiol. 12, 701.
20. Delcourt, A. and Deysson, G., 1975, Cytologia in the press.
21. Beer, C.T., 1955, Brit. Empire Cancer Campaign, 33 rd Ann. Rept. 487.
22. Cutts, J.H., Beer, C.T. and Noble, R.L., 1960, Cancer Res. 20, 1023.
23. Cardinali, Gius., Cardinali, Giul. and Blair, J., 1961, Cancer Res. 21, 1542.
24. Siebs, W., 1963, Z. Zellforsch. Mikroskop. Anat. 61, 231.
25. Sentein, P.,1964, Chromosoma 15, 416.
26. Deysson, G. and Truhaut, R., 1963, C.R. Acad. Sci. 257, 3512.
27. Kelly, M.G. and Hartwell, J.L., 1953, J. Nat. Cancer Inst. 14, 967.
28. Maturova, M., Malinsky, J. and Santavy, F., 1959, J. Nat. Cancer Inst. 22, 297.
29. Paget, G.E. and Walpole, A.L., 1958, Nature 182, 1320.
30. Deysson, G., 1964, Ann. Pharm. Fr. 22, 17.
31. Heilbrunn, L.V., 1956, The Dynamics of Living Protoplasm (Acad. Press, New York).
32. Deysson, G. and Longevialle, M., 1962, Biol. Med. 51, 42.
33. Marco, A. di, Necco, A. and Castegnaro, E., 1959, Biochem. Pharmacol.1, 179.
34. Deysson, G., 1965, C.R. Soc. Biol. 159, 125.
35. Harven, E. de and Dustin, P., 1960, Colloq. Intern. Centre Nat. Rech. Sci., Paris 88, 189.
36. Ledbetter, M.C. and Porter, K.R., 1963, J. Cell Biol. 19, 239.

37. Pickett-Heaps, J.D. and Northcote, D.H., 1966, J. Cell Sci. 1, 109.

38. Burgess, J. and Northcote, D.H., 1967, Planta 75, 319.

39. Deysson, G. and Benbadis, M.C., 1968, C.R. Soc. Biol. 162, 601.

40. Pickett-Heaps, J.D., 1969, J. Ultrastruct. Res. 27, 24.

41. Deysson, G. and Benbadis, M.C., 1969, C.R. Soc. Biol. 163, 1083.

42. Robbins, E. and Gonatas, N.K., 1964, J. Histochem. Cytochem. 12, 704.

43. Pickett-Heaps, J.D., 1967, Developmental Biol. 15, 206.

44. Mesquita, J.F., 1966, C.R. Acad. Sci. 263, 1827.

45. Rodkiewicz, B. and Mikulska, E., 1967, Bull. Acad. Polon. Sci. Cl. V, 15, 561.

46. Deysson, G., Benbadis, M.C. and Khessaïri, H., 1968, C.R. Soc. Biol. 162, 845.

47. Deysson, G. and Benbadis, M.C., 1971, C.R. Soc. Biol., 165, 1221.

48. Hanzely, L. and Olah, L.V., 1970, J. Cell Biol. 47, 82 a.

49. Ostergren, G., Koopmans, A. and Reitalu, J., 1953, Bot. Not. 4, 417.

50. Vannereau, A., Deysson, G., Benbadis, M.C., Brulfert, A. and Pareyre, C., 1972, C.R. Acad. Sci. 275, 2477.

51. Pareyre, C., 1973, Thesis, Fac. de Pharm., Univ. de Paris.

52. Benbadis, M.C., Levy, F. and Deysson, G., 1974, C.R. Acad. Sci. 278, 1353.

53. Deysson, G., 1960, Colloq. Intern. Centre Nat. Rech. Sci., Paris 88, 213.

54. Steinegger, E. and Levan, A., 1947, Hereditas 33, 385.

55. Steinegger, E. and Levan, A., 1948, Hereditas 34, 193.

56. Schindler, R., 1965, J. Pharmacol. Exper. Therap. 149, 409.

57. Wilson, L., Bamburg, J.R., Mizel, S.B., Grisham, L.M. and Creswell, K.M., 1974, Fed. Proc. 33, 158.

58. Deysson, G., 1952, Rev. Cytol. Biol. Veget. 13, 153.

59. Rapkine, L., 1931, Ann. Physiol. Physicochim. 7, 382.

60. Stern, H., 1960, Developping cell systems and their control (Ronald Press, New York), 135.

61. Sakai, H., 1968, Internat. Rev. Cytol. 23, 89.

62. Deysson, G. and Delcourt, A., 1970, C.R. Soc. Biol. 164, 2209.

63. Deysson, G., 1964, Ann. Pharm. Fr. 22, 89.

64. Deysson, G. and Delage, M., 1965, C.R. Acad. Sci. 261, 4490, 4811 and 5187.

65. Deysson, G. and Delcourt, A., 1970, C.R. Soc. Biol. 164, 2440.

66. Taylor, J.H., in Zimmerman, A.M., Padilla, G.M. and Cameron, I.L., 1973, Drugs and the Cell Cycle (Acad. Press, New York), 11.

67. Dustin, P., 1963, Pharmacol. Rev. 15, 449.

68. Allison, A.C. and Nunn, J.F., 1968, Lancet 1326.

69. Haschke, R.H., Byers, M.R. and Fink, B.R., 1974, J. Neurochem. 22, 837.

# ASSAY FOR ANTITUBULIN DRUGS IN LIVE CELLS: ORAL REGENERATION IN *STENTOR COERULEUS*

L. Margulis, Department of Biology, Boston University, Boston, MA. 02215; S. Banerjee, Department of Biology, Wellesley College, Wellesley, MA.; and J. K. Kelleher, Department of Biology, Boston University, Boston, MA.

Summary

This paper reviews our experience with the synchronous oral regeneration system of the large heterotrichous ciliate, *Stentor coeruleus*, as an assay for antitubulin agents. Certain compounds and treatments, known from other work to be antitubulins, prevent the polymerization of microtubule protein into normal functioning 250Å microtubules. We have found, over the past nine years, that such agents delay the migration of the ciliated mouth (≡ oral membranellar band) of stentor. We present a description including a film of normal stentor oral migration and a summary of the use of this regenerating system as a drug assay for small amounts of new compounds. We summarize the methods and major results in the hope that others will recognize and utilize the stentor system.

## I. Introduction

It is the purpose of this short paper to review our experience with the synchronous oral regeneration system of Stentor coeruleus as an assay for antitubulin agents. Certain compounds and treatments are known from other work to be antitubulins: they prevent the polymerization of microtubule protein into normal functioning 250Å microtubules. We have found, over the past nine years, that such agents delay the migration of the oral membranellar band of the large heterotrichous ciliate, Stentor coeruleus. We present here a description of this stentor oral migration assay and a summary of the drug results in the hope that others will recognize and utilize it. Although much of this work has been previously published, in this paper we summarize the methods and major conclusions in order to provide those completely unfamiliar with it a good inexpensive live assay system with which one can test small quantities of new potential antitubulins.

## II. Stentor oral morphogenesis

### A. Growth of Stentor

Stentor coeruleus is a fresh water single celled protist, a carnivorous ciliate that can easily be grown in culture bowls to a patchy density of about 50 cells per ml. In its extended form the cell may be about 700μ long (Fig. 1). Cultures are best started in 250 ml bowls, with several hundred organisms and two or so heat-treated wheat grains per bowl. A bit of teased nonabsorbent cotton reminds stentors of the fibrous plant material in their native ponds, providing attachment sites.

Water is very important: natural waters support the most rapid growth. Alas, native fresh water, deep well water, clear pond, lake or spring water work best for reasons we have never fully ascertained. Certain bottled european mineral waters support stentor growth as well. However, the much better defined tadpole solution and M-solution (see appendix) were adequate enough to permit growth and experimentation. Stentors are fed every two or three days (never overfed) by

---

Fig. 1. Drawing of Stentor coeruleus (top) and its oral ciliated membranellar band system (bottom right).

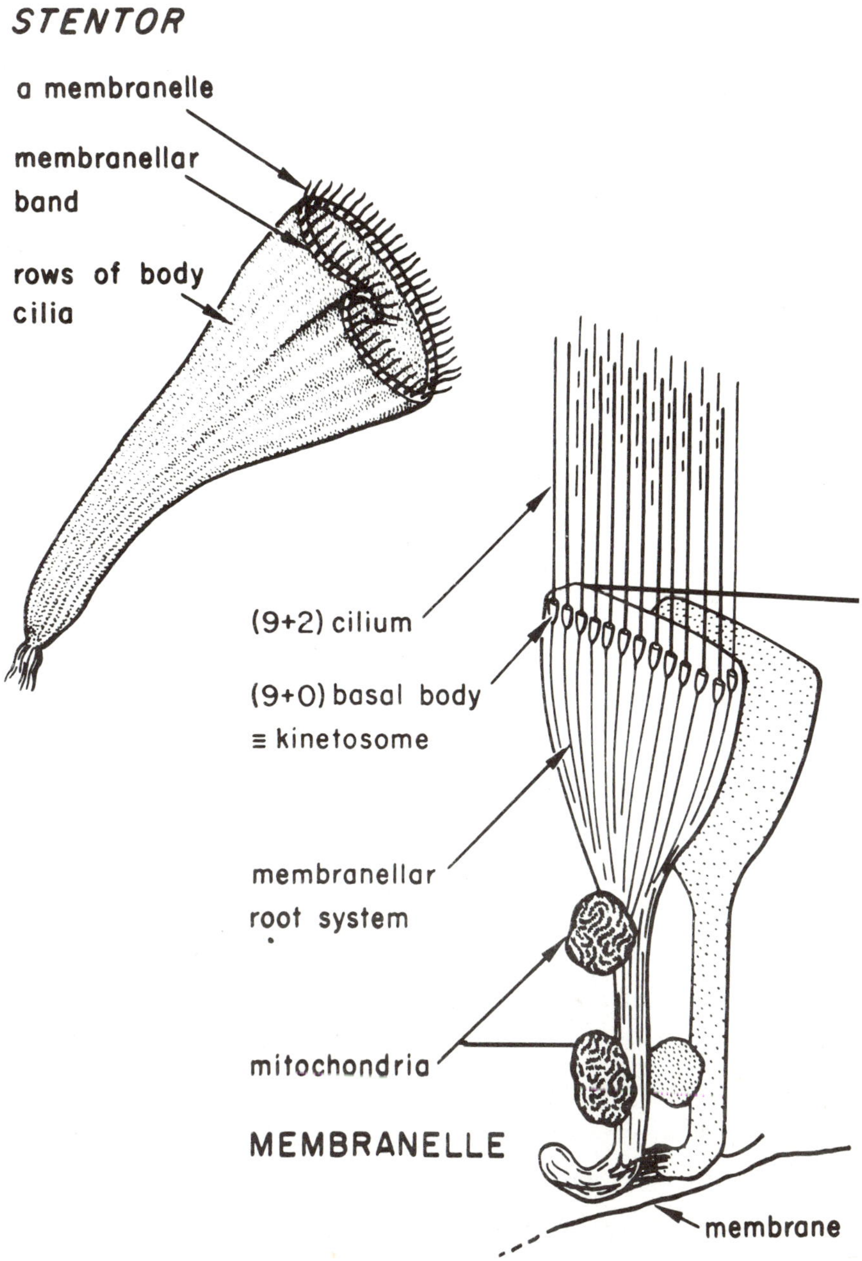
STENTOR
a membranelle
membranellar band
rows of body cilia
(9+2) cilium
(9+0) basal body
≡ kinetosome
membranellar root system
mitochondria
MEMBRANELLE
membrane

1

dropwise addition of flagellates and/or ciliates. Food preferred by stentor includes Colpidium (ciliate), Chilomonas (flagellate) and Tetrahymena (ciliate). The wheat grains decay providing a slow release of nutrients for the growth of bacteria which in turn are eaten by those ciliates and flagellates which provide food for stentors. In healthy cultures cells divide approximately every four days. Although this system of stentor stock maintenance and growth is messy, for regeneration experiments cells can be removed to far more clean and defined conditions: millipore filtered sterile pond water, tadpole or M-solution water. A wealth of experience concerning the growth, development and remarkable regenerative powers of this organism is available (Tartar,[1] 1961). Studies on stentor have a long history--although the organism may have first been seen by van Leeuwenhoek in the 17th century it was certainly well described and sketched by A. Trembley in the 1740's (H. Lenhoff,[2] 1973). Unfortunately genetic studies of S. coeruleus seem doomed: we agree with all of the experience since Ehrenberg in the 19th century: sex is lethal, at least in laboratory cultures of Stentor coeruleus--within four days or so after conjugation all exconjugants are dead.

B. Induction of oral morphogenesis

To our knowledge S. coeruleus is the only organism known in which basal body and ciliogenesis can be induced simultaneously in hundreds of cells in synchrony, Tartar,[3] 1957). Thousands of basal bodies and cilia per cell are released into the medium when the oral membranellar band is induced to shed by the experimenter (Fig. 2). Within about 30 seconds in certain toxic media such as urea, sucrose, glycerin or salt water the ciliated mouth bands are released. The shed cells are quickly washed through and removed to fresh filtered solutions of medium. In synchrony they then commence the regeneration sequence which is shown in Fig. 2. At room temperature they pass through these stages in an 8 hour period. If care is taken all cells will regenerate. New basal bodies always appear at a certain site on the cell surface; the striped contrast zone on the side of the cell

Fig. 2. Legend on figure. Macronucleus is beaded except in stage 6.

SHEDDING AND REGENERATION OF THE ORAL MEMBRANELLAR BAND OF

*STENTOR COERULEUS*

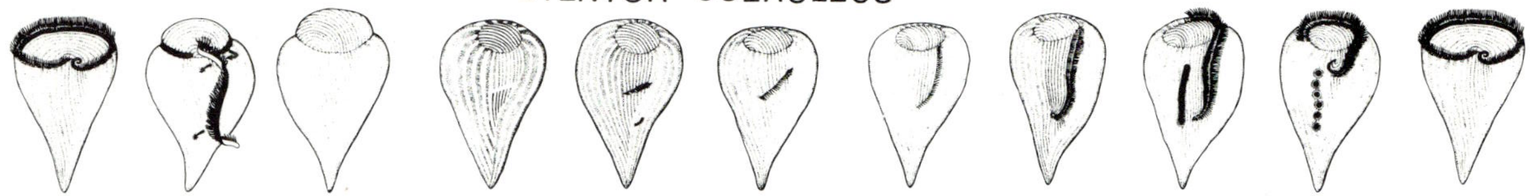


2

(Tartar,[1] 1961). This is the locus at which the large pigment stripes meet the small ones. New basal bodies appear precisely patterned in a regular asymmetric array: superficially basal bodies seem to "reproduce." DNA synthesis is not required for this regeneration of the approximately 20,000 basal bodies per cell (Younger et al.[4], 1972). However, evidence that new RNA, and protein synthesis as well as tubulin assembly accompany the normal regeneration sequence has been accumulated from many studies (Burchill[5], 1968; Frazier[6], 1970; Kelleher[7-8], 1971, 1975).

Shedding with 2% urea has been routinely used: one can see that the cilia themselves are sheared from the band with this treatment; in cytological preparations designed to retain the cilia on the band sucrose (4%) has been used (Fig. 3). Tartar's claim that upon shedding in this way the isolated bands contain the basal bodies was confirmed by electron microscopy (Van Wie[9], 1969).

The regeneration of the oral basal bodies, their cilia and associated structures which "package" to form rows of cilia (called membranelles) and the migration of these membranelles to the anterior end of the cell has been the subject of an elegant study by Paulin and Bussey[10] (1971). Two of their micrographs are reprinted here (Fig. 4 and 5): it can be seen that at stage 5 although the band is located along the side of the organism instead of at the anterior end the membranelles of the band are fully mature. In fact, from the ultrastructural point of view, a stage 5 band does not differ from a completely regenerated stage 8 band--except in its position. The later stages of oral regeneration simply involve a translocation of the band from the side to the anterior end of the organ-

---

Fig. 3. Oral membranellar bands of stentor isolated by different methods: 12.5% sucrose retains oral cilia (top); 2% urea shears cilia (middle); and since glycerol isolation (1.5%) tends to only partially and unpredictably remove cilia, its use was not extensive (bottom) (from Van Wie[9], 1969).

Fig. 4. Locus of stripe contrast 1-2 hours after shedding (Stage 2) showing the emergent newly formed cilia (from Paulin and Bussey[10], 1971).

Fig. 5. Stage 5 showing mature membranelles. The band is migrating anteriorly at this stage (from Paulin and Bussey[10], 1971).

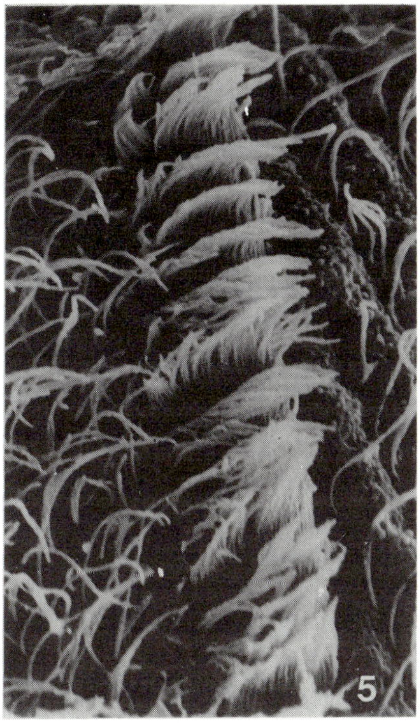
5

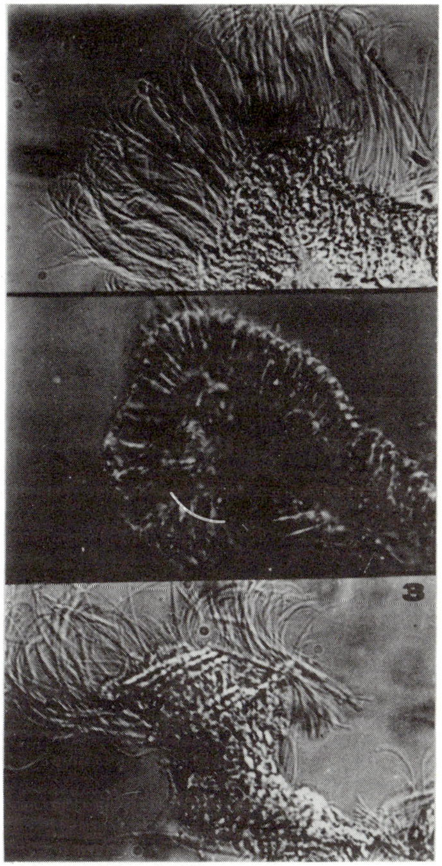
3

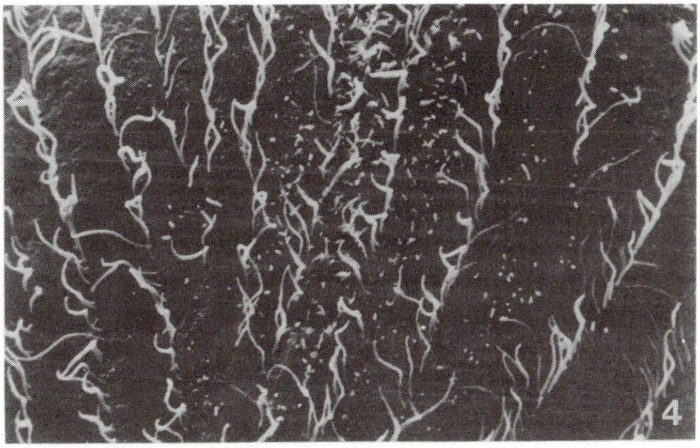
4

Table 1
Agents causing mitotic arrest and delay of oral regeneration in *Stentor coeruleus*

| TREATMENT | DELAY EQUATION | PROBIT VALUE 5 (concentration at which the velocity of regeneration is halved) | MECHANISM OF ACTION* |
|---|---|---|---|
| TRIFLURALIN | delay (hrs) = $4.1 \times 10^{19} [T]^{3.0}$ | $6 \times 10^{-7}$ M | binds to tubulin, not at colchicine site; (Hess et al.[29], Kelleher[8], 1975) |
| β-PELTATIN | delay (hrs) = $1.9 \times 10^{12} [\beta\text{-P}]^{1.8}$ | $8 \times 10^{-7}$ M | binds to tubulin at colchicine site (Kelleher[8], 1975) |
| β-PELTATIN A-METHYL ETHER | delay (hrs) = $1.1 \times 10^{7} [\beta\text{-P}_A]^{1.1}$ | $3 \times 10^{-6}$ M | binds to tubulin at colchicine site (Kelleher[8], 1975) |
| VINBLASTINE | delay (hrs) = $2.2 \times 10^{6} [V]^{1.0}$ | $6 \times 10^{-6}$ M | precipitates tubulin; binds at site other than colchicine site |
| PODOPHYLLOTOXIN | delay (hrs) = $1.1 \times 10^{9} [P]^{1.7}$ | $2 \times 10^{-5}$ M | binds to tubulin at colchicine site (Wilson[30], 1975) |
| COLCHICINE | delay (hrs) = $2.9 \times 10^{10} [C]^{2.1}$ | $3 \times 10^{-5}$ M | binds to tubulin at colchicine site |
| COLCEMID | delay (hrs) = $6.6 \times 10^{13} [CMD]^{3.2}$ | $9 \times 10^{-5}$ M | binds to tubulin at colchicine site |
| O-METHOXYTRYPTAMINE | delay (hrs) = $2.3 \times 10^{11} [O\text{-MT}]^{2.9}$ | $3 \times 10^{-4}$ M | at $10^{-3}$ M no specific effect on colchicine binding to tubulin |
| MELATONIN | delay (hrs) = $5.1 \times 10^{13} [M]^{3.9}$ | $7 \times 10^{-4}$ M | See Kelleher[8], 1975, for further data and discussion of tubulin-melatonin work |
| LOW TEMPERATURE | between 8-24°C Arrhenius activation energy Ea = 19.5 Kcal/mole | | (Kelleher[8], 1975) |

Table 1 cont.

| TREATMENT | DELAY EQUATION | PROBIT VALUE 5 (concentration at which the velocity of regeneration is halved) | MECHANISM OF ACTION* |
|---|---|---|---|
| $D_2O$ | delays induced in concentrations between 30 and 50% (at 22°C); cortical, membranellar and macronuclear abnormalities also induced. (55% lethal). Cold temperature lowered delay threshold of $D_2O$. | not meaningful | unpublished observations (Handler[31], 1970) |

*See Olmsted and Borisy[28], 1973, for review.

ism (Margulis et al.[11], 1975). This migration seems to depend on the continuing polymerization of microtubules. The actual basis of the observed migration is not known; however it may be due to the elongation of root fiber (nemadesmal) microtubules, and the accompanying stabilization of these microtubules (Paulin[12], 1974; Kelleher[8], 1975). Paulin is attempting to clarify this issue by continued direct ultrastructural studies of the late stages.

III. Summary of drug results

Regardless of the mechanism in both cases we have hypothesized that whatever forces underlie the anterior migration of the oral band of stentor also underlie the segregation of chromosomes in mitosis. Both these movements are slow (1 to a few μ/min) and are sensitive to disruption to low concentrations of drugs affecting microtubules. This hypothesis has been fruitful. It predicted mitotic arrest by melatonin (Banerjee and Margulis[13], 1973), the sensitivity of stentor regeneration to trifluralin (Banerjee et al.[14], 1975) and several other results. Our consistent finding is that agents known to affect microtubule-based processes in mitotic and other systems delay the migration of the oral band in stentor. With respect to the converse, compounds known to act on other cell systems: RNA synthesis inhibitors, radiation and so forth may delay the appearance of the new oral band but they do not slow the migration of the band as a function of concentration. For convenience, we list a summary of agents in Table 1, indicating the ones that are active in the stentor assay. It is these later that we expect will turn out to be antitubulins: agents that interact in some way to delay tubulin polymerization into microtubules.

Of the compounds tested (Table 2) we have identified nine drugs that at physiological concentrations seem to cause no other morphological effects except the delay of anterior migration of the oral membranellar band as an exponential function of concentration. At certain low concentrations characteristic for each drug, these drugs have no effect on stentor, at higher concentrations they delay migration relative to controls without any other visible cell abnormalities, at still higher concentrations no greater delays can be achieved: abnormalities

Table 2
Some compounds found inactive in the assay (do not delay stentor oral band migration as a function of concentration)

| Compound | Comments |
|---|---|
| PODOPHYLLOTOXIN DERIVATIVES | |
| 4-Demethylpodophyllotoxin | 2 hour delays; not a function of concentration between 3-6 x $10^{-6}$M |
| Podophyllic acid | Death at 3 x $10^{-6}$M; no delays without abnormalities |
| Picropodophyllic acid | Insoluble |
| β-Peltatin-β-D-glucopyranosid | No activity up to 4 x $10^{-4}$M |
| PUTATIVE SPINDLE POISONS | |
| Griseofulvin (mitotic spindle inhibitor)* | Delays but not as a function of concentration without other effects (Propst et al.[32], 1972) |
| Hexanitrodiphenylamine | Antimitotic (Deysson[22], 1968) |
| TRYPTOPHAN DERIVATIVES | |
| Serotonin (neurotransmitter) | Delays up to about 2.5 hours at 8 x $10^{-4}$M;not a function of concentration; death at 1 x $10^{-3}$M |
| N-acetyl serotonin | Delays up to about 3 hours at 6.0-6.4 x $10^{-3}$M; death at 6.5 x $10^{-3}$M; not a function of concentration |
| AMINOGLYCOSIDE ANTIBIOTICS | |
| Streptomycin, kanamycin, dihydrostreptomycin | Deshpande et al.[33], 1975; induce cilia membrane abnormalities |
| MISCELLANEOUS | |
| ATP, GTP, deoxyadenosine triphosphate, nicotinamide, riboflavin, cyclic AMP, high calcium, dinitrophenol, cytochalasin B, ethanol, dimethylsulfoxide, phenethyl alcohol, daunomycin, actinomycin D, chloramphenicol, cycloheximide, ethidium bromide, mitomycin C, hydroxyurea,urea | Inactive compounds, controls in various experiments; no synchronous delays in band migration a function of drug concentration |

*Also causes chromosome abnormalities (Banerjee and Margulis[13], 1973)

such as vacuolization, sparse bands, bleaching (loss of the characteristic deep blue green pigment) and so forth are induced. As the concentration is increased these appear in a greater and greater frequency in the cell population. As the concentrations are increased still higher cells begin to die at progressively earlier times. Eventually immediately lethal concentrations of these drugs are reached.

The curves summarizing the data are presented in Fig. 6. Delay in hours is an exponential function of concentration. Table 1 presents the delay equations for these nine active drugs. Only physiologically active concentrations are plotted: concentrations in which no morphological abnormalities, but only delay, can be observed. The maximum ranges are quite different for each of the drugs. As is often the case small chemical changes in the drugs may profoundly affect the potency in the drug assay. For example the switching of the -OH from $R_1$ to $R_2$ in β-peltatin relative to podophyllotoxin makes a significant difference in effect. In general the stentor assay results on podophyllotoxin derivatives correlate well with the ability of these drugs to bind tubulin, e.g., β-peltatin was found to be more effective than podophyllotoxin in the inhibition of tritiated colchicine binding to brain tubulin as well (Kelleher[8], 1975). Even though other direct studies with some of these compounds have shown clearly that different reactions with tubulin are involved (Column 4, Table 1), these nine all affect stentor in the same way: they delay band migration without abnormality as a function of concentration. One of us (S.B.) is presently testing the effect of certain other trifluralin derivatives on this stentor regeneration system.

IV. Comparison of the stentor assay with other biological systems

Microtubules of course are involved in the generation and maintenance of cell form in many organisms (see McIntosh[15], 1974, for an excellent short review and references). In some less well understood way the surfaces of microtubules seem to be involved in intracellular transport for example of pigment granules in teleost fishes (Murphy and Tilney[16], 1974) and in stentor (Blumberg et al.[17], 1973),

---

Fig. 6. Legend on figure.

DELAY OF ORAL REGENERATION IN *Stentor coeruleus* BY MITOTIC SPINDLE INHIBITORS

(delay in hrs) $y = k \propto$ (concentration of inhibitor in moles / liter)$^n$

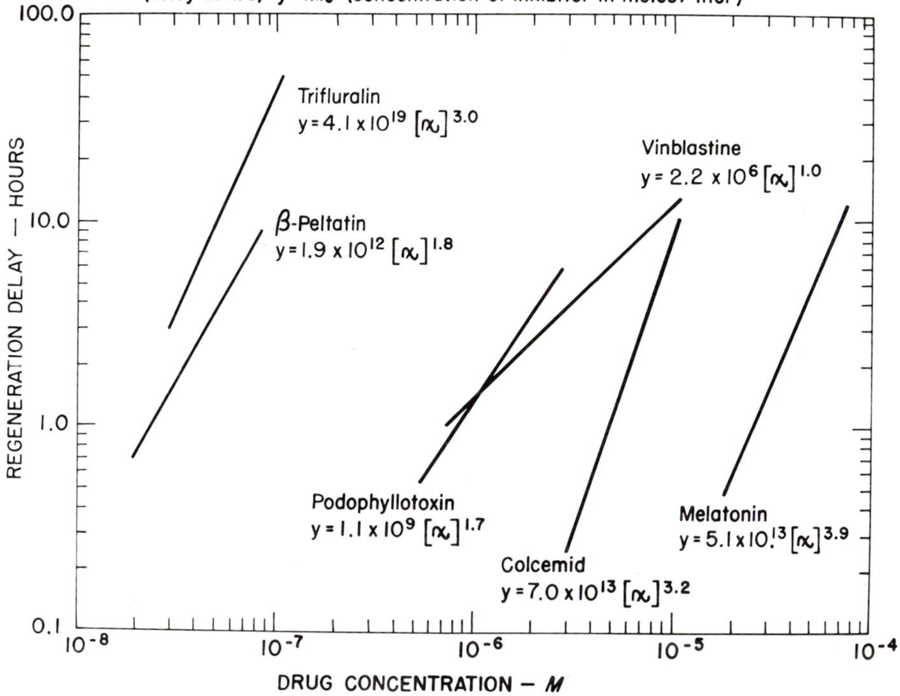


**other active derivatives:**
**β-Peltatin A-Methyl ether**
**Colchicine**
**O-Methoxytryptamine**

**6**

and of food particles and mitochondria in suctorians (Rudzinska[18], 1965) and even of intracellular symbiotic algae in Chlorohydra (Muscatine et al.[19], 1975). Microtubules are clearly involved in chromosome movement and perhaps in axoplasmic flow. They even seem to be used by some virions in certain plant cells perhaps for intracellular transport (Mayhew and Carroll[20], 1974). We currently have experiments underway that indicate microtubules may be involved in the migration of nuclei through the hyphae of fungi. In basidiomycetes, such as Schizophyllum pombe this process, which facilitates the coming together of dikaryon nuclei, is required in the formation of "clamp connections".

As the widespread use of colchicine has attested, antitubulins can be useful in the verification that a certain morphogenetic process is indeed microtubule based. The drugs listed in Table 1 may also be active against such processes. Thus established antitubulins can be used as probes, especially in systems in which electron microscopy is difficult and time-consuming and/or the number of microtubules low. We advocate the use and quantitation of antitubulin effects in other microtubule-based morphogenetic systems to determine the general utility and ultimate mechanism of action of these drugs. In some cases the active podophyllotoxin derivatives vinblastine and trifuluralin are to be preferred to colchicine derivatives in live cell assays because they seem to be more stable and less sensitive to interfering agents (e.g., colchicine is sensitive to light, tris buffer (Margulis et al.[21], 1969) and the two B vitamins riboflavin and nicotinamide (Deysson[22], 1968; Margulis[23], 1973). An unknown nontubulin protein in serum seems also to bind colchicine (Bamburg et al.[24], 1973; Wilson, L.[25], 1975). Stock solutions of colchicine can also serve as a carbon source for at least one species of fungi (Heath[26], 1975), and it is likely that fungi can grow on podophyllotoxin stock solutions too.

Because in some systems antitubulins have clearly documented other effects (e.g., inhibition of nucleoside uptake by colchicine and podophyllotoxin) and some bona fide microtubules are apparently insensitive to colchicine (e.g., Saprolegnia spindle tubules, Heath[27], 1975, in press) clearly studies with these compounds must be done carefully and interpreted with caution.

We anticipate then that these drugs will be useful as analytic tools in the understanding of the basis of a large variety of microtubule-based cellular processes. We also hope that others will recognize the advantages of the stentor assay.

Acknowledgments

Support for this research came from NASA NGR-004-025 (Microbial contributions to the Precambrian earth), the Boston University Graduate School and the Wellesley College Research Fund.

Appendix

ARTIFICIAL POND WATERS

I. TADPOLE SOLUTION

SOLUTION #2

| | | |
|---|---|---|
| 1) | Sodium bicarbonate ($NaHCO_3$) | 400g |
| 2) | Versene (EDTA) | 200g |
| 3) | Deionized, distilled water to make | 20 liters |

SOLUTION #1

| | | |
|---|---|---|
| 1) | Calcium chloride | 2 pounds |
| 2) | Magnesium chloride | 100g |
| 3) | Deionized, distilled water to make | 20 liters |

FINAL SOLUTION

| | | |
|---|---|---|
| 1) | #2 solution | 100 ml |
| 2) | #1 solution | 100 ml |
| 3) | Deionized, distilled water to make | 20 liters |

Dilute 100 ml of solution #2 with water bringing a 20 liter carbuoy almost to volume. Add 100 ml of solution #1 and bring to volume.

II. POND WATER - "M" SOLUTION

KMNT (6 ml/l) is added to $CaCl_2$ (1 ml/l) to make slightly over a liter of "pond water."

| Stock Solution Components (gm/liter) | | | Dilution | Final Molarity |
|---|---|---|---|---|
| Solution A (KMNT): | 1. | KCL - .016 M | 6/1000 | $10^{-4}$ |
| | 2. | $MgCl_2$ - .016 M | 6/1000 | $10^{-4}$ |
| | 3. | $NaHCO_3$ - 0.166 M | 6/1000 | $10^{-3}$ |
| | *4. | Trizma - 0.166 M (pH 7.4) | 6/1000 | $10^{-3}$ |
| Solution B: | 1. | $CaCl_2$ - 1.0 M (anhydrous) | 1/1000 | $10^{-3}$ |

PROCEDURE

1. Make a 3 liter batch of KMNT. This is prepared by mixing:

| | | gm/3 liters KMNT |
|---|---|---|
| a. | 1000 ml Trizma (0.5 M pH 7.4, MW 156) | 78 gm. |
| b. | 500 ml $NaHCO_3$ 1.0 M - prepare fresh each time (MW 84) | 42 gm. |
| c. | 50 ml KCl, 1.0 M (MW 74.6) | 3.73 gm. |
| d. | 50 ml $MgCl_2$ 1.0 M (MW $MgCl_2 \cdot 6H_2O$ is 203) | 10.16 gm. |

2. Make up and store $CaCl_2$ separately and use as separate reagent to prepare "M" solution.

3. To make "M" solution mix e into f.

   e. KNMT stock (6 ml/liter in distilled-deionized $H_2O$)

   f. $CaCl_2$, 1.0 M (MW 111) (1 ml/liter distilled-deionized water.)

References


1. Tartar, V. 1961. The Biology of Stentor. Pergamon Press, Oxford, 413 pp.

2. Lenhoff, H. Personal communication.

3. Tartar, V. 1957. Reactions of Stentor coeruleus to certain substances added to the medium. Exptl. Cell Res. 13: 317-332.

4. Younger, K. B., Banerjee, S., Kelleher, J. K., Winston, M. and Margulis, L. 1972. Evidence that the synchronized production of new basal bodies is not associated with DNA synthesis in Stentor coeruleus. J. Cell Science 11: 621-637.

5. Burchill, B. R. 1968. Synthesis of RNA and protein in relation to oral regeneration in the ciliate Stentor coeruleus. J. Exptl. Zool. 167: 427-438.


* Trizma buffer, Sigma Chemical Company, St. Louis, Missouri.

6. Frazier, E. A. J. 1970. Interactions between the nucleus and the cytoplasm in the ciliate Stentor coeruleus: An autoradiographic and microsurgical study of nucleic acid and protein metabolism during oral morphogenesis and macronuclear replication. Ph.D. Thesis, Department of Zoology, University of Washington.

7. Kelleher, J. K. 1971. Protein synthesis during oral regeneration in Stentor coeruleus. Masters Thesis, Boston University Graduate School, Department of Biology.

8. Kelleher, J. K. 1975. Interaction of antimicrotubule agents with tubulin in vitro and in Stentor coeruleus. Ph.D. Thesis, Boston University Graduate School, Department of Biology.

9. Van Wie, C. 1969. An electron microscopic investigation of the intact and shed membranellar band of Stentor coeruleus. Masters Thesis, Boston University Graduate School, Department of Biology.

10. Paulin, J. J. and Bussey, J. 1971. Oral regeneration in the ciliate Stentor coeruleus. A scanning and transmission electron optical study. J. Protozool. 18: 201-213.

11. Margulis, L., Banerjee, S. and Schaadt, J. G. 1975. Antitubulin compounds and induced oral cilia regeneration in Stentor coeruleus: Film. (American Society Cell Biology - Abstract, Puerto Rico).

12. Paulin, J. J. Personal communication.

13. Banerjee, S., and Margulis, L. 1973. Mitotic arrest by melatonin. Exp. Cell Research 78: 314-318.

14. Banerjee, S., Margulis, L. and Kelleher, J. K. 1975. The herbicide trifluralin is active against microtubule-based oral morphogenesis in Stentor coeruleus. Cytobios 11: (in press).

15. McIntosh, R. 1974. An introduction to microtubules. J. Supramolecular Structure 2: 385-392.

16. Murphy, D. G. and Tilney, L. G. 1974. The role of microtubules in the movement of pigment granules in teleost melanophores. J. Cell Biol. 61: 757-

17. Blumberg, S., Honjo, S., Otaka, T., Antanavage, J., Banerjee, S. and Margulis, L. 1973. Induced reversible pigment alteration in Stentor coeruleus. Trans. Amer. Microscop. Soc. 92: 557-569.

18. Rudzinska, M. 1965. Fine structure in function of the tentacles in Tokophyra infusorium. J. Cell. Biol. 25: 459-477.

19. Muscatine, L., Cook, C. B., Pardy, R. L. and Pool, R. R. 1975. Uptake, recognition and maintenance of symbiotic chlorella by Hydra viridis. Soc. Gen. Exp. Biol. Cambridge Univ. Press, D. Jennings, ed. volume 24 (in press).

20. Mayhew, D. E. and Carroll, T. W. 1974. Barley stripe mosaic virions associated with spindle microtubules. Science 185: 957-958.

21. Margulis, L., Banerjee, S. and White, T. 1969. Colchicine-inhibited cilia regeneration: Explanation of lack of effect in tris buffer medium. Science 164: 1177-1178.

22. Deysson, G. 1968. Antimitotic substances. Internatl. Rev. Cytol. 34: 99-148.

23. Margulis, L. 1973. Colchicine sensitive microtubules. Inter. Rev. Cytol. 34: 333-361, B. Bourne and J. F. Danielli, eds., Academic Press, New York.

24. Bamburg, J. R., Shooter, E. M., Wilson, L. 1973. Developmental changes in microtubule protein in chick brain. Biochem. 12: 1476-1482.

25. Wilson, L. 1975. Personal communication.

26. Heath, I. B. 1975. Department of Biology, York University, Toronto, Ontario, Personal communication.

27. Heath, I. B. 1975. In press.

28. Olmsted, J. B. and Borisy G. G. 1973. Microtubules. Ann. Rev. Biochem. 42: 507-540.

29. Hess, Daniel, 1975. Department of Botany, University of California, Davis, California, Personal communication.

30. Wilson, L. 1975. Microtubules as drug receptors: pharmacological properties of microtubule protein. New York Acad. Sci. 253: 213-231.

31. Handler, L. 1970. Unpublished observations.

32. Propst, S., Banerjee, S., Kelleher, J. K. and Margulis, L. 1972. Inhibition of cilia regeneration by antineoplastic agents: Delay of band migration by vinblastine (NSC-49842) and griseofulvin (NSC-34533). Cancer Chemotherapy Reports 56: 557-558.

33. Deshpande, K. L., Banerjee, S., Kelleher, J. K. and Margulis, L. 1975. Ciliary membrane abnormalities induced by streptomycin and other aminoglycoside antibiotics in Stentor. Cytobios 11: 1-15.

*Microtubules and Microtubule Inhibitors, eds. M. Borgers and M. de Brabander*



# ANTI-INFLAMMATORY ACTION OF COLCHICINE

K.Brune, P.Graf and M.Glatt

Department of Pharmacology, Biozentrum, Universität of Basel, Klingelbergstrasse 70, CH-4056 Basel, Switzerland.

## Summary

Colchicine in high doses has repeatedly been observed to be an active anti-inflammatory compound. To discover its possible mechanisms of action we measured the release of known mediators of acute inflammation (histamine, prostaglandins) in inflamed tissue. We found that colchicine in contrast to other nonsteroid anti-inflammatory drugs increases the amount of prostaglandins (PG) recovered from inflamed tissue. Three explanations for this effect appeared possible and were further investigated: colchicine could (1) increase the release, (2) reduce local metabolism and/or (3) reduce the disappearance of PG into the general circulation. Of these three explanations only the first one was found to be of importance. From these observations two mechanisms of action of colchicine in acute inflammation appear likely on the basis of our current knowledge; either the high concentrations of prostaglandins caused by colchicine in the inflamed tissue reduce the symptoms of inflammation as reported in the literature[1] or the increased concentrations of PG and the reduced signs of inflammation are independent effects of colchicine on cells involved in inflammation. Arguments for the latter explanation are presented.

## 1. Introduction

Among the so-called nonsteroid anti-inflammatory drugs (NSAID) colchicine assumes a unique position. It is the only substance for which a molecular receptor has clearly been isolated, namely the microtubules. Also, in contrast to all other NSAID colchicine is very effective in one specific disease, gouty arthritis, although high doses are required to observe anti-inflammatory effects in a variety of inflammatory states in humans and animals[2,3]. One would assume that these unique properties of colchicine would have led to

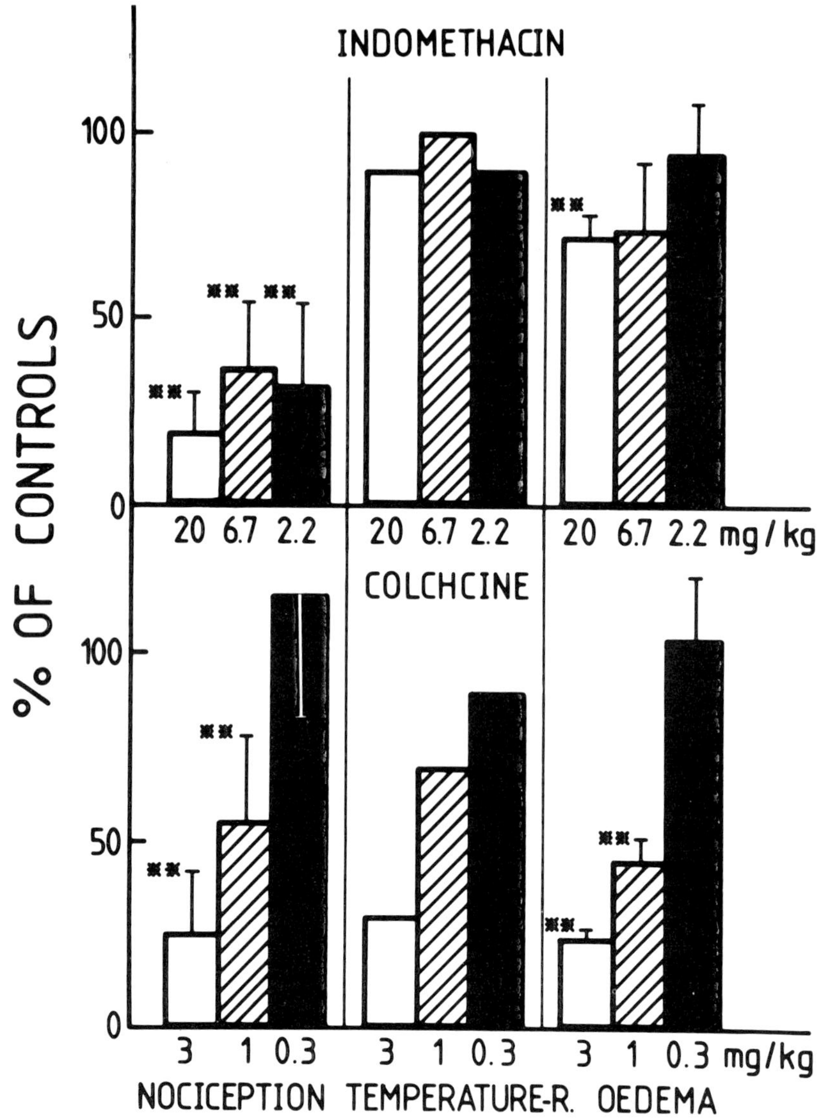


Fig. 1. Influence of colchicine and indomethacin on different parameters of inflammation. An inflammation was elicited by injecting urate crystals into an intertarsal joint of pigeons or chicken. The nociceptive reflex, the local warmth and the oedema formation were measured as described previously[6,7,8]. The drugs were administrated i.v. in the doses given one hour (indomethacin) or three hours (colchicine) before urate. The effects are expressed as % of non-drugged controls. Means and standard errors of 10 experiments are given. $*p < 0.05$, $**p < 0.001$.[6].

a complete understanding of its mode of action not only in gout but also in other inflammatory states. However, it rather appears as if the uniqueness of colchicine among other anti-inflammatory compounds has merely confused researchers and clinicians. Even today after almost a century of research in this field the mode of action of colchicine in gout is unresolved[4] and the role of microtubules in inflammation is a matter of speculations[5]. Using colchicine as an anti-inflammatory drug we have made some observations which might help towards a better understanding of colchicine and microtubules in inflammation.

## 2. Influence of Colchicine on Different Parameters of Inflammation

Three symptoms of an acute inflammation, i.e. nociception (an equivalent of pain), local warmth and oedema can be assessed using the microcrystal arthritis elicited in birds by the injection of urate crystals into intertarsal joints[6,7,8]. All three symptoms can be influenced, i.e. reduced by indomethacin and also by colchicine as may be seen from figure 1. However, the profile of action is different for both drugs. Surprisingly, the analgetic effect of indomethacin is more prominent than the other effects, whilst with colchicine the anti-oedema effect appears somewhat more accentuated. However, altogether there can be no doubt that both drugs are potent NSAID.

Table 1

Effect of indomethacin and colchicine on histamine- and PG-content in synovial fluid following injection of urate crystals.

| | INDOMETHACIN mg/kg | | | COLCHICINE mg/kg | | |
|---|---|---|---|---|---|---|
| | 2.2 | 6.7 | 20 | 0.3 | 1.0 | 3.3 |
| HISTAMINE | 105±28 | 86±21 | 48±13** | 31±14** | 28±16** | 10±3** |
| PG $F_{2\alpha}$ | 53±19** | 24±8** | 4±2** | 104±16 | 121±18 | 208±57** |

Indomethacin was given i.v. one hour and colchicine three hours before injection of urate crystals. Determinations of histamine and $PGF_{2\alpha}$-concentrations were performed after 40 min or 120 min respectively using standard methods for histamine[26] and prostaglandin[8] assays. Statistic evaluation as in fig.1.

Searching for an explanation of the different profile of anti-inflammatory action of colchicine in comparison to indomethacin we performed washes of inflamed joint cavities and assayed them for mediator content. According to the work of others ornithokinin and anaphylatoxins are not likely to be of relevance in this type of inflammation[9,10], whilst histamine[11] and prostaglandins (PG)[12] might be. Therefore we concentrated on these two mediators. The results of our experiments are summarized in table 1. As expected, indomethacin inhibits the build-up of measurable PG concentrations in a dose dependent manner. It also reduces the peak-concentrations of histamine as compared to untreated controls. In contrast, while colchicine also reduces the concentrations of histamine, we observed increased concentrations of prostaglandins in the inflamed joints of colchicine treated animals. From these results it is clear that colchicine certainly does not exert its anti-inflammatory action by inhibiting PG-synthesis as claimed for other NSAID[13].

3. Effects of Colchicine on Prostaglandin Metabolism, Absorption and Synthesis

The observed increased concentrations of PG in inflamed tissue following colchicine treatment could be the result of three different effects: (1) inhibition of local metabolism (2) reduced absorption of PG into the general circulation and/or (3) increased release. To test these possible mechanisms the following experiments were performed.

Table 2

$PGF_2$ concentration in joint washes further incubated in vitro

| INCUBATION (min) | 0 | 30 | 60 | 90 |
|---|---|---|---|---|
| CONTROLS | 1.1 $\pm$ 0.1 | 1.1 $\pm$ 0.2 | 1.2 $\pm$ 0.3 | 1.0 $\pm$ 0.2 |
| COLCHICINE-TREATED | 3.5 $\pm$ 0.3 | 3.3 $\pm$ 0.3 | 3.4 $\pm$ 0.4 | 3.4 $\pm$ 0.3 |

Joint washes were recovered 4 h after urate injection from non-drugged controls and colchicine-treated chicken (3 mg/kg, 3 h before UC-injection), diluted 1:2 with Hanks balanced salt solution containing 10 U Heparin (Liquemin,Roche), and further incubated in siliconized glass-tubes on a rotating disc at 40°C. After the time-intervals given, samples were removed and $PGF_2$ assayed in the cell-free supernatant, as described previously[8]. Means and S.E.M. from 5 experiments ($PGF_2$ in ng/ml) are given.

Firstly, joint washes were obtained from inflamed joints of colchicine treated chicken and controls, further incubated *in vitro* and PG-assays performed after different time intervals. The results are given in table 2. No significant inhibition of metabolism in the synovial fluid by colchicine could be detected. Also, the enzyme most likely to catalyse PG-breakdown in the body, namely PG-dehydrogenase[14] was not found to be inhibited by colchicine up to a concentration of $10^{-4}$M as may be seen from table 3.

Secondly, to test the possibility that colchicine interferes with the absorption of PG's from the inflamed tissue and its disappearance into the general circulation, radioactive PG $E_2$ was injected into the joint cavity of colchicine treated and control animals. After different time intervals joint fluid was recovered and the amount of label present at the site of inflammation defined. As may be seen from table 4 the disappearance of label was slightly lower in the colchicine treated animals as compared to the controls but nevertheless almost all label had disappeared after one hour. From these results it can be concluded that the observed long lasting[8] increased concentrations of PG's in inflamed joints cannot be solely explained by inhibited disappearance from the joint cavity.

Table 3

Influence of Colchicine on 15-Hydroxyprostaglandin-Dehydrogenase

| | CONTROLS | COLCHICINE $10^{-4}$M | COLCHICINE $10^{-5}$M |
|---|---|---|---|
| $\Delta$OD | 0.175 | 0.185 | 0.184 |
| | 0.172 | 0.185 | 0.189 |

15-Hydroxyprostaglandin-dehydrogenase was isolated according to the method of Tai et al.[26]. Using this enzyme the effect of colchicin on the enzymatic activity was assessed by the method of Anggard and Samuelsson[29]. The results show that the breakdown of $PGE_2$ as measured by the change of OD within 90 min. was not influenced by colchicin.

The third possible explanation for the increased PG-content, namely an increased release could not be tested *in vivo*. Since there is already evidence that colchicine might increase the formation of PG's in tissue culture cells[15] we decided to investigate the effects of microtubular inhibitors on PG-formation in 3T3-cell cultures.

We found that colchicine added to the tissue culture medium in concentrations as low as $10^{-7}$M in a dose dependent manner increased the PG-formation up to 300% of controls without significantly influencing the cell or protein content of the cultures (comp. table 5).

Table 4

Effect of Colchicine on the disappearance of PG from inflamed and not inflamed joints.

| | | % OF INJECTED PG $E_2$ RECOVERED AFTER (min). | | | | |
|---|---|---|---|---|---|---|
| | | 15 | 30 | 60 | 120 | 240 |
| NOT INFLAMED JOINTS | CONTROLS | 19.6 | 15.8 | 6.9 | 2.9 | 0.3 |
| | COLCHICINE | 25.8 | 18.1 | 10.2 | 5.9 | 1.6 |
| INFLAMED JOINTS | CONTROLS | 16.1 | 13.5 | 2.2 | 0.4 | 0.14 |
| | COLCHICINE | 21.4 | 18.7 | 6.4 | 3.0 | 0.7 |

Chicken (2kg body weight) received either saline (controls) or colchicine (3 mg/kg) three hours before either saline containing $^3H$ PG $E_2$ (100000 dpm) or urate crystals in saline with the same amount of labelled PG $E_2$ were injected (0-time). After different time intervals joint washes were performed in groups of each times 3 animals and the amount of activity recovered was assessed for each joint. Mean values of the three experiments expressed as percentage recovered of the amount of activity injected are taken as a means of the rates of disappearance of PG from the joint cavity.

Similar effects were observed when vinblastine was substituted, whilst lumicolchicine which does not interfere with microtubules was inactive. Also, the moderate activity of noradrenaline, isoprenaline and dibutyryl cAMP on PG-formation even in much higher concentrations stresses the specificity of the effect of microtubular inhibition (table 5). The argument that the increase in PG-content in tissue culture due to colchicine or vinblastine might be caused by inhibition of prostaglandindehydrogenase can be ruled out from our findings that this enzyme is not influenced by colchicine in concentrations up to $10^{-4}$M (table 4). In conclusion, we found that microtubular inhibitors act in an anti-inflammatory way and increase the formation of prostaglandins which are widely believed to be mediators of inflammation.

Table 5

Effects of microtubular inhibitors and other compounds on PG-formation in tissue culture.

| DRUG | CONCENTRATION [μM] | EFFECT ON PG $E_2$ FORMATION % [OF CONTROLS] |
|---|---|---|
| LUMICOLCHICINE | 1.0 | 98 |
| COLCHICINE | 0.015 | 70 |
| | 0.05 | 192 |
| | 0.15 | 294 |
| VINBLASTINE | 0.1 | 95 |
| | 0.3 | 166 |
| | 1.0 | 205 |
| NORADRENALIN | 10.0 | 156 |
| ISOPRENALIN | 10.0 | 172 |
| THEOPHYLLIN | 1000.0 | 264 |
| dBcAMP | 1000.0 | 87 |

Parent cultures of 3T3 cells were subcultured in Dulbecco's modified Eagle Medium,(GIBCO) with 10% calf serum (10% DMEM) for three days until layers were confluent. Then the medium was changed (3% DMEM). Each times three cultures were further incubated without drug and served as controls. Sets of each times three cultures received the medium containing the drugs in the concentrations given. Three days later the supernatants were separated and assayed for PG $E_2$ as described previously. The number of cells was counted (Coulter counter), the PG-content per cell in culture assessed and the relative change in comparison with the controls was calculated. Cultures showing more than 10% deviation in cell content as compared to controls were rejected.

## 4. Mechanism(s) of the Anti-inflammatory Action of Colchicine

From the results presented, two mechanisms of action of colchicine in acute inflammation appear likely on the base of our current knowledge. Either the high concentrations of PG effected by colchicine in inflamed tissue reduce the symptoms of inflammation as reported in the literature[1] or the increased concentration of PG and the reduced symptoms of inflammation are independent effects of the action of colchicine on cells involved in inflammation. Either possibility would be of interest for the understanding of the cellular processes and the role of prostaglandins in inflammation. To test the first explanatory hypothesis we injected high doses of

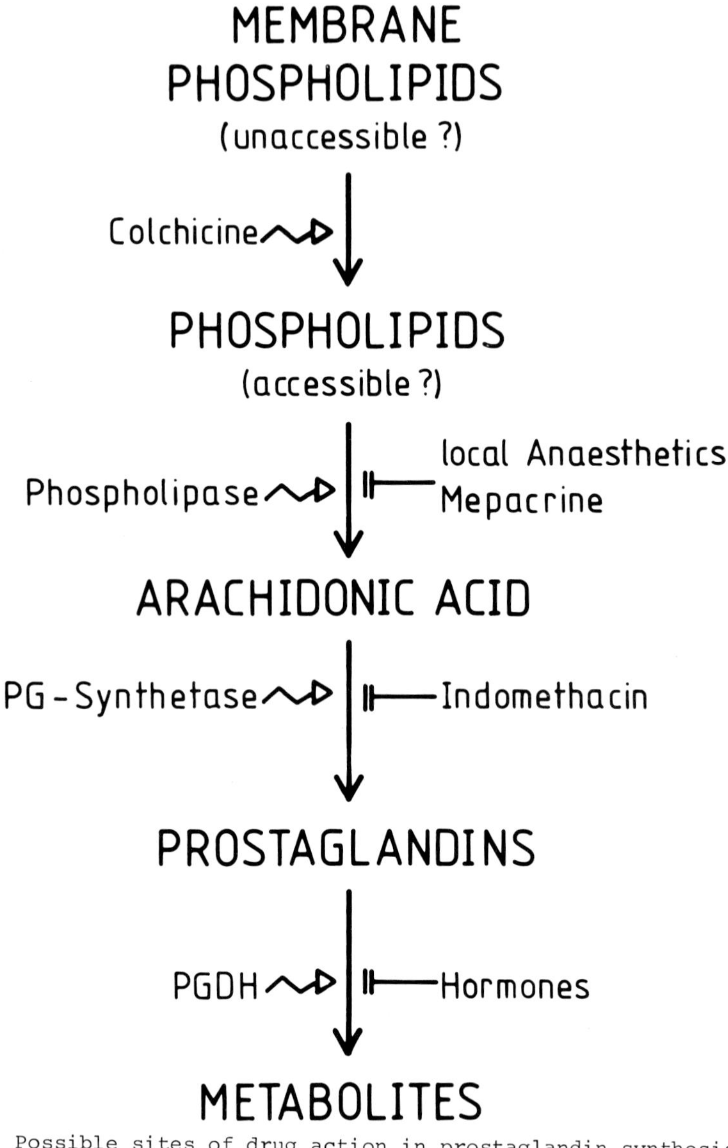


Fig.2. Possible sites of drug action in prostaglandin synthesis. a) Inhibition of phospholipases are achieved by mepacrine[28] and local anaesthetics[29]. b) Inhibition of the prostaglandin synthetase(s) is the probable target of indomethacin and other NSAID[13]. c) Hormones (thyroid hormones) are known to interfere with PG-metabolism[27]. d) Colchicine presumably increases PG-formation at a very early stage by interfering with organisation of cell membranes increasing the accessibility of phospholipids to phospholipases. This hypthesis is based on the observations of Edelman and coworkers[25].

$PG\ E_2$ and/or $PG\ F_{2\alpha}$ (100 ng of each per joint) together with urate crystals, the inflammatory stimulus, into pigeon joints and recorded the inflammatory symptoms, nociception and temperature rise, as reported previously[6]. Up to now, we could only observe a moderate anti-inflammatory action of $PG\ F_{2\alpha}$ whilst $PG\ E_2$ and a combination of both (as found in inflamed joints of colchicine treated animals) did not change the time course of the appearance or the intensity of the inflammatory symptoms. On the other hand, there is indirect evidence that intact microtubules are involved in the regulation of water, electrolyte and macromolecule transport across endothelial and epithelial cells[16,17,18]. Therefore we believe that colchicine and other microtubular inhibitors might exert their anti-inflammatory action by interfering with increased exudation of plasma compounds across vascular endothelia and/or synovial lining cells. Definite proof for this assumption however, is still lacking.

## 5. Conclusions

Two consequences of the observed effects of colchicine in inflammation appear worthy of consideration.

Firstly, colchicine apparently can inhibit vascular responses in inflammation, for example oedema formation, to a higher degree than other NSAID, e.g. indomethacin. Since this vascular effect can be demonstrated in the presence of even enhanced PG-concentrations, it is possible that colchicine exerts a direct action on vascular permeability by rendering endothelia unresponsive towards the effect of inflammatory mediators. In this context it is noteworthy that the observed action of colchicine could be useful for investigating the controversial question as to whether increased transendothelial transport, "cytopempsis"[20], contributes to the oedema formation in inflammation. It has always been claimed[21] that especially large molecules leave the intravascular space by "cytopempsis". Our findings are in support of this concept since colchicine inhibits oedema formation and there is morphological evidence that it reduces "cytopempsis"[23].

The second consequence of our observations seems possibly of great interest to the cell biologist working towards understanding how the organisation of cytoplasmic components is interconnected with characteristics of the cell membrane. The increased output of a membrane-derived mediator, PG, by interfering with the cytoplasmic microtubules may offer a tool for further studies in this direction. Since colchicine has not been reported to stimulate the PG-synthetase complex it is likely that this compound affects PG-synthesis at a very early stage. Possibly phospholipids become more accessible to phospholipase $A_2$ because colchicine and other microtubular inhibitors disorganize the pattern of protein distribution in the cell-membrane (Fig.2). This idea is supported by the findings of Ukena and Berlin[23] and Edelman et al.[24] who observed changes in the organisation of membrane proteins and lipids by colchicine.

References


1. Zurier,R.B., Hoffstein,S. and Weissmann,G., 1973, Arthritis Rheum. 16, 606.
2. Hibino,R., 1971, Gifu Daigaku Igahubu Kiyo 19, 739; Chem.Abstr. 77, 14112.
3. Fitzgerald,T.J., Williams,B. and Uyeki,E.M., 1971, Pharmacology 6, 265.
4. Wallace,S.L., 1974, Seminars in Arthritis and Rheumatism 3, 369.
5. Glatt,M., Wagner,K. and Brune,K., 1975, International Conference on Prostaglandins (Raven, New York). In press.
6. Brune,K., Walz,D. and Bucher,K.E., 1974, Agents and Actions 4,21.
7. Brune,K. and Glatt,M., 1974, Agents and Actions, 4, 95.
8. Glatt,M., Peskar,B. and Brune,K., 1974, Experientia 30, 1257.
9. Spilberg,I., 1974, Arthritis Rheum. 17, 143.
10. Spilberg,I. and Osterland,C.K., 1970, J.Lab.Clin.Med. 76, 472
11. Benzi,G., Crema,A. and Grigo,G.M., 1965, J.Pharm.Sci. 54, 1689.
12. Blackham,A., Farmer,J.B., Radzinowik,H. and Westwick,J., 1973, Brit.J.Pharmacol. 48, 343P.
13. Vane,J.R., 1971, Nature NB 231, 232.
14. Hamberg,M. and Samuelsson,B., 1971, J.Biol.Chem. 246, 6713.
15. Robinson,D.R., Smith,H. and Levine,L., 1973, Arthritis Rheum. 16, 129.
16. Taylor,A., Maffly,R., Wilson,L. and Reaven,E., 1975, Ann.N.Y. Acad.Sci., 253, 723.

17. Malaisse,W.J., Malaisse-Lagae,F., Van Obberghen,E., Somers,G., Devis,G., Ravazzola,M. and Orci,L., 1975, Ann.N.Y.Acad.Sci. 253, 630.

18. Poisner,M. and Cooke,P., 1975, Ann.N.Y.Acad.Sci. 253, 653.

19. Wooten,F., Kopin,I.J. and Axelrod,J., 1975, Ann.N.Y.Acad.Sci. 253, 528.

20. Moore,D.H. and Ruska,H., 1957, J.Biophys. Biochem.Cytol.9, 909.

21. Renkin,E.M., 1964, Physiologist 7, 13.

22. Casley-Smith,J.R., The University of Adelaide, Australia, pers. comm.

23. Ukena,T.E. and Berlin,R.D., 1972, J.exp.Med. 136, 1.

24. Edelman,M.G., Yahara,I. and Wang,J.L., 1973, Proc.Nat.Acad.Sci. USA, 70, 1442.

25. Lorenz,W., Benesch,L., Barth,H., Matejka,E., Meyer,R., Kusche,J., Hutzel,M. and Werle,E., 1970, Z. Anal.Chem. 252, 94.

26. Tai,H.H., Tai,C.L. and Hollander,C.S., 1974, Biochem. Biophys. Res.Commun. 57, 457.

27. Flower,R.J., 1974, The Wellcome Res. Laboratories, England, pers.comm.

28. Kunze,H., Bohn,E. and Vogt,W., 1974, Biochim. Biophys.Acta 360, 260.

29. Anggard,E. and Samuelsson,B., 1969, Methods Enzymol. 14, 215.

*Microtubules and Microtubule Inhibitors, eds. M. Borgers and M. de Brabander*



# ANTIMITOTIC ACTIVITY OF METHYL BENZIMIDAZOL-2-YLCARBAMATE IN FUNGI AND ITS BINDING TO CELLULAR PROTEIN

L.C. Davidse

Research Unit for Internal Therapy of Plants T.N.O.
Laboratory of Phytopathology, Agricultural University
Wageningen, The Netherlands

## 1. Summary

Several lines of evidence indicate that methyl benzimidazol-2-ylcarbamate (MBC or carbendazim)and some other benzimidazole derivatives may be considered as antimitotic agents, acting as spindle poisons. A macromolecular MBC receptor with tubulin-like properties appeared to be present in Aspergillus nidulans and other MBC-sensitive fungi.

The differences in response of fungi to MBC and the readiness with which changes in response in originally sensitive fungi can be induced may indicate a very specific action.

A high specificity of the mechanism of action may also account for the failure of MBC to inhibit both porcine brain tubulin polymerization in vitro and the copolymerization of fungal tubulin into microtubules in a heterologous assembly system with porcine brain tubulin.

In view of this, benzimidazole derivatives may prove to be valuable tools to compare biochemical and functional properties of tubulin from different sources.

Purification of tubulin in the heterologous assembly system did not involve copurification of the high molecular weight proteins as is normally observed. Incorporation of fungal tubulin into microtubules might prevent their association with structures, composed of these proteins.

The heterologous assembly system might provide a new approach by which the in vitro and in vivo control of microtubule assembly can be investigated.

## 2. Introduction

Methyl benzimidazole-2-ylcarbamate (MBC or carbendazim, fig. 1) is one of the benzimidazole derivatives which is highly toxic to some groups of fungi. Particularly fungi belonging to the Ascomycetes are very sensitive, whereas others belonging to the Oomycetes are resistant[1]. Other benzimidazole derivatives like 2-(thiazol-4-yl)benzimidazole (thiabendazole or TBZ, fig. 1) which is also known as an anthelminthic and 2-(2-furyl)benzimidazole (fuberidazole, fig. 1) have a similar spectrum of activity, although these compounds are less active than MBC.

In agricultural practice these compounds have become increasingly important for the control of plant diseases. In addition to their high toxicity their systemic behaviour within plants is an important property for effective disease control. Due to efficient translocation these compounds can eradicate established infections and cure the plant. In some cases protection of new growth may be achieved by these systemic compounds.

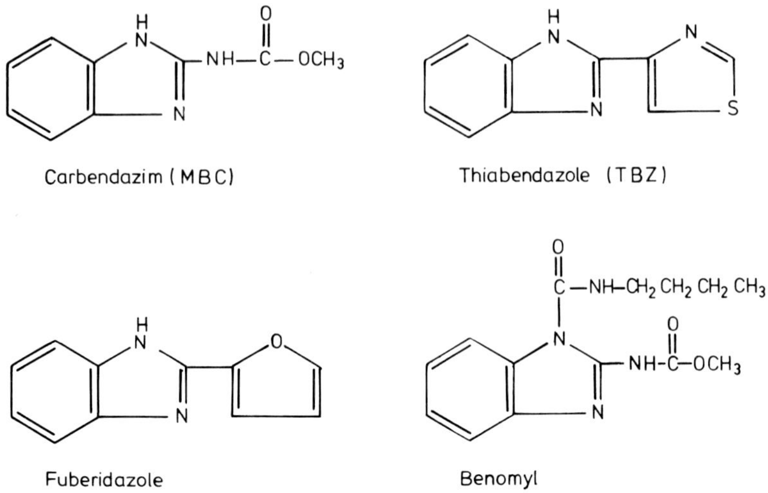


Fig. 1. Structural formulae of carbendazim (MBC), thiabendazole (TBZ), fuberidazole and benomyl.

Although MBC is used as such in spray formulations, methyl 1-(butylcarbamoyl) benzimidazol-2-ylcarbamate (benomyl, fig. 1) is more widely employed. The reason why benomyl which breaks down to MBC in aqueous solution, is used, is due to its slightly superior qualities in a number of applications[2,3].

However, soon after the introduction in practice of the benzimidazole derivatives which were occasionally considered as the final and decisive solution to the control of certain diseases, negative aspects became apparent. In some cases, disease control was no longer successful, because resistance developed in previously sensitive fungal populations. Also in laboratory experiments resistance could be induced by mutagenic agents. This phenomenon, well known in insect pest control, was seldom encountered in controlling fungal diseases until recently and might, therefore, be inherent to the use of the new systemic compounds[4].

Since the benzimidazole derivatives were assumed to be very specific in their action, scientists became interested in these compounds not only from the point of view of their fungicidal aspects, but also because of their possible use as investigative tools to study the physiology of fungi.

The purpose of this paper is to report recent findings on the mechanism of action of MBC and the identification of its site of action. Also the biochemical basis of resistance to this compound will be discussed.

3. Mechanism of action

The first report[5] concerning the mode of action of MBC drew attention to DNA and RNA synthesis as the possible targets of MBC action. However, inhibition of these processes became only evident several hours after addition of the compound to liquid cultures of the fungi studied, viz. Neurospora crassa and Ustilago maydis. The observed inhibition might therefore be considered as a secondary effect.

Since nuclear and cell division might be affected by MBC, these processes were studied separately from DNA synthesis using synchronous cultures of Saccharomyces cerevisiae and U. maydis[6] or by simply counting nuclei in growing germ tubes of Aspergillus nidulans[7]. These studies revealed that primarily mitosis was inhibited by MBC and that inhibition of DNA and RNA synthesis were secondary effects attributable to mitotic failure. In addition, light microscopic observations of MBC-treated hyphae revealed abnormal chromatin configurations.

The antimitotic activity of MBC in fungi resembles mitotic arrest caused by colchicine in mammalian cell cultures[8]. Because of this similarity MBC might be a member of a class of antimitotic agents commonly known as spindle poisons, of which colchicine can be considered as a prototype[9].

4. Induced resistance to MBC and TBZ in Aspergillus nidulans

In order to study more precisely the mechanism of action of the benzimidazole derivatives and the biochemical basis of resistance, resistance to MBC and TBZ was induced in A. nidulans[10]. It appeared that the TBZ-resistant strains could be distinguished into two categories. Strains belonging to the first category which comprised about 95 % of the total number of mutant strains examined, were also resistant to MBC. As distinct from these strains, strains belonging to the second category which comprised the remaining 5 %, showed an increase in sensitivity to MBC. Table 1 gives the inhibitory concentrations of TBZ or MBC causing a 50 % reduction in radial growth on agar ($ED_{50}$ values) for one representative strain of each category, viz. strain R and strain 186, as well as those for the wild type strain 003.

Genetic analysis proved both resistance and increased sensitivity to MBC to be caused by mutation in the same gene, located on linkage group VIII[10]. In addition to this a lower resistance to MBC was found to be caused by mutation of two other genes[11,12].

For biochemical studies, strain R, strain 186 and strain 003 were chosen, because of their characteristic behaviour with respect to MBC.

Table 1

Response of Aspergillus nidulans strains R, 186 and 003 to MBC and TBZ. Effects were estimated by measuring radial growth on malt agar medium containing MBC or TBZ at different concentrations. Values given are inhibitory concentrations causing a 50 % reduction in growth ($ED_{50}$ value)

| | $ED_{50}$ ( $10^{-6}$ M) | |
|---|---|---|
| Strain | TBZ | MBC |
| R | 800 | 95 |
| 186 | 195 | 1.5 |
| 003 | 50 | 4.5 |

5. Binding studies with $^{14}$C-MBC

The characteristic property of spindle poisons is their interference with the function of microtubules. Colchicine accomplishes this by binding tightly and very specifically to tubulin, the protein dimer of which the microtubule is composed.

In order to determine the involvement of a similar binding of MBC to fungal tubulin in the antimitotic action of this compound, cell-free extracts of mycelium of the three A. nidulans strains were incubated with $^{14}$C-MBC. Binding was determined by the standard gel filtration procedure using Sephadex G-100 columns. Typical elution patterns are shown in fig. 2.

From this figure it is evident that radioactivity was present in macromolecular fractions, when extracts of sensitive mycelium were used. The MBC sensitivity of the strains was reflected in the amount of bound radioactivity. Mycelial extracts of the resistant strain did not show binding activity.

Thin layer chromatographic analysis of the bound radioactivity showed that this was still present in $^{14}$C-MBC.

As distinct from colchicine binding, MBC binding did not require a high temperature and appeared to be rapidly reversible, as is suggested by the fact that the peak of bound radioactivity showed considerable tailing. In this respect the binding mechanism resembles that of the spindle poison podophyllotoxin[9].

The relation between binding of MBC and MBC sensitivity was further studied using normal sensitive strains of Penicillium species and resistant strains which emerged in benomyl-treated crops[14]. Also two naturally resistant fungi were included, viz. an Alternaria and a Pythium species. It appeared that only mycelial extracts of sensitive strains showed considerable MBC-binding activity, whereas mycelial extracts from resistant strains did not.

These data strongly suggest that MBC binding to fungal cellular material is essential in the antimitotic action of MBC.

According to the elution volume of the MBC complex it can be concluded that the molecular weight of the binding species is approximately 110,000 daltons. This value has also been obtained in a more accurate study using a Sephadex G-200 column, which was calibrated with proteins of known molecular weight.

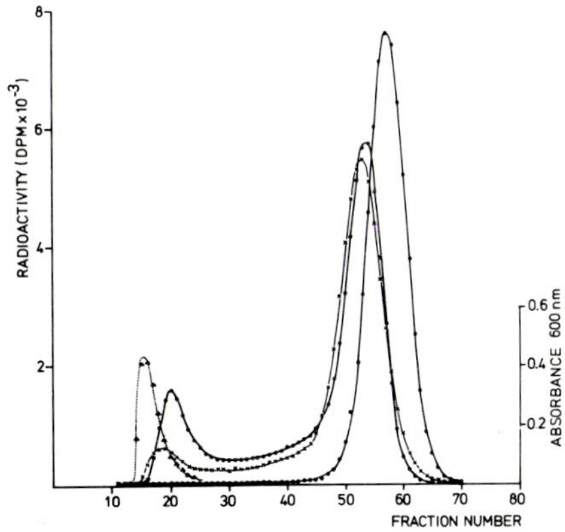


Fig. 2. Gel filtration of 40,000 x g supernatant mycelial extracts of Aspergillus nidulans strains, containing approximately 40 mg protein per ml, incubated with $2.7 \times 10^{-6}$ M 2-$^{14}$C-MBC (spec. act. 11.4 mCi/mmole) for 120 min. at $4^{\circ}$C. Extracts were prepared in 0.05 M potassiumphosphate buffer, pH 6.8, containing 0.1 M KCl, 5 mM $MgCl_2$, 0.1 mM GTP. Elution profiles of radioactivity of parallel experiments with extracts of strain 003 (x--x), strain 186 (o--o) and strain R (•--•) are shown. The elution profile of Blue Dextran is indicated by ▲...▲ (from Davidse[13]).

The MBC-binding substance was further characterized by chromatography on a DEAE-Sephadex A-50 anion exchanger. This method was employed because of the supposed identity of the MBC-binding substance with fungal tubulin. Elution profiles of radioactivity and protein in experiments with mycelial extracts of strain 186 and strain R, incubated with $^{14}$C-MBC are shown in fig 3A and 3B, respectively. With extracts of strain 186 radioactivity was eluted in two peaks, one with the bed volume of the column and the other at 0.6 M KCl, whereas with extracts of strain R the activity was eluted in one single peak with the bed volume of the column.

On the basis of the retention properties on DEAE-Sephadex ion exchangers and the estimated molecular weight of 110,000 daltons there is good reason to believe, that the MBC-binding macromolecule is identical with fungal tubulin. Assuming that binding prevents assembly of tubulin into functional microtubules, the mechanism of action of MBC might be similar with that of colchicine and MBC might be considered a spindle poison, especially active in certain fungi. Induced resistance and extra-sensitivity of A. nidulans strains might then be based on changes in affinity of the receptor site to MBC, caused by mutation of the coding gene. Mutations involving resistance and mutations involving increased sensitivity were shown to be about 16 nucleotides apart from each other[12], corresponding with a distance of about 5 amino acids in the protein. These data are consistent with

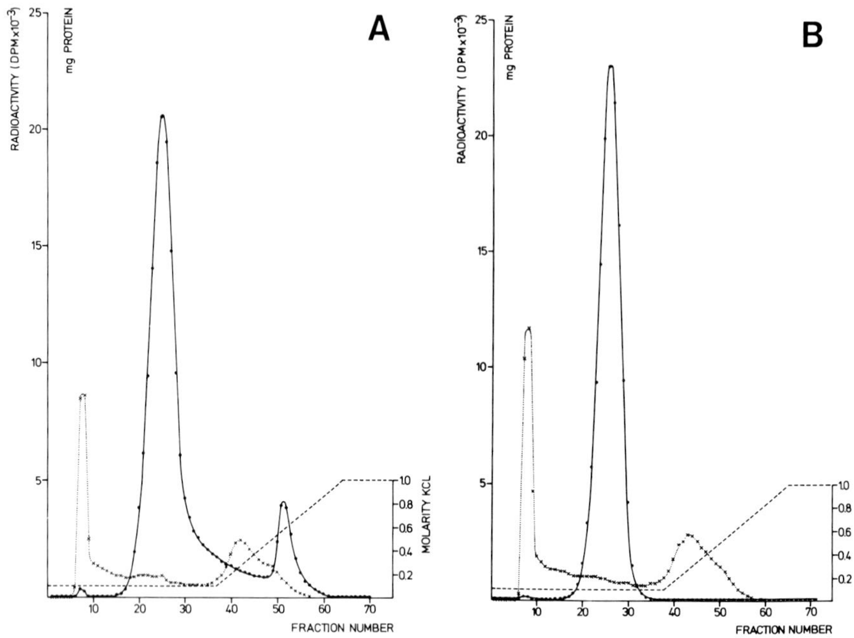


Fig. 3. DEAE-Sephadex A-50 column chromatography of 40,000 x g supernatant mycelial extracts of strain 186 (A) and strain R (B) of Aspergillus nidulans. A 1.5 x 23 column was loaded with 4 ml of an incubation mixture (for details see text fig. 2) and was eluted first with buffer containing 0.1 M KCl, followed by a linear gradient of 0.1 - 1.0 M KCl. Radioactivity (●--●) and protein (x--x) was determined in 1 ml of each 2 ml fraction (----) KCl concentration. (From Davidse[13]).

the hypothesis that changes in affinity might be associated with changes in amino acid composition of the MBC-binding site on tubulin, without interfering with its functional properties.

6. Partial purification of the MBC-binding protein

In order to know more about its biochemical properties an attempt was made to purify the MBC-binding protein, using $^{14}$C-MBC as a marker. Since the binding protein was assumed to be identical with fungal tubulin, a standard tubulin purification procedure[15] was employed. A 127,000 x g (60 min) supernatant mycelial extract was fractionated with a saturated ammonium sulphate solution in water. The fraction precipitating between 35 and 50 % saturation was taken up in buffer and incubated with $^{14}$C-MBC. The incubation mixture was run onto a DEAE-Sephadex A-50 column and the eluted fractions were assayed for radioactivity. Fractions eluted at c. 0.6 M KCl which contained bound MBC and, therefore, the binding protein were subjected to SDS-polyacrylamide electrophoresis (fig. 4). From this figure it is evident that in these fractions a protein is present with similar electrophoretic properties as brain tubulin. According to the specific tubulin purification procedure used, this protein might be considered as fungal tubulin. However, in addition to this protein several other proteins are present, so that from this experiment it can not be definitely concluded that fungal tubulin is responsible for MBC-binding.

Further purification of the MBC-binding protein was greatly hindered by the instability of the binding activity. In 40,000 x g supernatant mycelial extracts binding activity rapidly decayed in a first order manner with a half time of 280 min. at 4°C. Guanosine triphosphate (GTP), which is known to stabilize colchicine--binding activity of tubulin, did not affect the rate of decay. However, in these crude extracts, GTP might be rapidly hydrolyzed, so that no effect can be anticipated. MBC itself, at $10^{-5}$ M, stabilized the binding activity resulting in a half time of decay of 920 min.

Possible due to this rapid decay in the absence of MBC hardly any binding activity could be detected in combined and concentrated (vacuum dialysis or ammonium sulphate precipitation) DEAE-Sephadex A-50 fractions, eluted at c. 0.6 M KCl, when these were incubated with $^{14}$C-MBC and analyzed for bound radioactivity by gel filtration on Sephadex G-100 columns. However, the possibility that passage of an extract through this type of column might remove a factor required for the formation of the MBC-protein complex, can not be ruled out.

In mycelial extracts showing MBC-binding activity no $^{3}$H-colchicine-binding activity could be demonstrated. Although normally colchicine binding can be seen as an indication of the presence of tubulin, in fungi colchicine probably can not be used as a diagnostic tool. Fungal mitosis appeared to be insensitive to colchicine[18,19] presumably due to low affinity of fungal tubulin to this compound.

Although the definite proof has not yet been given the data presented support the idea that MBC is bound to fungal tubulin and that MBC might be used as a diagnostic tool for fungal tubulin.

## 7. Toxicity of MBC in mammalian cell cultures

As has been shown[20,21] MBC exhibits antimitotic activity in mammalian cell cultures. Studies of the mechanism of action in mammalian cell extracts might give more information about the character of MBC as a spindle poison. Binding studies with $^{14}$C-MBC and porcine brain extracts revealed that no binding occurred after incubation at 4°C nor at 37°C, although the preparations showed considerable $^{3}$H-colchicine-binding activity.

In vitro polymerization of porcine brain tubulin into neurotubules was not affected by MBC. Another benzimidazole derivative, methyl 5-(2-thenoyl)-

benzimidazol-2-ylcarbamate, (R 17934), however, proved to be very effective[22].

8. Effects of R 17934 on fungi

In preliminary studies with R 17934 it was shown that this compound could effectively inhibit the growth of fungi. It was found active against A. nidulans even at lower concentrations than MBC. The response of the three above-mentioned strains was similar to that towards MBC, viz. strain R was resistant compared with strain 003, whereas strain 186 appeared to be most sensitive.

In binding experiments with $^{14}$C-MBC, R 17934 could effectively reduce the amount of MBC bound to the protein. These facts indicate that at least in fungi R 17934 and MBC have a similar mechanism of action.

9. Heterologous assembly of fungal and porcine brain tubulin

Assuming interference of MBC with microtubule formation during mitosis in fungi, a study of its effect on fungal tubulin polymerization in vitro might add to our understanding of the ability of MBC to act as a spindle poison.

Preliminary polymerization experiments with fungal tubulin preparations, purified by DEAE-Sephadex A-50 chromatography were not successful. This could be due to the low amount of protein which could be purified and which was probably not enough to reach the critical concentration essential for tubulin polymeri-

Fig. 4. SDS-polyacrylamide electrophoretic analysis[16] of DEAE-Sephadex A-50 fractions containing MBC-binding protein. Five ml of a 35 - 50 % ammonium sulphate preparation in extraction buffer (see text fig. 1) of a 127,000 x g (60 min) supernatant mycelial extract were incubated with $^{14}$C-MBC(final concentration 9 x $10^{-6}$ M; 226,000 dpm/ml) for 120 min. at 4$^{o}$C. The incubation mixture was loaded onto a 20 ml DEAE-Sephadex column, which was subsequently eluted with 15 ml of a linear gradient of 0.1 - 0.4 M KCl, 15 ml of 0.4 M KCl, 30 ml of a linear gradient of 0.4 - 1.0 M KCl and 20 ml 1.0 M KCl, respectively, all made up in buffer. Fractions of 2.5 ml were collected and assayed for radioactivity. Fractions from the 0.6 M region which contained resp. 1670, 2160, 4170, 7230, 5570, 2590 and 1070 dpm bound $^{14}$C-MBC per ml were subjected to electrophoresis (gel A - G). As reference tubulin purified by two cycles of assembly[17] from a porcine brain extract was run on a separate gel (T). Densitometer tracings were made of gel T and gel D (peak fraction of MBC-binding protein). Samples were prepared for electrophoresis as described[16]. Gels consisted of 7.5 % acrylamide, 0.13 % bisacrylamide, 0.1 % SDS in 0.1 M sodium phosphate buffer (pH 7.0). Gels were stained for 60 min. in 0.1 % Coomassie Brilliant Blue in a methanol, water, acetic acid mixture (227:227:46). Destaining was performed by diffusion in 7 % acetic acid in water.

Fig. 5. SDS-polyacrylamide gel electrophoresis of tubulin, partially purified by one cycle of assembly[17] in porcine brain extract (1) and in a mixture of porcine brain and fungal extract (2). For electrophoretic and staining conditions see fig. 4.

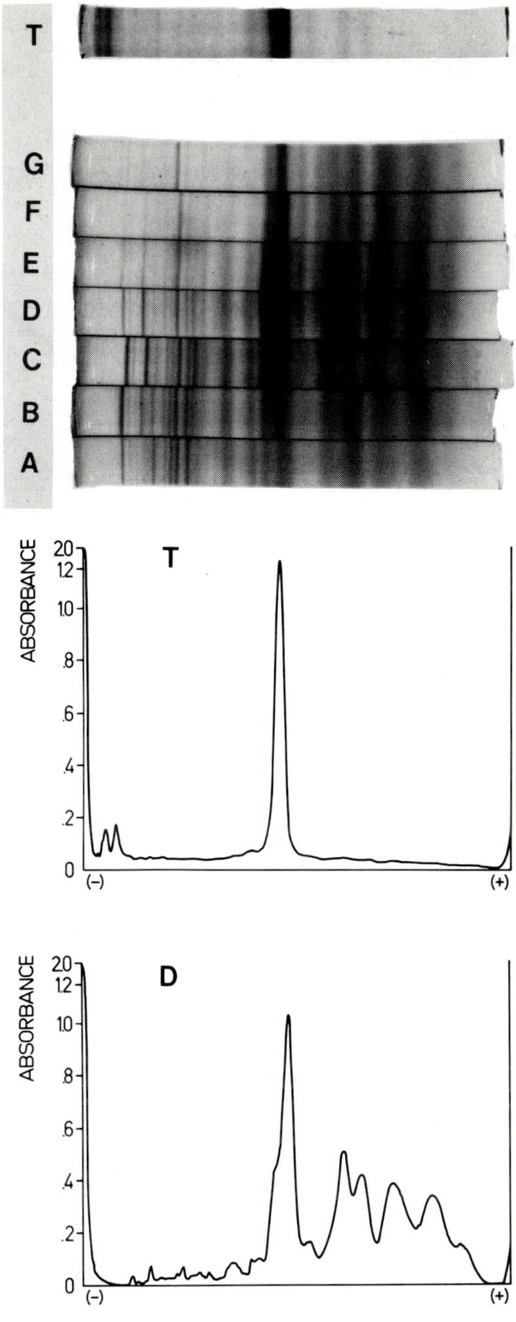

T
G
F
E
D
C
B
A
ABSORBANCE
2.0
1.2
1.0
.8
.6
.4
.2
0
T
(–)
(+)
ABSORBANCE
2.0
1.2
1.0
.8
.6
.4
.2
0
D
(–)
(+)


Fig. 4

1 2


Fig. 5

zation[23]. On the other hand also cofactors essential for polymerization might have been removed during the purification procedure[24,25].

Possible interference of MBC with the assembly of fungal tubulin might also be studied in a heterologous assembly system. Literature data[23,26] report the successful assembly of brain tubulin onto flagellar microtubules from Chlamydomonas and sea urchin sperm. However, tubulin from the cystoplasmatic pool of Chlamydomonas did not copolymerize with brain tubulin, but prevented assembly[27].

In preliminary experiments using 40,000 x g (60 min) supernatant porcine brain extracts and 127,000 x g (120 min) supernatant mycelial extracts of A. nidulans no polymerization could be obtained, as was shown by analysis of the various fractions by SDS-polyacrylamide gel electrophoresis. Partial purification of the mycelial extract by 35 - 50 % ammonium sulphate fractionation and passage of the preparation over a Sephadex G-25 column, equilibrated with reassembly buffer was found necessary to obtain polymerization equal to that in the porcine brain extract, as was revealed by protein determinations in corresponding fractions. That indeed microtubule assembly occurred in the mixture was established by electron microscopy. Microtubules being assembled in the mixture appeared to be similar to those in porcine brain extracts. Fig. 5 shows the results of electrophoretic analysis of the tubulin preparations after one cycle of assembly and disassembly.

It is evident that the assembly product, obtained in the mixture does not contain the two high molecular weight proteins which were found to be always present in tubulin preparations[23,28,29,30], purified by the assembly procedure. These proteins were reported to be the components of filaments attached to in vitro assembled brain microtubules (neurotubules) and neurotubules in situ[31].

Besides the tubulin band several other minor bands were visible on the gels loaded with the assembly product of the mixture which did not appear on the gels of the control preparation. These components persisted through a further second assembly and disassembly step.

These results prompted us to carry out a heterologous assembly experiment, in which the proteins of the fungal extract were labeled. Proteins were labeled by growing mycelium during the last three hours of the growing period in the presence of 25 $\mu$Ci $^{14}$C-phenylalanine per l of medium. About 90 % of the label was taken up and proteins were labeled at about 44,850 dpm/mg protein as was determined in the 35 - 50 % ammonium sulphate preparation after gel filtration on the Sephadex G-25 column.

After one cycle of assembly and disassembly MBC ($10^{-5}$ M) was added to a sample of the preparation, whereas a similar sample served as control, and a second cycle of assembly was performed.

The various fractions were analyzed by SDS-polyacrylamide gel electrophoresis. Gels were sliced and the amount of radioactivity was determined in each segment.

The results of this experiment are given in table 2.

Table 2

Heterologous assembly of fungal and porcine brain tubulin. Two ml of a partially purified mycelial extract (13 mg protein per ml; 44,850 dpm/mg protein) of *Aspergillus nidulans* strain 003 were mixed with 8 ml of a porcine brain extract (5.5 mg protein per ml). Two cycles of assembly and disassembly were performed[17]. In cyle II the effect of MBC ($10^{-5}$ M) was studied. Total radioactivity (dpm) and its distribution (%) among the various fractions are given. Also the percentages of radioactivity actually present in tubulin subunits as revealed by SDS-polyacrylamide gel electrophoretic analysis are given.

| Assembly | | Radioactivity of the fractions (dpm) | | |
|---|---|---|---|---|
| | | mixture | $H_1S$ | $C_1S$ |
| cycle I | | 1,166,000 (100 %) | 1,150,000 (98.6 %) | 15,750 (1.4 %) |
| | | $C_1S$ | $H_2S$ | $H_2P$ |
| cycle II | control | 10,500 (100 %) | 7,680 (73 %) | 2,410 (23 %) |
| | MBC | 10,500 (100 %) | 8,000 (76 %) | 2,390 (23 %) |

| SDS-polyacrylamide gel electrophoresis | | Radioactivity of tubulin (% of total radioactivity of the fraction) | | |
|---|---|---|---|---|
| | | $C_1S$ | $H_2S$ | $H_2P$ |
| cycle II | control | 49 | 42 | 73 |
| | MBC | 49 | 47 | 72 |

$H_1S$ : 45,000 x g (30 min. 25 °C) supernatant from first assembly
$C_1S$ : 45,000 x g (30 min. 4 °C) supernatant from resuspended pellet (30 min. 0 °C) from first assembly
$H_2S$ : 45,000 x g (30 min. 25 °C) supernatant from second assembly
$H_2P$ : resuspended pellet (30 min. 0 °C) from second assembly

As is evident from this table, copolymerization of radiolabeled fungal protein with porcine brain tubulin under microtubule assembly conditions strongly suggests that heterologous assembly has occurred in a mixture of brain and mycelial extract. Although in the first cycle only 1.4 % of the radioactivity could be recovered in the tubulin preparation, in the second cycle this percentage increased to 23 % indicating that actual temperature dependent polymerization of labeled protein has occurred. Gel electrophoresis proved that tubulin was mainly responsible for this phenomenon, because the amount of radioactivity present in tubulin increased from 49 % in $C_1S$ to 73 % in $H_2P$.

At the concentration used MBC did not have any effect on polymerization of radioactive protein, although the fungal preparation used did bind MBC, as was determined in similar preparations.

An explanation for the inability of MBC to inhibit the assembly of fungal tubulin in a heterologous system can only be tentative, as long as the processes of MBC binding and fungal and procine brain tubulin copolymerization are not completely clear. The inability of MBC to interfere with the heterologous assembly might be explained by supposing that an MBC-fungal tubulin complex is the result of interaction of MBC with the suggested Y-state of the tubulin molecule[25,32] and that Y-state fungal tubulin molecules do not take part in the heterologous assembly process.

At present experiments along these lines are in progress.

## 10. Discussion

Although the exact mechanism of action of MBC has not yet been elucidated, several lines of evidence indicate that it acts as a spindle poison in fungi. Its similarity with colchicine in binding to a protein with several biochemical characteristics of tubulin is one of these. The structural relationship of MBC with R 17934, a compound which certainly belongs to this class of antimitotic agents[22] and which probably displays a similar mechanism of action in fungi, also supports this idea.

Whether MBC binding to fungal tubulin is comparable with colchicine binding to mammalian tubulin or whether MBC binding requires a specific state of the tubulin molecule is not yet clear.

Changes in amino acid composition of the MBC-binding site which might result in affinity changes might explain the phenomena of resistance and increased sensitivity to the benzimidazole derivatives. However, mechanism underlying these phenomena might be more complicated and might also involve changes in amounts of tubulin molecules being able to bind MBC.

Although it has been shown that part of the primary structure of the tubulin molecule is strongly conserved during evolution[33], the colchicine-binding site might have evolved along different lines according to the differential colchicine--binding properties of fungal and mammalian tubulin. In a comparable way, a binding site for benzimidazole derivatives might have been evolved leading to a differential response of mammalian and fungal cells to MBC. However, modification of the molecule might strongly influence its interaction with tubulin. Benzimidazole derivatives, therefore, might be valuable tools to study differences in biochemical and functional properties of tubulin from different sources.

Purification of tubulin in a mixture of porcine brain and mycelial extract did not involve copurification of high molecular weight proteins as is normally observed. Supposing polymerization of these proteins being independent of that of microtubules, this phenomenon might be due to an inhibitory action of components of the mycelial extract. However, according to literature data[29,31] it is more likely that presence of the high molecular weight proteins in microtubule preparations is caused by a structural linkage of these proteins to microtubules.

Their absence in tubulin preparations purified from mixtures of brain and mycelial extracts, might, therefore, indicate that in this case linkage did not occur. This might be due to incorporation of fungal tubulin into microtubules making these less suitable for association with high molecular weight proteins.

The incorporation of radioactive fungal protein in microtubule preparations in the heterologous assembly system suggests copolymerization of fungal and brain tubulin. This means that despite the above mentioned differences between the two tubulin species, binding sites involved in polymerization should be highly conserved during evolution.

The heterologous assembly system might provide a new approach by which the in vitro and in vivo control of microtubule assembly can be investigated.

The author wishes to thank Miss Willy Flach for excellent assistance with the experimental work, and Dr. A. Fuchs for help and advice in the preparation of the manuscript. He feels indebted to Dr.Ir. J. Visser (Department of Genetics, Agricultural University) and Prof.Dr.Ir. J. Dekker for a critical reading of the manuscript.

References

1. Bollen, G.J. and Fuchs, A., 1970, Neth. J. Pl. Path. 76, 299.
2. Fuchs, A. et al., 1972, Pestic. Biochem. Physiol. 2, 191.
3. Upham, P.M. and Delp, C.J., 1973, Phytopathology 63, 814.
4. Dekker, J., 1973, OEPP/EPPO Bul. 10, 47.
5. Clemons, G.P. and Sisler, H.D., 1971, Pestic. Biochem. Physiol. 1, 32.
6. Hammerschlag, R.S. and Sisler, H.D., 1971, Pestic. Biochem. Physiol. 3, 42.
7. Davidse, L.C., 1973, Pestic. Biochem. Physiol. 3, 317.
8. Taylor, E.W., 1965, J. Cell. Biol. 25, 145.
9. Wilson, L. et al., 1974, Federation Proc. 33, 158.
10. Van Tuyl, J.M. et al., 1974, Neth. J. Pl. Path. 80, 165.
11. Hastie, A.C. and Georgopoulos, S.G., 1971, J. Gen. Microbiol. 67, 371.
12. Van Tuyl, J.M., 1975, Meded. Fak. Landb. Wetensch. Gent, 40, in press.
13. Davidse, L.C., 1974, Proc. IV Reinhardsbrunner Symp. 5 - 11 May Reinhardsbrunn (DDR), in press.
14. Bollen, G.J., 1971, Neth. J. Pl. Path. 77, 187.
15. Weisenberg, R.C. et al., 1968, Biochemistry 7, 4466.
16. Weber, K. and Osborn, M., 1969, J. Biol. Chem. 244, 4406.
17. Shelanski, M.L. et al., 1973, Proc. Nat. Acad. Sci. USA 70, 765.
18. Haber, J.E. et al., 1972, J. Cell. Biol. 55, 355.
19. Burns, R.G., 1973, Exptl. Cell Res. 81, 285.
20. Styles, J.A. and Garner, R., 1974, Mutation Res. 26, 177.
21. De Brabander, M. personal communication.
22. De Brabander, M. et al., these proceedings.
23. Borisy, G.G. et al., 1974, Federation Proc. 33, 167.
24. Murphy, D.B. and Borisy, G.G., 1974, J. Cell. Biol. 63, 236a.
25. Weingarten, M.D. et al., 1975, Proc. Nat. Acad. Sci. USA 72, 1858.
26. Binder, L.I. et al., 1975, Proc. Nat. Acad. Sci. USA 72, 1122.
27. Farrell, K.W. and Burns, R.G., 1974, J. Cell Biol. 63, 98a.
28. Dentler, W.L. et al., 1974, Proc. Nat. Acad. Sci. USA 71, 1710.
29. Gaskin, F. et al., 1974, FEBS Letters 40, 281.
30. Kuriyama, R., 1975, J. Biochem. 77, 23.
31. Dentler, W.L. et al., 1975, J. Cell Biol. 65, 237.
32. Kirschner, M.W. et al., 1974, Proc. Nat. Acad. Sci. USA 70, 3594.
33. Luduena, R.F. and Woodward, D.O., 1973, Proc. Nat. Acad. Sci. USA 70, 3594.

*Microtubules and Microtubule Inhibitors, eds. M. Borgers and M. de Brabander*
© 1975, North-Holland Publishing Company - Amsterdam



# EFFECTS OF NEW ANTHELMINTICS ON THE MICROTUBULAR SYSTEM OF PARASITES

M. Borgers, S. De Nollin, A. Verheyen, M. De Brabander and D. Thienpont
Janssen Pharmaceutica, Research Laboratories
2340 BEERSE, Belgium.

## SUMMARY

The successive and time-related changes occurring in the absorptive organs of three parasite species after treatment of their hosts with the anthelmintics mebendazole and R 17889 are reported.

Though *Ascaris suum*, *Syngamus trachea* and cysticerci of *Taenia taeniaeformis* inhabit different organs and live under totally different conditions, the response towards these drugs is very similar.

The ultrastructural alterations in the intestinal cells of the two nematodes and the tegument of the cestode appear shortly after the onset of treatment and concern the disappearance of cytoplasmic microtubules. Subsequently, a block in the movement of subcellular organelles with in particular a block in the transport of secretory granules results in an excessive accumulation of these structures in the cytoplasm. After the prolonged storage of the secretory substances, the cytoplasm becomes lytic and the absorptive cells degenerate completely. The damage is irreversible to this "permanent" type of cells and is lethal to the parasites. In contrast to the extreme vulnerability towards these drugs of the microtubules of the parasite cells, those of the hosts (cytoplasmic and spindle) remain intact.

## INTRODUCTION

Mebendazole (methyl(5-benzoyl-1H-benzimidazol-2-yl)carbamate) is a broad-spectrum anthelmintic, active against a series of nematodal and cestodal species[1-7].

Preliminary data obtained with one of its analogues, R 17889 (methyl[5-(4-fluorobenzoyl)-1H-benzimidazol-2-yl]carbamate) have shown that this drug possesses anthelmintic properties that are comparable to those of mebendazole as far as potency and spectrum are concerned.

In spite of a tremendous amount of biochemical and histological work that has been done in the past, the mode of action of the drug, i.e. the mechanism by which the parasites are actually killed, has never been explained in a satisfactory way. Our current work in this field is therefore concerned with the morphologic identification of the target organ(s) and the target subcellular organelle(s) hoping to gain more substantial information on the basic problems : How do helminths get killed and what makes them so susceptible to this type of drugs ? Recently reported data[8], describing the ultrastructural changes in *Ascaris suum* after mebendazole treatment, have shown that the first morphologic alterations are observed in the intestinal epithelium and that the death of the parasites is most probably caused by irreversible degeneration of this organ. The aim of the present study is to report the successive and time-related changes in the absorptive compartment of helminths after treating their hosts with mebendazole or R 17889.

In order to obtain a more complete evaluation of these drugs on the morphology of parasites, observations were made on three different species living under totally different conditions, i.e. *Ascaris suum* (roundworm of the pig), *Syngamus trachea* (gapeworm of the turkey) and cysticerci of *Taenia taeniaeformis* (larval form of the tapeworm of the cat). To establish the specificity of action of these drugs, host tissues were examined as well.

*ASCARIS SUUM*

This nematode settles in the jejunal part of the pig small intestine. Pigs, naturally infected with adult *A. suum*, were fed *ad libitum* a medicated food containing 30 or 125 ppm mebendazole or R 17889. They were sacrificed 6, 9, 15 or 24 hours after the onset of medication. Infected pigs, fed *ad libitum* a normal pig food, served as controls. The preparation procedure of the parasite samples for electron microscopy has been detailed before[8].

As described previously[9,10] the intestinal tract of *A. suum* is composed of a single layer of cylindrical cells which do not undergo continuous replacement. Characteristic for these "permanent" epithelial cells is the presence of a large number of secretory granules, derived from the Golgi apparatus and transported through the cytoplasm to reach the terminal web area where they dissolve or fuse with the apical plasmalemma (Fig. 1a). It is assumed that these granules are the carriers of microvillous coat material and that at least part of this material constitutes in fact hydrolytic and proteolytic enzymes, needed for the extracellular digestion of nutrients. Microtubules, which are present all over the cytoplasm and run mostly in apico-basal direction, are particularly abundant in the Golgi region (Fig. 1b).

The changes observed after 6 hours of drug medication were twofold and, as far as one can judge morphologically, they occurred simultaneously : microtubules disappeared from the apical half of the cytoplasm and secretory granules accumulated in the Golgi areas (Fig. 1c). In cells where accumulation in the Golgi area was apparent, granules were absent in the terminal web region. After 9 hours of medication, all intestinal cells showed massive accumulation and fusion of granules at the site of final synthesis. While in the cells of untreated *Ascaris*, secretory granules were devoid of acid phosphatase (ACPase) activity, this enzyme could be demonstrated in the fused granules after 9 hours of treatment.

Figure 1 : *Ascaris suum* intestine.
a) Control. Apical part of an epithelial cell showing multiple secretion granules (arrowheads) in the terminal web (tw) and the Golgi-apparatus (g) ; microvilli (mv). b) Control. Detail of a Golgi apparatus showing the formation of secretory granules (arrowheads) and the presence of multiple microtubules (arrows) in their vicinity. c) After 6 hours of mebendazole treatment. Area comparable to b. Early signs of accumulation and fusion of granules (arrowheads) are seen in the Golgi. Note the absence of microtubules. d) After 9 hours of treatment. The terminal web (tw) is devoid of granules whereas accumulation of single and fused granules is present in the Golgi areas. e) 24 hours after mebendazole treatment. Intestinal cells showing severe cytoplasmic autolysis.

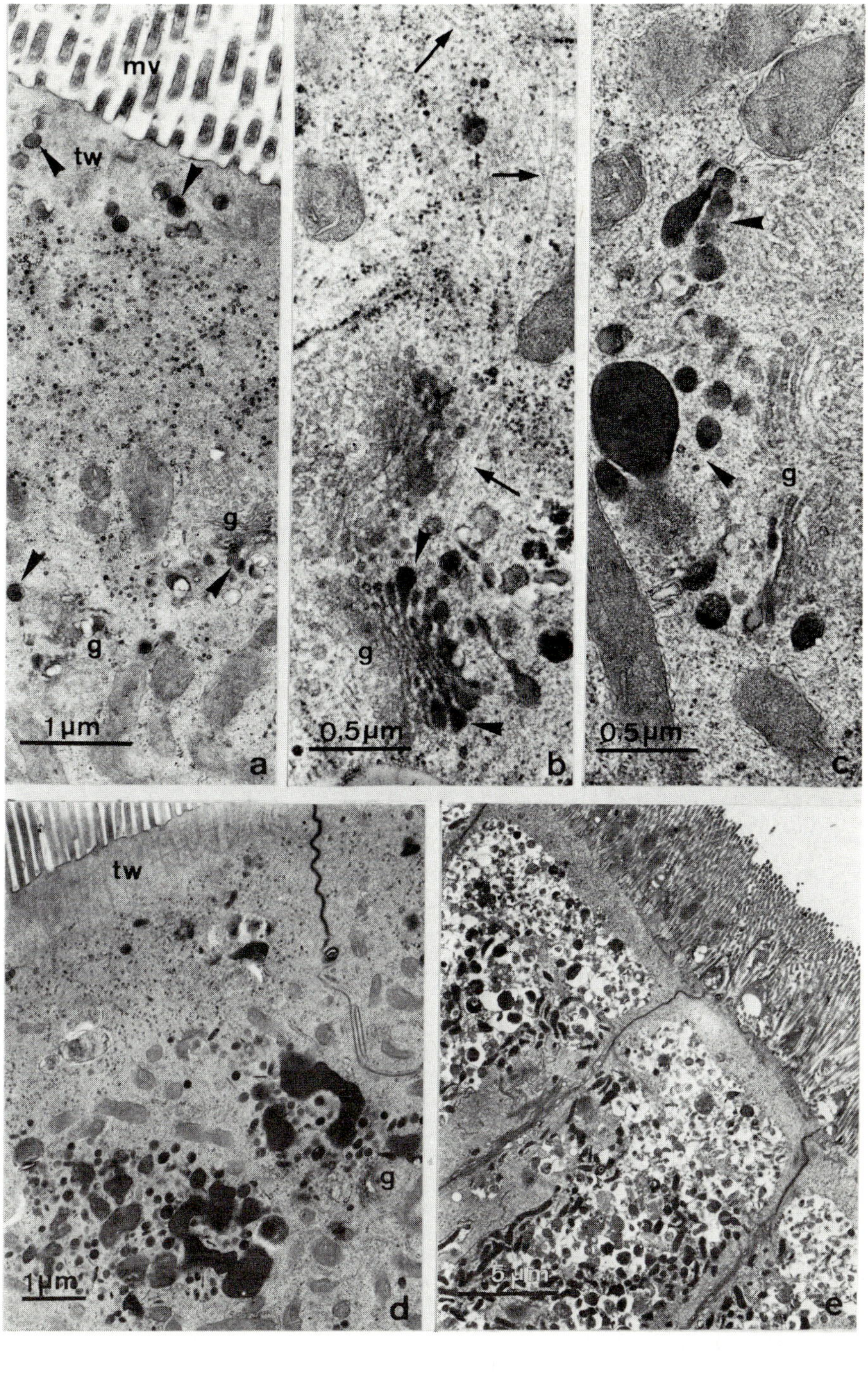
mv
tw
g
g
1 µm
a
g
0.5 µm
b
g
0.5 µm
c
tw
g
1 µm
d
5 µm
e

After 15 and 24 hours of medication, cellular injury was predominent in every cell and was characterized by swelling, fusion or even complete loss of the microvilli, lysis of the cytoplasm with excessive accumulation of vacuoles containing most probably remnants of secretory substances and degeneration of subcellular organelles (Fig. 1c). ACPase activity was found distributed over the total height of the necrotic areas.

The degenerative changes of the intestinal cells of *A. suum* which are irreversible for such a non-proliferative cell type and which lead to the death of the parasite, are interpreted as follows : mebendazole and R 17889 cause the disintegration of the cytoplasmic microtubular system, which results in a block in the movement of subcellular organelles including a block in the transport of secretory granules from the Golgi towards the apical cell surface. This leads to an excessive accumulation of secretory products in the cell cytoplasm. The deleterious effects on the cell of a prolonged storage of such materials in the cytoplasm may be twofold : 1) deprivation of the absorptive surface from protective coat substances and from enzymes needed for digestion and absorption of nutrients, and 2) proteolytic and hydrolytic enzymes, which are normally transported in inactive form[11] to avoid autolysis, may become active upon prolonged storage (cfr. presence of ACPase) and cause cellular autolysis as seen after 15 and 24 hours of treatment.

Examination of the host intestine reveals no changes in distribution and ultrastructure of the various subcellular organelles, including cytoplasmic and spindle microtubules (Fig. 5a).

*SYNGAMUS TRACHEA.*

From the above described results on *A. suum*, it appears that there is a striking difference in sensitivity towards this drug between host and parasite. To see whether this discrepancy can be explained by a differential uptake of the drugs by host and parasite, we studied the effects of oral treatment on *S. trachea*, a sanguivorous nematode that inhabits the trachea of turkeys. Since the drug can reach the parasite by systemic route only, the host cells such as intestinal-, tracheal- and blood cells are exposed to the same drug concentrations as the parasite, unless accumulation occurs.

Turkeys, carriers of adult *S. trachea*, received medicated food containing 125 ppm mebendazole and were sacrificed 6, 12, 24, 48 or 72 hours after the onset of

Figure 2 : *Syngamus trachea* intestine.
a) Control. Section through the apical part of a gut syncytial cell showing the continuity between the fibrillar core of the microvilli (mv) and the fibrillar network of the terminal web (tw). Vesicular structures (v) are intermingled with numerous microtubules (arrows). b) After 6 hours of mebendazole treatment. Area comparable to figure 5. Microtubules (arrows) are rare. c) After 24 hours of mebendazole treatment. Microvilli no longer contact the fibrillar terminal web (tw). The space in between is occupied by dislocated organelles such as a Golgi apparatus (g), an autophagic vacuole (av), a lipid globule (l) and a mitochondrion (m). Microtubules are no longer visible. d) After 72 hours of mebendazole treatment. A severely altered microvillous border (mv) and a heavily vacuolated lytic cytoplasm are seen.

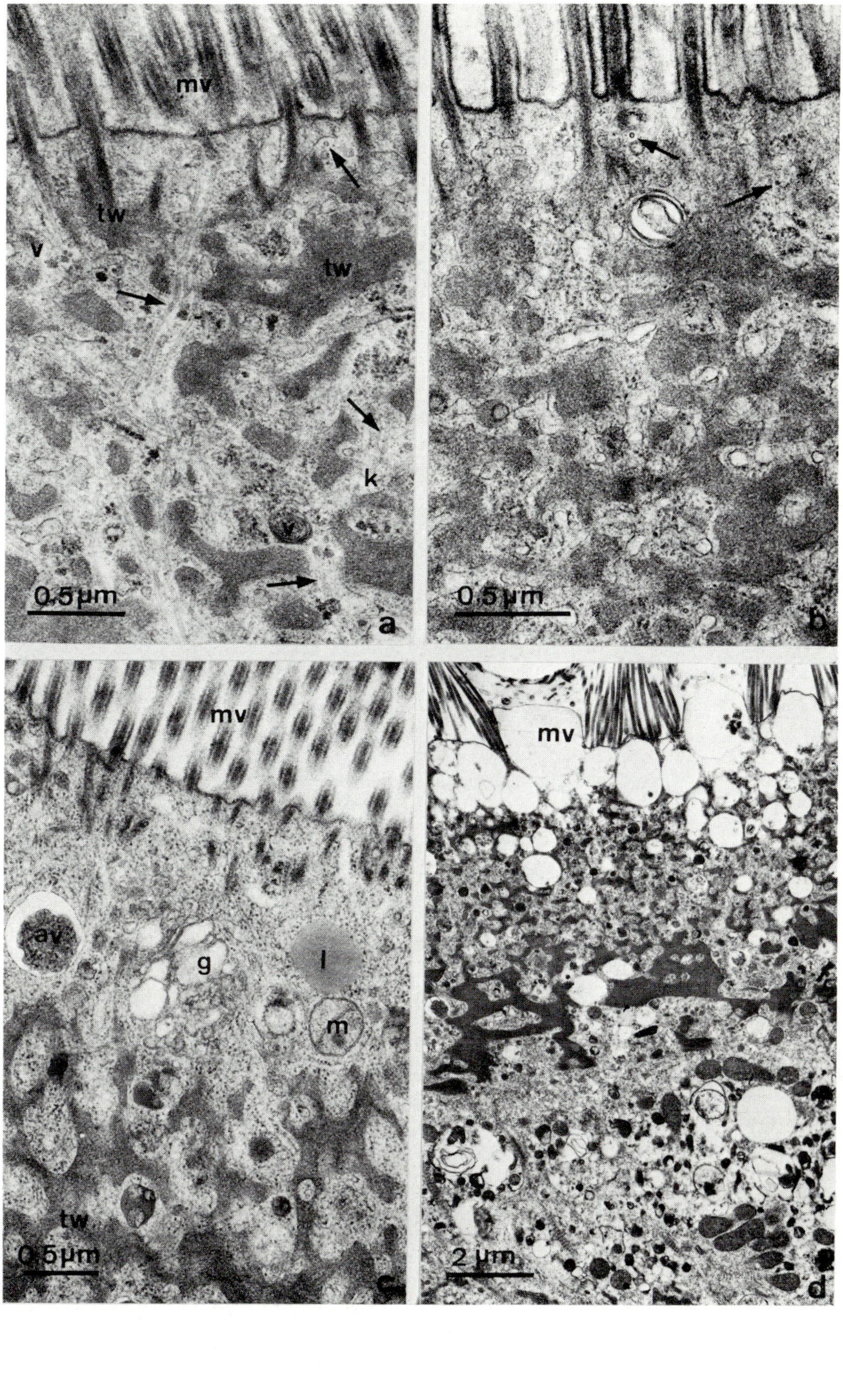
mv
tw
v
tw
k
v
0.5µm
a
0.5µm
b
mv
av
g
l
m
tw
0.5µm
c
mv
2 µm
d

continuous medication. Untreated, infected animals served as controls. The parasites and host tissues (small intestine and trachea) were prepared for ultrastructural investigation.

The intestinal tract of *S. trachea* is a syncytial structure. The terminal web region consists of a densely packed network of filaments which are directly connected with the microvillous core. In between these filaments and in the area just beneath, microtubules are abundant (Fig. 2a). Like for *A. suum*, granules derived from the Golgi apparatus are found intermingled with these structures.

The morphologic changes induced by mebendazole were very well comparable with those found in *A. suum*, though the response in time is different. Microtubules became very scarce already 6 and 12 hours after treatment, however, no abnormalities were visible in other organelles at these time-periods (Fig. 2b). After 24 hours of medication, when microtubules were completely absent from the syncytial cell, the first changes occurred in other organelles. Autophagic vacuoles were formed just beneath the microvillous border and secretory substances accumulated in the central cytoplasm. Very peculiar at that moment was the frequent occurrence of dislocated organelles. Individual Golgi-organelles and mitochondria were found in the uppermost part of the terminal web in close apposition to the microvilli (Fig. 2c).

A lytic degeneration of the gut, quite similar to that observed in *A. suum*, became prominent after 48 and 72 hours of continuous medication (Fig. 2d).

As for the host tissues, the general as well as the micromorphology of the mucosal and submucosal cells of the small intestine and the trachea in all treated animals remained unchanged. Mitoses in the glands of Lieberkühn proceeded normally (Fig. 5b).

### *TAENIA TAENIAEFORMIS*

To verify whether the same vulnerability of the "absorptive organ" applies to parasites lacking a proper intestinal tract, we chose to extend our work to a cestodal species which feeds solely by taking up nutrients through its syncytial tegumental surface.

Swiss mice, infected orally with eggs of *T. taeniaeformis* were used in this study after 1 year of infection. The larvae, measuring about 20 cm at that period of maturation, are encysted in a host capsule that separates them from the surrounding liver parenchyma. Mebendazole or R 17889 was administered as a single parenteral injection of 10 - 40 mg/kg or mixed with the food in a 250 ppm concentration. The

---

Figure 3 : *Taenia taeniaeformis* cysticerci.
a) Control. Survey showing the general organization of the syncytial tegument and the surrounding cells. The anucleated surface of the tegument is studded with microtriches (mt). The nucleated region (tc) is connected with the surface (tg) through tiny cytoplasmic processes (arrows). Muscle cells (mu), parenchymal cells (pc).
b) Control. Detail of the elongated cytoplasmic bridge connecting the nucleated with the anucleated tegument, filled with microtubules (arrows). c) After 6 hours of parenteral R 17889 treatment. Comparable area as shown in b. The tiny cytoplasmic connections are devoid of microtubules.

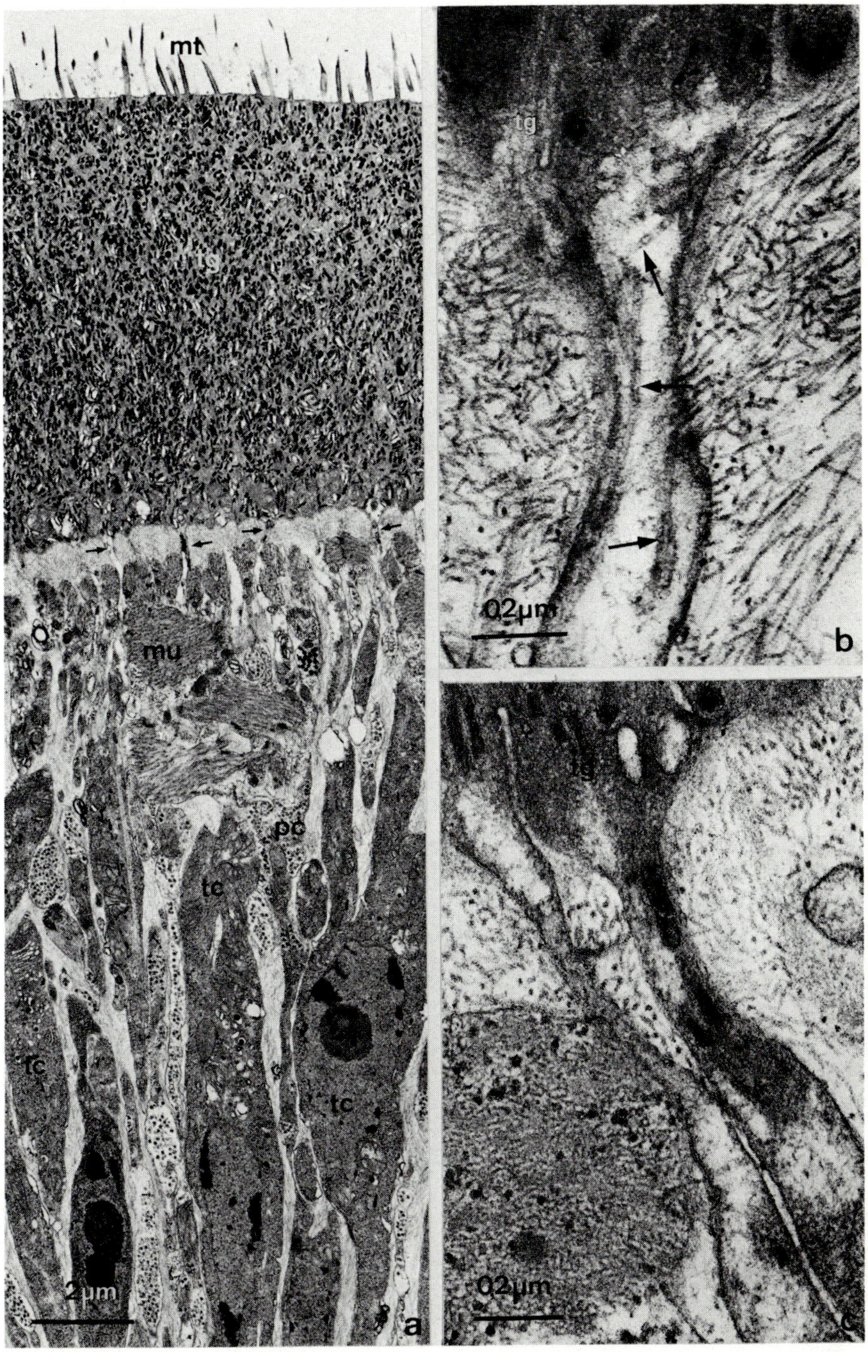
mt
mu
pc
tc
tc
tc
2μm
a
tg
0.2μm
b
tg
0.2μm
c

cestodal larvae and host tissues (intestine, liver and capsule wall) were processed for electron microscopy 6, 12, 24, 48, 72 hours, 1 and 2 weeks after the single parenteral injection or after continuous feeding of the medicated diet.

The tegument of the cysticercus is a syncytial structure bordered by a microvillous surface, called microtriches, through which nutrients are taken up. The anucleated region of the tegument (surface part) is connected with the nucleated region through tiny cytoplasmic processes (Fig. 3a). The anucleated region is filled with electron-lucent round and electron-dense ovoid vesicles. Both types of vesicles derive from the Golgi cisternae in the nucleated region[12]. In the untreated cysticerci microtubules are seen in cells of the tegument, especially in the nucleated region and in the cytoplasmic processes connecting the nucleated part with the surface part of the tegument (Fig. 3b). Large amounts of microtubules are also present in cells of the parenchyma, the excretory channel, the muscle and the sensory organ.

From 6 hours after parenteral administration of the anthelmintics on, microtubules had disappeared from the tegument (Fig. 3c). The progressive changes occurring from 12 hours after the onset of treatment on concerned mainly the tegument. In the nucleated region, accumulation of vesicles took place subsequent to the disappearance of microtubules (Fig. 4a). The number of microtriches decreased considerably ; they were totally absent after 24 hours, leaving an irregularly lined surface (Fig. 4b). After 24 and 48 hours, all nucleated regions were balloon-shaped because of the excessive accumulation of vesicles. After 72 hours, such cells became lytic and burst (Fig. 4c). The tegumental surface at that time was very irregular in thickness and heavily vacuolated. Changes in other cells were noted after 72 hours only. Even at this late period intact microtubules remained visible in sensory-, parenchymal-, excretory duct- and muscle cells. After one week, structural organization was completely lost, tegumental cells were no longer observed and only a fibrous skeleton intermingled with cell residues of unknown origin remained. As for the oral treatment, after 2 weeks continuous medication with mebendazole or R 17889 at 250 ppm concentration, no surviving cysticerci were found.

Examination of the host tissues revealed no changes in microtubule distribution in the intestinal-, liver parenchymal- and capsular wall cells in any of the treated animals (Fig. 5c).

Like in nematodes described above, the microtubular system in the cells functioning as "nutrient-transporting cells" are most vulnerable towards these drugs. The

Figure 4 : *Taenia taeniaeformis* cysticerci.
a) After 12 hours of parenteral R 17889 treatment. Nucleated region of the tegument showing accumulation of electron-lucent and electron dense vesicles (v) in the Golgi area (g). Microtubules are absent. b) After 24 hours of parenteral R 17889 treatment. The tegumental surface (tg) is devoid of microtriches and becomes irregularly lined. c) After 24 hours of parenteral mebendazole treatment. The thinned tegument surface is strongly altered, whereas all nucleated areas of the tegument (tc) are balloon-shaped and filled in excess with vesicles. Some of them become lytic and show bursts (arrows).

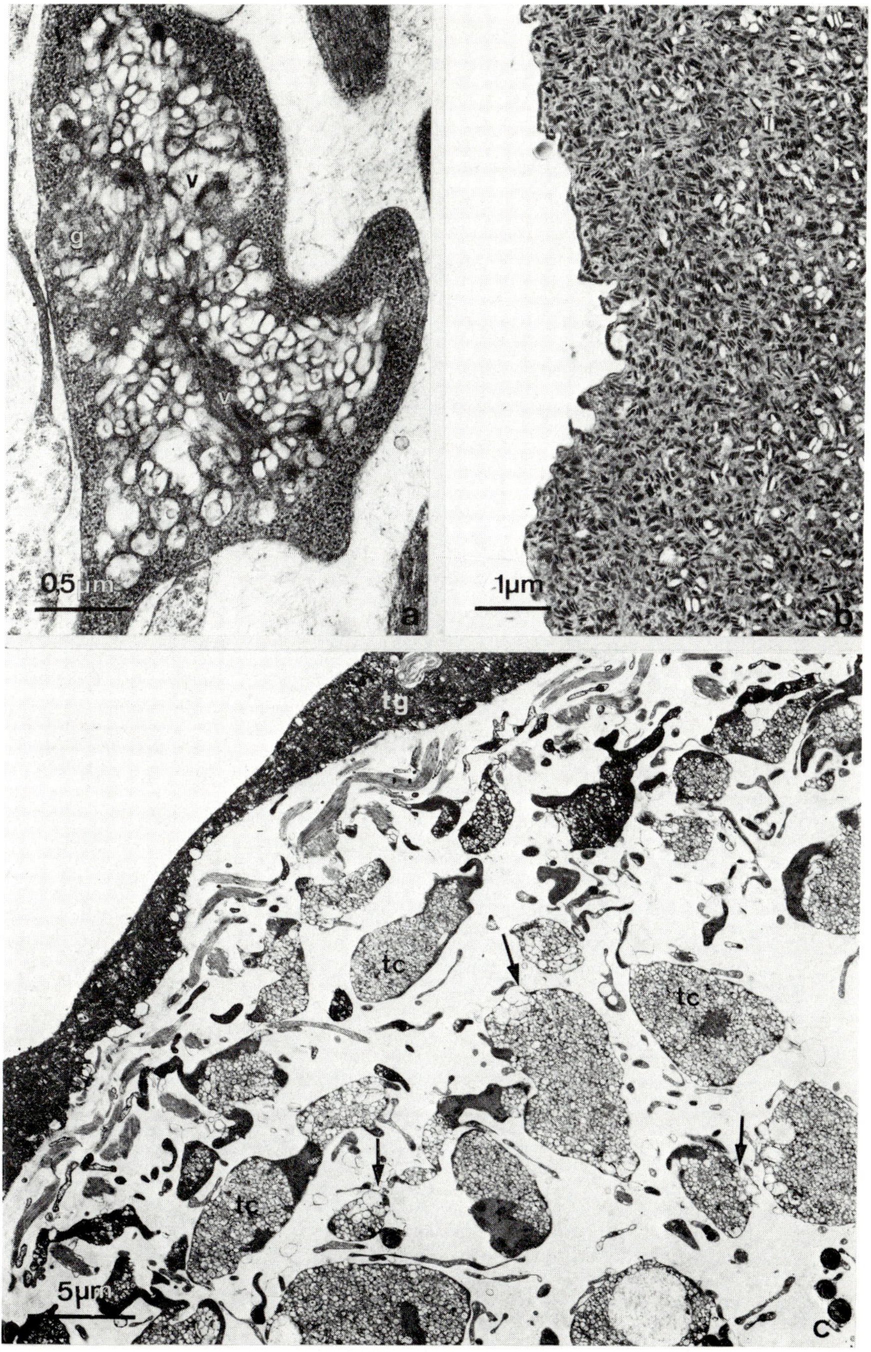
v
tg
v
0.5 µm
a
1µm
b
tg
tc
tc
tc
5µm
c

subsequently disturbed mechanism of transport of vesicles, which have been claimed by Lumsden et al.[12] to play an important role in the formation of the microtrix border covering the worm surface, causes a progressive breakdown of this important organ. The observation that microtriches disappear shortly after a block in transport of the vesicles has been evoked by drug treatment indicates that such vesicles are not only involved with the formation of microtriches, but may be important for their maintenance as well. Another consequence of the disintegration of microtubules is its deleterious effect on the nucleated parts of the tegument. These parts which apparently continue to produce secretory vesicles, accumulate these structures till they burst.

## CONCLUSIONS

The successive morphologic changes brought about by mebendazole and R 17889 in the absorptive organs of three parasites living under different environmental conditions are quite comparable to each other. The impact of the primary interaction of these drugs with cytoplasmic microtubules is of great importance and is interpreted as the initial cause of the subsequently observed cytoplasmic disorganization and lytic degeneration of the absorptive cells leading to the death of the parasites. The involvement of microtubules in the process of intracellular organelle-movement has been reported for various cells[11,13-15]. Unlike most reports dealing with the action of anti-microtubular agents on proliferative cells where cellular necrosis results from an irreversible block of the mitosis, the anthelmintics mebendazole and R 17889 induce necrosis of non-proliferative cells in parasites through interference with their cytoplasmic microtubular apparatus. The extreme difference in sensitivity of microtubules towards these drugs between host and parasite remains unexplained. Although there is, as yet, no indication of any structural difference, chemically or morphologically, between microtubules of parasites and higher vertebrates, one may not rule out this possibility. On the other hand, the divergence might be due to either a different metabolism of the drugs, a different interaction with nucleation sites, or a different susceptibility of the regulatory mechanisms responsible for polymerization and depolymerization of microtubules.

---

Figure 5 : *Host tissues.*
a) Pig small intestine after 24 hours of continuous oral feeding with mebendazole. Survey of the absorptive epithelium indicating a completely normal cell organization and preservation of subcellular organelles. b) Turkey small intestine after 72 hours of continuous oral feeding with mebendazole. Detail of a mitotic cell in the crypts of Lieberkühn. The spindle microtubules are tangentially or transversely cut (arrows). Chromosome (chr). c) Mouse capsule surrounding *T. taeneaeformis* cysticerci after continuous feeding of R 17889 for 14 days. Detail of a mononuclear phagocyte. Section through the centriolar - Golgi region showing numerous well-preserved microtubules (arrows).

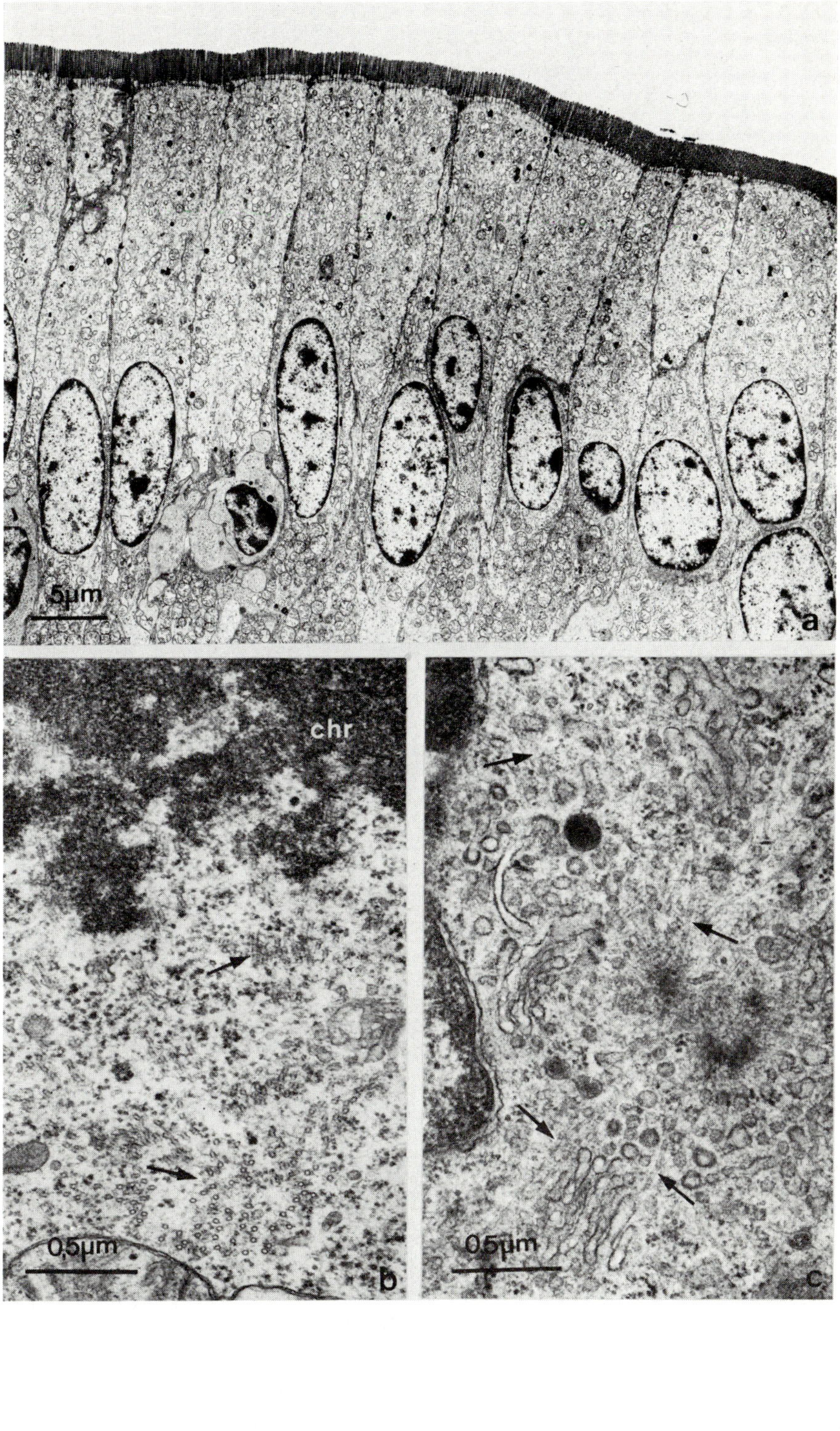
5µm
a
chr
0,5µm
b
0,5µm
c

REFERENCES


1. Brugmans, J., Thienpont, D., Van Wijngaarden, I., Vanparijs, O., Schuermans, V. and Lauwers, H., 1971, J. amer. med. Ass. 217, 313.
2. Chaia, G., Métene, F., Chiari, L., Araujo, S. de M. and de Abreu, I.B., 1972, Folia med. 64, 139.
3. Thienpont, D.C., Vanparijs, O.F.J. and Hermans, L.C., 1973, Poultry Sci. 52, 1712.
4. Thienpont, D., Vanparijs, O. and Hermans, L., 1974, J. Parasit. 60, 1052.
5. Heath, D.D. and Chevis, R.A.F., 1974, Lancet 2, 218.
6. Heath, D.D., Christie, M.J. and Chevis, R.A.F., 1975, Parasitology 70, 273.
7. Campbell, W.C., McCracken, R.O. and Blair, L.S., 1974, Jama 230, 825.
8. Borgers, M. and De Nollin, S., 1975, J. Parasit. 61, 110.
9. Sheffield, H.G., 1964, J. Parasit. 50, 365.
10. Borgers, M. and De Nollin, S., 1974, J. Parasit. 60, 953.
11. Colam, J.B., 1971, Parasitology 62, 259.
12. Lumsden, R.D., Oaks, J.A. and Mueller, J.F., 1974, J. Parasit. 60, 209.
13. Bikle, D., Filney, L.G. and Porter, K.R., 1966, Protoplasma 61, 322.
14. Bhisey, A.N. and Freed, J.J., 1971, Exptl. Cell Res. 64, 430.
15. Stein, O., Sanger, L. and Stein, Y., 1974, J. Cell Biol. 62, 90.

*Microtubules and Microtubule Inhibitors, eds. M. Borgers and M. de Brabander*



# ONCODAZOLE (R 17934) : A NEW ANTI-CANCER DRUG INTERFERING WITH MICROTUBULES. EFFECTS ON NEOPLASTIC CELLS CULTURED IN VITRO AND IN VIVO.

M. De Brabander, R. Van de Veire, F. Aerts, G. Geuens,
M. Borgers, L. Desplenter and J. De Crée
Janssen Pharmaceutica, Research Laboratories
2340 BEERSE, Belgium.

## SUMMARY

R 17934 is a new synthetic compound showing antitumoral activity in many experimental and human neoplasms. Ultrastructural investigations on tissue cultured cells show that the activity of R 17934 resides, most probably, in its microtubule-dissoluting properties. This results in the complete disorganization of both mitotic and interphase cells as a consequence of the arrest of ordered directional organelle-movements. Prolonged treatment results in the appearance of bundles composed of 100 Å filaments and of annulated lamellae, as is the case in cells treated with other antimicrotubular substances, suggesting a close functional link between these structures and microtubules.

*In vivo* experiments on the L1210 system and preliminary clinical trials in human patients were done with a micronized suspension of R 17934, which is poorly soluble in aquous media. Ultrastructural observations show that the compound is avidly phagocytosed by neoplastic cells and normal phagocytes. The antimicrotubular activity of the compound remains present when given in this formulation, resulting in disorganization of dividing and non-dividing neoplastic cells. The microtubules of non-dividing normal cells (phagocytes, peritoneal mesothelial cells, liver parenchymal cells, vascular endothelial cells etc.) are apparently unaffected by doses many times higher than those which attack microtubules of neoplastic cells in the same animals.

From an experimental, cell biological point of view, R 17934 offers an additional tool in microtubule research.

From a therapeutic point of view R 17934 seems to be a promising new anti-neoplastic agent. It should be stressed that the antitumoral activity of this and related compounds is most probably not solely related to its mitoclastic properties since non-dividing neoplastic cells are also completely disorganized. Both its rather high degree of specificity and its "phagosomotropic" properties may be factors favoring its activity.

## INTRODUCTION

R 17934 (methyl[5-(2 thienylcarbonyl)-1H-benzimidazol-2-yl] carbamate; MW. 301,31; Janssen Pharmaceutica) is a new antineoplastic compound that shows activity against many experimental tumors and leukemias (1,2). Antitumoral activity was also shown in preliminary clinical trials. This report deals with our investigations on the subcellular site of action of this compound. For this purpose we used tissue cultured cells of various origin. The effects of R 17934 were assessed by using time-lapse phase contrast observation, quantitative cell kinetic measurements and ultrastructural observation. In order to find out whether the phenomena observed *in vitro* had any relevance to the antitumoral effect *in vivo*, we examined the altera-

Supported by a grant from the I.W.O.N.L. (Instituut tot aanmoediging van het wetenschappelijk onderzoek in nijverheid en landbouw) Belgium

tions of tumoral cells after treatment of experimental animals or human patients.

EFFECTS ON TISSUE CULTURED CELLS

The experiments were performed using a variety of different cell lines. These included : $MO_4$ (a virally transformed mouse embryonal cell line) and its untransformed variant MO; primary embryonal fibroblast cultures derived by trypsinization from C3H mouse embryos; melanoma B16 cells; HeLa cells; and a human embryonal fibroblast cell line.

Most effects were largely the same in all the cells studied. Differences in reaction will be stressed whenever appropriate.

When subconfluent tissue cultured cells are treated with R 17934 at doses ranging from 0.04 - 10 µg/ml, starting from a solution of the compound in dimethyl sulfoxide (5 mg/ml) they loose their stretched bipolar character within 10-20 minutes (3). Undulating membrane activity, which is normally confined to the leading pole of the cell becomes distributed all over the cell periphery. This results in the loss of net translational movement of the cells over the culture surface. The centripetal movement of endocytic vacuoles is also immediately halted, as is the saltatory movement of the perinuclear lysosome-like structures. Subsequently the gelated aspect of the cytoplasm is replaced by periodic bursts of cytoplasmic mass streaming which ensues in a randomization of the granular structures surrounding the nucleus within 2-4 hrs. (3).

At the ultrastructural level treatment with R 17934 (0.04 - 10 µg/ml) induces the loss of most of the cytoplasmic microtubules within 10 minutes. The microtubular disintegration is completed within 120 minutes. Concomitantly a total loss of the normal structural organization and compartmentalization becomes apparent. The centrosphere area - the Golgi organelles surrounding the centrioles - normally located in a perinuclear zone, becomes "exploded" with individual Golgi organelles distributed all over the cytoplasm (fig. 1a) and with dislocated centriolar complexes, often present near the plasma membrane. The perinuclear rim of lysosomal structures assumes a randomized patchy distribution too (fig. 1a).

The foregoing phenomena (and the effects on mitotic cells as described below) are identical to the effects of the standard antimicrotubular substances on the same cells and confirm the many previous publications on this subject (4, 5, 6, 7, 8, 9, 10). Since the behavior and ultrastructure of tissue cultured cells after prolonged treatment with antimicrotubular substances seemed to be a somewhat neglected area we investigated this using R 17934 and colchicine, vin-

Figure 1

a) MO cell treated for 5 hrs. with 1 µg/ml of R 17934. The Golgi organelles (arrows) are dispersed and often surrounded by accumulations of smooth endoplasmic profiles. A large accumulation of smooth endoplasmic vesicles is denoted by an asterisk. Small filament bundles are already present. Microtubules are completely absent.
b) Perinuclear (n) region of a MO cell treated for 6 days with 1 µg/ml of R 17934. A large bundle of 100 Å filaments (f) and annulated lamellae (al) are shown.

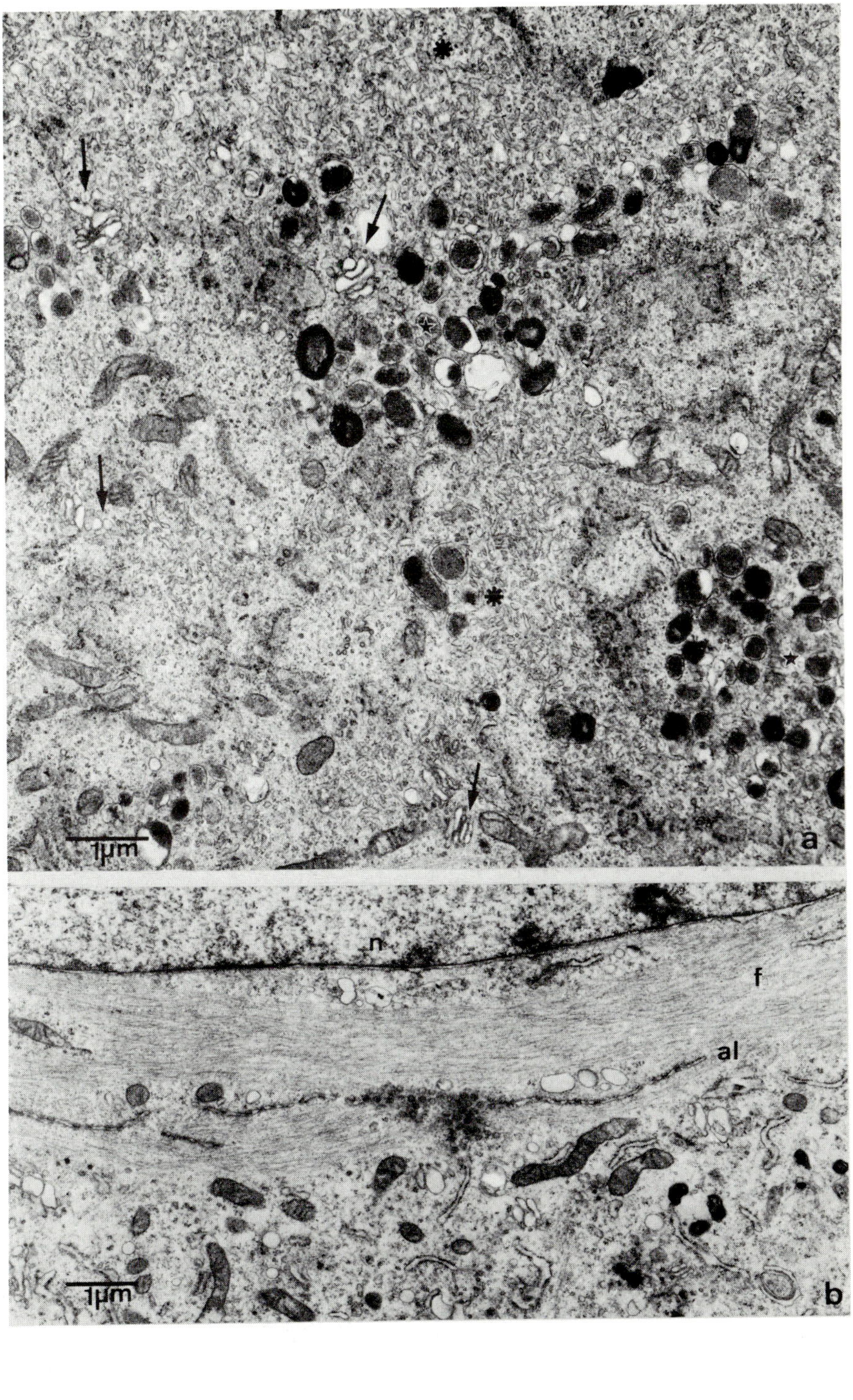
1µm
a
n
f
al
1µm
b

blastine and vincristine as reference tools. The following gives a description of these observations which were identical with all the compounds used except for the production of tubulin crystals with high dosages ( 10 μg/ml) of the vinca alkaloids.

The loss of intracellular organisation, which is accomplished after approximately 24 hrs. is followed by a gradual accumulation of smooth endoplasmic vesicular material, apparently starting around the dispersed Golgi organelles (fig. 1a). This reaches a maximum after approximately 5 hrs. and subsides afterwards when a general hypertrophy of the rough endoplasmic reticulum becomes apparent. After 24-48 hrs. accumulations of two peculiar structures appear in all cell lines tested with all microtubule disintegrating doses of all the above mentioned compounds. These structures are : bundles of 100 Å filaments curving throughout the cytoplasm and annulated lamellae, parallel stacks of endoplasmic reticulum characterized by periodic annuli which closely resemble the pores in the nuclear membrane (fig. 1b).

We have previously speculated that the delayed appearance of these structures, after microtubule disintegration, could be the result of feedback mechanisms that would induce a reactive hypersyntheses of structures which are presumed to be functionally linked with microtubules (11,12).

The effects of R 17934 on mitotic cells were also identical to those of the standard antimicrotubular substances and will be described shortly. Untreated cells round up completely before mitosis and show the normal meta- ana- telophase transition of the chromosomes followed by cytokinesis and readhesion on the culture surface within 45 minutes. In cultures treated with R 17934 (0.04 - 10 μg/ml) the mitotic cells round up in a normal fashion. The ordered chromosome distribution and movement however fails to occur. At the ultrastructural level, the unseparated chromatids are seen to be scattered orderless within the cytoplasm which is completely devoid of microtubules (fig. 2a). At this point an essential difference in response becomes apparent according to the cell line used. The $MO_4$ and MO cells, after having spent approximately 5-6 hrs. in the rounded phase, form multiple furrows which show vigorous movements on time lapse microfilms. After an additional 5-6 hrs. new stacks of nuclear membranes appear appositioned to the scattered chromosomes which are enveloped separately, resulting in the formation of polyploid multimicronucleated cells after readhesion onto the substratum (fig. 2b). Other cell types become polyploid too, (human and murine embryonal fibroblasts). However, in these cells the chromatin is enveloped mostly in 1 large nucleus. HeLa cells and melanoma B16 cells rapidly degenerate by pyknosis during the abortive mitotic phase

Figure 2

a) A mitotic MO cell treated for 5 hours with 1 μg/ml of R 17934. Spindle microtubules are completely absent. The double chromosomes (chr) show two prominent centromeres (arrowheads).

b) A mitotic MO cell treated for 24 hours with 1 μg/ml of R 17934. The scattered chromosomes (chr) are being enveloped individually into a new nuclear membrane (arrowhead). The asterisk denotes a chromosome which is still naked.

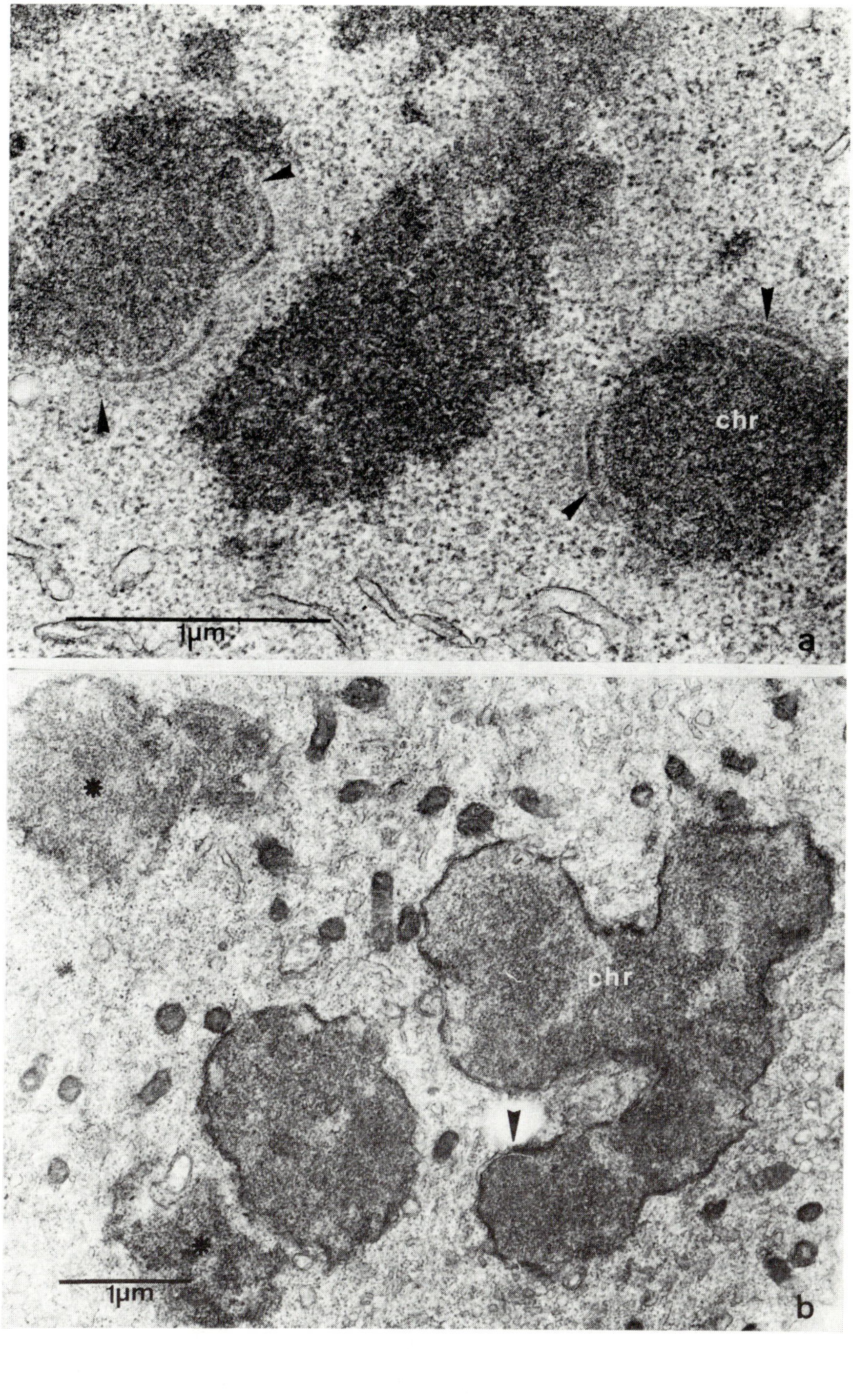
chr
1μm
a
chr
1μm
b

and are apparently unable to resume the interphase stage.

Quantitative experiments on cell growth correlate with the mode of action as described above and with the different kinds of reaction of individual cell lines. Cultures of cells which become endopolyploid show an immediate arrest of growth expressed as cell number. Actual cell death begins after approximately 48 hrs. of continuous treatment. Resumption of growth after withdrawal of the treatment is immediate after 2,4 and 8 hrs. of treatment. The reversibility becomes hampered in correlation with the duration of the treatment time if this is larger than 24 hrs.

Cultures of cells which show mitotic necrosis during treatment (e.g. HeLa) become totally exterminated within ± 48 hrs. However, this cytotoxicity cannot solely be ascribed to mitotic death since cell death occurs also in these cultures when they are blocked in a premitotic stage with antimetabolites.

Since R 17934 is injected in experimental animals or human patients as a micronized suspension we checked whether the above described phenomena could be reproduced by the compound in that formulation. This proved indeed to be the case. The micronized particles dissolved completely in the culture medium at a concentration below 1 µg/ml. At higher concentrations, the remaining particles were actively phagocytosed by the tissue cultured cells. The only noticeable, but important difference with the soluble formulation was the loss of the reversibility after short incubation times.

## EFFECTS ON MALIGNANT AND NON-MALIGNANT CELLS IN VIVO

Tissue culture methods can give excellent information on the subcellular site of action of chemotherapeutic agents. Their predictability for the therapeutic outcome in a complex in vivo situation, however, is practically nihil. Moreover, little information can be gained on the eventual difference in sensitivity between malignant and non-malignant cells. Indeed, most so-called "normal" cell lines are in fact of embryonic origin and, by the routine passage methods, they are in some way forced to behave as malignant cells.

In order to test the relevance of the findings on cultured cells for the in vivo situation we have studied the ultrastructure of L1210 cells after treatment of the host with R 17934 in the micronized form. In one experiment the animals were inoculated with $10^5$ L1210 cells intraperitoneally and treated after 5 days with 160, 80, 40 or 0 mg/kg of R 17934. Ascitic fluid (containing L1210 cells and host leukocytes) peritoneal serosal membrane, and intestinal tissue of animals sacrificed 5, 10 and 32 hrs. after treatment, were prepared for electron microscopy. In another experiment L1210 cells were inoculated intravenously and the

---

Figure 3
A mitotic HeLa cell treated for 24 hrs. with 1 µg/ml of R 17934. The clumped chromosomes (chr) are intermingled with prenecrotic accumulations of lipid vesicles (asterisk) and parallel layers of endoplasmic cisternae. Part of a completely necrotic HeLa cell is shown in the upper left corner.

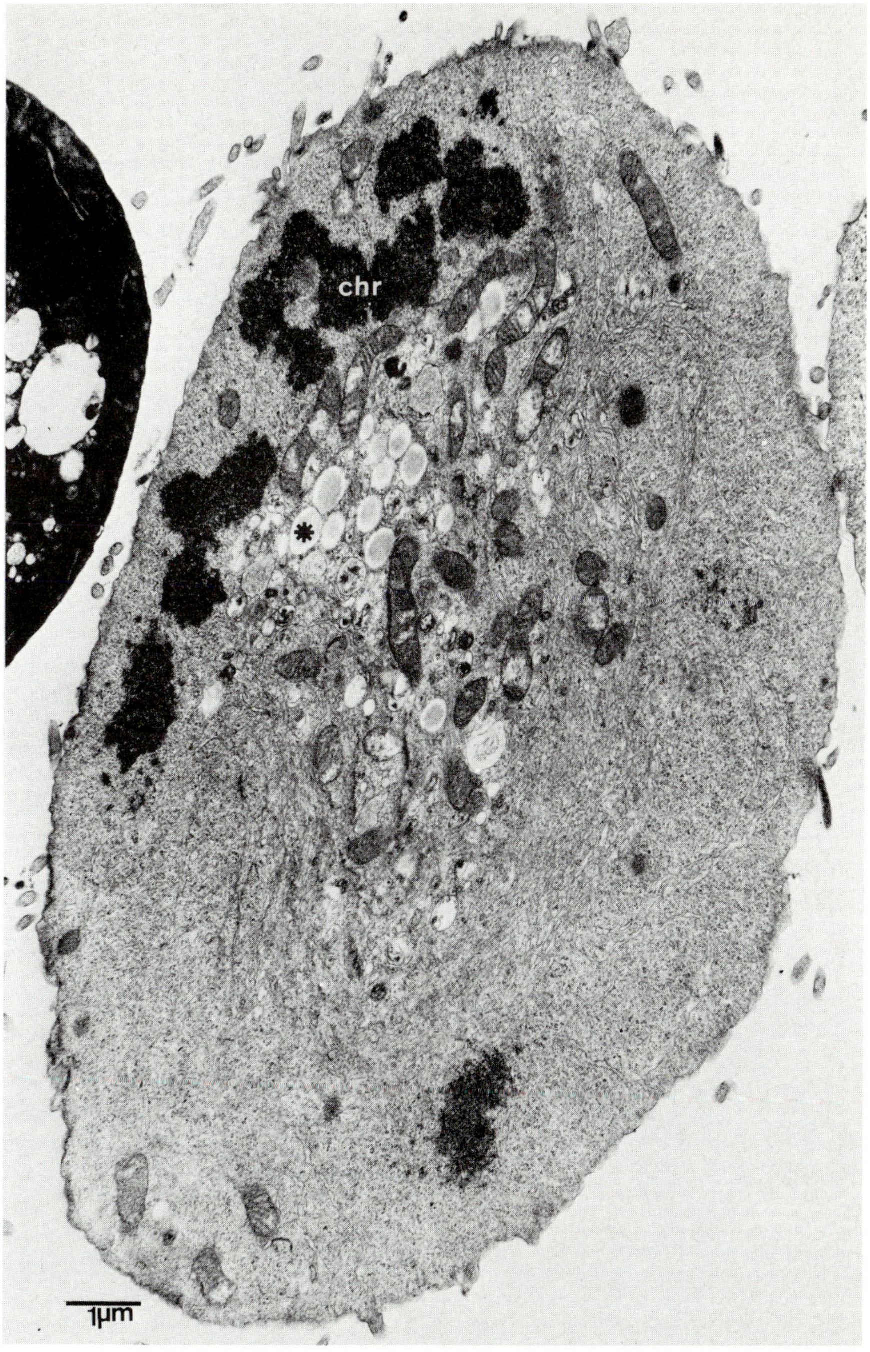
chr
*
1μm

mice were treated intravenously after 4 days with 160, 80, 40 or 0 mg/kg of R 17934. Spleen and liver tissue (containing nests of L1210 cells) and intestinal tissue of animals sacrificed 5, 10 and 32 hrs. after treatment, were prepared for electron microscopy. The optimal therapeutic dose, in both systems, is between 80 - 160 mg/kg.

In both systems, it was obvious that a substantial amount of the micronized particles of R 17934 had been phagocytosed by L1210 cells and normal phagocytes (granulocytes, monocytic phagocytes, Küpffer cells). This was more prominent with higher doses and at the shorter sampling times although we have no exact quantitative data. Microtubules were completely absent after 5 and 10 hrs. in both systems with all doses, in all L1210 cells, although they were present in 100 % of untreated cells. After 32 hrs. a beginning recuperation was apparent in the 40 mg/kg groups, while most cells of the animals treated with the higher doses were still devoid of microtubules. The overall result was similar to that in tissue-cultured cells : disorganization of both mitotic and interphase cells. Especially after 32 hrs. many necrotic cells were present but also a large number of multinucleated L1210 cells. Interestingly, many L1210 cells showed bundles of 100 Å filaments and annulated lamellae. The leukocytes and the serosal mesothelial cells in the mice with the intraperitoneal ascites and the liver parenchymal cells, blood cells, Küpffer cells, and vascular endothelial cells in the animals with the systemic leukemia, had a normal complement of microtubules and a normal intracellular organization, even in the animals treated with 160 mg/kg, and regardless of whether they contained R 17934 particles or not. Annulated lamellae or filament bundles were never observed in these cells.

Since the possibility exists that the solubilization of R 17934 would be different in normal phagocytes compared with L1210 cells, we incubated spleen cell suspensions of leukemic mice (5 days after i.v. inoculation) with a solution of R 17934. Here again L1210 cells were affected while normal cells remained apparently unchanged with doses up to 1.6 µg/ml.

This shows that there is an intrinsic difference in sensitivity, towards the antimicrotubular-properties of R 17934, between interphase malignant cells and non-malignant cells. Mitotic normal cells however, (in the intestinal crypts) were often seriously affected. A mitotic block was most prominent at 5 and 10 hrs. after intravenous treatment with 160-80 mg/kg. Although some microtubules were

---

Figure 4

a) A malignant ascitic cell derived from a gastric parietal cell carcinoma, 24 hrs. after treatment with 3 mg/kg of R 17934 as a micronized suspension. A large phagosome contains the phagocytosed crystals of R 17934. In this cell and all the other malignant ascitic cells, (both mitotic and non-dividing) microtubules were completely absent.

b) Mitotic ascitic cell of the same patient 24 hrs. after intraperitoneal treatment with 3 mg/kg of a micronized suspension of R 17934. Chromosomes (chr) are scattered in the cytoplasm which is completely devoid of spindle microtubules.

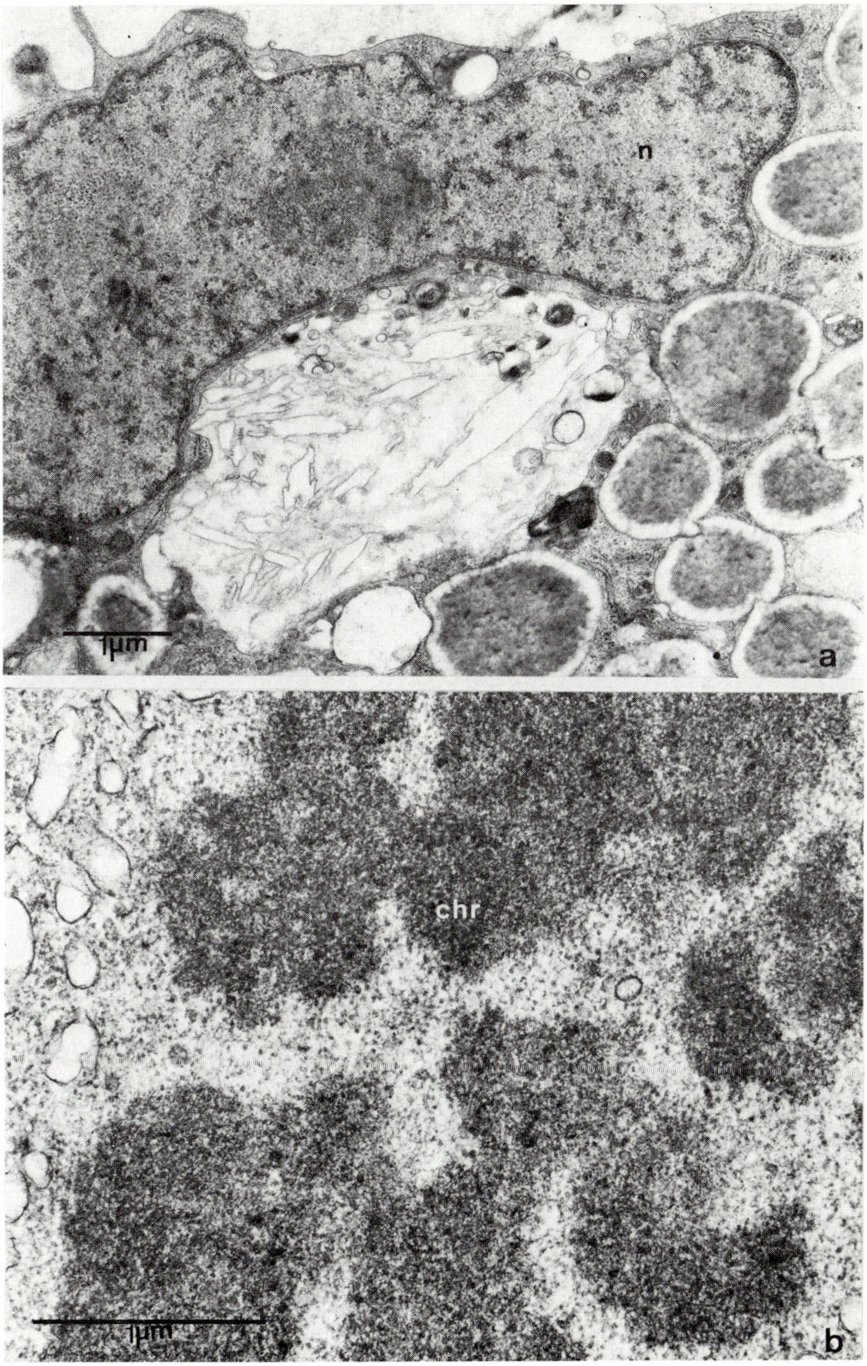
n
1µm
a
chr
1µm
b

still present, centriolar migration had mostly failed resulting in ring mitoses. Further studies, with more narrow dose ranges and sampling times are needed in order to find out whether any specificity exists for malignant mitotic cells.

Finally, we wanted to see whether the described activity of R 17934 - disintegration of microtubules - has any relevance to its effects on human neoplasms. For this purpose we investigated the ultrastructure of malignant ascitic cells before and after treatment of a patient suffering from a carcinoma of the stomach. A sample of the ascitic fluid was withdrawn before and 24 hrs. after intraperitoneal treatment with 3 mg/kg of a micronized suspension of R 17934. As in the murine leukemia, the micronized particles, were actively engulfed both by the malignant cells and by normal phagocytes. Moreover, a similar difference in sensitivity was obvious since all malignant cells, in interphase or mitosis, had completely lost their microtubules, while the normal phagocytes, even those containing the compound, (as microcrystals) had apparently normal microtubules and a normal intracellular organization.

Both mitotic necrosis and multinucleation were present in the treated samples. However, also many interphase cells became strongly vacuolated and necrotic. This could be related to the fact that these neoplastic cells were apparently derived from the gastric parietal cells, since they showed several morphological characteristics of these. The disintegration of the microtubules could have been responsible for the lack of migration of the secretory vesicles ensuing in intracellular accumulation and autolytic necrosis.

Continuing intermittent treatment of this (and other patients) resulted in a gradual fall in malignant cells and in the disappearance of the ascitic fluid.

## CONCLUSION

From an experimental point of view R 17934 apparently provides a new, synthetic substance with a rather specific antimicrotubular activity. This was confirmed by the ability of R 17934 to inhibit the in vitro polymerization of rat brain tubulin (own observations and 13). The existence of a simple chemical structure with such an activity could prove to be of value in cell biological research. Especially the eventual elucidation of its biochemical mode of action, could give new clues with regard to the intricate mechanism regulating microtubule function.

Figure 5

a) A mononuclear phagocyte in the ascitic fluid of the same patient as in figure 4 after intraperitoneal treatment with 3 mg/kg of a micronized suspension of R 17934. The vacuoles (v) are filled with R 17934 crystals. However, microtubules are present around the centrioles (arrow) and the cellular organization is normal. Indeed the Golgi organelles (arrowheads) are fixed in a pericentriolar position.
b) An eosinophylic leukocyte in a L1210 ascites 5 hrs. after intraperitoneal treatment with 160 mg/kg of a micronized suspension of R 17934. Note the abundant presence of microtubules in the vicinity of the centriolar satelites (arrowhead) and the normal pericentriolar location of the Golgi organelles (arrows).

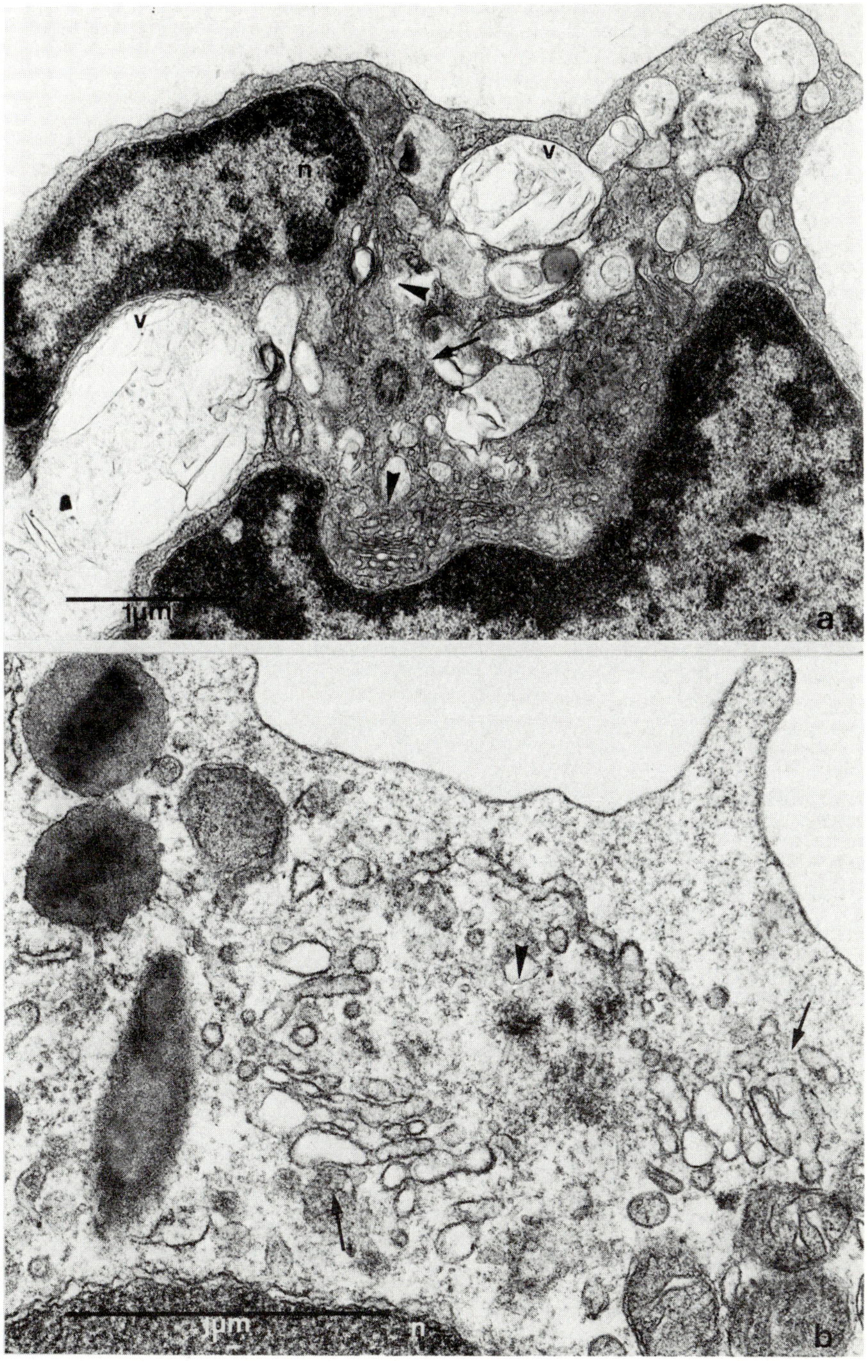
n
v
v
1µm
a
1µm
n
b

From the antitumoral chemotherapeutic point of view, R 17934 seems to offer and additional promising tool. It is probably worth stressing that, as for the vinca alkaloids (14), the antitumoral activity of this compound cannot entirely be explained by the oversimplifying term "mitotic block". The total disorganization of non-mitotic cells should strongly affect all the vital processes which depend on ordered subcellular movements or on an adequate compartmentalization. The autolytic necrosis of the neoplastic cells described is an extreme example of this. The interference with directional cell migration could have important consequences for the invasive and metastatic properties of tumoral cells. At last, both the alteration of the fluid mosaic behavior of the cell membrane (15) and the altered turnover of the cell coat (16) could seriously change the interaction with immune and non-immune host defense mechanisms. In view of this, the apparent difference in sensitivity between interphase malignant and non-malignant cells towards R 17934 could be a favorable factor in determining its therapeutic efficacy.

The use of a micronized suspension as the formulation for intravenous or local injections makes this compound a new "phagosomotropic" drug (17). This property may favor an accumulation of the compound in actively phagocytosing cancer cells and the formation of a slow release depot which ensures prolonged active blood levels. Preliminary clinical observations have shown that this can be favorably exploited in the local treatment of ascitic or pleural effusions where regressions can be obtained without any systemic toxicity.

REFERENCES


1. Atassi, G. and Tagnon, H., July 1975, 9th International Congress of Chemotherapy, London, England.
2. De Brabander, M., Van de Veire, R., Aerts, F., Geuens, G., Desplenter, L., De Crée, J., Borgers, M. and Janssen, P.A.J., July 1975, 9th International Congress of Chemotherapy, London, England.
3. De Brabander, M.J., Van de Veire, R.M.L., Aerts, F.E.M., Borgers, M.J.M. and Janssen, P.A.J. Submitted.
4. Bhisey, A.N. and Freed, J.J., 1971, Exp. Cell Res. 64, 430.
5. Bhisey, A.N. and Freed, J.J., 1971, Exp. Cell Res. 64, 419.
6. Buckley, I.K., 1974, Tissue & Cell 6, 1.
7. Deig, E.F., Starr, T.J. and Church, K.K., 1962, Proc. Soc. exp. Biol. Med. 111, 248.
8. Freed, J.J. and Lebowitz, M.M., 1970, J. Cell Biol. 45, 334.
9. Robbins, E. and Gonatas, N.K., 1964, J. Histochem. Cytochem. 12, 704.
10. Starr, T.J., 1963, Texas Rep. Biol. Med. 21, 412
11. De Brabander, M., Aerts, F., Van de Veire, R. and Borgers, M., 1975, 253, 119.
12. De Brabander, M. and Borgers, M., 1975, J. Cell Sciences, in press.
13. Hoebeke, J. and Van Nijen, G., 1975, Proceedings of the International Symposium on Microtubules and Microtubular Inhibitors, Beerse (Belgium) in press.

14. Rosner, F., Hirshaut, Y., Grünwald, H.W. and Dietrich, M., 1975, Cancer Res. 35, 700.

15. Berlin, R.D., Oliver, J.M., Ukena, T.E. and Yin, H.H., 1974, Nature 247, 45.

16. Ginsel, L.A., Daems, W.Th. and Debets, W., 1975, Proceedings of the International Symposium on Microtubules and Microtubular Inhibitors, Beerse (Belgium) in press.

17. Trouet, A., Deprez-De Campeneere, D. and De Duve, C., 1972, Nature 239, 110.

ACKNOWLEDGEMENTS

The skilful assistance of L. Leyssen in preparation of the photographs, of M. Van de Ven in typing and of H. Vanhove in reviewing the manuscript is greatly appreciated.

*Microtubules and Microtubule Inhibitors, eds. M. Borgers and M. de Brabander*



# EFFECTS OF SPINDLE POISONS ON CELL SURVIVAL IN THE MOUSE

L.M. van Putten, T. Smink and P. Lelieveld.

Radiobiological Institute TNO, Rijswijk, The Netherlands.

## Summary

Cell survival studies in vivo in the mouse are reported after exposure to a number of microtubular inhibiting agents. The survival curves of spleen colony forming haemopoietic stem cells both at rest in normal mice and in rapid proliferation in radiation depleted mice were compared with the survival curve of L1210 leukemia cells. The resulting survival curves have been interpreted both for their kinetic and their pharmacokinetic implications. The results obtained with the recently developed agent R17934 suggest that it is likely to be clinically most effective upon i.v. administration when given in repeated doses or in a slow infusion.

## Introduction

Different spindle poisons have found use in the treatment of cancer. Especially the vinca alkaloids are applied widely for this purpose, but earlier also a colchicine derivative (demecolcin) was used and more recently podophyllotoxin derivatives (VM 26 and VP 16-213) have been shown to be effective agents for the treatment of malignant disease. Presently various new microtubular inhibitors are produced by chemical synthesis and it would be most helpful if laboratory methods could be used to predict the clinical effectiveness of cytostatic microtubular inhibitors. In the present report cell survival in vivo after single and fractionated exposure of mice to various agents have been compared in an attempt to uncover differences and similarities between five cytostatic microtubular inhibitors. Studies of a similar type had been performed earlier by Valeriote and Bruce (1, 2) on vinblastine and vincristine and in earlier papers from our institute widely different cytostatic agents have been compared (3, 4). For a review on the interaction of cytostatic drugs with stem cells see (5).

Materials and Methods

The survival of the following cells was assessed after in vivo exposure in the mouse:

1. Normally resting haemopoietic stem cells from the mouse as identified by the spleen colony assay of Till and McCulloch (6).
2. Rapidly proliferating haemopoietic stem cells in the spleen of a mouse of which 7 days before assay the bone marrow had been depleted by a lethal radiation dose followed by injection of about one quarter femur's content of bone marrow cells. In the spleen of these mice the repopulating haemopoietic stem cells are in rapid proliferation (cycle time about 12 hours).
3. Cells of transplantable leukemia L1210 present in the femur of a mouse 7 days after i.v. inoculation of $10^4$ cells.

Haemopoietic stem cells: for 1) and 2) 10 - 14 week old male (C57BL/Rij x CBA/Rij)F1 hybrid mice were used. Groups of five treated animals and untreated controls were killed 16 hours after injection of the drug and the bone marrow was quantitatively collected from the femurs and pooled for each group. A single cell suspension was prepared by pushing the marrow gently through 6 layers of nylon gauze and washing the filter with Hanks' solution. Spleens were minced with scissors and similarly made into a single cell suspension in Hanks'. Suitable dilutions of these suspensions were then injected i.v. in 0.5 ml volume into lethally irradiated isologous recipient mice (950 rad gamma irradiation from a $^{137}Cs$ source). Nine days later the recipients were sacrificed, their spleens were fixed in Telleyesniczki solution and the number of colonies was counted. By reference to the fraction of the donor's femur or spleen injected per recipient, the average colony forming units (CFU) content of the donor femur or spleen could be determined and comparison with the simultaneously assayed untreated control group permitted to calculate the relative survival of CFU in the femurs and spleens of the treated mice.

L1210 cell survival was similarly followed in 10 - 15 week old (DBA2 x C57BL/RIJ) F1 mice carrying about $10^6$ leukemic cells per femur. Assay was performed by plating known fractions of pooled femur contents of groups of 5 treated or control mice in tissue culture dishes in vitro. The culture medium consisted of Dulbecco's medium, 20 % horse serum and 0.3 % agar supplemented with 60 µmol/l of 2 mercapto-ethanol. The dishes were incubated at 37° C in an atmosphere of 10 % $CO_2$ and 100 % humidity. After 7 days colonies of more than 50 cells were counted and the number of leukemic CFU-C (colony forming units-culture) per femur was calculated. Survival was expressed as the fraction of the average number of colonies per femur counted in at least four dishes prepared from the pooled femur contents of at least four donor mice per treatment group to that of an untreated control group.

Drugs: The vinca alkaloids were dissolved in saline. VM 26* and VP 16-213* were used dissolved as supplied, and administered subcutaneously. R 17934* was administered i.v. in the form of a micronized suspension.

## RESULTS

The survival curves of normal resting stem cells, of rapidly proliferating stem cells and of L1210 leukemic cells after different doses of the five microtubular inhibitors are presented in figure 1 and 2. Most of the curves are characterized by a plateau at higher drug doses indicating the presence of a drug resistant subpopulation of cells. Since the level of this plateau is lower (i.e. the quantity of the subpopulation is smaller) in the rapidly proliferating cell population, it is assumed that this population consists of the cells which have not passed through the drug-sensitive phase of the cell cycle (1). In all cases the leukemic cells are more sensitive than either of the normal stem cell populations. In figure 3 the results for the rapidly proliferating stem cells are compared for four of the drugs under two conditions of administration; either

---

*VM 26 is 4-demethyl-epipodophyllotoxin-β-D-thenylidene glucoside (NSC 122819)
VP 16-213 is 4-demethyl-epipodophyllotoxin-β-D-ethylidene glucoside (NSC 141540)
R 17934 is methyl(5 -(2-thienylcarbonyl)-1H-benzimidazol-2-yl)carbamate (NSC 238159).

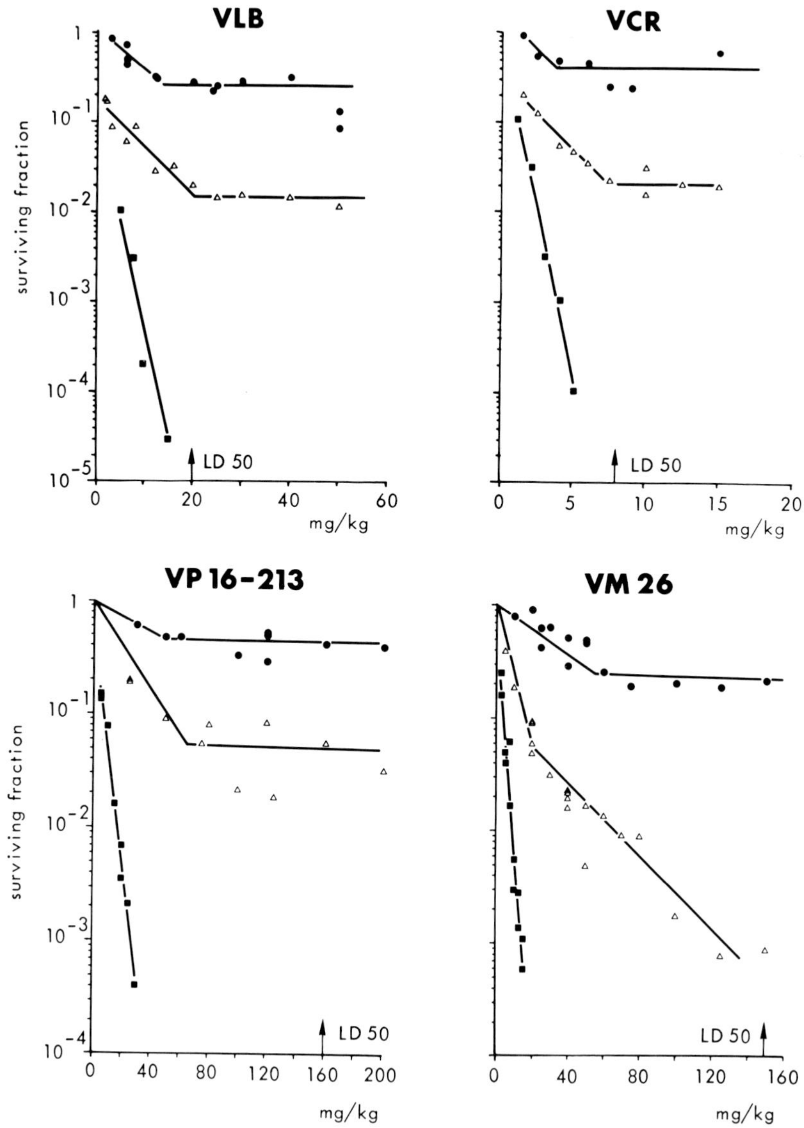


Figure 1: Dose-effect curves for survival in vivo of normal resting CFU, rapidly proliferating CFU and L1210 leukemia cells after exposure to single doses of vinca alkaloids (VLB = vinblastine; VCR = vincristine) and epipodophyllotoxin derivatives (VP 16-213 and VM 26), administered s.c.

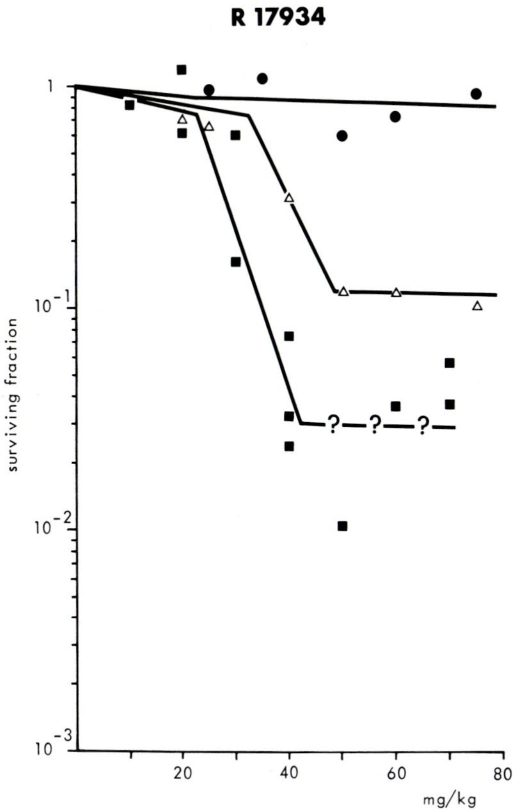


Figure 2:

Dose-effect curves for survival <u>in vivo</u> of normal resting CFU, rapidly proliferating CFU and L1210 leukemia cells after exposure to R 17934 in a micronized suspension, administered i.v.

as a single dose, as also given earlier or in three divided doses at 4 hour intervals. This time period is expected to cover a full cycle time of the rapidly repopulating haemopoietic stem cells tested under these conditions.

## Discussion

Cell kinetic aspects: In the bone marrow of normal mice only about 10 % of the cells are in cell cycle; the remainder of the cells are in a state of prolonged rest from which they can be recruited to re-enter the cycle under suitable

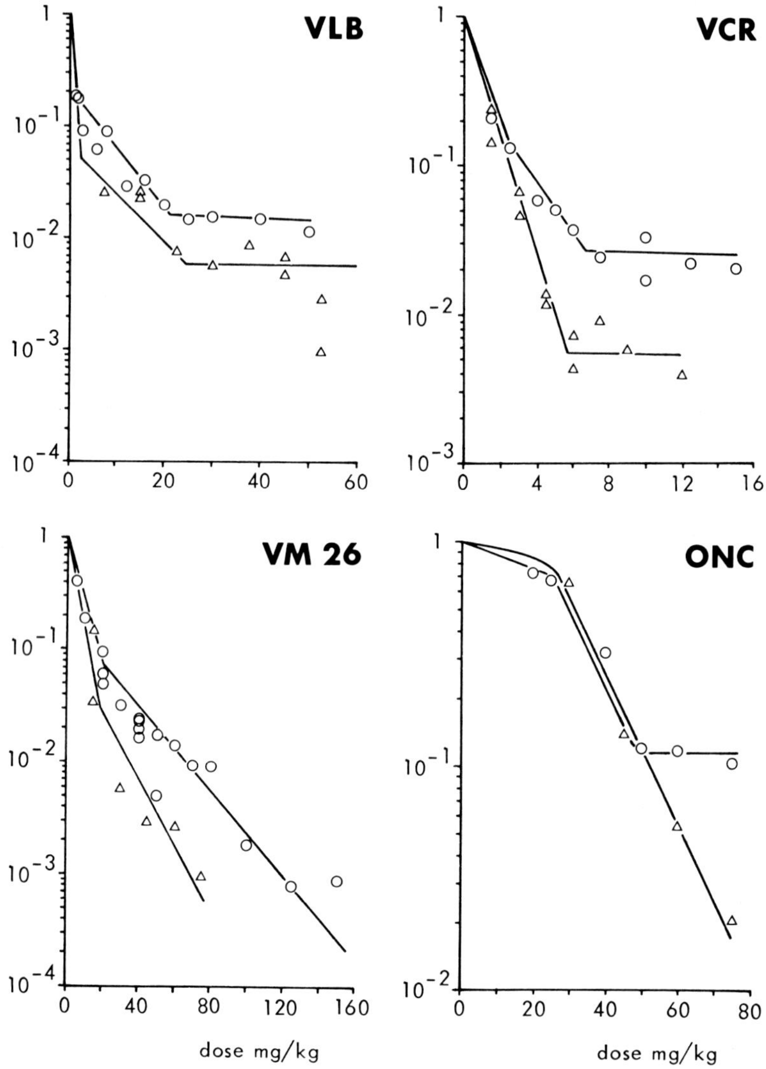


Figure 3: Comparison of the dose effect curves for survival of rapidly proliferating CFU to a single dose, and to three doses at 4 hour intervals of different microtubular inhibitors. VLB = vinblastine s.c.; VCR = vincristine s.c.; VM 26 s.c.; ONC = R 17934 micronized suspension i.v. The indicated dose represents the total dose.

conditions of stimulation, e.g. in our spleen colony assay, when they are stimulated to repopulate the depleted haemopoietic system of an irradiated host. Nevertheless we find that more than 10 % of the stem cells seem to be killed by at least four of the spindle poisons. The most likely explanation lies in the fact that the drugs under these conditions may kill other cells in the bone marrow which are in cell cycle such as the erythroblasts, myeloblasts and other proliferating precursor cells. This may lead to depletion of these cell systems which in turn may stimulate some stem cells to differentiate in order to compensate for this depletion. By such differentiation the stem cells lose the capacity to be recognized as spleen colony forming cells and appear to be killed. This early differentiation is well documented (5) and it makes it difficult to assess cell killing directly for the resting cells. In the repopulating rapidly proliferating CFU population such differentiation is much less likely since in these depleted mice there is presumably already a maximal stimulus to proliferate and differentiate, which can hardly be affected by additional cell depletion. Under these conditions the assays thus seem to reflect true cell survival. The level of the plateau indicates that for vinblastine and vincristine a large majority of the proliferating stem cells may be killed by a single administration of the drug and a plateau is reached in the curve. This seems also to be the case for VP 16-213 but for VM 26 there is not a true plateau but only a break in the curve. Since the level of the plateau for rapidly proliferating stem cells exposed during a full cell cycle should indicate the fraction of the cells which are not in cycle, these levels should be similar for the different drugs. Additional evidence obtained with other agents such as radiotoxic doses of tritiated thymidine, procarbazine and cytosine arabinoside indicate that this fraction comprises about 1 to 2 % of these CFU. It is evident that for the leukemic L1210 cells the plateau is not reached; apparently with leukemia a yet larger fraction of the cells is in cycle.

Pharmacokinetic aspects: As mentioned earlier it can be expected that all cells in cycle are killed by agents of this type if effective drug levels are maintained for the period of one full cell cycle. However, we know that some drugs are bound in all phases of the cell cycle but released slowly. Such drugs could be bound even in resting cells and not completely released, thus permitting a toxic effect when the cells are later drawn into cycle in the assay system. VM 26 seems a typical example. Obviously the resting cells are not completely resistant to this drug and this could well be explained if the cells accumulate the drug even when resting, then releasing it slowly so that exposure of resting cells to high concentrations could lead to increased cell killing due to greater retention of effective amounts of drug which kill the cells when they enter the cell cycle and pass through the critical phase. This hypothesis is for VM 26 in agreement with the data obtained by Stähelin on the poor reversibility of binding of this agent to cells. He found (7) in contrast to experience with other microtubular inhibitors that cells exposed to VM 26 released the agent only very slowly upon washing and since it does not lead to accumulation of metaphases but prevents cells from entering mitosis it may not only act as a true microtubular inhibitor.

For the drugs which bind reversibly to the microtubular target in the cell our survival curves do indicate whether there is a persistently effective cell killing level during the whole cell cycle. This is clear from a comparison of the survival level of rapidly proliferating CFU exposed for single and for fractionated drug doses. From figure 2 it is evident that there is little difference in plateau cell killing for vinblastine. Apparently this drug is present for a full cell cycle duration after a single injection. Vincristine is more effective after repeated administration, indicating that a single dose leads to too rapid excretion to permit maximal cell killing. After R 17934 (ONC) the difference is also rather marked; after a single dose, cell survival at the plateau level is depressed even less than after vincristine, suggesting that the duration of effective drug levels may have been even shorter. This is rather surprising for an insoluble agent that has been administered as a micronized suspension. It indicates that the drug in a soluble form may be eliminated even faster and could be completely inactive after single doses. This was indeed confirmed to be the case (De Brabander, personal communication). Furthermore, this agent is peculiar in another aspect, namely the presence of a clear threshold effect for cell killing. Both for the effects on leukemia (figure 2) and for the effects of single and divided doses on rapidly proliferating stem cells a threshold total dose of 25 to 30 mg/kg must be exceeded to produce significant cell killing. A similar critical serum level has been described for the cellular effects of vinblastine (8) but in this case the threshold is much lower compared to toxic drug levels. These findings indicate the possibility that some of the soluble synthetic microtubular inhibitors may appear inactive when tested as cytostatic agents *in vivo* due to extremely rapid elimination. Continuous infusion or use of a delayed release formulation may identify the cytostatic activity of such drugs. Furthermore the optimal dosage schedule found for mice treated i.p. with a single daily dose of R 17934 might well be suboptimal for human application if patients are treated by intravenous administration. In that case it seems that a better effectiveness of this agent may be expected if administered by repeated injection or by slow infusion.

## Acknowledgements

The capable technical assistance of W. Akkerman and M. Pijpers-de Bruin is gratefully acknowledged.

Gifts of VM 26 and VP 16-213 were received from Sandoz A.G., Basel, Switzerland, and of R 17934 from Janssen Pharmaceutica, Beersse, Belgium.

## References

1. Valeriote, F.A. and Bruce, R.W., Comparison of the sensitivity of hematopoietic colony-forming cells in different proliferative states to vinblastine. J. Nat. Cancer Inst. 38, 393 - 399, 1967.
2. Bruce, W.R. et al., Comparison of the dose- and time-survival curves for normal hematopoietic and lymphoma colony-forming cells exposed to vinblastine, vincristine, arabinosylcytosine and amethopterin. J. Nat. Cancer Inst. 42, 1015 - 1023, 1969.

3. Van Putten, L.M. et al., Cell cycle specificity and therapeutic effectiveness of cytostatic agents.
Cancer Chemotherapy Rep. 56, 691 - 700, 1972.

4. Van Putten, L.M. et al., Abhängigkeit der Wirksamkeit Phasenspezifischer Zytostatica vom Zellzyklus; Interpretation von Dosis Wirkungs-Kurven.
In: Aktuelle Probleme der Therapie maligner Tumoren. Editor G. Wüst.
Published by George Thieme Verlag, Stuttgart, 1973.

5. Valeriote, F.A. and Van Putten, L.M., Proliferation dependent cytotoxicity of anticancer-agents.
Cancer Research, to be published 1975.

6. Till, J.E. and McCulloch, E.A., A direct measurement of the radiation sensitivity of normal mouse bone marrow cells.
Radiat. Res. 14, 213 - 222, 1961.

7. Stähelin, R., Reversibility of the cytostatic effect of the podophyllotoxin derivatives VP 16-213.
In: Progress in chemotherapy; proceedings of the 8th International congress of chemotherapy. K. Daikos, Ed., Athens, 1974, Volume III, 819 - 823.

8. Valeriote, F.A. et al., A model for the action of vinblastine in vivo.
Biophysical Journal 6, 145 - 152, 1966.

# CONCLUDING REMARKS

*Microtubules and Microtubule Inhibitors, eds. M. Borgers and M. de Brabander*
© 1975, North-Holland Publishing Company - Amsterdam 

# NEW PERSPECTIVES OF MICROTUBULE RESEARCH

## CONCLUDING REMARKS

P. Dustin
Department of Pathology
Université Libre de Bruxelles
Brussels
Belgium.

One billion years ago, a momentous change took place in Evolution : the birth of the Eucaryotes was not only the separation of the nucleus from the cytoplasm by the nuclear membrane ; it entailed also considerable changes in the structure of the ribosomes, and the acquisition of a new machinery to separate the genomes, more or less condensed in the form of chromosomes, at cell division. The microtubules (MT) appear to have been at the start closely related to cell and chromosome movements, as pointed out by Lynn MARGULIS. One thinks immediately of a first unsolved problem : were the first MT intranuclear, as they are still to be found in some Unicellulars, or extranuclear, the chromosomes being attached to the nuclear membrane by the kinetochores ? Whatever their origin, symbiotic or not, microtubules diversified rapidly into the surprising structures which we have been studying this week : spindle fibers, centrioles, cilia, structural supports, organs of transport of cytoplasmic organelles, of secretion, of sensory transduction. As Jelle ATEMA has suggested, MT have played a prominent part in the development of the nervous system and of cellular transmission : their grouping into cilia and flagella was first necessary for cell movement ; later, these structures were put to use in most sensory organs. Neurotubules and specialized cilia are the base of all knowledge of the external world and have remained so in the most complex Metazoa.

One could certainly not expect a symposium such as this to cover more than a small part of the so rapidly increasing knowledge about MT. The recently published volume of last year's New York conference (1) gives a good idea of the quantity of information gained since the last ten years, although it did not attempt to cover the whole field of MT research. It may be useful to-day to underline the principal developments which have taken place since last year, and were presented here.

An important aspect of MT physiology is that of the regulation of MT formation and disassembly. It has always been surprising to see how cells grow MT exactly at the place and the time when they appear to be needed : how can this be controlled ? The necessity of such control is evident, as a protection against the assembly of all the cell tubulin into MT, as it is known that much, if not most, of the tubulin remains in the cell as a mono- or a dimer. Joseph BRYAN has isolated heat-stable substances present in non-neural cell extracts, which act by sequestering a heat-stable, basic protein, which is indispensable for MT assembly. This may explain the contradictory results obtained in vitro with solutions of tubulins from neural and non-neural tissues. The ease with which neural tubulins can assemble in vitro under the proper ionic and temperature conditions contrasts with the impossibility of assembling other tubulins without some nucleation centers - microtubule organizing centers (MTOC), centrioles, preexisting MT, or ring-structures of tubulin. The extensive growth of neurotubules in neurons is necessary for axoplasmic transport : it may perhaps explain why neurons lose so rapidly, in embryonic growth, their capacity to divide by mitosis. This problem has been well studied by NUNEZ and collaborators, who clearly indicated that the activation of the growth of neurotubules took place at a definite time in the embryonic growth of the brain. This suggested either the existence of different populations of tubulins, or the presence of a factor necessary for the propagation of MT assembly.

Another aspect of the regulation of MT is the formation of microtubule organizing centers (MTOC), centrioles and basal bodies. The recent findings of Weisenberg (2) showing that centrioles can be formed de novo in the isolated cytoplasm of activated eggs opens a new path in the study of the very old problem of activation. Apparently, the egg is not the only cell in which such changes can take place : the classical studies on amoeboflagellates such as Naegleria gruberi had demonstrated that at short notice a cell could grow flagellae, with all the complexity of MT assembly which is needed for this. The results we have observed, with Mrs. Jacqueline FLAMENT-DURAND and J.-P. HUBERT indicate that the pituicyte, this neglected glial cell which plays some role in the regulation of the posterior pituitary secretion, can also be activated and grow multiple cilia at short notice. Like Weisenberg, it was found that this was not prevented, on the contrary, by the MT poisons colchicine and vincristine, indicating that these drugs do not necessarily prevent pools of tubulin to assemble into complex structures. The same experimental model - the hypothalamic-posterior pituitary pathway - has been further studied by SLOPER and Felicity GRAINGER. They have found that in mice with congenital diabetes insipidus, the number of MT increases shortly after birth, as a consequence of the functional, although ineffective, overloading of the neurosecretory neurons. It will be interesting to understand the regulatory processes involved, and the site of assembly of the newly-formed MT.

Lynn MARGULIS gave us a splendid demonstration of the capacity of the cell - in this occurrence, the giant ciliate, Stentor - to regenerate in a few hours thousands of cilia of the mouth region. Among the drugs which interfere actively with the MT-associated migration of the newly-formed buccal ciliature, are many MT poisons, and the most active is the widely-used herbicide, trifuralin. Protists are excellent tools for MT research and many important facts, it should be remembered, about MT assembly, ciliary growth and the mechanisms of mitosis are to be found in Protozoology textbooks.

The study of the relations of tubulins with contractile proteins in the cytoplasm is necessary to understand the mechanisms of tubulin-associated movements. Since dynein was discovered in cilia, relations between actin and tubulin have been suggested. Linda AMOS, who has studied the diffraction patterns of negatively stained flagellar outer doublet tubules, has presented a tentative model based on the symetries of MT and actin, indicating the possibility of actin becoming wrapped around MT as a paracrystalline coating. CRAWFORD and CASTLE, who isolated tubulin from blood platelets, found that it was associated with a protein with the properties of actin, and that myosin was also present in these cells, which are an excellent model for further studies on the possible interrelations between tubulins and actin-myosin complexes.

MT may now be studied by immuno-fluorescence methods. A great advance - rather surprising, as the molecular structure of all tubulins was thought not long ago to be identical or very closely related - has been the preparation of specific antibodies against tubulins. These demonstrate, as beautifully illustrated by BRINKLEY and collaborators, the large amount of tubulin present in the cytoplasm as the "cytoplasmic microtubule complex" (CMTC). Observations of transformed cells appear to indicate that the level of CMTC may have some relation with their behaviour. Klaus WEBER who presented remarkable documents of immunofluorescent staining of tubulin, actin and myosin, illustrated in a striking way the complexity of the cytoplasmic fibrillar structures. These two papers clearly indicated that the use of immunofluorescent techniques for the study of MT has opened a new field in MT research.

The association of a protein-kinase with MT has been much debated. It is not yet clear which substance this enzyme phosphorylates, whether it is tubulin itself or another protein, and what exactly is the role of cAMP. FULLER and his group have presented experiments which link the role of $Ca^{2+}$ and cAMP : the liberation of $Ca^{2+}$ by altered mitochondria may prevent the assembly of MT by stimulating the efflux of $Ca^{2+}$. It remains, however, to be demonstrated whether the intracellular changes in $Ca^{2+}$ concentration are, in normal conditions, sufficient to explain the formation and disparition of MT.

The role of the guanine nucleotides in the assembly of tubulin has been studied by JACOBS and collaborators. GTP must occupy the easily dissociating (E) site of tubulin for a rapid assembly of MT. This GTP-dependent assembly is antagonized by GDP. In the presence of ATP, GDP is phosphorylated to GTP and assembly can take place. These authors have described also a new type of tubulin assembly, which takes place in the presence of protamine : the "duplex-microtubule". This could act as a nucleating center. It shows once more that the tubulin monomers are capable of many types of assembly, apart from the classical MT with its 13 subunits. Recently, MT with 11 or 15 subunits have been described, and ring and helical forms are well known. Guy DEYSSON,who gave us a detailed review of the action of MT poisons in Plant cells (where they are playing such an important role in the creation of new polyploids and amphidiploids of use in agriculture) described the formation of a peculiar type of macrotubules under the influence of colchicine. The molecular structure of macrotubules, observed after many drugs (vincristine, halothane) remains poorly understood and the model proposed last year by Fujiwara and Tilney (3) does not seem to provide a satisfactory explanation of this assembly of MT subunits into the various types of macrotubules.

Returning to the role of phosphorylation in MT assembly, LAGNADO and KIRAZOV, using $^{32}P$, demonstrated the incorporation of phosphorus in high molecular weight proteins associated with tubulins, and also in phospholipids. These were present in twice purified MT fractions, and their significance may be important. Lipids may be present in some MT structures which have not yet received a good explanation, such as the clear "exclusion" zone which surrounds MT. These results should be compared to those of P.J. QUINN who found that a phospholipase, associated with β-tubulin extracted from brain and thyroid, is inhibited by vinblastine and chlorpromazine, but not by colchicine.

An interesting question must be answered to understand the assembly of MT : if these are helical structures made of dimers of α and β tubulins, are they assembled from dimers or from monomers ? Working with the cross linking agent dimethyl-3,3'-(tetramethylenedioxy)-dipropionimidate, LUDUENA, WILSON and SHOOTER colclude that solubilized tubulin consists mainly of αβ dimers. The results presented by ENGELBORGHS, HEREMANS and HOEBEKE on the effects of pressure and temperature on nucleation and on propagation (i.e. the growth of the MT polymer) should be considered in relation to this problem.

The study of the Vinca alcaloids may help to explain the assembly of MT : L. WILSON and his group have presented important data on the action of vinblastine, which binds to tubulin relatively independently of temperature, contrary to colchicine. Vinblastine acts by combining to two high affinity binding sites on the tubulin molecule, and preventing its assembly in MT. A quantitative analysis demonstrates the surprising fact that at the low concentrations which inhibit mitosis, only a very small fraction of the tubulin molecules are combined with vinblastine. This may be explained by the fact that blocking of one dimer of tubulin may prevent any further elongation of a MT.

Few papers on mitosis have been presented at this meeting although this remains one of the central themes of MT research. BAJER and MOLE-BAJER (who's book on the ultrastructure of mitosis is well known and admired by all of us) (4), in collaboration with LAMBERT, have refined their "zipper" theory of chromosome movements. Lateral interactions between coupled MT would in their opinion explain how the MT move the chromosomes, and experiments with MT poisons agree with this hypothesis. It remains doubtful however, that these remarkable analytical studies, based on excellent electron microscopical observations and a most beautiful movie, may explain all types of mitosis, and in particular those where at anaphase, a parallel bundle of MT is seen to elongate considerably, as in some Unicellulars. But should one expect to find a single theory for all mitoses, as cell division has been diversifying since the birth of the Eucaryotes ?

While the mechanism of chromosome transport remains a perplexing problem, the transport of many other cell structures, secretion granules, organelles, viruses, takes clearly place in relation with MT, and several papers presented at this conference demonstrate the advances in this field. The transport of melanin granules and axonal flow are two closely related activities, as melanin-carrying cells are often of neural crest origin. However, the mechanisms may be quite different, as neurotubules appear to be static structures, while in pigment cells, as in mitosis, assembly-disassembly of MT plays a major role. This has been clearly demonstrated by SCHLIWA in the anglefish, Pterophyllum scalare : the MT disassemble during the aggregation of pigment granules. However, the intervention of another, maybe contractile protein of the cytoplasm, is suggested, perhaps in relation to lateral expansions of the MT. It is known that in other species, melanin granules may move to and fro without any apparent change in the number of MT. More information is clearly needed about the lateral expansions of MT and their relations with the "sleeve" element surrounding the MT and studied by STEBBINGS in the ovaries of insects, where ribosomes are transported along the MT.

The role played by MT in secretion and exocytosis has been illustrated by many experiments showing that MT poisons interfere with these cellular activities. Other cell organelles, in particular the microfilaments located in the periphery of the cell ("cell web") control also secretion, and the action of MT is more on the transport of granules than on their opening at the cell surface. MALAISSE and collaborators have pursued their work on the mechanism of secretion by the pancreatic islets. In this model, colchicine and vinblastine have now been found to enhance glucagon secretion by $A_2$-cells, which is inhibited by heavy water and facilitated by cytochalasin B. In Spiny mice, the impaired insulin release has been related to a defect in the amount of MT protein. Colchicine, vinblastine and deuterium oxide have been used as tools by LE MARCHAND and collaborators in a detailed study of the secretion by liver cells of very low density lipoproteins, albumin and globulins. This is observed in perfused animals and the doses of poisons are relatively high. In normal mice, the injection of colchicine leads however to decrease in plasma triglyceride and protein levels ; at the same time, the MT of hepatocytes are no longer visible. These reversible changes underline once more the complexity of colchicine pharmacology. The MT of hepatocytes have also been studied by PATZELT and collaborators : they found that about 40 % of the tubulin was in this form. The tubulin was isolated from the liver of mice after centrifugation and its colchicine-binding activity measured.

The liberation of lipoprotein-lipase by endothelial cells is decreased after colchicine treatment, while the synthesis is not modified, as shown by CHAJEK and collaborators : the liberation of this enzyme obeys laws quite comparable to those demonstrated in true secretory cells.

In the thyroid, secretion begins by the phagocytosis of colloid, and this activity is prevented by colchicine and other MT poisons. Pierre NEVE and Mrs. KETELBANT-BALASSE have studied this engulfment of colloid within the scanning electron microscope. They have also found that chronic stimulation by TSH secretion apparently takes place by a vincristine-insensitive mechanism. The movements of leukocyte granules are also related to MT : Stephen MALAWISTA found that colchicine decreased the degranulation of polymorphonuclear leukocytes during phagocytosis. MT may help to bring together the granules (lysosomes) and the phagosome (phagocytic vacuole) containing bacteria. It would be interesting to know whether this action of colchicine has any relation to the colchicine-leukocytosis known since Dixon and Malden (5), nearly seventy years ago, and which still remains without proper explanation. In the Chediak-Higashi anomaly, an hereditary disorder which exists in man, mink, cattle and mice, the leukocyte granules are abnormally large. The polymorphonuclears show disturbances of chemotaxis and degranulation : Janet OLIVER found that these defects were related to MT and could perhaps be explained by an inadequate generation of cGMP. Cholinergic drugs normalize the size of the granules of Chediak-Higashi fibroblasts in tissue culture. The study of this disease provides also a model for the study of the role of MT in cell capping and its possible relation with cyclic GMP.

Movements of the cell membrane may be MT-dependent. Colchicine-binding proteins have been isolated from membranes. In the nervous system, for instance, the synaptosomes fix strongly colchicine, although it remains to be proved that this results from the presence of tubulin in the membrane, from the close association of MT with cell membranes, or from another unidentified colchicine-binding protein. The surface receptors for lectins have been studied by Richard BERLIN, and specific MT-membrane associations demonstrated. Their internalization plays a fundamental role in phagocytosis and is prevented by colchicine. BHISEY and collaborators have presented data on modifications of membrane ruffling of mouse macrophages under the effects of MT poisons. This ruffling should be compared to the surface changes of the thyroid cells but here the membrane movements are more active in colchicinized cells. The results of GINSEL, DAEMS and DEBETS, on the inhibition by colchicine of the intracellular transport of glycocalyx material show also a close relation between MT and cell surface.

Resistance to colchicine and podophyllotixin is known since long in the Syrian hamster (6, 7). Jerome FREED and Betsy OHLSSON-WILHELM have isolated a culture of haploid frog cells by selection for vinblastine and podophyllin resistance. The strain resistant to podophyllin became resistant to colchicine while remaining sensitive to vinblastine indicating that the resistance is related to the fixation of the drugs on the MT. This is an important finding, as other types of resistance described in the literature are usually explained by changes in cell permeability.

Another new aspect of colchicine pharmacology is the antagonism between $Li^+$ and colcemid, as observed in rat glioma cells in culture, by REISER and collaborators. This is explained by a stabilisation of MT by $Li^+$, although no similar antagonism is found with vinblastine.

Although colchicine and the Vinca alkaloids are the most widely used MT poisons, many other chemicals have similar effects, and it is often forgotten that the typical metaphase mitotic arrest (stathmokinesis) was observed in mice, several years before the rediscovery of the effects of colchicine, after injection of several arsenical compounds (8) which, in the light of present research, combine probably with the SH group of tubulin and prevent its assembly into MT.

Many problems related to the pharmacology of colchicine remain poorly understood, and it is premature to think that any action of this alcaloid is necessarily mediated through MT. Kay BRUNE has studied the mechanism of the anti-inflammatory action of colchicine, a property which led to the wide use of this drug in medicine at the beginning of the last century. The amount of prostaglandins in inflamed tissue is increased by colchicine : it is difficult to understand how this could decrease inflammation, and it is more probable that the release of prostaglandins and the reduced inflammation are independent results of the action of colchicine, probably through its effects of MT or cell membranes.

Several new uses of MT inhibitors in chemotherapy have been presented, and the results obtained so far with vinblastine and vincristine in cancer therapy show the importance of this approach. DE BRABANDER and the group of workers of Janssen Pharmaceutica have studied a MT inhibitor, of an original structure, R 17934. This is more toxic to neoplastic than to normal cells, and may destroy even non-dividing cells. VAN PUTTEN and collaborators have tested this new compound on hematopoietic cells of the mouse where its action is clearly related to its antimitotic effects. It should be recalled, however, that the therapeutic effects of the Vinca alcaloids in cancer chemotherapy are difficult to explain by mitotic poisoning alone.
DAVIDSE has studied the effects on fungi of several benzimidazole derivatives, which act as spindle poisons and may be more specific than colchicine in this regard. *In vitro*, these compounds do not prevent the assembly of porcine brain tubulin into MT, nor the heterologous assembly of porcine and fungal tubulins.

The fields of application of MT inhibitors are varied, and the surprising action of two antihelminthics, mebendazole and R 17889, as reported by BORGERS and his group, are of great practical interest : the MT of the nutrient-absorptive cells of three species of parasitic worms are specifically destroyed, leading to an irreversible cell damage and to death of the parasite, without any effect on the MT of the intestinal wall of the host.

Looking through all these contributions and the poster demonstrations which I do not have the time to comment to-day, one senses the general biological significance of tubulins, and the varied use of the drugs which specifically bind to these proteins. The causes of this specificity remain obscure, and we will be waiting in the years to come for a molecular model of tubulin which may explain how colchicine combines to this protein : the work of the chemists, as shown by T.N. MARGULIS earlier at this meeting, on the structure of colchicine should help to find why this molecule blocks the tubulin assembly into MT.

Much remains to be learnt, for if it is agreed that colchicine prevents the assembly of subunits into MT, the apparent destructive action of colchicine and other inhibitors on already formed MT, remains poorly explained. It may be related to a rapid turn-over of MT. However, cold and high hydrostatic pressures are known to destroy already formed MT, and we will be expecting the chemists and thermodynamicists to explain us how colchicine could have the same action. MT research is being carried on with great activity throughout the world, and in many fields : this meeting, which I had the great pleasure and honour to preside, will, I am sure, have helped to further research and has fostered new links between workers in this thrilling field of biology.

## REFERENCES


1. Soifer, D. (Edt.), 1975, The biology of cytoplasmic microtubules. Ann. N.Y. Acad. Sci. 253.
2. Weisenberg, R.C. and Rosenfeld, A.C., 1975, J. Cell Biol. 64, 42-53.
3. Fujiwara, K. and Tilney, L.G., 1975, in : The biology of cytoplasmic microtubules. Ann. N.Y. Acad. Sci. 253, 27-50.
4. Bajer, A.S. and Mole-Bajer, J., 1972, Spindle dynamics and chromosome movements (Academic Press, New York, London).
5. Dixon, W. and Malden, W., 1908, J. Physiol. 37, 50-76.
6. Orsini, M.W. and Pansky, B., 1952, Science 115, 88-89.
7. de Harven, E., 1955, Bull. Acad. roy. Méd. Belg. Cl. Sci. 5e série, 41, 1056-1060.
8. Piton, R., 1929, Arch. intern. Méd. exp. 5, 355-411.

# AUTHOR INDEX

The names and page numbers printed in *italic* face refer to the articles of the contributors of these Proceedings. Names and page numbers appearing in normal print refer to the cited authors.